Linear Algebra for Computational Sciences and Engineering

Ferrante Neri

Linear Algebra for Computational Sciences and Engineering

Second Edition

Foreword by Alberto Grasso

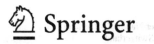 Springer

Ferrante Neri
School of Computer Science
University of Nottingham
Nottingham, UK

ISBN 978-3-030-21323-7   ISBN 978-3-030-21321-3   (eBook)
https://doi.org/10.1007/978-3-030-21321-3

This Springer imprint is published by the registered company Springer Nature Switzerland AG.
The registered company address is: Gewerbestrasse 11, 6330 Cham, Switzerland

Счастье - это когда тебя понимают.

Happiness is to be understood.

– Георгий Полонский–
(Доживём до понедельника)

– Georgi Polonsky –
(We'll Live Till Monday)

Foreword

Linear Algebra in Physics

The history of linear algebra can be viewed within the context of two important traditions.

The first tradition (within the history of mathematics) consists of the progressive broadening of the concept of number so to include not only positive integers but also negative numbers, fractions and algebraic and transcendental irrationals. Moreover, the symbols in the equations became matrices, polynomials, sets and, permutations. Complex numbers and vector analysis belong to this tradition. Within the development of mathematics, the one was concerned not so much about solving specific equations but mostly about addressing general and fundamental questions. The latter were approached by extending the operations and the properties of sum and multiplication from integers to other linear algebraic structures. Different algebraic structures (lattices and Boolean algebra) generalized other kinds of operations, thus allowing to optimize some non-linear mathematical problems. As a first example, lattices were generalizations of order relations on algebraic spaces, such as set inclusion in set theory and inequality in the familiar number systems ($\mathbb{N}$, $\mathbb{Z}$, $\mathbb{Q}$, and $\mathbb{R}$). As a second example, Boolean algebra generalized the operations of intersection and union and the principle of duality (De Morgan's relations), already valid in set theory, to formalize the logic and the propositions' calculus. This approach to logic as an algebraic structure was much similar as the Descartes' algebra approach to the geometry. Set theory and logic have been further advanced in the past centuries. In particular, Hilbert attempted to build up mathematics by using symbolic logic in a way that could prove its consistency. On the other hand, Gödel proved that in any mathematical system, there will always be statements that can never be proven either true or false.

The second tradition (within the history of physical science) consists of the search for mathematical entities and operations that represent aspects of the physical reality. This tradition played a role in the Greek geometry's bearing and its following application to physical problems. When observing the space around us, we always suppose the existence of a reference frame, identified with an ideal 'rigid body', in the part of the universe in which the system we want to study evolves (e.g. a three axes system having the sun as their origin and direct versus three fixed stars). This is modelled in the so-called Euclidean affine space. A reference frame's choice is purely descriptive at a purely kinematic level. Two reference frames have to be intuitively considered distinct if the correspondent 'rigid bodies' are in relative motion. Therefore, it is important to fix the links (*linear transformations*) between the kinematic entities associated with the same motion but relatives to two different reference frames (*Galileo's relativity*).

In the seventeenth and eighteenth centuries, some physical entities needed a new representation. This necessity made the above-mentioned two traditions converge by adding quantities as velocity, force, momentum and acceleration (*vectors*) to the traditional quantities as mass and time (*scalars*). Important ideas led to the vectors' major systems: the forces' parallelogram concept by Galileo, the situations' geometry and calculus concepts by Leibniz and by Newton and the complex numbers' geometrical representation. Kinematics studies the motion of bodies in space and in time independently on the causes which provoke it. In classical physics, the role of time is reduced to that of a parametric independent variable. It needs also to choose a model for the body (or bodies) whose motion one wants to study. The fundamental and simpler model is that of point (useful only if the body's extension is smaller than the extension of its motion and of the other important physical quantities considered in a particular problem). The motion of a point is represented by a curve in the tridimensional Euclidean affine space. A second fundamental model is the "rigid body" one, adopted for those extended bodies whose component particles do not change mutual distances during the motion.

Later developments in electricity, magnetism and optics further promoted the use of vectors in mathematical physics. The nineteenth century marked the development of vector space methods, whose prototypes were the three-dimensional geometric extensive algebra by Grassmann and the algebra of quaternions by Hamilton to, respectively, represent the orientation and rotation of a body in three dimensions. Thus, it was already clear how a simple algebra should meet the needs of the physicists in order to efficiently describe objects in space and in time (in particular their dynamical symmetries and the corresponding conservation laws) and the properties of space-time itself. Furthermore, the principal characteristic of a simple algebra had to be its linearity (or at most its multi-linearity). During the latter part of the nineteenth century, Gibbs based his three-dimensional vector algebra on some ideas by Grassmann and by Hamilton, while Clifford united these systems into a single geometric algebra (direct product of quaternions' algebras). Afterwards, the Einstein's description of the four-dimensional continuum space-time (special and general relativity theories) required a tensor algebra. In the 1930s, Pauli and Dirac introduced Clifford algebra's matrix representations for physical reasons: Pauli for describing the electron spin while Dirac for accommodating both the electron spin and the special relativity.

Each algebraic system is widely used in contemporary physics and is a fundamental part of representing, interpreting and understanding the nature. Linearity in physics is principally supported by three ideas: Superposition Principle, Decoupling Principle and Symmetry Principle.

Superposition Principle. Let us suppose to have a linear problem where each O_k is the fundamental output (linear response) of each basic input I_k. Then, both an arbitrary input as its own response can be written as a linear combination of the basic ones, i.e. $I = c_1 I_1 + \ldots + c_k I_k$ and $O = c_1 O_1 + \ldots + c_k O_k$.

Decoupling Principle. If a system of coupled differential equations (or difference equations) involves a diagonalizable square matrix $\mathbf{A}$, then it is useful to consider new variables $\mathbf{x'_k} = \mathbf{U}\mathbf{x_k}$ with ($k \in \mathbb{N}; 1 \le k \le n$), where $\mathbf{U}$ is a unitary matrix and $\mathbf{x'_k}$ is an orthogonal eigenvectors set (basis) of $\mathbf{A}$. Rewriting the equations in terms of the $\mathbf{x'_k}$, one discovers that each eigenvector's evolution is independent on the others and that the form of each equation depends only on the corresponding eigenvalue of $\mathbf{A}$. By solving the equations so to get each $\mathbf{x'_k}$ as a function of time, it is also possible to get $\mathbf{x_k}$ as a function of time ($\mathbf{x_k} = \mathbf{U}^{-1}\mathbf{x'_k}$). When $\mathbf{A}$ is not diagonalizable (not normal), the resulting equations for $\mathbf{x'}$ are not completely decoupled (Jordan canonical form), but are still relatively easy (of course, if one does not take into account some deep problems related to the possible presence of resonances).

Symmetry Principle. If $\mathbf{A}$ is a diagonal matrix representing a linear transformation of a physical system's state and $\mathbf{x'_k}$ its eigenvectors' set, each unitary transformation satisfying the matrix equation $\mathbf{U}\mathbf{A}\mathbf{U}^{-1} = \mathbf{A}$ (or $\mathbf{U}\mathbf{A} = \mathbf{A}\mathbf{U}$) is called "symmetry transformation" for the considered physical system. Its deep meaning is to eventually change each eigenvector without changing the whole set of eigenvectors and their corresponding eigenvalues.

Thus, special importance in computational physics is assumed by the standard methods for solving systems of linear equations: the procedures suited for symmetric real matrices and the iterative methods converging fast when applied to matrix having its nonzero elements concentrated near the main diagonal (diagonally dominated matrix).

Physics has a very strong tradition about tending to focus on some essential aspects while neglecting other important issues. For example, Galileo founded the mechanics neglecting friction, despite its important effect on mechanics. The statement of Galileo's inertia law (Newton's first law, i.e. 'An object not affected by forces moves with constant velocity') is a pure abstraction, and it is approximately valid. While modelling, a popular simplification has been for centuries the search of a linear equation approximating the nature. Both ordinary and partial linear differential equations appear through classical and quantum physics, and even where the equations are non-linear, linear approximations are extremely powerful. For example, thanks to Newton's second law, much of classical physics is expressed in terms of second-order ordinary differential equations' systems. If the force is a position's linear function, the resulting equations are linear ($m\frac{d^2 x}{dt^2} = -\mathbf{A}\mathbf{x}$, where $\mathbf{A}$ matrix not depending on $\mathbf{x}$). Every solution may be written as a linear combination of the special solutions (oscillation's normal modes) coming from eigenvectors of the $\mathbf{A}$ matrix. For non-linear problems near equilibrium, the force can always be expanded as a Taylor's series, and the leading (linear) term is dominant for small oscillations. A detailed treatment of coupled small oscillations is possible by obtaining a diagonal matrix of the coefficients in N coupled differential equations by finding the eigenvalues and the eigenvectors of the Lagrange's equations for coupled oscillators. In classical mechanics, another example of linearization consists of looking for the principal moments and principal axes of a solid body through solving the eigenvalues' problem of a real symmetric matrix (inertia tensor). In the theory of continua (e.g. hydrodynamics, diffusion and thermal conduction, acoustic, electromagnetism), it is (sometimes) possible to convert a partial differential equation into a system of linear equations by employing the finite difference formalism. That ends up with a diagonally dominated coefficients' matrix. In particular, Maxwell's equations of electromagnetism have an infinite number of degrees of freedom (i.e. the value of the field at each point), but the Superposition Principle and the Decoupling Principle still apply. The response to an arbitrary input is obtained as the convolution of a continuous basis of Dirac δ functions and the relevant Green's function.

Even without the differential geometry's more advanced applications, the basic concepts of multilinear mapping and tensor are used not only in classical physics (e.g. inertia and electromagnetic field tensors) but also in engineering (e.g. dyadic).

In particle physics, it was important to analyse the problem of neutrino oscillations, formally related both to the Decoupling and the Superposition Principles. In this case, the three neutrino mass matrices are not diagonal and not normal in the so-called gauge states' basis. However, through a bi-unitary transformation (one unitary transformation for each "parity" of the gauge states), it is possible to get the eigenvalues and their own eigenvectors (mass states) which allow to render it diagonal. After this transformation, it is possible to obtain the Gauge States as a superposition (linear combination) of mass states.

Schrödinger's linear equation governs the nonrelativistic quantum mechanics, and many problems are reduced to obtain a diagonal Hamiltonian operator. Besides, when studying the quantum angular momentum's addition, one considers Clebsch-Gordon coefficients related to an unitary matrix that changes a basis in a finite-dimensional space.

In experimental physics and statistical mechanics (stochastic methods' framework), researchers encounter symmetric, real positive definite and thus diagonalizable matrices (so-called covariance or dispersion matrix). The elements of a covariance matrix in the i, j positions are the covariances between ith and jth elements of a random vector (i.e. a vector of random variables, each with finite variance). Intuitively, the variance's notion is so generalized to multiple dimension.

The geometrical symmetry's notion played an essential part in constructing simplified theories governing the motion of galaxies and the microstructure of matter (quarks' motion confined inside the hadrons and leptons' motion). It was not until the Einstein's era that the discovery of the space-time symmetries of the fundamental laws and the meaning of their relations to the conservation laws were fully appreciated, for example, Lorentz transformations, Noether's theorem and Weyl's covariance. An object with a definite shape, size, location and orientation constitutes a state whose symmetry properties are to be studied. The higher its "degree of symmetry" (and the number of conditions defining the state is reduced), the greater is the number of transformations that leave the state of the object unchanged.

While developing some ideas by Lagrange, by Ruffini and by Abel (among others), Galois introduced important concepts in group theory. This study showed that an equation of order $n \geq 5$ cannot, in general, be solved by algebraic methods. He did this by showing that the functional relations among the roots of an equation have symmetries under the permutations of roots. In the 1850s, Cayley showed that every finite group is isomorphic to a certain permutation group (e.g. the crystals' geometrical symmetries are described in finite groups' terms). Fifty years after Galois, Lie unified many disconnected methods of solving differential equations (evolved over about two centuries) by introducing the concept of continuous transformation of a group in the theory of differential equations. In the 1920s, Weyl and Wigner recognized that certain group theory's methods could be used as a powerful analytical tool in quantum physics. In particular, the essential role played by Lie's groups, e.g. rotation isomorphic groups $SO(3)$ and $SU(2)$, was first emphasized by Wigner. Their ideas have been used in many contemporary physics' branches which range from the theory of solids to nuclear physics and particle physics. In classical dynamics, the invariance of the equations of motion of a particle under the Galilean transformation is fundamental in Galileo's relativity. The search for a linear transformation leaving "formally invariant" the Maxwell's equations of electromagnetism led to the discovery of a group of rotations in space-time (Lorentz transformation).

Frequently, it is important to understand why a symmetry of a system is observed to be broken. In physics, two different kinds of symmetry breakdown are considered. If two states of an object are different (e.g. by an angle or a simple phase rotation) but they have the same energy, one refers to 'spontaneous symmetry breaking'. In this sense, the underlying laws of a system maintain their form (Lagrange's equations are invariant) under a symmetry transformation, but the system as a whole changes under such transformation by distinguishing between two or more fundamental states. This kind of symmetry breaking, for example, characterizes the ferromagnetic and the superconductive phases, where the Lagrange function (or the Hamiltonian function, representing the energy of the system) is invariant under rotations (in the ferromagnetic phase) and under a complex scalar transformation (in the superconductive phase). On the contrary, if the Lagrange function is not invariant under particular transformations, the so-called 'explicit symmetry breaking' occurs. For example, this happens when an external magnetic field is applied to a paramagnet (Zeeman's effect).

Finally, by developing the determinants through the permutations' theory and the related Levi-Civita symbol, one gains an important and easy calculation tool for modern differential geometry, with applications in engineering as well as in modern physics. This is the case in general relativity, quantum gravity, and string theory.

We can therefore observe that the concepts of *linearity* and *symmetry* aided to solve many physical problems. Unfortunately, not the entire physics can be straightforwardly modelled by linear algebra. Moreover, the knowledge of the laws among the elementary constituents of a system does not implicate an understanding of the global behaviour. For example, it is not easy at all to deduce from the forces acting between the water's molecules because the ice is lighter than water. Statistical mechanics, which was introduced between the end of the nineteenth century and the beginning of the twentieth century (the work by Boltzmann and Gibbs), deals with the problem of studying the behaviour of systems composed of many particles without determining each particle's trajectory but by probabilistic methods. Perhaps, the most interesting result of statistical mechanics consists of the emergence of collective behaviours: while the one cannot say whether the water is in the solid or liquid state and which is the transition temperature by observing a small number of atoms, clear conclusions can be easily reached if a large number of atoms are observed (more precisely when the number of atoms tends to infinity). Phase transitions are therefore created as a result of the collective behaviour of many components.

The latter is an example of a physical phenomenon which requires a mathematical instrument different from linear algebra. Nonetheless, as mentioned above, linear algebra and its understanding is one of the basic foundations for the study of physics. A physicist needs algebra either to model a phenomenon (e.g. classical mechanics) or to model a portion of phenomenon (e.g. ferromagnetic phenomena) or to use it as a basic tool to develop complex modern theories (e.g. quantum field theory).

This book provides the readers the basics of modern linear algebra. The book is organized with the aim of communicating to a wide and diverse range of backgrounds and aims. The book can be of great use to students of mathematics, physics, computer science, and engineering as well as to researchers in applied sciences who want to enhance their theoretical knowledge in algebra. Since a prior rigorous knowledge about the subject is not assumed, the reader may easily understand how linear algebra aids in numerical calculations and problems in different and diverse topics of mathematics, physics, engineering and computer science.

I found this book a pleasant guide throughout linear algebra and an essential vademecum for the modern researcher who needs to understand the theory but has also to translate theoretical concepts into computational implementations. The plethora of examples make the topics, even the most complex, easily accessible to the most practical minds. My suggestion is to read this book, consult it when needed and enjoy it.

Catania, Italy   Alberto Grasso
April 2016

Preface to the Second Edition

The first edition of this book has been tested in the classroom over three academic years. As a result, I had the opportunity to reflect on my communication skills and teaching clarity.

Besides correcting the normal odd typos and minor mistakes, I decided to rewrite many proofs which could be explained in a clearer and more friendly way. Every change to the book has been made by taking into great consideration the reactions of the students as well as their response in terms of learning. Many sections throughout the book have been rephrased, some sections have been reformulated, and a better notation has been used. The second edition contains over 150 pages of new material, including theory, illustrations, pseudocodes and examples throughout the book summing up to more than 500.

New topics have been added in the chapters about matrices, vector spaces and linear mapping. However, numerous additions have been included throughout the text. The section about Euclidean spaces has been now removed from the chapter about vector spaces and placed in a separated introductory chapter about inner product spaces. Finally, a section at the end of the book showing the solutions to the exercises placed at the end of each chapter has been included.

In its new structure, this book is divided still into two parts: Part I illustrates basic topics in algebra, while Part II presents more advanced topics.

Part I is composed of six chapters. Chapter 1 introduces the basic notation, concepts and definitions in algebra and set theory. Chapter 2 describes theory and applications about matrices. Chapter 3 analyses systems of linear equation by focussing on analytic theory as well as numerical methods. Chapter 4 introduces vectors with a reference to the three-dimensional space. Chapter 5 discusses complex numbers and polynomials as well as the fundamental theorem of algebra. Chapter 6 introduces the conics from the perspective of algebra and matrix theory.

Part II is composed of seven chapters. Chapter 7 introduces algebraic structures and offers an introduction to group and ring theories. Chapter 8 analyses vector spaces. Chapter 9 introduces inner product spaces with an emphasis on Euclidean spaces. Chapter 10 discusses linear mappings. Chapter 11 offers a gentle introduction to complexity and algorithm theory. Chapter 12 introduces graph theory and presents it from the perspective of linear algebra. Finally, Chap. 13 provides an example on how all linear algebra studied in the previous chapters can be used in practice in an example about electrical engineering.

In Appendix A, Boolean algebra is presented as an example of non-linear algebra. Appendix B outlines some proofs to theorems stated in the book chapters but where the proof was omitted since it required a knowledge of calculus and mathematical analysis which was beyond the scope of this book.

I feel that the book, in its current form, is a substantially improved version of the first edition. Although the overall book structure and style remained broadly the same, the new way to present and illustrate the concept makes the book accessible to a broad audience, guiding them towards a higher education level of linear algebra.

As the final note, the second edition of this book has been prepared with the aim of making Algebra accessible and easily understandable to anybody who has an interest in mathematics and wants to devote some effort to it.

Nottingham, UK   Ferrante Neri
April 2019

Preface to the First Edition

Theory and practice are often seen as entities in opposition characterizing two different aspects of the world knowledge. In reality, applied science is based on the theoretical progress. On the other hand, the theoretical research often looks at the world and practical necessities to be developed. This book is based on the idea that theory and practice are not two disjointed worlds and that the knowledge is interconnected matter. In particular, this book presents, without compromising on mathematical rigour, the main concepts of linear algebra from the viewpoint of the computer scientist, the engineer, and anybody who will need an in depth understanding of the subject to let applied sciences progress. This book is oriented to researchers and graduate students in applied sciences but is also organized as a textbook suitable to courses of mathematics.

Books of algebra are either extremely formal, thus will not be enough intuitive for a computer scientist/engineer, or trivial undergraduate textbooks, without an adequate mathematical rigour in proofs and definitions. "Linear Algebra for Computational Sciences and Engineering" aims at maintaining a balance between rigour and intuition, attempting to provide the reader with examples, explanations, and practical implications of every concept introduced and every statement proved. When appropriate, topics are also presented as algorithms with associated pseudocode. On the other hand, the book does not contain logical skips or intuitive explanations to replace proofs.

The "narration" of this book is thought to flow as a story of (a part of) the mathematical thinking. This story affected, century after century, our brain and brought us to the modern technological discoveries. The origin of this knowledge evolution is imagined to be originated in the stone age when some caveman/cavewoman had the necessity to assess how many objects he/she was observing. This conceptualization, happened at some point in our ancient history, has been the beginning of mathematics, but also of logics, rational thinking, and somehow technology.

The story narrated in this book is organized into two parts composed of six chapters each, thus twelve in total. Part I illustrates basic topics in algebra which could be suitable for an introductory university module in Algebra while Part II presents more advanced topics that could be suitable for a more advance module. Furthermore, this book can be read as a handbook for researchers in applied sciences as the division into topics allows an easy selection of a specific topic of interest.

Part I opens with Chap. 1 which introduces the basic concepts and definitions in algebra and set theory. Definitions and notation in Chap. 1 are used in all the subsequent chapters. Chapter 2 deals with matrix algebra introducing definitions and theorems. Chapter 3 continues the discussion about matrix algebra by explaining the theoretical principles of systems of linear equations as well as illustrating some exact and approximate methods to solve them. Chapter 4, after having introduced vectors in an intuitive way as geometrical entities, progressively abstracts and generalizes this concept leading to algebraic vectors which essentially require the solution of systems of linear equations and are founded on matrix theory. The narration about vectors leads to Chap. 5 where complex numbers and polynomials are discussed. Chapter 5 gently introduces algebraic structures by providing statement and interpretation for the fundamental theorem of algebra. Most of knowledge achieved during the first five chapters is proposed again in Chap. 6 where conics are introduced and explained. It is shown how a conic has, besides its geometrical meaning, an algebraic interpretation and is thus equivalent to a matrix.

In a symmetric way, Part II opens with an advanced introduction to algebra by illustrating basic algebraic structures in Chap. 7. Group and ring theories are introduced as well as the concept of field which constitutes the basics for Chap. 8 where vector spaces are presented. Theory of vector spaces is described from a theoretical viewpoint as well as with reference to their physical/geometrical meaning. These notions are then used within Chap. 10 which deals with linear mappings, endomorphism, and eigenvalues. In Chaps. 8 and 10 the connections with matrix and vector algebra is self-evident. The narration takes a break in Chap. 11 where some logical instruments for understanding the final chapters are introduced. These concepts are the basics of complexity theory. While introducing these concepts it is shown that algebra is not only an abstract subject. On the contrary, the implementation of algebraic techniques has major practical implications which must be taken into account. Some simple algebraic operations are revisited as instructions to be executed within a machine. Memory and operator representations are also discussed. Chapter 12 discusses graph theory and emphasizes the equivalence between a graph and a matrix/vector space. Finally Chap. 13 introduces electrical networks as algebraic entities and shows how an engineering problem is the combination of multiple mathematical (in this case algebraic) problems. It is emphasized that the solution of an electric network incorporates graph theory, vector space theory, matrix theory, complex numbers, and systems of linear equations, thus covering all the topics presented within all the other chapters.

I would like express my gratitude to my long-standing friend Alberto Grasso who inspired me with precious comments and useful discussions. As a theoretical physicist, he offered me a different perspective of Algebra which is more thoroughly explained in the Foreword written directly by himself.

Furthermore, I would like to thank my colleagues of the Mathematics Team at De Montfort University, especially Joanne Bacon, Michéle Wrightham, and Fabio Caraffini for support and feedback.

Last but not least, I wish to thank my parents, Vincenzo and Anna Maria, for the continued patience and encouragement during the writing of this book.

As a final note, I hope this book will be a source of inspiration for young minds. To the youngest readers who are approaching Mathematics for the first time with the present book I would like to devote a thought. The study of Mathematics is similar to running of a marathon: it requires intelligence, hard work, patience, and determination, where the latter three are as important as the first one. Understanding mathematics is a lifetime journey which does not contain short-cuts but can be completed only mile by mile, if not step by step. Unlike the marathon, the study of mathematics does not have a clear and natural finish line. However, it has the artificial finish lines that the society imposes to us such as an exam, the publication of an article, a funding bid, a national research exercise etc. Like in a marathon, the study of mathematics contains easier and harder stretches, comfortable downhill and nasty uphill bends. In a marathon, like in the study of mathematics, the most important point is the focus on the personal path, the passion towards the goal, and to persevere despite of the difficulties.

This book is meant to be a training guide towards an initial understanding of linear and abstract algebra and possibly a first or complementary step towards better research in Computational Sciences and Engineering.

I wish to readers a fruitful and enjoyable time reading this book.

Leicester, UK   Ferrante Neri
April 2016

Contents

Contents

About the Author

Ferrante Neri received his master's degree and a PhD in Electrical Engineering from the Technical University of Bari, Italy, in 2002 and 2007 respectively. In 2007, he also received a PhD in Scientific Computing and Optimization from the University of Jyväskylä, Finland. From the latter institution, he received the DSc degree in Computational Intelligence in 2010. He was appointed Assistant Professor in the Department of Mathematical Information Technology at the University of Jyväskylä, Finland, in 2007, and in 2009 as a Research Fellow with the Academy of Finland. Dr. Neri moved to De Montfort University, United Kingdom, in 2012, where he was appointed Reader in Computational Intelligence and, in 2013, promoted to Full Professor of Computational Intelligence Optimization. Since 2019 Ferrante Neri moved to the School of Computer Science, University of Nottingham, United Kingdom. Ferrante Neri's teaching expertise lies in mathematics for computer science. He has a specific long lasting experience in teaching linear and abstract algebra. His research interests include algorithmics, hybrid heuristic-exact optimization, scalability in optimization and large-scale problems.

Part I
Foundations of Linear Algebra

Chapter 1
Basic Mathematical Thinking

1.1 Introduction

Mathematics, from the Greek word "mathema", is simply translated as science or expression of the knowledge. Regardless of the fact that mathematics is something that exists in our brain as well as in the surrounding nature and we discover it little by little or an invention/abstraction of a human brain, mathematics has been with us with our capability of thinking and is the engine of human progress.

Although it is impossible to determine the beginning of mathematics in the history, at some point in the stone-age some caveman/cavewoman probably asked to himself how many objects (stones, trees, fruits) he/she was looking at. In order to mark the amount, he/she used a gesture with the hands where each gesture corresponded to a certain amount. That caveman who made this for first, is the one who invented/discovered the concept of *enumeration*, that is closely related to the concepts of *set* and its cardinality. The most natural gesture I can think is to lift a finger for each object taken into account. This is probably a common opinion and, since we totally have ten fingers in our hands, is the reason why our *numeral system* is *base 10*.

Obviously, mathematics is much more than enumeration as it is a logical thinking system in which the entire universe can potentially be represented and lives aside, at an abstract level, even when there is no physically meaningful representation.

This book offers a gentle introduction to mathematics and especially to linear algebra (from Arabic al-gabr = connection), that traditionally is the discipline that connects quantities (numbers) to symbol (letters) in order to extract general rules. Algebra, as well as mathematics is based on a set of initial rules that are considered basic system of truth that is the basis of all the further discoveries. This system is named *Axiomatic System*.

© Springer Nature Switzerland AG 2019
F. Neri, *Linear Algebra for Computational Sciences and Engineering*,
https://doi.org/10.1007/978-3-030-21321-3_1

1.2 Axiomatic System

A concept is said to be *primitive* when it cannot be rigorously defined since its meaning is intrinsically clear. An *axiom* or *postulate* is a premise or a starting point for reasoning. Thus, an axiom is a statement which appears unequivocally true and that does not require any proof to be verified but cannot be, in any way, falsified.

Primitive concepts and axioms compose the axiomatic system. The axiomatic system is the ground onto the entire mathematics is built. On the basis of this ground, a *definition* is a statement that introduces a new concept/object by using previously known concepts (and thus primitive concepts are necessary for defining new ones). When the knowledge can be extended on the basis of previously established statements, this knowledge extension is named *theorem*. The previously known statements are the *hypotheses* while the extension is the *thesis*. A theorem can be expressed in the form: "if the hypotheses are verified then the thesis occurs". In some cases, the theorem is symmetric, i.e. besides being true that "if the hypotheses are verified then the thesis occurs" it is also true that "if the thesis is verified then the hypotheses occur". More exactly, if A and B are two statements, a theorem of this kind can be expressed as "if A is verified than B occurs and if B is verified then A occurs". In other words the two statements are *equivalent* since the truth of one of them automatically causes the truth of the other. In this book, theorems of this kind will be expressed in the form "A is verified *if and only if* B is verified".

The set of logical steps to deduce the thesis on the basis of the hypotheses is here referred as *mathematical proof* or simply *proof*. A large number of proof strategies exist. In this book, we will use only the *direct proof*, i.e. from the hypotheses we will logically arrive to the thesis or *by contradiction* (or *reductio ad absurdum*), i.e. the negated thesis will be new hypothesis that will lead to a paradox. A successful proof is indicated with the symbol □. It must be remarked that a theorem that states the equivalence of two facts requires two proofs. More specifically, a theorem of the kind 'A is verified *if and only if* B is verified' is essentially two theorems in one. Hence, the statements "if A is verified than B occurs" and "if B is verified then A occurs" require two separate proofs.

A theorem that enhances the knowledge by achieving a minor result that is then usable to prove a major result is called *lemma* while a minor result that uses a major theorem to be proved is called *corollary*. A proved result that is not as important as a theorem is called *proposition*.

1.3 Basic Definitions in Set Theory

The first important primitive concept of this book is the *set* that without mathematical rigour is here defined as a collection of objects that share a common feature. These objects are the *elements* of the set. Let us indicate with A a generic set and with x its element. In order to indicate that x is an element of A, we will write $x \in A$ (otherwise $x \notin A$).

Definition 1.1. Two sets A and B are said to be *coincident* if every element of A is also an element of B and every element of B is also an element of A.

Definition 1.2. The *cardinality* of a set A is the number of elements contained in A.

Definition 1.3. A set A is said *empty* and indicated with $\emptyset$ when it does not contain any element.

Definition 1.4. Universal Quantifier. In order to indicate all the elements x of a set A, we will write $\forall x \in A$.

If a proposition is applied to all the elements of the set, the statement "for all the elements of A it follows that" is synthetically written as $\forall x \in A :$.

Definition 1.5. Existential Quantifier. In order to indicate that it exists at least one element x of a set A, we will write $\exists x \in A$. If we want to specify that only one element exists we will use the symbol $\exists!$.

If we want to state that "it exists at least one element of A such that" we will write $\exists x \in A \ni$ '.

The statement $\forall x \in \emptyset$ is perfectly meaningful (and it is equivalent to "for no elements") while the statement $\exists x \in \emptyset$ is always wrong.

Definition 1.6. Let m be the cardinality of a set B and n be the cardinality of A. If $m \leq n$ and all the elements of B are also elements of A, then B is contained in A (or is a subset of A) and is indicated $B \subseteq A$.

Definition 1.7. Let A be a set. The set composed of all the possible subsets of A (including the empty set and A itself) is said *power set*.

Definition 1.8. For two given sets, A and B, the *intersection set* $C = A \cap B$ is that set containing all the elements that are in both the sets A and B.

Definition 1.9. For two given sets, A and B, the *union set* $C = A \cup B$ is that set containing all the elements that are in either or both the sets A and B.

Although proofs of set theory do not fall within the objectives of this book, in order to have a general idea of the mathematical reasoning the proof of the following property is provided.

Proposition 1.1. Associativity of the Intersection.

$$(A \cap B) \cap C = A \cap (B \cap C)$$

Proof. Let us consider a generic element x such that $x \in (A \cap B) \cap C$. This means that $x \in (A \cap B)$ and $x \in C$ which means that $x \in A$ and $x \in B$ and $x \in C$. Hence the element x belongs to the three sets. This fact can be re-written by stating that $x \in A$ and $x \in (B \cap C)$ that is $x \in A \cap (B \cap C)$. We can repeat the same operation $\forall x \in (A \cap B) \cap C$ and thus find out that all the elements of $(A \cap B) \cap C$ are also elements of $A \cap (B \cap C)$. Hence, $(A \cap B) \cap C = A \cap (B \cap C)$. $\square$

Definition 1.10. For two given sets, A and B, the *difference set* $C = A \setminus B$ is that set containing all the elements that are in A but not in B.

Definition 1.11. For two given sets, A and B, the *symmetric difference* $C = A \Delta B = (A \setminus B) \cup (B \setminus A) = (A \cup B) \setminus (A \cap B)$.

The symmetric difference set is, thus, the set of those elements that belong either to A or B (elements that do not belong to their intersection).

Definition 1.12. For a given set A, the *complement* of a set A is the set of all the elements not belonging to A. The complement of A is indicated as A^c. Complement set: $A^c = \{x | x \notin A\}$.

Proposition 1.2. Complement of a Complement.

$$(A^c)^c = A$$

Proof. Let us consider a generic element $x \in (A^c)^c$. By definition, $x \notin A^c$. This means that $x \in A$. We can repeat the reasoning $\forall x \in (A^c)^c$ and find out that $x \in A$. Hence, $(A^c)^c = A$. $\square$

Definition 1.13. Cartesian Product. Let A and B be two sets with n and m their respective cardinalities. Let us indicate each set by its elements as $A = \{a_1, a_2, \ldots, a_n\}$ and $B = \{b_1, b_2, \ldots, b_n\}$. The Cartesian product C is a new set generated by all the possible pairs

$$\begin{aligned} C = A \times B = \{ &(a_1, b_1), (a_1, b_2), \ldots, (a_1, b_m), \\ &(a_2, b_1), (a_2, b_2), \ldots, (a_2, b_m), \ldots, \\ &(a_n, b_1), (a_n, b_2), \ldots, (a_n, b_m) \} \end{aligned}$$

The Cartesian product $A \times A$ is indicated with A^2 or in general $A \times A \times A \times \ldots \times A = A^n$ if A is multiplied n times.

Example 1.1. Let us consider the following two sets

$$A = \{1, 5, 7\}$$
$$B = \{2, 3\}.$$

Let us calculate the Cartesian product $A \times B$:

$$C = A \times B = \{(1,2),(1,3),(5,2),(5,3),(7,2),(7,3)\}.$$

Example 1.2. In order to better understand the Cartesian product let us give a graphical example. Let us interpret the following as a set A composed of the points on the paper needed to draw it.

The Cartesian product $C = A \times A = A^2$ would be graphically represented by the area below (all the points composing the area).

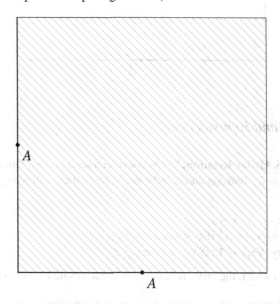

Definition 1.14. Let $C = A \times B$ be a Cartesian product. A *relation* on C is an arbitrary subset $\mathscr{R} \subseteq C$. This subset means that some elements of A relates to B according to a certain criterion $\mathscr{R}$. The set A is said *domain* while B is said *codomain*. If x is the generic element of A and y the generic element of B. The relation can be written as $(x,y) \in \mathscr{R}$ or $x\mathscr{R}y$.

Example 1.3. With reference to the graphical example above, a relation $\mathscr{R}$ would be any subarea contained in the striped area $C = A \times A = A^2$. For example, a relation $\mathscr{R}$ is the starred ellipse below.

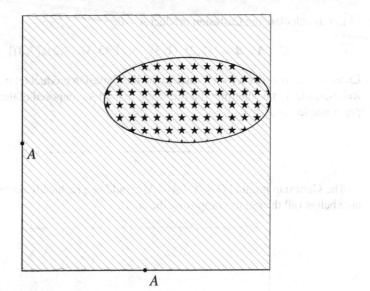

1.3.1 Order and Equivalence

Definition 1.15. Order Relation. Let us consider a set A and a relation $\mathscr{R}$ on A. This relation is said *order relation* and is indicated with $\preceq$ if the following properties are verified.

- reflexivity: $\forall x \in A : x \preceq x$
- transitivity: $\forall x, y, z \in A$: if $x \preceq y$ and $y \preceq z$ then $x \preceq z$
- antisymmetry: $\forall x, y \in A$: if $x \preceq y$ then $y \not\preceq x$

The set A, on which the order relation $\preceq$ is valid, is said *totally ordered set*.

Example 1.4. If we consider a group of people we can always sort them according theirs age. Hence the relation "to not be older than" (i.e. to be younger or to have the same age) with a set of people is a totally ordered set since every group of people can be fully sorted on the basis of their age.

From the definition above, the order relation can be interpreted as a predicate to be defined over the elements of a set. Although this is not wrong, we must recall that, rigorously, a relation is a set and an order relation is a set with some properties. In order to emphasise this fact, let us give again the definition of order relation by using a different notation.

Definition 1.16. Order Relation (Set Notation). Let us consider a set A and the Cartesian product $A \times A = A^2$. Let $\mathscr{R}$ be a relation on A, that is $\mathscr{R} \subseteq A^2$. This relation is said *order relation* if the following properties are verified for the set $\mathscr{R}$.

- reflexivity: $\forall x \in A : (x,x) \in \mathscr{R}$
- transitivity: $\forall x,y,z \in A :$ if $(x,y) \in \mathscr{R}$ and $(y,z) \in \mathscr{R}$ then $(x,z) \in \mathscr{R}$
- antisymmetry: $\forall x,y \in A :$ if $(x,y) \in \mathscr{R}$ then $(y,x) \notin \mathscr{R}$

If the properties above are valid for all the elements of A, that is $\mathscr{R}$ and A^2 are coincident, then A is a *totally ordered set*.

Definition 1.17. Partially Ordered Set. A set A where the order relation $\preceq$ is verified to some of its elements, that is $\mathscr{R} \subset A^2$, is a *partially ordered set* (also named as *poset*) and is indicated with $(A, \preceq)$.

Intuitively, a partially ordered set is a set whose elements (at least some of them) can be sorted according to a certain criterion, that is the relation $\preceq$.

Example 1.5. If we consider a group of people we can identify the relation "to be successor of". It can be easily verified that this is an order relation as the properties are verified. Furthermore, in the same group of people, some individuals can also be not successors of some others. Thus, we can have groups of individuals that are in relation with respect to the "to be successor of" order relation and some others that are unrelated to each other. The partially ordered set can be seen as a relation that allows the sorting of groups within a set.

Definition 1.18. Let $(Y, \preceq)$ be a subset of a poset $(X, \preceq)$. An element u in X is an *upper bound* of Y if $y \preceq u$ for every element $y \in Y$. If u is an upper bound of Y such that $u \preceq v$ for every other upper bound v of Y, then u is called a *least upper bound* or *supremum* of Y (sup Y).

Example 1.6. Let us consider the set $X = \{1,3,5,7,9\}$ and the set $Y \subset X = \{1,3,5\}$. Let us now consider the relation "to be less or equal" $\leq$.

We can easily verify that $(Y, \leq)$ is a poset:

- reflexivity: $\forall x \in Y : x \leq x$. In terms of set notation, $(x,x) \in \mathscr{R}$. For example, $1 \leq 1$ or $(1,1) \in \mathscr{R}$.
- transitivity: $\forall x,y,z \in Y :$ if $x \leq y$ and $y \leq z$ then $x \leq z$. For example, we can see that $1 \leq 3$ and $3 \leq 5$. As expected, $1 \leq 5$.
- antisymmetry: $\forall x,y \in Y :$ if $x \leq y$ then $y \nleq x$ (which means $y > x$). For example, we can see that $1 \leq 3$ and $3 > 1$.

The elements $5,7,9$ are all upper bounds of Y since they are all the element of B are $\leq$ than them. However, only 5 is the supremum of Y since $5 \leq 7$ and $5 \leq 9$.

Definition 1.19. Let $(Y, \preceq)$ be a subset of a poset $(X, \preceq)$. An element l in X is said to be a *lower bound* of Y if $l \preceq y$ for all $y \in Y$. If l is a lower bound of Y such that $k \preceq l$ for every other lower bound k of Y, then l is called a *greatest lower bound* or *infimum* of Y (inf Y).

Theorem 1.1. *Let Y be a nonempty subset of a poset X ($Y \subset X$). If Y has a supremum, then this supremum is unique.*

Proof. Let us assume by contradiction that a set Y has two suprema, u_1 and u_2, respectively, such that $u_1 \neq u_2$. By definition of supremum $\forall u \in X$ upper bound it follows that $u_1 \preceq u$. Analogously, $\forall u \in Y$ upper bound it follows that $u_2 \preceq u$. Since both u_1 and u_2 are upper bounds, we would have $u_1 \preceq u_2$ and $u_2 \preceq u_1$. This is impossible due to the antisymmetry of the order relation. Thus, it must occur that $u_1 = u_2$. □

Example 1.7. With reference to Example 1.6, 5 is the only supremum.

The same proof can be done for the uniqueness of the infima.

Theorem 1.2. *If Y has an infimum, this infimum is unique.*

Definition 1.20. Equivalence Relation. A relation $\mathscr{R}$ on set A is an *equivalence relation* and is indicated with $\equiv$ if the following properties are verified.

- reflexivity: $\forall x \in A$ it happens that $x \equiv x$
- symmetry: $\forall x, y \in A$ if $x \equiv y$ then it also happens that $y \equiv x$
- transitivity: $\forall x, y, z \in A$ if $x \equiv y$ and $y \equiv z$ then $x \equiv z$

The equivalence relation is also given in the following alternative definition.

Definition 1.21. Equivalence Relation (Set Notation). Let us consider a set A and the Cartesian product $A \times A = A^2$. Let $\mathscr{R}$ be a relation on A, that is $\mathscr{R} \subseteq A^2$. This relation is said *equivalence relation* if the following properties are verified for the set $\mathscr{R}$.

- reflexivity: $\forall x \in A : (x, x) \in \mathscr{R}$
- symmetry: $\forall x, y \in A :$ if $(x, y) \in \mathscr{R}$ then $(y, x) \in \mathscr{R}$
- transitivity: $\forall x, y, z \in A :$ if $(x, y) \in \mathscr{R}$ and $(y, z) \in \mathscr{R}$ then $(x, z) \in \mathscr{R}$

Example 1.8. Let us consider the set $A = \{1, 2, 3\}$ and the relation $\mathscr{R} \subset A^2$:

$$\mathscr{R} = \{(1,1), (1,2), (2,1), (2,2)\}.$$

The relation $\mathscr{R}$ is not an equivalence since it is not reflexive: $(3,3) \notin \mathscr{R}$.

Example 1.9. Let us give a graphical representation of the equivalence relation. Let us consider the Cartesian product $A \times A = A^2$

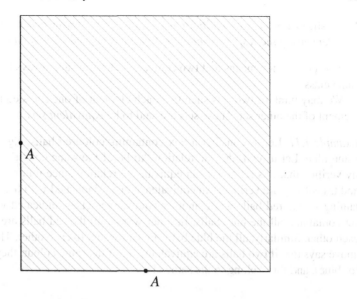

an example of equivalence relation is the subset composed of the diagonal points and the bullet points in figure.

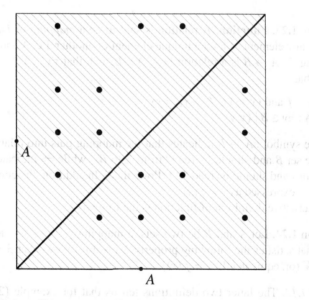

Definition 1.22. Let $\mathscr{R}$ be an equivalence relation defined on A. The *equivalence class* of an element a is a set defined as

$$[a] = \{x \in A | x \equiv a\}.$$

Example 1.10. The set composed of diagonal elements and bullet points of the graphic example above (Example 1.9) is an equivalence class.

Proposition 1.3. *Let* [a] *and* [b] *be two equivalence classes and* $x \in [a]$ *and* $y \in [b]$ *two elements of the respective classes. It follows that* [a] = [b] *if and only if* $x \equiv y$.

This proposition means that two equivalent elements are always belonging to the same class.

We may think of two sets such that each element of one set is equivalent to one element of the other set. These sets are said to be *equivalent sets*.

Example 1.11. Let us consider a box containing coloured balls, e.g. some red and some blue. Let us consider the relation "to be of the same colour". It can be easily verified that this relation is an equivalence relation since reflexivity, symmetry, and transitivity are verified. An equivalence class, indicated with [r], is the set containing all the red balls while another equivalence class, indicated with [b], is the set containing all the blue balls. In other words, all the red balls are equivalent to each other. Similarly, all the blue balls are equivalent to each other. The proposition above says that if two balls are equivalent, i.e. of the same colour, they are both red (or blue), and thus belong to the same class.

1.4 Functions

Definition 1.23. Function. A relation is said to be a *mapping* or *function* when it relates to any element of a set a unique element of another. Let A and B be two sets, a mapping $f : A \rightarrow B$ is a relation $\mathscr{R} \subseteq A \times B$ such that $\forall x \in A$, $\forall y_1$ and $y_2 \in B$ it follows that

- $(x, y_1) \in f$ and $(x, y_2) \in f \Rightarrow y_1 = y_2$
- $\forall x \in A : \exists y \in B \mid (x, y) \in f$

where the symbol $: A \rightarrow B$ indicates that the mapping puts into relationship the set A and the set B and should be read "from A to B", while $\Rightarrow$ indicates the material implications and should be read "it follows that". In addition, the concept $(x, y) \in f$ can be also expressed as $y = f(x)$.

An alternative definition of function is the following.

Definition 1.24. Let A and B be two sets, a mapping $f : A \rightarrow B$ is a relation $\mathscr{R} \subseteq A \times B \mid$ that satisfies the following property: $\forall x \in A$ it follows that $\exists! y \in B$ such that $(x, y) \in \mathscr{R}$ (or, equivalently $y = f(x)$).

Example 1.12. The latter two definitions tell us that for example $(2, 3)$ and $(2, 6)$ cannot be both element of a function. We can express the same concept by stating that if $f(2) = 3$ then it cannot happen that $f(2) = 6$. In other words, if we fix $x = 2$ then we can have only one y value such that $y = f(x)$.

Thus, although functions are often interpreted as "laws" that connect two sets, mathematically, a function is any set (subset of a Cartesian product) for which the property in Definition 1.24 is valid.

If we consider again the caveman example, he generates a mapping between a physical situation and the position of the fingers. More precisely, this mapping allows the relation of only one amount of object to only one position of the fingers. This means that enumeration is a special mapping $f\ A \to B$ that simultaneously satisfies the two properties described in the following two definitions.

Definition 1.25. Let $f\ A \to B$ be a mapping. This mapping is said to be an *injection* (or the function is injective) if the function values of two different elements is always different: $\forall x_1$ and $x_2 \in A$ if $x_1 \neq x_2$ then $f(x_1) \neq f(x_2)$.

Example 1.13. Let us consider the set $A = \{0,1,2,3,4\}$ and the Cartesian product $A \times A = A^2$. Let us consider the following function $f \subset A^2$:

$$f = \{(0,0),(1,1),(2,4),(3,4),(4,3)\}.$$

The function f is not injective since it contains both $(2,4)$ and $(3,4)$. We can rewrite this statement as: although $x_1 \neq x_2$ it happens that $f(x_1) = f(x_2)$ where $x_1 = 2$ and $x_2 = 3$. Clearly, $f(2) = f(3) = 4$.

Definition 1.26. Let $f\ A \to B$ be a mapping. This mapping is said to be a *surjection* (or the function is surjective) if all the elements of B are mapped by an element of A: $\forall y \in B$ it follows that $\exists x \in A$ such that $y = f(x)$.

Example 1.14. Let us consider the sets $A = \{0,1,2,3,4\}$ and $B = \{0,1,2,7\}$. Let us consider the Cartesian product $A \times B$ and the following function $f \subset A \times B$:

$$f = \{(0,0),(1,0),(2,1),(3,1),(4,2)\}.$$

The function f is not surjective since $7 \in B$ is not mapped by the function. In other words, there is no element $x \in A$ such that $(x,7) \in f$ (equivalently there is no element $x \in A$ such that $f(x) = 7$).

Definition 1.27. Let $f\ A \to B$ be a mapping. This mapping is said to be a *bijection* (or the function is bijective) when both injection and surjection are verified, i.e. when the function f is both injective and surjective.

Example 1.15. Let us consider the set $A = \{0,1,2,3,4\}$ and the Cartesian product $A \times A = A^2$. Let us consider the following function $f \subset A^2$ ($f : A \to A$):

$$f = \{(0,0),(1,1),(2,2),(3,4),(4,3)\}.$$

- Since this function uses all the elements of the codomain, the function is surjective.
- Since the function never takes the same value for two distinct elements of the domain, it is injective.

Thus, the function is bijective.

Within our example, surjection says that each physical situation can be potentially represented (by means of a position of the fingers if up to ten objects are involved). Injection says that each representation is unique (and hence unambiguous).

It must be remarked that two sets are equivalent if a bijection between them exists. In our example, the position of the fingers is thus a symbol that univocally identifies a quantity, i.e. the set of quantities and the set of finger positions (symbols) are equivalent. These symbols represent another important primitive concept in mathematics and is called *number*. This concept was previously used when the cardinality of sets was introduced. The concept of a enumeration was also intuitively introduced above as a special function which is injective and surjective (and thus bijective). The following proposition gives a formal justification of this fact.

Proposition 1.4. *Let $f A \rightarrow B$ be a mapping. Let n be the cardinality of A and m be the cardinality of B. If f is bijective then $n = m$.*

Proof. If f is bijective, then f is injective and surjective.

Since f is injective

$$\forall x_1, x_2 \in A \text{ with } x_1 \neq x_2 \text{ it follows that } f(x_1) \neq f(x_2).$$

This means that it cannot happen that for two distinct x_1 and x_2 there is only one $y = f(x_1) = f(x_2)$. On the contrary, for each pair of distinct elements of A we have a pair of distinct elements of B. Thus, if we map the n elements of A we get n elements of B. This happens only if B contains at least n elements. This means that the number of elements of A cannot be greater than the number of elements of B or, in other words, $n \leq m$.

Since f is surjective

$$\forall y \in B \; \exists x \in A \text{ such that } y = f(x).$$

This means that there are no elements of B which are not mapped by an element of A. In other words, $m \leq n$.

Since both the properties are verified, it follows that $m = n$. $\quad \square$

This proposition is important since it is one useful mathematical tool: in order to prove that two sets have the same cardinality, we need to find a bijection between them.

1.5 Number Sets

A set can be composed of numbers, i.e. the elements of a set are numbers. In this case it will be a *number set*. Before introducing number set we need one more primitive concept which will be the last in in this book. This concept is the *infinity*, indicated as ∞. We will consider ∞ as a special number that is larger than any possible number we can think. Any other number but ∞ is said *finite*. Thanks to this introduction, we can further characterize the sets by these definitions.

Definition 1.28. A set is said *finite* if its cardinality is a finite number. Conversely, a set is said *infinite* if its cardinality is ∞.

Definition 1.29. Let A be a number set. A is said *continuous* if $\forall x_0 \in A : \exists x \mid |x - x_0| < \varepsilon$, regardless how small ε is taken.

This means that a number in a continuous set is contiguously surrounded by neighbours. In other words, in a continuous set we cannot identify a minimum neighbourhood radius $\varepsilon > 0$ that separates two neighbour numbers. On the contrary, when an $\varepsilon > 0$ can be found to separate two neighbour numbers the set is said *discrete*. These definitions implicitly state that continuous sets are always infinite. More specifically, these sets are *uncountably infinite*. A discrete set composed of infinite elements is still an infinite set but in a different way as it is *countably infinite*.

Definition 1.30. Countability. A set A is said to be *countably infinite* if A can be put into a bijective relation with the set of natural numbers $\mathbb{N}$. If the set A is infinite but cannot be put into a bijective relation with the set of natural numbers, the set A is *uncountably infinite*.

The set of natural numbers $\mathbb{N}$ can be defined as a discrete set $\{0, 1, 2, \ldots\}$.

Other relevant sets are:

- relative numbers $\mathbb{Z} = \{\ldots, -2, -1, 0, 1, 2 \ldots\}$

- rational numbers $\mathbb{Q}$: the set containing all the possible fractions $\frac{x}{y}$ where x and $y \in \mathbb{Z}$ and $y \neq 0$

- real numbers $\mathbb{R}$: the set containing $\mathbb{Q}$ and all the other numbers that cannot be expressed as fractions of relative numbers

- complex numbers $\mathbb{C}$: the set of numbers that can be expressed as $a + jb$ where $a, b \in \mathbb{R}$ and the imaginary unit $j = \sqrt{-1}$, see Chap. 5.

It can be easily seen that all the number sets above except from $\mathbb{C}$ are totally ordered sets with respect to the relation $\leq$, i.e. $\forall x, y$ we can always assess whether $x \leq y$ or $y \leq x$.

Definition 1.31. An *interval* is a continuous subset of $\mathbb{R}$ delimited by infimum and supremum. Let a be the infimum and b the supremum, respectively. The interval is indicated as $[a, b]$ to denote that a and b belong to the interval. Conversely, the notation $]a, b[$ denotes that infimum and supremum do not belong to the interval.

Example 1.16. Let us consider the set $X \subset \mathbb{N}$, $\{2, 4, 7, 12\}$ and the relation "to be less or equal" $\leq$. It follows that 14 is an upper bound while 12 is the supremum. The infimum is 2 while a lower bound is 1.

Example 1.17. Let us consider now the same relation and the set $X \subset \mathbb{R}$ defined as $\forall x \in \mathbb{R}$ such that $0 \leq x \leq 1$. The supremum and infimum are 1 and 0, respectively. If the set is defined as $\forall x \in \mathbb{R}$ such that $0 < x < 1$, still the upremum and infimum are 1 and 0, respectively. Finally, if we consider the set $X \subset \mathbb{R}$ defined as $\forall x \in \mathbb{R}$ such that $0 \leq x$, this set has no supremum.

Example 1.18. Let us consider $x, y \in \mathbb{Z} \setminus \{0\}$ and the following relation $x \mathscr{R} y$: $xy > 0$.

- Let us check the reflexivity of this relation: $x \mathscr{R} x$ means $xx = x^2 > 0$ which is true.
- Let us check the symmetry: if $xy > 0$ then $yx > 0$ which is true.
- Let us check the transitivity: if $xy > 0$ and $yz > 0$ then x has the same sign of y and y has the same sign of z. It follows that x has the same sign of z. Hence $xz > 0$, i.e. the transitivity is verified.

This means that the relation above is an equivalence relation.

Example 1.19. Let us now consider the set $\mathbb{N} \setminus \{0\}$. Let us define the relation $\mathscr{R}$ "to be a divisor".

- Let us check the reflexivity of this relation: a number is always a divisor of itself. Hence the relation is reflexive.
- Let us check the symmetry: if $\frac{x}{y} = k \in \mathbb{N}$ then $\frac{y}{x} = p = \frac{1}{k}$ which is surely not $\in \mathbb{N}$. Hence this relation is antisymmetric.
- Let us check the transitivity: if $\frac{x}{y} = k \in \mathbb{N}$ and $\frac{y}{z} = h \in \mathbb{N}$ then $\frac{x}{z} = \frac{kyh}{y} = kh$. The product of two natural numbers is a natural number. Hence, the relation is also transitive.

This means that the relation above is of partial order.

Example 1.20. Let us consider $x, y \in \mathbb{Z}$ and the following relation $x \mathscr{R} y$: $x - y$ is dividable by 4.

- Let us check the reflexivity of this relation: $\frac{x-x}{4} = 0 \in \mathbb{Z}$. Hence the relation is reflexive.
- Let us check the symmetry: if $\frac{x-y}{4} = k \in \mathbb{Z}$ then $\frac{y-x}{4} = p = -k$ which is $\in \mathbb{Z}$. Hence this relation is symmetric.
- Let us check the transitivity: if $\frac{x-y}{4} = k \in \mathbb{Z}$ and $\frac{y-z}{4} = h \in \mathbb{Z}$ then $\frac{x-z}{4} = \frac{x-y+y-z}{4} = k+h$. The sum of these two numbers is $\in \mathbb{Z}$. Hence, the relation is also transitive.

This means that the relation above is an equivalence relation.

Example 1.21. Let us consider the following set

$$E = \{(x,y) \in \mathbb{R}^2 \,|\, (x \geq 0) \, AND \, (0 \leq y \leq x)\}$$

and the relation

$$(a,b) \mathscr{R} (c,d) : (a-c, b-d) \in E.$$

- Let us check the reflexivity of this relation: $(a-a, b-b) = (0,0) \in E$. Hence the relation is reflexive.

- Let us check the symmetry: if $(a-c, b-d) \in E$ then $a-c \geq 0$. If the relation is symmetric $(c-a, d-b) \in E$ with $c-a \geq 0$. Hence if it is symmetric $a-c = c-a$. This is possible only when $a=c$. Moreover, $0 \leq d-b \leq c-a = 0$. This means that $b=d$. In other words, the symmetry occurs only if the two solutions are coincident. In all the other cases this relation is never symmetric. It is then antisymmetric.

- Let us check the transitivity. We know that $(a-c, b-d) \in E$ and if $(c-e, d-e) \in E$. For the hypothesis $a-c \geq 0$, $c-e \geq 0$, $0 \leq b-d \leq a-c$, and $0 \leq d-f \leq c-e$. If we sum positive numbers we obtain positive numbers. Hence, $(a-c+c-e) = (a-e) \geq 0$ and $0 \leq b-d+d-f \leq a-c+c-e$ that is $0 \leq b-f \leq a-e$. Hence, the transitivity is valid.

The relation above is of partial order.

1.6 A Preliminary Introduction to Algebraic Structures

If a set is a primitive concept, on the basis of a set, algebraic structures are sets that allow some operations on their elements and satisfy some properties. Although an in depth analysis of algebraic structures is out of the scopes of this chapter, this section gives basic definitions and concepts. More advanced concepts related to algebraic structures will be given in Chap. 7.

Definition 1.32. An *operation* is a function $f : A \rightarrow B$ where $A \subset X_1 \times X_2 \times \ldots \times X_k$, $k \in \mathbb{N}$. The k value is said *arity* of the operation.

Definition 1.33. Let us consider a set A and an operation $f : A \rightarrow B$. If A is $X \times X \times \ldots \times X$ and B is X, i.e. the result of the operation is still a member of the set, the set is said to be *closed* with respect to the operation f.

Definition 1.34. Ring. A *ring* R is a set equipped with two operations called *sum* and *product*. The sum is indicated with a + sign while the product operator is simply omitted (the product of x_1 by x_2 is indicated as $x_1 x_2$). Both these operations process two elements of R and return an element of R (R is closed with respect to these two operations). In addition, the following properties must be valid.

- commutativity (sum): $x_1 + x_2 = x_2 + x_1$
- associativity (sum): $(x_1 + x_2) + x_3 = x_1 + (x_2 + x_3)$
- neutral element (sum): $\exists$ an element $0 \in R$ such that $\forall x \in R : x + 0 = x$
- inverse element (sum): $\forall x \in R : \exists (-x) \,|\, x + (-x) = 0$
- associativity (product): $(x_1 x_2) x_3 = x_1 (x_2 x_3)$
- distributivity 1: $x_1 (x_2 + x_3) = x_1 x_2 + x_1 x_3$
- distributivity 2: $(x_2 + x_3) x_1 = x_2 x_1 + x_3 x_1$
- neutral element (product): $\exists$ an element $1 \in R$ such that $\forall x \in R \, x1 = 1x = x$

The inverse element with respect to the sum is also named *opposite element*.

Definition 1.35. Let R be a ring. If in addition to the ring properties also the

- commutativity (product): $x_1 x_2 = x_2 x_1$

is valid, then the ring is said to be *commutative*.

Definition 1.36. Field. A *field* is a commutative ring which contains an inverse element with respect to the product for every element of the field except 0. In other words, for a field, indicated here with F, besides the commutative ring properties, also the

- inverse element (product): $\forall x \in F \setminus \{0\} : \exists \left(x^{-1}\right) | xx^{-1} = 1$

is valid.

For example, if we consider the set of real numbers $\mathbb{R}$, and associate to it the sum and product operations, we obtain the *real field*.

Exercises

1.1. Prove the following statement

$$A \cup (A \cap B) = A.$$

1.2. Prove the associativity of union operation

$$(A \cup B) \cup C = A \cup (B \cup C).$$

1.3. Calculate $A \times B$ where

$$A = \{a, b, c\}$$

and

$$B = \{x \in \mathbb{Z} | x^2 - 2x - 8 = 0\}.$$

1.4. Let us consider the following sets

$$A = \{1, 2, 3\}$$
$$B = \{1, 2, 3\}$$

Let the relation $\mathscr{R} \subset A \times B$ be a set defined as

$$\mathscr{R} = \{(1,1), (1,2), (1,3), (2,2), (2,3), (3,3)\}.$$

Verify reflexivity, symmetry and transitivity of $\mathscr{R}$ and then (1) assess whether or not $\mathscr{R}$ is an order relation; (2) assess whether or not $\mathscr{R}$ is an equivalence relation.

1.5. Let us consider the following sets

$$A = \{0,1,2,3\}$$
$$B = \{1,2,3\}$$

Let $f \subset A \times B$ be a set defined as

$$f = \{(0,1),(1,1),(2,2),(3,3)\}$$

Assess whether or not f is a function and whether or not f is injective. Justify your answers by using the relevant definitions.

Chapter 2
Matrices

2.1 Numeric Vectors

Although this chapter intentionally refers to the set of real numbers $\mathbb{R}$ and its sum and multiplication operations, all the concepts contained in this chapter can be easily extended to the set of complex numbers $\mathbb{C}$ and the complex field. This fact is further remarked in Chap. 5 after complex numbers and their operations are introduced.

Definition 2.1. Numeric Vector. Let $n \in \mathbb{N}$ and $n > 0$. The set generated by the Cartesian product of $\mathbb{R}$ by itself n times ($\mathbb{R} \times \mathbb{R} \times \mathbb{R} \times \mathbb{R} \ldots$) is indicated with $\mathbb{R}^n$ and is a set of ordered n-tuples of real numbers. The generic element $\mathbf{a} = (a_1, a_2, \ldots, a_n)$ of this set is named *numeric vector* or simply vector of order n on the real field and the generic $a_i \; \forall i$ from 1 to n is said the ith *component* of the vector $\mathbf{a}$.

Example 2.1. The n-tuple

$$\mathbf{a} = \left(1, 0, 56.3, \sqrt{2}\right)$$

is a vector of $\mathbb{R}^4$.

Definition 2.2. Scalar. A numeric vector $\lambda \in \mathbb{R}^1$ is said *scalar*.

Definition 2.3. Let $\mathbf{a} = (a_1, a_2, \ldots, a_n)$ and $\mathbf{b} = (b_1, b_2, \ldots, b_n)$ be two numeric vectors $\in \mathbb{R}^n$. The *sum* of these two vectors is the vector $\mathbf{c} = (a_1 + b_1, a_2 + b_2, \ldots, a_n + b_n)$ generated by the sum of the corresponding components.

Example 2.2. Let us consider the following vectors of $\mathbb{R}^3$

$$\mathbf{a} = (1, 0, 3)$$
$$\mathbf{b} = (2, 1, -2).$$

© Springer Nature Switzerland AG 2019
F. Neri, *Linear Algebra for Computational Sciences and Engineering*,
https://doi.org/10.1007/978-3-030-21321-3_2

The sum of these two vectors is

$$\mathbf{a}+\mathbf{b} = (3,1,1).$$

Definition 2.4. Let $\mathbf{a} = (a_1,a_2,\ldots,a_n)$ be a numeric vector $\in \mathbb{R}^n$ and λ a number $\in \mathbb{R}$. The *product of a vector by a scalar* is the vector $\mathbf{c} = (\lambda a_1, \lambda a_2, \ldots, \lambda a_n)$ generated by the product of λ by each corresponding component.

Example 2.3. Let us consider the vector $\mathbf{a} = (1,0,4)$ and the scalar $\lambda = 2$. The product of this scalar by this vector is

$$\lambda\mathbf{a} = (2,0,8).$$

Definition 2.5. Let $\mathbf{a} = (a_1,a_2,\ldots,a_n)$ and $\mathbf{b} = (b_1,b_2,\ldots,b_n)$ be two numeric vectors $\in \mathbb{R}^n$. The *scalar product* of $\mathbf{a}$ by $\mathbf{b}$ is a real number

$$\mathbf{ab} = c = a_1b_1 + a_2b_2, \ldots, a_nb_n$$

generated by the sum of the products of each pair of corresponding components.

Example 2.4. Let us consider again

$$\mathbf{a} = (1,0,3)$$
$$\mathbf{b} = (2,1,-2).$$

The scalar product of these to vectors is

$$\mathbf{ab} = (1\cdot2) + (0\cdot1) + (3\cdot(-2)) = 2+0-6 = -4.$$

Let $\mathbf{a},\mathbf{b},\mathbf{c} \in \mathbb{R}^n$ and $\lambda \in \mathbb{R}$. It can be proved that the following properties of the scalar product are valid.

- symmetry: $\mathbf{ab} = \mathbf{ba}$
- associativity: $\lambda(\mathbf{ba}) = (\lambda\mathbf{a})\mathbf{b} = \mathbf{a}(\lambda\mathbf{b})$
- distributivity: $\mathbf{a}(\mathbf{b}+\mathbf{c}) = \mathbf{ab}+\mathbf{ac}$

2.2 Basic Definitions About Matrices

Definition 2.6. Matrix. Let $m,n \in \mathbb{N}$ and both $m,n > 0$. A matrix $(m \times n)$ $\mathbf{A}$ is a generic table of the kind:

$$\mathbf{A} = \begin{pmatrix} a_{1,1} & a_{1,2} & \ldots & a_{1,n} \\ a_{2,1} & a_{2,2} & \ldots & a_{2,n} \\ \ldots & \ldots & \ldots & \ldots \\ a_{m,1} & a_{m,2} & \ldots & a_{m,n} \end{pmatrix}$$

where each *matrix element* $a_{i,j} \in \mathbb{R}$. If $m = n$ the matrix is said *square* while it is said *rectangular* otherwise.

The numeric vector $\mathbf{a_i} = (a_{i,1}, a_{i,2}, \ldots, a_{i,n})$ is said generic i^{th} *row vector* while $\mathbf{a^j} = (a_{1,j}, a_{2,j}, \ldots, a_{m,j})$ is said generic j^{th} *column vector*.

The set containing all the matrices of real numbers having m rows and n columns is indicated with $\mathbb{R}_{m,n}$.

Definition 2.7. A matrix is said null $\mathbf{O}$ if all its elements are zeros.

Example 2.5. The null matrix of $\mathbb{R}_{2,3}$ is

$$\mathbf{O} = \begin{pmatrix} 0\,0\,0 \\ 0\,0\,0 \end{pmatrix}$$

Definition 2.8. Let $\mathbf{A} \in \mathbb{R}_{m,n}$. The *transpose* matrix of $\mathbf{A}$ is a matrix $\mathbf{A^T}$ whose elements are the same of $\mathbf{A}$ but $\forall i, j$: $a_{j,i} = a_{i,j}^T$.

Example 2.6.

$$\mathbf{A} = \begin{pmatrix} 2\,7\,3.4\ \sqrt{2} \\ 5\,0\ 4\ \ 1 \end{pmatrix}$$

$$\mathbf{A^T} = \begin{pmatrix} 2 & 5 \\ 7 & 0 \\ 3.4 & 4 \\ \sqrt{2} & 1 \end{pmatrix}$$

It can be easily proved that the transpose of the transpose of a matrix is the matrix itself: $\left(\mathbf{A^T}\right)^{\mathbf{T}}$.

Definition 2.9. A matrix $\mathbf{A} \in \mathbb{R}_{n,n}$ is said n *order square matrix*.

Definition 2.10. Let $\mathbf{A} \in \mathbb{R}_{n,n}$. The *diagonal of a matrix* is the ordered n-tuple that displays the same index twice: $\forall i$ from 1 to n $a_{i,i}$.

Definition 2.11. Let $\mathbf{A} \in \mathbb{R}_{n,n}$. The *trace of a matrix* $\text{tr}(\mathbf{A})$ is the sum of the diagonal elements: $\text{tr}(\mathbf{A}) = \sum_{i=1}^{n} a_{i,i}$.

Example 2.7. The diagonal of the matrix

$$\begin{pmatrix} 1\,3\,0 \\ 9\,2\,1 \\ 0\,1\,2 \end{pmatrix}$$

is $(1, 2, 2)$ and the trace is $1 + 2 + 2 = 5$.

Definition 2.12. Let $\mathbf{A} \in \mathbb{R}_{n,n}$. The matrix $\mathbf{A}$ is said *lower triangular* if all its elements above the diagonal (the elements $a_{i,j}$ where $j > i$) are zeros. The matrix $\mathbf{A}$ is said *upper triangular* if all its elements below the diagonal (the elements $a_{i,j}$ where $i > j$) are zeros.

Example 2.8. The following matrix is lower triangular:

$$\mathbf{L} = \begin{pmatrix} 5 & 0 & 0 \\ 4 & 1 & 0 \\ 3 & 1 & 8 \end{pmatrix}.$$

Definition 2.13. An *identity matrix* $\mathbf{I}$ is a square matrix whose diagonal elements are all ones while all the other extra-diagonal elements are zeros.

Example 2.9. The identity matrix of $\mathbb{R}_{3,3}$ is

$$\mathbf{I} = \begin{pmatrix} 1 & 0 & 0 \\ 0 & 1 & 0 \\ 0 & 0 & 1 \end{pmatrix}$$

Definition 2.14. A matrix $\mathbf{A} \in \mathbb{R}_{n,n}$ is said *symmetric* when $\forall i, j : a_{i,j} = a_{j,i}$.

Example 2.10. The following matrix is symmetric:

$$\mathbf{A} = \begin{pmatrix} 2 & 3 & 0 \\ 3 & 1 & 2 \\ 0 & 2 & 4 \end{pmatrix}$$

Proposition 2.1. *Let $\mathbf{A}$ be a symmetric matrix. It follows that $\mathbf{A}^\mathbf{T} = \mathbf{A}$.*

Proof. From the definition of symmetric matrix $\mathbf{A} \in \mathbb{R}_{n,n}$:

$$\forall i, j : a_{i,j} = a_{j,i}.$$

If we transpose the matrix we have

$$\forall i, j : a_{i,j}^T = a_{j,i} = a_{i,j},$$

that is $\mathbf{A}^\mathbf{T} = \mathbf{A}$. □

Example 2.11. Let us consider again the symmetric matrix $\mathbf{A}$ from Example 2.10 and let us calculate the transpose

$$\mathbf{A}^\mathbf{T} = \begin{pmatrix} 2 & 3 & 0 \\ 3 & 1 & 2 \\ 0 & 2 & 4 \end{pmatrix} = \mathbf{A}.$$

2.3 Matrix Operations

Definition 2.15. Let $\mathbf{A}, \mathbf{B} \in \mathbb{R}_{m,n}$. The *matrix sum* $\mathbf{C}$ is defined as: $\forall i, j : c_{i,j} = a_{i,j} + b_{i,j}$.

Example 2.12. The sum of two matrices is shown below

$$\begin{pmatrix} 1 & 2 & 3 \\ 2 & 2 & 0 \\ 1 & 0 & 1 \end{pmatrix} + \begin{pmatrix} 0 & 5 & 1 \\ 0 & 3 & 1 \\ 2 & 0 & 1 \end{pmatrix} = \begin{pmatrix} 1 & 7 & 4 \\ 2 & 5 & 1 \\ 3 & 0 & 2 \end{pmatrix} \tag{2.1}$$

The following properties can be easily proved for the sum operation amongst matrices.

- commutativity: $\mathbf{A} + \mathbf{B} = \mathbf{B} + \mathbf{A}$
- associativity: $(\mathbf{A} + \mathbf{B}) + \mathbf{C} = \mathbf{A} + (\mathbf{B} + \mathbf{C})$
- neutral element: $\mathbf{A} + \mathbf{O} = \mathbf{A}$
- opposite element: $\forall \mathbf{A} \in \mathbb{R}_{m,n} : \exists! \mathbf{B} \in \mathbb{R}_{m,n} | \mathbf{A} + \mathbf{B} = \mathbf{O}$

The proof of these properties can be carried out simply considering that commutativity and associativity are valid for the sum between numbers. Since the sum between two matrices is the sum over multiple numbers the properties are still valid for matrices.

Definition 2.16. Let $\mathbf{A} \in \mathbb{R}_{m,n}$ and $\lambda \in \mathbb{R}$. The *product of a scalar by a matrix* is a matrix $\mathbf{C}$ defined as: $\forall i, j : c_{i,j} = \lambda a_{i,j}$.

Example 2.13. The product of a scalar $\lambda = 2$ by the matrix

$$\begin{pmatrix} 2 & 1 & 0 \\ 1 & -1 & 4 \end{pmatrix}$$

is

$$\lambda \mathbf{A} = \begin{pmatrix} 4 & 2 & 0 \\ 2 & -2 & 8 \end{pmatrix}.$$

The following properties can be easily proved for the product of a scalar by a matrix.

- associativity: $\forall \mathbf{A} \in \mathbb{R}_{m,n}$ and $\forall \lambda, \mu \in \mathbb{R} : (\lambda \mu) \mathbf{A} = (\mathbf{A} \mu) \lambda = (\mathbf{A} \lambda) \mu$
- distributivity of the product of a scalar by the sum of two matrices: $\forall \mathbf{A}, \mathbf{B} \in \mathbb{R}_{m,n}$ and $\forall \lambda \in \mathbb{R} \lambda (\mathbf{A} + \mathbf{B}) = \lambda \mathbf{A} + \lambda \mathbf{B}$
- distributivity of the product of a matrix by the sum of two scalars: $\forall \mathbf{A} \in \mathbb{R}_{m,n}$ and $\forall \lambda, \mu \in \mathbb{R} : (\lambda + \mu) \mathbf{A} = \lambda \mathbf{A} + \mu \mathbf{A}$

Definition 2.17. Let $\mathbf{A} \in \mathbb{R}_{m,r}$ and $\mathbf{B} \in \mathbb{R}_{r,n}$. The *product of matrices* $\mathbf{A}$ *and* $\mathbf{B}$ is a matrix $\mathbf{C} = \mathbf{AB}$ whose generic element $c_{i,j}$ is defined in the following way:

$$c_{i,j} = \mathbf{a_i b^j} = \sum_{k=1}^{n} a_{i,k} b_{k,j} = a_{i,1} b_{1,j} + a_{i,2} b_{2,j} + \ldots + a_{i,n} b_{m,j}.$$

The same definition with a different notation can be expressed in the following way.

Definition 2.18. Let us consider two matrices $\mathbf{A} \in \mathbb{R}_{m,r}$ and $\mathbf{B} \in \mathbb{R}_{r,n}$. Let us express the matrices as vectors of vectors: row vectors for the matrix $\mathbf{A}$, column vectors for the matrix $\mathbf{B}$. The matrices $\mathbf{A}$ and $\mathbf{B}$ are

$$\mathbf{A} = \begin{pmatrix} a_{1,1} & a_{1,2} & \dots & a_{1,r} \\ a_{2,1} & a_{2,2} & \dots & a_{2,r} \\ \dots & \dots & \dots & \dots \\ a_{m,1} & a_{m,2} & \dots & a_{m,r} \end{pmatrix} = \begin{pmatrix} \mathbf{a_1} \\ \mathbf{a_2} \\ \dots \\ \mathbf{a_m} \end{pmatrix}$$

and

$$\mathbf{B} = \begin{pmatrix} b_{1,1} & b_{1,2} & \dots & b_{1,n} \\ b_{2,1} & b_{2,2} & \dots & b_{2,n} \\ \dots & \dots & \dots & \dots \\ b_{r,1} & b_{r,2} & \dots & b_{r,n} \end{pmatrix} = \begin{pmatrix} \mathbf{b^1} & \mathbf{b^2} & \dots & \mathbf{b^n} \end{pmatrix}.$$

The *product of matrices* $\mathbf{A}$ *and* $\mathbf{B}$ is the following matrix $\mathbf{C}$ where each element is the scalar product of a row vector of $\mathbf{A}$ and a column vector of $\mathbf{B}$:

$$\mathbf{C} = \begin{pmatrix} \mathbf{a_1 b^1} & \mathbf{a_1 b^2} & \dots & \mathbf{a_1 b^n} \\ \mathbf{a_2 b^1} & \mathbf{a_2 b^2} & \dots & \mathbf{a_2 b^n} \\ \dots & \dots & \dots & \dots \\ \mathbf{a_m b^1} & \mathbf{a_m b^2} & \dots & \mathbf{a_m b^n} \end{pmatrix}.$$

Example 2.14. Let us multiply the matrix $\mathbf{A}$ by the matrix $\mathbf{B}$.

$$\mathbf{A} = \begin{pmatrix} 2 & 7 & 3 & 1 \\ 5 & 0 & 4 & 1 \end{pmatrix}$$

$$\mathbf{B} = \begin{pmatrix} 1 & 2 \\ 2 & 5 \\ 8 & 0 \\ 2 & 2 \end{pmatrix}$$

$$\mathbf{C} = \mathbf{AB} = \begin{pmatrix} \mathbf{a_1 b^1} & \mathbf{a_1 b^2} \\ \mathbf{a_2 b^1} & \mathbf{a_2 b^2} \end{pmatrix} = \begin{pmatrix} 42 & 41 \\ 39 & 12 \end{pmatrix}$$

The following properties can be easily proved for the product between two matrices.

- left distributivity: $\mathbf{A}(\mathbf{B} + \mathbf{C}) = \mathbf{AB} + \mathbf{AC}$
- right distributivity: $(\mathbf{B} + \mathbf{C})\mathbf{A} = \mathbf{BA} + \mathbf{CA}$
- associativity: $\mathbf{A}(\mathbf{BC}) = (\mathbf{AB})\mathbf{C}$

- transpose of the product: $(\mathbf{AB})^{\mathbf{T}} = \mathbf{B}^{\mathbf{T}}\mathbf{A}^{\mathbf{T}}$
- neutral element: $\forall \mathbf{A} : \mathbf{AI} = \mathbf{A}$
- absorbing element: $\forall \mathbf{A} : \mathbf{AO} = \mathbf{O}$

Theorem 2.1. *Let us consider two compatible matrices*

$$\mathbf{A} \in \mathbb{R}_{m,r}$$
$$\mathbf{B} \in \mathbb{R}_{r,n}.$$

It follows that

$$(\mathbf{AB})^{\mathbf{T}} = \mathbf{B}^{\mathbf{T}}\mathbf{A}^{\mathbf{T}}.$$

Proof. Without a loss of generality, let us give the proof for $\mathbf{A}$ and $\mathbf{B}$ square matrices

$$\mathbf{A} \in \mathbb{R}_{n,n}$$
$$\mathbf{B} \in \mathbb{R}_{n,n}.$$

Let us consider the matrices

$$\mathbf{A} = \begin{pmatrix} a_{1,1} & a_{1,2} & \ldots & a_{1,n} \\ a_{2,1} & a_{2,2} & \ldots & a_{2,n} \\ \ldots & \ldots & \ldots & \ldots \\ a_{n,1} & a_{n,2} & \ldots & a_{n,n} \end{pmatrix} = \begin{pmatrix} \mathbf{a_1} \\ \mathbf{a_2} \\ \ldots \\ \mathbf{a_n} \end{pmatrix}$$

and

$$\mathbf{B} = \begin{pmatrix} b_{1,1} & b_{1,2} & \ldots & b_{1,n} \\ b_{2,1} & b_{2,2} & \ldots & b_{2,n} \\ \ldots & \ldots & \ldots & \ldots \\ b_{n,1} & b_{n,2} & \ldots & b_{n,n} \end{pmatrix} = \begin{pmatrix} \mathbf{b^1} & \mathbf{b^2} & \ldots & \mathbf{b^n} \end{pmatrix}.$$

Let us calculate the product of matrices $\mathbf{A}$ and $\mathbf{B}$:

$$\mathbf{AB} = \begin{pmatrix} \mathbf{a_1 b^1} & \mathbf{a_1 b^2} & \ldots & \mathbf{a_1 b^n} \\ \mathbf{a_2 b^1} & \mathbf{a_2 b^2} & \ldots & \mathbf{a_2 b^n} \\ \ldots & \ldots & \ldots & \ldots \\ \mathbf{a_n b^1} & \mathbf{a_n b^2} & \ldots & \mathbf{a_n b^n} \end{pmatrix}.$$

Let us now calculate the transpose $(\mathbf{AB})^{\mathbf{T}}$ and place its elements in a matrix $\mathbf{C}$:

$$\mathbf{C} = (\mathbf{AB})^{\mathbf{T}} = \begin{pmatrix} \mathbf{a_1 b^1} & \mathbf{a_2 b^1} & \ldots & \mathbf{a_n b^1} \\ \mathbf{a_1 b^2} & \mathbf{a_2 b^2} & \ldots & \mathbf{a_n b^2} \\ \ldots & \ldots & \ldots & \ldots \\ \mathbf{a_1 b^n} & \mathbf{a_2 b^n} & \ldots & \mathbf{a_n b^n} \end{pmatrix}.$$

Let us consider an arbitrary element of $\mathbf{C}$, e.g.

$$c_{2,1} = \mathbf{a}_1 \mathbf{b}^2 = \sum_{k=1}^{n} = a_{1,k} b_{k,2}.$$

The generic element $c_{i,j}$ of the matrix $\mathbf{C}$ is

$$c_{i,j} = \mathbf{a}_j \mathbf{b}^i = \sum_{k=1}^{n} a_{j,k} b_{k,i}.$$

Let us now calculate

$$\mathbf{B}^{\mathbf{T}} = \begin{pmatrix} b_{1,1} & b_{2,1} & \cdots & b_{n,1} \\ b_{1,2} & b_{2,2} & \cdots & b_{n,2} \\ \cdots & \cdots & \cdots & \cdots \\ b_{1,n} & b_{2,n} & \cdots & b_{n,n} \end{pmatrix} = \begin{pmatrix} \mathbf{b}^1 \\ \mathbf{b}^2 \\ \cdots \\ \mathbf{b}^n \end{pmatrix}$$

and

$$\mathbf{A}^{\mathbf{T}} = \begin{pmatrix} a_{1,1} & a_{2,1} & \cdots & a_{n,1} \\ a_{1,2} & a_{2,2} & \cdots & a_{n,2} \\ \cdots & \cdots & \cdots & \cdots \\ a_{1,n} & a_{2,n} & \cdots & a_{n,n} \end{pmatrix} = \begin{pmatrix} \mathbf{a}_1 & \mathbf{a}_2 & \cdots & \mathbf{a}_n \end{pmatrix}.$$

The product $\mathbf{B}^{\mathbf{T}} \mathbf{A}^{\mathbf{T}}$ is given by

$$\mathbf{C} = \mathbf{B}^{\mathbf{T}} \mathbf{A}^{\mathbf{T}} = \begin{pmatrix} \mathbf{b}^1 \mathbf{a}_1 & \mathbf{b}^1 \mathbf{a}_2 & \cdots & \mathbf{b}^1 \mathbf{a}_n \\ \mathbf{b}^2 \mathbf{a}_1 & \mathbf{b}^2 \mathbf{a}_2 & \cdots & \mathbf{b}^2 \mathbf{a}_n \\ \cdots & \cdots & \cdots & \cdots \\ \mathbf{b}^n \mathbf{a}_1 & \mathbf{b}^n \mathbf{a}_1 & \cdots & \mathbf{b}^n \mathbf{a}_n \end{pmatrix}.$$

We can already see that $(\mathbf{AB})^{\mathbf{T}} = \mathbf{B}^{\mathbf{T}} \mathbf{A}^{\mathbf{T}}$. However, let us consider an arbitrary element of $\mathbf{C}$, e.g.

$$c_{2,1} = \mathbf{b}^2 \mathbf{a}_1 = \sum_{k=1}^{n} b_{k,2} a_{1,k}.$$

The generic element $c_{i,j}$ of the matrix $\mathbf{C}$ is

$$c_{i,j} = \mathbf{a}_j \mathbf{b}^i = \sum_{k=1}^{n} b_{k,i} a_{j,k} = \sum_{k=1}^{n} a_{j,k} b_{k,i}.$$

It follows that $(\mathbf{AB})^{\mathbf{T}} = \mathbf{B}^{\mathbf{T}} \mathbf{A}^{\mathbf{T}}$. $\square$

Example 2.15. Let us consider the following two matrices:

$$\mathbf{A} = \begin{pmatrix} 1 & 0 & 1 \\ 2 & 1 & 0 \\ 3 & 1 & 0 \end{pmatrix}$$

and

$$\mathbf{B} = \begin{pmatrix} 2 & 2 & 4 \\ 0 & 0 & 1 \\ 2 & 1 & 0 \end{pmatrix}.$$

Let us calculate

$$(\mathbf{AB})^{\mathbf{T}} = \begin{pmatrix} 4 & 4 & 6 \\ 3 & 4 & 6 \\ 4 & 9 & 13 \end{pmatrix}$$

and

$$\mathbf{B}^{\mathbf{T}}\mathbf{A}^{\mathbf{T}} = \begin{pmatrix} 2 & 0 & 2 \\ 2 & 0 & 1 \\ 4 & 1 & 0 \end{pmatrix} \begin{pmatrix} 1 & 2 & 3 \\ 0 & 1 & 1 \\ 1 & 0 & 0 \end{pmatrix} = \begin{pmatrix} 4 & 4 & 6 \\ 3 & 4 & 6 \\ 4 & 9 & 13 \end{pmatrix}.$$

We have verified that $(\mathbf{AB})^{\mathbf{T}} = \mathbf{B}^{\mathbf{T}}\mathbf{A}^{\mathbf{T}}$.

It must be observed that the commutativity with respect to the matrix product is generally not valid. It may happen in some cases that $\mathbf{AB} = \mathbf{BA}$. In these cases the matrices are said *commutable* (one with respect to the other). Every matrix $\mathbf{A}$ is commutable with $\mathbf{O}$ (and the result is always $\mathbf{O}$) and with $\mathbf{I}$ (and the result is always $\mathbf{A}$).

Since the commutativity is not valid for the product between matrices, the set $\mathbb{R}_{m,n}$ with sum and product is not a commutative ring.

Example 2.16. Let us consider again the matrices $\mathbf{A}$ and $\mathbf{B}$ from Example 2.15. The product $\mathbf{AB}$ is

$$\mathbf{AB} = \begin{pmatrix} 4 & 3 & 4 \\ 4 & 4 & 9 \\ 6 & 6 & 13 \end{pmatrix}.$$

The product $\mathbf{BA}$ is

$$\mathbf{BA} = \begin{pmatrix} 18 & 6 & 2 \\ 3 & 1 & 0 \\ 4 & 1 & 2 \end{pmatrix}.$$

We can verify that in general $\mathbf{AB} \neq \mathbf{BA}$.

If we interpret a vector as a matrix of n rows and one column (or one row and n columns).

Definition 2.19. If we consider the following two vectors:

$$\mathbf{x} = \begin{pmatrix} x_1 \\ x_2 \\ \dots \\ x_n \end{pmatrix}$$

$$\mathbf{y} = \begin{pmatrix} y_1 \\ y_2 \\ \dots \\ y_n \end{pmatrix}.$$

The scalar product **xy** can be redefined as

$$\mathbf{x}^T\mathbf{y} = \sum_{i=1}^{n} x_i y_i$$

where the transpose is calculated to ensure that the two matrices are compatible.

Example 2.17. Let us consider the following two vectors:

$$\mathbf{x} = \begin{pmatrix} 1 \\ 3 \\ 5 \end{pmatrix}$$

$$\mathbf{y} = \begin{pmatrix} 1 \\ 0 \\ 2 \end{pmatrix}.$$

If we consider **x** and **y** as matrices, the product **xy** would not be viable since the number of columns of **x**, i.e. one, is not equal to the number of rows of **y**, i.e. three. Thus to have the matrices compatible we need to write

$$(1\ 3\ 5) \begin{pmatrix} 1 \\ 0 \\ 2 \end{pmatrix} = \mathbf{x}^T\mathbf{y} = 11.$$

Proposition 2.2. *Let* $\mathbf{A}, \mathbf{B} \in \mathbb{R}_{n,n}$. *The* $\mathrm{tr}(\mathbf{AB}) = \mathrm{tr}(\mathbf{BA})$.

Example 2.18. Let us verify Proposition 2.2 If we consider again the matrices from Example 2.16,

$$\mathrm{tr}(\mathbf{AB}) = 4 + 4 + 13 = 21$$

and

$$\mathrm{tr}(\mathbf{BA}) = 18 + 1 + 2 = 21.$$

2.4 Determinant of a Matrix

Definition 2.20. Let us consider n objects. We will call *permutation* every grouping of these objects. For example, if we consider three objects a, b, and c, we could group them as $a - b - c$ or $a - c - b$, or $c - b - a$ or $b - a - c$ or $c - a - b$ or $b - c - a$. In this case, there are totally six possible permutations. More generally, it can be checked that for n objects there are $n!$ (n factorial) permutations where $n! = (n)(n-1)(n-2)\ldots(2)(1)$ with $n \in \mathbb{N}$ and $(0)! = 1$.

We could fix a reference sequence (e.g. $a - b - c$) and name it *fundamental permutation*. Every time two objects in a permutation follow each other in a reverse order with respect to the fundamental we will call it *inversion*. Let us define *even class permutation* a permutation undergone to an even number of inversions and *odd class permutation* a permutation undergone to an odd number of inversions, see also [1].

In other words, a sequence is an even class permutation if an even number of swaps is necessary to obtain the fundamental permutation. Analogously, a sequence is an odd class permutation if an odd number of swaps is necessary to obtain the fundamental permutation.

Example 2.19. Let us consider the fundamental permutation $a - b - c - d$ associated with the objects a, b, c, d. The permutation $d - a - c - b$ is of even class since two swaps are required to reconstruct the fundamental permutation. At first we swap a and d to obtain $a - d - c - b$ and then we swap d and b to obtain the fundamental permutation $a - b - c - d$.

On the contrary, the permutation $d - c - a - b$ is of odd class since three swaps are necessary to reconstruct the fundamental permutation. Let us reconstruct the fundamental permutation step-by-step. At first we swap d and b and obtain $b - c - a - d$. Then, let us swap b and a to obtain $a - c - b - d$. Eventually, we swap c and b to obtain the fundamental permutation $a - b - c - d$.

Definition 2.21. Associate Product of a Matrix. Let us consider a matrix $\mathbf{A} \in \mathbb{R}_{n,n}$ (a square matrix),

$$\mathbf{A} = \begin{pmatrix} a_{1,1} & a_{1,2} & \cdots & a_{1,n} \\ a_{2,1} & a_{2,2} & \cdots & a_{2,n} \\ \cdots & \cdots & \cdots & \cdots \\ a_{n,1} & a_{n,2} & \cdots & a_{n,n} \end{pmatrix}.$$

From this matrix, we can select n elements that do not belong neither to the same row nor to the same column (the selected elements have no common indices). The product of these n elements is here referred to as *associated product* and indicated with the symbol $\varepsilon(c)$. If we order the factors of the associated product according to the row index, the generic associated product can be expressed in the form

$$\varepsilon(c) = a_{1,c1} a_{2,c2} a_{3,c3} \ldots a_{n,cn}.$$

Example 2.20. Let us consider the matrix

$$\begin{pmatrix} 2 & 1 & 4 \\ 3 & 1 & 0 \\ 2 & 1 & 2 \end{pmatrix}.$$

An associated product is $\varepsilon(c) = a_{1,2} a_{2,1} a_{3,3} = (1)(3)(2) = 6$. Two other examples of associated product are $\varepsilon(c) = a_{1,1} a_{2,3} a_{3,2} = (2)(0)(1) = 0$ and $\varepsilon(c) = a_{1,2} a_{2,3} a_{3,1} = (1)(0)(2) = 0$.

Proposition 2.3. *Totally, $n!$ associated products can be extracted from a matrix $\mathbf{A} \in \mathbb{R}_{n,n}$.*

Example 2.21. A matrix $\mathbf{A} \in \mathbb{R}_{3,3}$, i.e. a square matrix with three rows and three columns has six associated products.

Definition 2.22. Let us consider $1 - 2 - \ldots - n$ as the fundamental permutation, the scalar η_k is defined as:

$$\eta_k = \begin{cases} 1 \text{ if } c1\text{-}c2\text{-}\ldots\text{-}cn \text{ is an even class permutation} \\ -1 \text{ if } c1\text{-}c2\text{-}\ldots\text{-}cn \text{ is an odd class permutation} \end{cases}.$$

Example 2.22. Let us consider again the matrix

$$\begin{pmatrix} 2 & 1 & 4 \\ 3 & 1 & 0 \\ 2 & 1 & 2 \end{pmatrix}$$

and the associated product $\varepsilon(c) = a_{1,2}a_{2,1}a_{3,3} = (1)(3)(2) = 6$. The column indices $c1, c2, c3$ and $2, 1, 3$. If we express this fact in terms of permutations, the column indices are the permutation $2 - 1 - 3$. Since one swap is needed to obtain the natural permutation (we need to swap 1 and 2) then the permutation is of odd order. Thus the corresponding coefficient $\eta_k = -1$.

In the case of the associated product $\varepsilon(c) = a_{1,1}a_{2,3}a_{3,2} = (2)(0)(1) = 0$ the permutation associated with the column indices is $1 - 3 - 2$. Thus, also in this case only one swap (3 and 2) is needed to reconstruct the fundamental permutation and consequently $\eta_k = -1$.

In the case of the associated product $\varepsilon(c) = a_{1,2}a_{2,3}a_{3,1} = (1)(0)(2) = 0$, the permutation associated with the column indices is $2 - 3 - 1$. Since we need to swaps to reconstruct the fundamental permutation, i.e. 2 and 3 and then 3 and 1, the permutation is of even class. Thus, the corresponding coefficient is $\eta_k = 1$.

Definition 2.23. Determinant of a Matrix. Let $\mathbf{A} \in \mathbb{R}_{n,n}$, the *determinant* of the matrix $\mathbf{A}$, indicated as $\det \mathbf{A}$, is the function

$$\det : \mathbb{R}_{n,n} \to \mathbb{R}$$

defined as the sum of the $n!$ associated products where each term is weighted by the corresponding η_k:

$$\det \mathbf{A} = \sum_{k=1}^{n!} \eta_k \varepsilon_k(c).$$

Example 2.23. If $n = 1$, $\det \mathbf{A}$ is equal to the only element of the matrix. This can be easily seen considering that

$$\mathbf{A} = (a_{1,1}).$$

It follows that only associated product can be found that is $a_{1,1}$. In this case $\eta_k = 1$ (only one element following the fundamental permutation). Thus

$$\det \mathbf{A} = a_{1,1}.$$

Example 2.24. If $n = 2$, the matrix $\mathbf{A}$ appears as

$$\mathbf{A} = \begin{pmatrix} a_{1,1} & a_{1,2} \\ a_{2,1} & a_{2,2} \end{pmatrix}$$

and its $\det \mathbf{A} = a_{1,1}a_{2,2} - a_{1,2}a_{2,1}$.

Example 2.25. If $n = 3$, the matrix $\mathbf{A}$ appears as

$$\mathbf{A} = \begin{pmatrix} a_{1,1} & a_{1,2} & a_{1,3} \\ a_{2,1} & a_{2,2} & a_{2,3} \\ a_{3,1} & a_{3,2} & a_{3,3} \end{pmatrix}$$

and its $\det \mathbf{A} = a_{1,1}a_{2,2}a_{3,3} + a_{1,2}a_{2,3}a_{3,1} + a_{1,3}a_{2,1}a_{3,2} - a_{1,3}a_{2,2}a_{3,1} - a_{1,2}a_{2,1}a_{3,3} - a_{1,1}a_{2,3}a_{3,2}$. Looking at the column indices it can be noticed that in $(1,2,3)$ is the fundamental permutation, it follows that $(2,3,1)$ and $(3,1,2)$ are even class permutation because two swaps are required to obtain $(1,2,3)$. On the contrary, $(3,2,1)$, $(2,1,3)$, and $(1,3,2)$ are odd class permutations because only one swap allows to be back to $(1,2,3)$.

Example 2.26. Let us consider again the matrix

$$\begin{pmatrix} 2 & 1 & 4 \\ 3 & 1 & 0 \\ 2 & 1 & 2 \end{pmatrix}$$

and let us calculate its determinant:

$$\begin{aligned} \det \mathbf{A} = \\ = (2)(1)(2) + (1)(0)(2) + (4)(3)(1) - (4)(1)(2) - (1)(3)(2) - (2)(0)(1) = \\ = 4 + 0 + 12 - 8 - 6 - 0 = 2. \end{aligned}$$

2.4.1 Linear Dependence of Row and Column Vectors of a Matrix

Definition 2.24. Let $\mathbf{A}$ be a matrix. The ith row is said *linear combination* of the other rows if each of its element $a_{i,j}$ can be expressed as weighted sum of the other elements of the jth column by means of the same scalars $\lambda_1, \lambda_2, \ldots, \lambda_{i-1}, \lambda_{i+1}, \ldots \lambda_n$:

$$\mathbf{a_i} = \lambda_1 \mathbf{a_1} + \lambda_2 \mathbf{a_2} + \cdots + \lambda_{i-1} \mathbf{a_{i-1}} + \lambda_{i+1} \mathbf{a_{i+1}} + \ldots + \lambda_n \mathbf{a_n}.$$

Equivalently, we may express the same concept by considering each row element:

$$\forall j : \exists \lambda_1, \lambda_2, \ldots, \lambda_{i-1}, \lambda_{i+1}, \ldots \lambda_n |$$
$$a_{i,j} = \lambda_1 a_{1,j} + \lambda_2 a_{2,j} + \ldots \lambda_{i-1} a_{i-1,j} + \lambda_{i+1} a_{i+1,j} + \ldots \lambda_n a_{n,j}.$$

Example 2.27. Let us consider the following matrix:

$$\mathbf{A} = \begin{pmatrix} 0 & 1 & 1 \\ 3 & 2 & 1 \\ 6 & 5 & 3 \end{pmatrix}.$$

The third row is a linear combination of the first two by means of scalars $\lambda_1, \lambda_2 = 1, 2$, the third row is equal to the weighted sum obtained by multiplying the first row by 1 and summing to it the second row multiplied by 2:

$$(6, 5, 3) = (0, 1, 1) + 2(3, 2, 1)$$

that is

$$\mathbf{a_3} = \mathbf{a_1} + 2\mathbf{a_2}.$$

Definition 2.25. Let $\mathbf{A}$ be a matrix. The jth column is said *linear combination* of the other column if each of its element $a_{i,j}$ can be expressed as weighted sum of the other elements of the ith row by means of the same scalars $\lambda_1, \lambda_2, \ldots, \lambda_{j-1}, \lambda_{j+1}, \ldots \lambda_n$:

$$\mathbf{a^j} = \lambda_1 \mathbf{a^1} + \lambda_2 \mathbf{a^2} + \cdots + \lambda_{j-1} \mathbf{a^{j-1}} + \lambda_{j+1} \mathbf{a^{j+1}} + \ldots + \lambda_n \mathbf{a^n}.$$

Equivalently, we may express the same concept by considering each row element:

$$\forall i : \exists \lambda_1, \lambda_2, \ldots, \lambda_{j-1}, \lambda_{j+1}, \ldots \lambda_n |$$
$$a_{i,j} = \lambda_1 a_{i,1} + \lambda_2 a_{i,2} + \ldots \lambda_{i-1} a_{i,j-1} + \lambda_{i+1} a_{i,j+1} + \ldots \lambda_n a_{i,n}.$$

Example 2.28. Let us consider the following matrix:

$$\mathbf{A} = \begin{pmatrix} 1 & 2 & 1 \\ 2 & 2 & 4 \\ 1 & 3 & 0 \end{pmatrix}.$$

The third column is a linear combination of the first two by means of scalars $\lambda_1, \lambda_2 = 3, -1$, the third column is equal to the weighted sum obtained by multiplying the first column by 3 and summing to it the second row multiplied by -1:

$$\begin{pmatrix} 1 \\ 4 \\ 0 \end{pmatrix} = 3 \begin{pmatrix} 1 \\ 2 \\ 1 \end{pmatrix} - \begin{pmatrix} 2 \\ 2 \\ 3 \end{pmatrix}.$$

that is

$$\mathbf{a^3} = 3\mathbf{a^1} - \mathbf{a^2}.$$

Definition 2.26. Let $\mathbf{A} \in \mathbb{R}_{m,n}$ be a matrix. The m rows (n columns) are *linearly dependent* if a row (column) composed of all zeros $\mathbf{o} = (0, 0, \ldots, 0)$ can be expressed as the linear combination of the m rows (n columns) by means of nun-null scalars.

In the case of linearly dependent rows, if the matrix $\mathbf{A}$ is represented as a vector of row vectors:

$$\mathbf{A} = \begin{pmatrix} \mathbf{a_1} \\ \mathbf{a_2} \\ \dots \\ \mathbf{a_m} \end{pmatrix}$$

the rows are linearly dependent if

$$\exists \lambda_1, \lambda_2, \dots, \lambda_m \neq 0, 0, \dots, 0$$

such that

$$\mathbf{o} = \lambda_1 \mathbf{a_1} + \lambda_2 \mathbf{a_2} + \dots + \lambda_m \mathbf{a_m}.$$

Example 2.29. The rows in the following matrix

$$\mathbf{A} = \begin{pmatrix} 1 & 2 & 1 \\ 2 & 2 & 4 \\ 4 & 6 & 6 \end{pmatrix}$$

are linearly dependent since

$$\mathbf{o} = -2\mathbf{a_1} - \mathbf{a_2} + \mathbf{a_3}$$

that is a null row can be expressed as the linear combination of the row vector by means of $\lambda_1, \lambda_2, \lambda_3 = -2, -1, 1$.

Definition 2.27. Let $\mathbf{A} \in \mathbb{R}_{>,\kappa}$ be a matrix. The m rows (n columns) are *linearly independent* if the only way to express a row (column) composed of all zeros $\mathbf{o} = (0, 0, \dots, 0)$ as the linear combination of the m rows (n columns) is by means of null scalars.

Example 2.30. The rows in the following matrix

$$\mathbf{A} = \begin{pmatrix} 1 & 2 & 1 \\ 0 & 1 & 4 \\ 0 & 0 & 1 \end{pmatrix}$$

are linearly independent.

Proposition 2.4. *Let $\mathbf{A} \in \mathbb{R}_{>,\kappa}$ be a matrix. Let r be a row index such that $r \leq m$. r rows are linearly dependent if and only if at least one row can be expressed as the linear combination of the others.*

Proof. Let us indicate with $\mathbf{o}$ a row vector composed of all zeros and let us represent the matrix $\mathbf{A}$ as a vector of row vectors:

$$\mathbf{A} = \begin{pmatrix} \mathbf{a_1} \\ \mathbf{a_2} \\ \dots \\ \mathbf{a_m} \end{pmatrix}.$$

If r rows are linearly dependent, we can write

$$\mathbf{o} = \lambda_1 \mathbf{a_1} + \lambda_2 \mathbf{a_2} + \cdots + \lambda_r \mathbf{a_r}$$

with $\lambda_1, \lambda_2, \ldots, \lambda_r \neq 0, 0, \ldots 0$.

Since at least one coefficient is non-null, let us assume that the coefficient $\lambda_i \neq 0$ and let us write

$$-\lambda_i \mathbf{a_i} = \lambda_1 \mathbf{a_1} + \lambda_2 \mathbf{a_2} + \cdots + \lambda_{i-1} \mathbf{a_{i-1}} + \lambda_{i+1} \mathbf{a_{i+1}} + \cdots + \lambda_r \mathbf{a_r}.$$

Since $\lambda_i \neq 0$ we can write

$$\mathbf{a_i} = -\frac{\lambda_1}{\lambda_i} \mathbf{a_1} - \frac{\lambda_2}{\lambda_i} \mathbf{a_2} + \cdots - \frac{\lambda_{a-1}}{\lambda_i} \mathbf{a_{i-1}} - \frac{\lambda_{i+1}}{\lambda_i} \mathbf{a_{i+1}} - \frac{\lambda_r}{\lambda_i} \mathbf{a_r}. \square$$

If one vector can be expressed as the linear combination of the others then

$$\mathbf{a_i} = \mu_1 \mathbf{a_1} + \mu_2 \mathbf{a_2} + \cdots + \mu_{i-1} \mathbf{a_{i-1}} + \mu_{i+1} \mathbf{a_{i+1}} + \mu_r \mathbf{a_r}.$$

We can now rearrange the equation as

$$\mathbf{o} = \mu_1 \mathbf{a_1} + \mu_2 \mathbf{a_2} + \cdots + \mu_{i-1} \mathbf{a_{i-1}} + \mu_{i+1} \mathbf{a_{i+1}} + \mu_r \mathbf{a_r} - \mathbf{a_i}$$

that is a row vector composed of all zeros expressed as linear combination of the r row vectors by means of $\mu_1, \mu_2, \ldots, \mu_{i-1}, \mu_{i+1}, \ldots, \mu_r, -1$. This means that the row vectors are linearly dependent. $\square$

Example 2.31. If we consider again

$$\mathbf{A} = \begin{pmatrix} 1 & 2 & 1 \\ 2 & 2 & 4 \\ 4 & 6 & 6 \end{pmatrix}$$

whose rows are linearly dependent since

$$\mathbf{o} = -2\mathbf{a_1} - \mathbf{a_2} + \mathbf{a_3}$$

we can write

$$\mathbf{a_3} = 2\mathbf{a_1} + \mathbf{a_2}.$$

The same concept is valid for column vectors.

Proposition 2.5. *Let $\mathbf{A} \in \mathbb{R}_{>,\times}$ be a matrix. Let s be a column index such that $s \leq n$. s columns are linearly dependent if and only if at least one column can be expressed as the linear combination of the others.*

2.4.2 Properties of the Determinant

For a given matrix $\mathbf{A} \in \mathbb{R}_{n,n}$, the following properties about determinants are valid.

Proposition 2.6. *The determinant of a matrix* $\mathbf{A}$ *is equal to the determinant of its transpose matrix:* $\det \mathbf{A} = \det \mathbf{A}^{\mathbf{T}}$.

Example 2.32. Let us consider the matrix

$$\mathbf{A} = \begin{pmatrix} 1,1,0 \\ 2,1,0 \\ 1,1,1 \end{pmatrix}.$$

The determinant of this matrix is

$$\det \mathbf{A} = 1 + 0 + 0 - 0 - 0 - 2 = -1.$$

The transpose of the matrix $\mathbf{A}$ is

$$\mathbf{A}^{\mathbf{T}} = \begin{pmatrix} 1\ 2\ 1 \\ 1\ 1\ 1 \\ 0\ 0\ 1 \end{pmatrix}$$

and its determinant is

$$\det \mathbf{A}^{\mathbf{T}} = 1 + 0 + 0 - 0 - 0 - 2 = -1.$$

As shown $\det \mathbf{A} = \det \mathbf{A}^{\mathbf{T}}$.

Proposition 2.7. *The determinant of a triangular matrix is equal to the product of the diagonal elements.*

Example 2.33. Let us consider the matrix

$$\mathbf{A} = \begin{pmatrix} 1\ 2\ 37 \\ 0\ 1\ 144 \\ 0\ 0\ 2 \end{pmatrix}.$$

The determinant of $\mathbf{A}$ is

$$\det \mathbf{A} = (1)\,(1)\,(2) = 2.$$

Proposition 2.8. *Let* $\mathbf{A}$ *be a matrix and* $\det \mathbf{A}$ *its determinant. If two rows (columns) are swapped the determinant of the modified matrix* $\mathbf{A}_s$ *is* $-\det \mathbf{A}$.

Example 2.34. Let us consider the following matrix:

$$\det \mathbf{A} = \begin{pmatrix} 1\ 1\ 1 \\ 2\ 1\ 2 \\ 1\ 1\ 3 \end{pmatrix} = 3 + 2 + 2 - 1 - 2 - 6 = -2.$$

Let us swap the second row with the third one. The new matrix $\mathbf{A}_s$ is such that

$$\det \mathbf{A}_s = \begin{pmatrix} 1 & 1 & 1 \\ 1 & 1 & 3 \\ 2 & 1 & 2 \end{pmatrix} = 2+6+1-2-3-2 = 2.$$

Proposition 2.9. *If two rows (columns) of a matrix* $\mathbf{A}$ *are identical then the* det $\mathbf{A} = 0$.

Proof. Let us write the matrix $\mathbf{A}$ as a vector of row vectors

$$\mathbf{A} = \begin{pmatrix} \mathbf{a}_1 \\ \mathbf{a}_2 \\ \dots \\ \mathbf{a}_k \\ \dots \\ \mathbf{a}_k' \\ \dots \\ \mathbf{a}_n \end{pmatrix}$$

where the rows indicated as $\mathbf{a}_k$ and $\mathbf{a}_k'$ are identical. The determinant of this matrix is $\det \mathbf{A}$. Let us swap the rows $\mathbf{a}_k$ and $\mathbf{a}_k'$ and let us generate the matrix $\mathbf{A}_s$:

$$\mathbf{A}_s = \begin{pmatrix} \mathbf{a}_1 \\ \mathbf{a}_2 \\ \dots \\ \mathbf{a}_k' \\ \dots \\ \mathbf{a}_k \\ \dots \\ \mathbf{a}_n \end{pmatrix}.$$

Since two rows have been swapped $\det \mathbf{A}_s = -\det \mathbf{A}$. On the other hand, since the rows $\mathbf{a}_k$ and $\mathbf{a}_k'$ are identical, also the matrices $\mathbf{A}_s$ and $\mathbf{A}_s$ are identical. Thus $\det \mathbf{A}_s = \det \mathbf{A}$. The only number that is equal to its opposite is 0, hence $\det \mathbf{A} = 0$. $\square$

Proposition 2.10. *Let* $\mathbf{A}$ *be a matrix and* $\det \mathbf{A}$ *its determinant. If to the elements of a row (column) the elements of another row (column) all multiplied by the same scalar* λ *are added, the determinant remains the same.*

Example 2.35. Let us consider the following matrix:

$$\mathbf{A} = \begin{pmatrix} 1 & 1 & 1 \\ 1 & 2 & 1 \\ 0 & 0 & 1 \end{pmatrix}$$

whose determinant $\det \mathbf{A} = 2 - 1 = 1$.

Let us now add to the third row the first row multiplied by $\lambda = 2$:

$$\mathbf{A_n} = \begin{pmatrix} 1\ 1\ 1 \\ 1\ 2\ 1 \\ 2\ 2\ 3 \end{pmatrix}.$$

The determinant of this new matrix is $\det \mathbf{A_n} = 6 + 2 + 2 - 4 - 2 - 3 = 1$, i.e. it remained the same.

Proposition 2.11. *Let $\mathbf{A}$ be a matrix and $\det \mathbf{A}$ its determinant. if a row (column) is proportional to another row (column) then the determinant is zero, $\det \mathbf{A} = 0$: if $\exists i, j$ such that $\mathbf{a_i} = k\mathbf{a_j}$, with $k \in \mathbb{R}$.*

Two equal rows (columns) ($k = 1$) and a row(column) composed only zeros ($k = 0$) are special cases of this property .

Example 2.36. Let us consider the matrix

$$\mathbf{A} = \begin{pmatrix} 8\ 8\ 0 \\ 1\ 1\ 0 \\ 1\ 1\ 1 \end{pmatrix}.$$

It can be observed that $\mathbf{a_1} = 8\mathbf{a_2}$ and $\det \mathbf{A} = 0$.

Proposition 2.12. *Let $\mathbf{A} \in \mathbb{R}_{n,n}$ be a matrix and $\det \mathbf{A}$ its determinant. The determinant of the matrix is zero if and only if at least one row (column) is a linear combination of the other rows (columns).*

In the case of the rows, $\det \mathbf{A} = 0$ if and only if $\exists$ index i such that

$$\mathbf{a_i} = \lambda_1 \mathbf{a_1} + \lambda_2 \mathbf{a_2} + \cdots + \lambda_{i-1} \mathbf{a_{i-1}} + \lambda_{i+1} \mathbf{a_{i+1}} + \lambda_n \mathbf{a_n}$$

with the scalars $\lambda_1, \lambda_2, \ldots, \lambda_{i-1}, \lambda_{i+1}, \ldots, \lambda_n$.

This proposition is essentially a generalisation of Proposition 2.11. The propositions are presented as separate results for the sake of clarity but the first is a special case of the second.

An equivalent way to express Proposition 2.12 is the following.

Proposition 2.13. *Let $\mathbf{A} \in \mathbb{R}_{n,n}$ be a matrix and $\det \mathbf{A}$ its determinant. The determinant of the matrix is zero if and only if the rows (columns) are linearly dependent.*

Example 2.37. The following matrix

$$\mathbf{A} = \begin{pmatrix} 5\ 2\ 3 \\ 8\ 6\ 2 \\ 2\ 0\ 2 \end{pmatrix}$$

has the first column equal to the sum of the other two. In other words, the matrix $\mathbf{A}$ is such that $\mathbf{a}^1 = \lambda_1 \mathbf{a}^2 + \lambda_2 \mathbf{a}^3$ where $\lambda_1 = 1$ and $\lambda_2 = 1$. It is easy to verify that

$$\det \mathbf{A} = \det \begin{pmatrix} 5 & 2 & 3 \\ 8 & 6 & 2 \\ 2 & 0 & 2 \end{pmatrix} = 60 + 8 + 0 - 36 - 0 - 32 = 0.$$

Proposition 2.14. *Let $\mathbf{A}$ be a matrix and $\det \mathbf{A}$ its determinant. If a row (column) is multiplied by a scalar λ its determinant is $\lambda \det \mathbf{A}$.*

Example 2.38. For the following matrix,

$$\det \mathbf{A} = \det \begin{pmatrix} 2 & 2 \\ 1 & 2 \end{pmatrix} = 2$$

while, if we multiply the second row by 2,

$$\det \mathbf{A} = \det \begin{pmatrix} 2 & 2 \\ 2 & 4 \end{pmatrix} = 4.$$

Proposition 2.15. *Let $\mathbf{A}$ be a matrix and $\det \mathbf{A}$ its determinant. If λ is a scalar, $\det(\lambda \mathbf{A}) = \lambda^n \det \mathbf{A}$.*

Example 2.39. For the following matrix,

$$\det \mathbf{A} = \det \begin{pmatrix} 2 & 2 \\ 1 & 2 \end{pmatrix} = 2$$

while, if we multiply all the elements of the matrix by $\lambda = 2$,

$$\det \lambda \mathbf{A} = \det \begin{pmatrix} 4 & 4 \\ 2 & 4 \end{pmatrix} = 16 - 8 = 8 = \lambda^2 \det \mathbf{A}.$$

Proposition 2.16. *Let $\mathbf{A}$ and $\mathbf{B}$ be two matrices and $\det \mathbf{A}$, $\det \mathbf{B}$ their respective determinants. The determinant of the product between two matrices is equal to the products of the determinants:* $\det(\mathbf{AB}) = \det \mathbf{A} \det \mathbf{B} = \det \mathbf{B} \det \mathbf{A} = \det(\mathbf{BA})$.

Example 2.40. Let us consider the following matrices,

$$\mathbf{A} = \begin{pmatrix} 1 & -1 \\ 1 & 2 \end{pmatrix}$$

and

$$\mathbf{B} = \begin{pmatrix} 1 & 2 \\ 0 & 2 \end{pmatrix}$$

whose determinants are $\det \mathbf{A} = 3$ and $\det \mathbf{B} = 2$, respectively.

If we calculate $\mathbf{AB}$ we obtain the product matrix:

$$\mathbf{AB} = \begin{pmatrix} 1 & 0 \\ 1 & 6 \end{pmatrix}$$

whose determinant is 6 that is equal to $\det \mathbf{A} \det \mathbf{B}$. If we calculate $\mathbf{BA}$ we obtain the product matrix:

$$\mathbf{BA} = \begin{pmatrix} 3 & 3 \\ 2 & 4 \end{pmatrix}$$

whose determinant is 6 that is equal to $\det \mathbf{A} \det \mathbf{B}$.

2.4.3 Submatrices, Cofactors and Adjugate Matrices

Definition 2.28. Submatrices. Let us consider a matrix $\mathbf{A} \in \mathbb{R}_{m,n}$. Let r, s be two positive integer numbers such that $1 \leq r \leq m$ and $1 \leq s \leq n$. A *submatrix* is a matrix obtained from $\mathbf{A}$ by cancelling $m - r$ rows and $n - s$ columns.

Example 2.41. Let us consider the following matrix:

$$\mathbf{A} = \begin{pmatrix} 3 & 3 & 1 & 0 \\ 2 & 4 & 1 & 2 \\ 5 & 1 & 1 & 1 \end{pmatrix}.$$

The submatrix obtained by cancelling the second row, the second and fourth columns is

$$\begin{pmatrix} 3 & 1 \\ 5 & 1 \end{pmatrix}.$$

Definition 2.29. Let us consider a matrix $\mathbf{A} \in \mathbb{R}_{m,n}$ and one of its square submatrices. The determinant of this submatrix is said *minor*. If the submatrix is the largest square submatrix of the matrix $\mathbf{A}$, its determinant is said *major determinant* or simply *major*.

It must be observed that a matrix can have multiple majors. The following example clarifies this fact.

Example 2.42. Let us consider the following matrix $\mathbf{A} \in \mathbb{R}_{4,3}$:

$$\mathbf{A} = \begin{pmatrix} 1 & 2 & 0 \\ 2 & 2 & 3 \\ 0 & 1 & 0 \\ 1 & 1 & 0 \end{pmatrix}.$$

An example of minor is

$$\det \begin{pmatrix} 1 & 0 \\ 2 & 3 \end{pmatrix}$$

obtained after cancelling the second column as well as the third and fourth rows.

Several minors can be calculated. Furthermore, this matrix has also several major. For example, one major is

$$\det \begin{pmatrix} 1 & 2 & 0 \\ 2 & 2 & 3 \\ 1 & 1 & 0 \end{pmatrix}$$

obtained by cancelling the third row and another major is

$$\det \begin{pmatrix} 2 & 2 & 3 \\ 0 & 1 & 0 \\ 1 & 1 & 0 \end{pmatrix}$$

obtained by cancelling the first row.

Definition 2.30. Let us consider a matrix $\mathbf{A} \in \mathbb{R}_{n,n}$. The submatrix is obtained by cancelling only the ith row and the jth column from $\mathbf{A}$ is said *complement submatrix* to the element $a_{i,j}$ and its determinant is here named *complement minor* and indicated with $M_{i,j}$.

Example 2.43. Let us consider a matrix $\mathbf{A} \in \mathbb{R}_{3,3}$:

$$\mathbf{A} = \begin{pmatrix} a_{1,1} & a_{1,2} & a_{1,3} \\ a_{2,1} & a_{2,2} & a_{2,3} \\ a_{3,1} & a_{3,2} & a_{3,3} \end{pmatrix}.$$

The complement submatrix to the element $a_{1,2}$ is

$$\mathbf{A} = \begin{pmatrix} a_{2,1} & a_{2,3} \\ a_{3,1} & a_{3,3} \end{pmatrix}$$

while the complement minor $M_{1,2} = a_{2,1}a_{3,3} - a_{1,3}a_{3,1}$.

Definition 2.31. Let us consider a matrix $\mathbf{A} \in \mathbb{R}_{n,n}$, its generic element $a_{i,j}$ and corresponding complement minor $M_{i,j}$. The *cofactor* $A_{i,j}$ of the element $a_{i,j}$ is defined as $A_{i,j} = (-1)^{i+j} M_{i,j}$.

Example 2.44. From the matrix of the previous example, the cofactor $A_{1,2} = (-1)M_{1,2}$.

Definition 2.32. Adjugate Matrix. Let us consider a matrix $\mathbf{A} \in \mathbb{R}_{n,n}$:

$$\mathbf{A} = \begin{pmatrix} a_{1,1} & a_{1,2} & \dots & a_{1,n} \\ a_{2,1} & a_{2,2} & \dots & a_{2,n} \\ \dots & \dots & \dots & \dots \\ a_{n,1} & a_{n,2} & \dots & a_{n,n} \end{pmatrix}.$$

Let us compute the transpose matrix $\mathbf{A}^T$:

$$\mathbf{A}^T = \begin{pmatrix} a_{1,1} & a_{2,1} & \dots & a_{n,1} \\ a_{1,2} & a_{2,2} & \dots & a_{n,2} \\ \dots & \dots & \dots & \dots \\ a_{1,n} & a_{2,n} & \dots & a_{n,n} \end{pmatrix}.$$

Let us substitute each element of the transpose matrix with its corresponding cofactor $A_{i,j}$. The resulting matrix is said *adjugate matrix* (or adjunct or adjoint) of the matrix $\mathbf{A}$ and is indicated with adj($\mathbf{A}$):

$$\text{adj}(\mathbf{A}) = \begin{pmatrix} A_{1,1} & A_{2,1} & \dots & A_{n,1} \\ A_{1,2} & A_{2,2} & \dots & A_{n,2} \\ \dots & \dots & \dots & \dots \\ A_{1,n} & A_{2,n} & \dots & A_{n,n} \end{pmatrix}.$$

Example 2.45. Let us consider the following matrix $\mathbf{A} \in \mathbb{R}_{3,3}$:

$$\mathbf{A} = \begin{pmatrix} 1 & 3 & 0 \\ 5 & 3 & 2 \\ 0 & 1 & 2 \end{pmatrix}$$

and compute the corresponding Adjugate Matrix. In order to achieve this purpose, let us compute $\mathbf{A}^T$:

$$\mathbf{A}^T = \begin{pmatrix} 1 & 5 & 0 \\ 3 & 3 & 1 \\ 0 & 2 & 2 \end{pmatrix}.$$

Let us compute the nine complements minors: $M_{1,1} = 4$, $M_{1,2} = 6$, $M_{1,3} = 6$, $M_{2,1} = 10$, $M_{2,2} = 2$, $M_{2,3} = 2$, $M_{3,1} = 5$, $M_{3,2} = 1$, and $M_{3,3} = -12$. The Adjugate Matrix adj $(\mathbf{A})$ is:

$$\text{adj}(\mathbf{A}) = \begin{pmatrix} 4 & -6 & 6 \\ -10 & 2 & -2 \\ 5 & -1 & -12 \end{pmatrix}.$$

2.4.4 Laplace Theorems on Determinants

Theorem 2.2. I Laplace Theorem *Let* $\mathbf{A} \in \mathbb{R}_{n,n}$. *The determinant of* $\mathbf{A}$ *can be computed as the sum of each row (element) multiplied by the corresponding cofactor:*
$\det \mathbf{A} = \sum_{j=1}^{n} a_{i,j} A_{i,j}$ *for any arbitrary i and*
$\det \mathbf{A} = \sum_{i=1}^{n} a_{i,j} A_{i,j}$ *for any arbitrary j.*

The I Laplace Theorem can be expressed in the equivalent form: *the determinant of a matrix is equal to scalar product of a row (column) vector by the corresponding vector of cofactors.*

Example 2.46. Let us consider the following $\mathbf{A} \in \mathbb{R}_{3,3}$:

$$\mathbf{A} = \begin{pmatrix} 2 & -1 & 3 \\ 1 & 2 & -1 \\ -1 & -2 & 1 \end{pmatrix}$$

The determinant of this matrix is $\det \mathbf{A} = 4 - 1 - 6 + 6 + 1 - 4 = 0$. Hence, the matrix is singular. Let us now calculate the determinant by applying the I Laplace Theorem. If we consider the first row, it follows that $\det \mathbf{A} = a_{1,1} A_{1,1} + a_{1,2}(-1)A_{1,2} + a_{1,3} A_{1,3}$, $\det \mathbf{A} = 2(0) + 1(0) + 3(0) = 0$. We arrive to the same conclusion.

Example 2.47. Let us consider the following $\mathbf{A} \in \mathbb{R}_{3,3}$:

$$\mathbf{A} = \begin{pmatrix} 1 & 2 & 1 \\ 0 & 1 & 1 \\ 4 & 2 & 0 \end{pmatrix}$$

The determinant of this matrix is $\det \mathbf{A} = 8 - 4 - 2 = 2$. Hence, the matrix is nonsingular. Let us now calculate the determinant by applying the I Laplace Theorem. If we consider the second row, it follows that $\det \mathbf{A} = a_{2,1}(-1)A_{2,1} + a_{2,2} A_{2,2} + a_{2,3}(-1)A_{2,3}$, $\det \mathbf{A} = 0(-1)(-2) + 1(-4) + 1(-1)(-6) = 2$. The result is the same.

Let us prove the I Laplace Theorem in the special case of an order 3 matrix:

$$\mathbf{A} = \begin{pmatrix} a_{1,1} & a_{1,2} & a_{1,3} \\ a_{2,1} & a_{2,2} & a_{2,3} \\ a_{3,1} & a_{3,2} & a_{3,3} \end{pmatrix}.$$

Proof. By applying the definition

$$\det \mathbf{A} = a_{1,1} a_{2,2} a_{3,3} + a_{1,2} a_{2,3} a_{3,1} + a_{1,3} a_{2,1} a_{3,2} - a_{1,3} a_{2,2} a_{3,1} - a_{1,2} a_{2,1} a_{3,3} - a_{1,1} a_{2,3} a_{3,2}.$$

By applying the I Laplace Theorem we obtain

$$\det \mathbf{A} = a_{1,1} \det \begin{pmatrix} a_{2,2} & a_{2,3} \\ a_{3,2} & a_{3,3} \end{pmatrix} - a_{1,2} \det \begin{pmatrix} a_{2,1} & a_{2,3} \\ a_{3,1} & a_{3,3} \end{pmatrix} + a_{1,3} \det \begin{pmatrix} a_{2,1} & a_{2,2} \\ a_{3,1} & a_{3,2} \end{pmatrix} =$$
$$= a_{1,1} (a_{2,2}a_{3,3} - a_{2,3}a_{3,2}) - a_{1,2} (a_{2,1}a_{3,3} - a_{2,3}a_{3,1}) + a_{1,3} (a_{2,1}a_{3,2} - a_{2,2}a_{3,1}) =$$
$$a_{1,1}a_{2,2}a_{3,3} - a_{1,1}a_{2,3}a_{3,2} - a_{1,2}a_{2,1}a_{3,3} + a_{1,2}a_{2,3}a_{3,1} + a_{1,3}a_{2,1}a_{3,2} - a_{1,3}a_{2,2}a_{3,1}$$

that is the determinant calculated by means of the definition. $\square$

Theorem 2.3. II Laplace Theorem. *Let* $\mathbf{A} \in \mathbb{R}_{n,n}$ *with* $n > 1$. *The sum of the elements of a row (column) multiplied by the corresponding cofactor related to another row (column) is always zero:*
$\sum_{j=1}^{n} a_{i,j} A_{k,j} = 0$ *for any arbitrary* $k \neq i$ *and*
$\sum_{i=1}^{n} a_{i,j} A_{i,k} = 0$ *for any arbitrary* $k \neq j$.

The II Laplace Theorem can be equivalently stated as: *the scalar product of a row (column) vector by the vector of cofactors associated with another row (column) is always null.*

Example 2.48. Let us consider again the matrix

$$\mathbf{A} = \begin{pmatrix} 1 & 2 & 1 \\ 0 & 1 & 1 \\ 4 & 2 & 0 \end{pmatrix}$$

and the cofactors associated with second row: $A_{2,1} = (-1)(-2)$, $A_{2,2} = (-4)$, and $A_{2,3} = (-1)(-6)$. Hence, $A_{2,1} = 2$, $A_{2,2} = -4$, and $A_{2,3} = 6$. Let us apply now the II Laplace Theorem by applying the scalar product of the first row vector by the vector of the cofactors associated with the second row:

$$a_{1,1}A_{2,1} + a_{1,2}A_{2,2} + a_{1,3}A_{2,3} = 1(2) + 2(-4) + 1(6) = 0.$$

It can been seen that if we multiply the third row vector by the same vector of cofactors the result still is 0:

$$a_{3,1}A_{2,1} + a_{3,2}A_{2,2} + a_{3,3}A_{2,3} = 4(2) + 2(-4) + 0(6) = 0.$$

Now, let us prove the II Laplace Theorem in the special case of an order 3 matrix:

$$\mathbf{A} = \begin{pmatrix} a_{1,1} & a_{1,2} & a_{1,3} \\ a_{2,1} & a_{2,2} & a_{2,3} \\ a_{3,1} & a_{3,2} & a_{3,3} \end{pmatrix}.$$

Proof. Let us calculate the elements of the scalar product of the element of the second row by the cofactors associated with the first row:

$$a_{2,1} \det \begin{pmatrix} a_{2,2} & a_{2,3} \\ a_{3,2} & a_{3,3} \end{pmatrix} - a_{2,2} \det \begin{pmatrix} a_{2,1} & a_{2,3} \\ a_{3,1} & a_{3,3} \end{pmatrix} + a_{2,3} \det \begin{pmatrix} a_{2,1} & a_{2,2} \\ a_{3,1} & a_{3,2} \end{pmatrix} =$$
$$= a_{2,1} \left(a_{2,2}a_{3,3} - a_{2,3}a_{3,2} \right) - a_{2,2} \left(a_{2,1}a_{3,3} - a_{2,3}a_{3,1} \right) + a_{2,3} \left(a_{2,1}a_{3,2} - a_{2,2}a_{3,1} \right) =$$
$$a_{2,1}a_{2,2}a_{3,3} - a_{2,1}a_{2,3}a_{3,2} - a_{2,2}a_{2,1}a_{3,3} + a_{2,2}a_{2,3}a_{3,1} + a_{2,3}a_{2,1}a_{3,2} - a_{2,3}a_{2,2}a_{3,1} = 0. \square$$

2.5 Invertible Matrices

Definition 2.33. Let $\mathbf{A} \in \mathbb{R}_{n,n}$. If $\det \mathbf{A} = 0$ the matrix is said singular. If $\det \mathbf{A} \neq 0$ the matrix is said non-singular.

Definition 2.34. Let $\mathbf{A} \in \mathbb{R}_{n,n}$. The matrix $\mathbf{A}$ is said *invertible* if $\exists$ a matrix $\mathbf{B} \in \mathbb{R}_{n,n} | \mathbf{AB} = \mathbf{I} = \mathbf{BA}$. The matrix $\mathbf{B}$ is said *inverse* matrix of the matrix $\mathbf{A}$.

Theorem 2.4. *If $\mathbf{A} \in \mathbb{R}_{n,n}$ is an invertible matrix and $\mathbf{B}$ is its inverse. It follows that the inverse matrix is unique:* $\exists! \mathbf{B} \in \mathbb{R}_{n,n} | \mathbf{AB} = \mathbf{I} = \mathbf{BA}$.

Proof. Let us assume by contradiction that the inverse matrix is not unique. Thus, besides $\mathbf{B}$, there exists another inverse of $\mathbf{A}$, indicated as $\mathbf{C} \in \mathbb{R}_{n,n}$.

This would mean that for the hypothesis $\mathbf{B}$ is inverse of $\mathbf{A}$ and thus

$$\mathbf{AB} = \mathbf{BA} = \mathbf{I}.$$

For the contradiction hypothesis also $\mathbf{C}$ is inverse of $\mathbf{A}$ and thus

$$\mathbf{AC} = \mathbf{CA} = \mathbf{I}.$$

Considering that $\mathbf{I}$ is the neutral element with respect to the product of matrices ($\forall \mathbf{A} : \mathbf{AI} = \mathbf{IA} = \mathbf{A}$) and that the product of matrices is associative, it follows that

$$\mathbf{C} = \mathbf{CI} = \mathbf{C}(\mathbf{AB}) = (\mathbf{CA})\mathbf{B} = \mathbf{IB} = \mathbf{B}.$$

In other words, if $\mathbf{B}$ is an inverse matrix of $\mathbf{A}$ and another inverse matrix $\mathbf{C}$ exists, then $\mathbf{C} = \mathbf{B}$. Thus, the inverse matrix is unique. $\square$

The only inverse matrix of the matrix $\mathbf{A}$ is indicated with $\mathbf{A}^{-1}$.

Theorem 2.5. *Let $\mathbf{A} \in \mathbb{R}_{n,n}$ and $A_{i,j}$ its generic cofactor. The inverse matrix $\mathbf{A}^{-1}$ is*

$$\mathbf{A}^{-1} = \frac{1}{\det \mathbf{A}} \operatorname{adj}(\mathbf{A}).$$

Proof. Let us consider the matrix $\mathbf{A}$,

$$\mathbf{A} = \begin{pmatrix} a_{1,1} & a_{1,2} & \dots & a_{1,n} \\ a_{2,1} & a_{2,2} & \dots & a_{2,n} \\ \dots & \dots & \dots & \dots \\ a_{n,1} & a_{n,2} & \dots & a_{n,n} \end{pmatrix}$$

and its adjugate matrix,

$$\mathrm{adj}\,(\mathbf{A}) = \begin{pmatrix} A_{1,1} & A_{2,1} & \dots & A_{n,1} \\ A_{1,2} & A_{2,2} & \dots & A_{n,2} \\ \dots & \dots & \dots & \dots \\ A_{1,n} & A_{2,n} & \dots & A_{n,n} \end{pmatrix}.$$

Let us compute the product matrix of $\mathbf{A}$ by $\mathrm{adj}\,(\mathbf{A})$, i.e. $(\mathbf{A})\,(\mathrm{adj}\,(\mathbf{A}))$,

$$(\mathbf{A})\,(\mathrm{adj}\,(\mathbf{A})) =$$

$$= \begin{pmatrix} (a_{1,1}A_{1,1}+a_{1,2}A_{1,2}+\dots+a_{1,n}A_{1,n}) & (\dots) & (a_{1,1}A_{n,1}+a_{1,2}A_{n,2}+\dots+a_{1,n}A_{n,n}) \\ (a_{2,1}A_{1,1}+a_{2,2}A_{1,2}+\dots+a_{2,n}A_{2,n}) & (\dots) & (a_{2,1}A_{n,1}+a_{2,2}A_{n,2}+\dots+a_{2,n}A_{n,n}) \\ \dots & \dots & \dots \\ (a_{n,1}A_{1,1}+a_{n,2}A_{1,2}+\dots+a_{n,n}A_{1,n}) & (\dots) & (a_{n,1}A_{n,1}+a_{n,2}A_{n,2}+\dots+a_{n,n}A_{n,n}) \end{pmatrix}.$$

The matrix can also be written as

$$\begin{pmatrix} \sum_{j=1}^{n} a_{1,j}A_{1,j} & \sum_{j=1}^{n} a_{1,j}A_{2,j} & \dots & \sum_{j=1}^{n} a_{1,j}A_{n,j} \\ \sum_{j=1}^{n} a_{2,j}A_{1,j} & \sum_{j=1}^{n} a_{2,j}A_{2,j} & \dots & \sum_{j=1}^{n} a_{2,j}A_{n,j} \\ \dots & \dots & \dots & \dots \\ \sum_{j=1}^{n} a_{n,j}A_{1,j} & \sum_{j=1}^{n} a_{n,j}A_{2,j} & \dots & \sum_{j=1}^{n} a_{n,j}A_{n,j} \end{pmatrix}$$

For the I Laplace Theorem, the diagonal elements are equal to $\det \mathbf{A}$:

$$a_{i,1}A_{i,1}+a_{i,2}A_{i,2}+\dots+a_{i,n}A_{i,n} = \sum_{j=1}^{n} a_{i,j}A_{i,j} = \det \mathbf{A}.$$

for all the rows i.

For the II Laplace Theorem, the extra-diagonal elements are equal to zero:

$$a_{i,1}A_{k,1}+a_{i,2}A_{k,2}+\dots+a_{i,n}A_{k,n} = \sum_{j=1}^{n} a_{i,j}A_{k,j} = 0.$$

with $i \neq k$.

The result of the multiplication is then

$$(\mathbf{A})\,(\mathrm{adj}\,(\mathbf{A})) = \begin{pmatrix} \det\mathbf{A} & 0 & \ldots & 0 \\ 0 & \det\mathbf{A} & \ldots & 0 \\ \ldots & \ldots & \ldots & \ldots \\ 0 & 0 & \ldots & \det\mathbf{A} \end{pmatrix}.$$

Thus,

$$(\mathbf{A})\,(\mathrm{adj}\,(\mathbf{A})) = (\det\mathbf{A})\,\mathbf{I}$$

and

$$\mathbf{A}^{-1} = \frac{1}{\det\mathbf{A}}\mathrm{adj}\,(\mathbf{A}).\ \square$$

Example 2.49. Let us calculate the inverse of the matrix

$$\mathbf{A} = \begin{pmatrix} 2 & 1 \\ 1 & 1 \end{pmatrix}.$$

The determinant of this matrix is $\det\mathbf{A} = 1$. The transpose of this matrix is

$$\mathbf{A}^{\mathbf{T}} = \begin{pmatrix} 2 & 1 \\ 1 & 1 \end{pmatrix},$$

which, in this case, is equal to $\mathbf{A}$.
   The adjugate matrix is

$$\mathrm{adj}\,(\mathbf{A}) = \begin{pmatrix} 1 & -1 \\ -1 & 2 \end{pmatrix}.$$

The inverse of the matrix $\mathbf{A}$ is then

$$\mathbf{A}^{-1} = \frac{1}{\det\mathbf{A}}\mathrm{adj}\,(\mathbf{A}) = \begin{pmatrix} 1 & -1 \\ -1 & 2 \end{pmatrix}.$$

Example 2.50. Let us calculate the inverse of the matrix

$$\mathbf{A} = \begin{pmatrix} 1 & 1 \\ -2 & 1 \end{pmatrix}.$$

The determinant of this matrix is $\det\mathbf{A} = 3$. The transpose of this matrix is

$$\mathbf{A}^{\mathbf{T}} = \begin{pmatrix} 1 & -2 \\ 1 & 1 \end{pmatrix}.$$

The adjugate matrix is

$$\text{adj}(\mathbf{A}) = \begin{pmatrix} 1 & -1 \\ 2 & 1 \end{pmatrix}.$$

The inverse of the matrix $\mathbf{A}$ is then

$$\mathbf{A}^{-1} = \tfrac{1}{\det \mathbf{A}} \text{adj}(\mathbf{A}) = \tfrac{1}{3} \begin{pmatrix} 1 & -1 \\ 2 & 1 \end{pmatrix} = \begin{pmatrix} \frac{1}{3} & -\frac{1}{3} \\ \frac{2}{3} & \frac{1}{3} \end{pmatrix}.$$

The latter two examples suggest introduce the corollary of Theorem 2.5.

Corollary 2.1. *Let* $\mathbf{A} \in \mathbb{R}_{2,2}$:

$$\mathbf{A} = \begin{pmatrix} a_{1,1} & a_{1,2} \\ a_{2,1} & a_{2,2} \end{pmatrix},$$

then

$$\mathbf{A}^{-1} = \tfrac{1}{\det \mathbf{A}} \begin{pmatrix} a_{2,2} & -a_{1,2} \\ -a_{2,1} & a_{1,1} \end{pmatrix}.$$

Proof. Let us calculate the transpose of $\mathbf{A}$:

$$\mathbf{A}^{\mathsf{T}} = \begin{pmatrix} a_{1,1} & a_{2,1} \\ a_{1,2} & a_{2,2} \end{pmatrix}.$$

the adjugate matrix is

$$\text{adj}(\mathbf{A}) = \begin{pmatrix} a_{2,2} & -a_{1,2} \\ -a_{2,1} & a_{1,1} \end{pmatrix}$$

and the inverse is

$$\mathbf{A}^{-1} = \tfrac{1}{\det \mathbf{A}} \begin{pmatrix} a_{2,2} & -a_{1,2} \\ -a_{2,1} & a_{1,1} \end{pmatrix}.$$

Example 2.51. Let us consider the matrix

$$\mathbf{A} = \begin{pmatrix} 1 & 2 \\ 3 & 4 \end{pmatrix}.$$

The determinant is $\det \mathbf{A} = 4 - 6 = -2$. By applying Corollary 2.1, the inverse matrix of $\mathbf{A}$ is

$$\mathbf{A}^{-1} = -\frac{1}{2} \begin{pmatrix} 4 & -2 \\ -3 & 1 \end{pmatrix}.$$

Example 2.52. Let us now invert a matrix $\mathbf{A} \in \mathbb{R}_{3,3}$:

$$\mathbf{A} = \begin{pmatrix} 2 & 1 & 1 \\ 0 & 1 & 0 \\ 1 & 3 & 1 \end{pmatrix}.$$

The determinant of this matrix is $\det \mathbf{A} = 2 - 1 = 1$. The transpose of this matrix is

$$\mathbf{A}^{\mathrm{T}} = \begin{pmatrix} 2 & 0 & 1 \\ 1 & 1 & 3 \\ 1 & 0 & 1 \end{pmatrix}.$$

The adjugate matrix is

$$\mathrm{adj}\,(\mathbf{A}) = \begin{pmatrix} 1 & 2 & -1 \\ 0 & 1 & 0 \\ -1 & -5 & 2 \end{pmatrix}$$

and the corresponding inverse matrix is

$$\mathbf{A}^{-1} = \tfrac{1}{\det \mathbf{A}} \mathrm{adj}\,(\mathbf{A}) = \begin{pmatrix} 1 & 2 & -1 \\ 0 & 1 & 0 \\ -1 & -5 & 2 \end{pmatrix}.$$

In all the previous examples, only one inverse matrix can be found in accordant with Theorem 2.4. Furthermore, all the inverted matrices are non-singular, i.e. $\det \mathbf{A} \neq 0$. Intuitively, it can be observed that for a singular matrix the inverse cannot be calculated since the formula $\mathbf{A}^{-1} = \frac{1}{\det \mathbf{A}} \mathrm{adj}\,(\mathbf{A})$ cannot be applied (as it would require a division by zero). The following theorem introduces the theoretical foundation of this intuition.

Theorem 2.6. *Let* $\mathbf{A} \in \mathbb{R}_{n,n}$. *The matrix* $\mathbf{A}$ *is invertible if and only if* $\mathbf{A}$ *is non-singular.*

Proof. If $\mathbf{A}$ is invertible then

$$\exists \mathbf{A}^{-1} \text{ such that } \mathbf{A}\mathbf{A}^{-1} = \mathbf{I} = \mathbf{A}^{-1}\mathbf{A}$$

where $\mathbf{A}^{-1}$, for Theorem 2.4, is unique ($\mathbf{A}^{-1}$ is the only inverse matrix of $\mathbf{A}$).

Since two identical matrices have identical determinants it follows that

$$\det\left(\mathbf{A}\mathbf{A}^{-1}\right) = \det \mathbf{I} = 1.$$

Since for the properties of the determinant, the determinant of the product of two (square) matrices is equal to the product of the determinants of these two matrices, it follows that

$$\det\left(\mathbf{A}\mathbf{A}^{-1}\right) = (\det\mathbf{A})\left(\det\mathbf{A}^{-1}\right) = 1.$$

Hence, $\det\mathbf{A} \neq 0$, i.e. $\mathbf{A}$ is non-singular. □

If $\mathbf{A}$ is non-singular, then $\det\mathbf{A} \neq 0$. Thus, we can consider a matrix $\mathbf{B} \in \mathbb{R}_{n,n}$ taken as

$$\mathbf{B} = \frac{1}{\det\mathbf{A}}\,\mathrm{adj}\,(\mathbf{A}).$$

We could not have taken $\mathbf{B}$ in this way if $\det\mathbf{A} = 0$. We know from Theorem 2.5 that $\mathbf{B} = \mathbf{A}^{-1}$ and thus $\mathbf{A}$ is invertible. □

Example 2.53. Let us consider the following matrix

$$\mathbf{A} = \begin{pmatrix} 1 & 3 & 4 \\ 0 & 2 & 2 \\ 4 & -2 & 2 \end{pmatrix}.$$

It can be easily seen that $\det\mathbf{A} = 0$. The matrix is singular and not invertible. We could not calculate the inverse

$$\mathbf{A}^{-1} = \frac{1}{\det\mathbf{A}}\,\mathrm{adj}\,(\mathbf{A})$$

since its calculation would lead to a division by zero.

Corollary 2.2. *Let $\mathbf{A} \in \mathbb{R}_{n,n}$ be an invertible matrix. It follows that $\det\mathbf{A}^{-1} = \frac{1}{\det\mathbf{A}}$.*

Proof. From the proof of Theorem 2.6 we know that

$$\det\left(\mathbf{A}\mathbf{A}^{-1}\right) = \det\mathbf{I} = 1.$$

It follows that

$$\det\mathbf{A}^{-1} = \frac{1}{\det\mathbf{A}}. □$$

Example 2.54. Let us consider the following matrix

$$\mathbf{A} = \begin{pmatrix} 2 & 4 & 1 \\ 0 & 1 & 0 \\ 4 & 0 & 4 \end{pmatrix},$$

whose determinant is $\det\mathbf{A} = 4$.

The inverse matrix of $\mathbf{A}$ is

$$\mathbf{A} = \begin{pmatrix} 1 & -4 & -0.25 \\ 0 & 1 & 0 \\ -1 & 4 & 0.5 \end{pmatrix},$$

whose determinant is $\det\mathbf{A}^{-1} = 0.25 = \frac{1}{\det\mathbf{A}}$.

Corollary 2.3. *Let* $\mathbf{A} \in \mathbb{R}_{1,1}$, *i.e.* $\mathbf{A} = (a_{1,1})$, *be a matrix composed of only one element. The inverse matrix of* $\mathbf{A}$ *is*

$$\mathbf{A}^{-1} = \frac{1}{\det \mathbf{A}} = \frac{1}{a_{1,1}}.$$

Proof. For the I Laplace Theorem

$$a_{1,1} A_{1,1} = \det(\mathbf{A})$$

with $A_{1,1}$ cofactor of $a_{1,1}$.

Since, as shown in Example 2.23, the $\det \mathbf{A} = a_{1,1}$ it follows that $A_{1,1} = 1$. Since the only cofactor is also the only element of the adjugate matrix:

$$\mathrm{adj}(\mathbf{A}) = A_{1,1} = 1$$

it follows that

$$\mathbf{A}^{-1} = \frac{1}{\det \mathbf{A}} \mathrm{adj}(\mathbf{A}) = \frac{1}{\det \mathbf{A}} = \frac{1}{a_{1,1}}. \square$$

The last corollary simply shows how numbers can be considered as special cases of matrices and how the matrix algebra encompasses the number algebra.

Example 2.55. If $\mathbf{A} = (5)$, its inverse matrix would be $\mathbf{A}^{-1} = \left(\frac{1}{5}\right)$. If $\mathbf{A} = (0)$ the matrix would be singular and thus not invertible, in accordance with the definition of field, see Definition 1.36.

The following illustration represents and summarizes the matrix theory studies in the past sections. The empty circles represent singular matrices which are not linked to other matrices since they have no inverse. Conversely, filled circles represent non-singular matrices. Each of them has one inverse matrix to which is linked in the illustration. Non-singular matrices are all paired and connected by a link. There are no isolated filled circles (every non-singular matrix has an inverse) and every filled circle has only one link (the inverse matrix is unique).

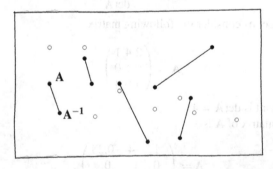

Proposition 2.17. *Let* $\mathbf{A}$ *and* $\mathbf{B}$ *be two square and invertible matrices. it follows that*

$$(\mathbf{AB})^{-1} = \mathbf{B}^{-1}\mathbf{A}^{-1}.$$

Proof. Let us calculate

$$(\mathbf{AB})\left(\mathbf{B}^{-1}\mathbf{A}^{-1}\right) = \mathbf{AIA}^{-1} = \mathbf{AA}^{-1} = \mathbf{I}.$$

Thus, the inverse of $(\mathbf{AB})$ is $\left(\mathbf{B}^{-1}\mathbf{A}^{-1}\right)$, i.e.

$$(\mathbf{AB})^{-1} = \mathbf{B}^{-1}\mathbf{A}^{-1}.\square$$

Example 2.56. Let us consider the following matrices

$$\mathbf{A} = \begin{pmatrix} 2 & 0 & 1 \\ 1 & 1 & 0 \\ 3 & 2 & 1 \end{pmatrix}$$

and

$$\mathbf{B} = \begin{pmatrix} 1 & 0 & 1 \\ 1 & 1 & 0 \\ 2 & 2 & 2 \end{pmatrix}.$$

We can easily verify that

$$\mathbf{AB} = \begin{pmatrix} 4 & 2 & 4 \\ 2 & 1 & 1 \\ 7 & 4 & 5 \end{pmatrix}$$

and

$$\mathbf{AB}^{-1} = \begin{pmatrix} 0.5 & 3 & -1 \\ -1.5 & -4 & 2 \\ 0.5 & -1 & 0 \end{pmatrix}.$$

Also, let us calculate the inverse matrices

$$\mathbf{A}^{-1} = \begin{pmatrix} 1 & 2 & -1 \\ -1 & -1 & 1 \\ -1 & -4 & 2 \end{pmatrix}$$

and

$$\mathbf{B}^{-1} = \begin{pmatrix} 1 & 1 & -0.5 \\ -1 & 0 & 0.5 \\ 0 & -1 & 0.5 \end{pmatrix}.$$

Let us now calculate their product

$$\mathbf{B}^{-1}\mathbf{A}^{-1} = \begin{pmatrix} 0.5 & 3 & -1 \\ -1.5 & -4 & 2 \\ 0.5 & -1 & 0 \end{pmatrix}.$$

2.6 Orthogonal Matrices

Definition 2.35. A matrix $\mathbf{A} \in \mathbb{R}_{n,n}$ is said *orthogonal* if the product between it and its transpose is the identity matrix:

$$\mathbf{A}\mathbf{A}^{\mathbf{T}} = \mathbf{I} = \mathbf{A}^{\mathbf{T}}\mathbf{A}.$$

Theorem 2.7. *An orthogonal matrix is always non-singular and its determinant is either* 1 *or* -1.

Proof. Let $\mathbf{A} \in \mathbb{R}_{n,n}$ be an orthogonal matrix. Then,

$$\mathbf{A}\mathbf{A}^{\mathbf{T}} = \mathbf{I}.$$

Thus, the determinants are still equal:

$$\det\left(\mathbf{A}\mathbf{A}^{\mathbf{T}}\right) = \det\mathbf{I}.$$

For the properties of the determinant

$$\det\left(\mathbf{A}\mathbf{A}^{\mathbf{T}}\right) = \det\mathbf{A}\det\mathbf{A}^{\mathbf{T}}$$
$$\det\mathbf{A} = \det\mathbf{A}^{\mathbf{T}}$$
$$\det\mathbf{I} = 1$$

Thus,

$$(\det\mathbf{A})^2 = 1.$$

This can happen only when $\det\mathbf{A} = \pm 1$.   □

Theorem 2.8. Property of Orthogonal Matrices. *A matrix* $\mathbf{A} \in \mathbb{R}_{n,n}$ *is orthogonal if and only if the sum of the squares of the element of a row (column) is equal to* 1 *and the scalar product of any two arbitrary rows (columns) is equal to* 0:

$$\sum_{j=1}^{n} a_{i,j}^2 = 1$$
$$\forall i, j \; \mathbf{a_i}\mathbf{a_j} = 0.$$

Proof. The proof will be given for row vectors. The proof for column vectors is analogous. Let us consider a matrix $\mathbf{A} \in \mathbb{R}_{n,n}$

$$\mathbf{A} = \begin{pmatrix} a_{1,1} & a_{1,2} & \dots & a_{1,n} \\ a_{2,1} & a_{2,2} & \dots & a_{2,n} \\ \dots & \dots & \dots & \dots \\ a_{n,1} & a_{n,2} & \dots & a_{n,n} \end{pmatrix} = \begin{pmatrix} \mathbf{a_1} \\ \mathbf{a_2} \\ \dots \\ \mathbf{a_n} \end{pmatrix}$$

and its transpose

$$\mathbf{A^T} = \begin{pmatrix} a_{1,1} & a_{2,1} & \cdots & a_{n,1} \\ a_{1,2} & a_{2,2} & \cdots & a_{n,2} \\ \cdots & \cdots & \cdots & \cdots \\ a_{1,n} & a_{2,n} & \cdots & a_{n,n} \end{pmatrix} = \begin{pmatrix} \mathbf{a_1} & \mathbf{a_2} & \cdots & \mathbf{a_n} \end{pmatrix}.$$

Since $\mathbf{A}$ is orthogonal it follows that

$$\mathbf{A}\mathbf{A^T} = \mathbf{I}.$$

Thus,

$$\mathbf{A}\mathbf{A^T} = \begin{pmatrix} \mathbf{a_1 a_1} & \mathbf{a_1 a_2} & \cdots & \mathbf{a_1 a_n} \\ \mathbf{a_2 a_1} & \mathbf{a_2 a_2} & \cdots & \mathbf{a_2 a_n} \\ \cdots & \cdots & \cdots & \cdots \\ \mathbf{a_n a_1} & \mathbf{a_n a_2} & \cdots & \mathbf{a_n a_n} \end{pmatrix} = \begin{pmatrix} 1 & 0 & \cdots & 0 \\ 0 & 1 & \cdots & 0 \\ \cdots & \cdots & \cdots & \cdots \\ 0 & 0 & \cdots & 1 \end{pmatrix}.$$

This means that $\forall i, j$

$$\mathbf{a_i a_i} = \sum_{j=1}^{n} a_{i,j}^2 = 1$$

and

$$\mathbf{a_i a_j} = 0. \square$$

The scalar product of every two row vectors is 0. This condition as shown in Chap. 4 means that the vectors are orthogonal, which is the reason why these matrices are called orthogonal matrices.

Example 2.57. The following matrices are orthogonal:

$$\begin{pmatrix} sin(\alpha) & cos(\alpha) \\ cos(\alpha) & -sin(\alpha) \end{pmatrix}$$

and

$$\begin{pmatrix} sin(\alpha) & 0 & cos(\alpha) \\ 0 & 1 & 0 \\ -cos(\alpha) & 0 & sin(\alpha) \end{pmatrix}.$$

Example 2.58. The following matrix is orthogonal

$$\mathbf{A} = \frac{1}{\sqrt{2}} \begin{pmatrix} 1 & 1 \\ 1 & -1 \end{pmatrix}.$$

Let us verify that the matrix is orthogonal by calculating $\mathbf{AA^T}$:

$$\mathbf{AA^T} = \frac{1}{\sqrt{2}}\begin{pmatrix} 1 & 1 \\ 1 & -1 \end{pmatrix} \frac{1}{\sqrt{2}}\begin{pmatrix} 1 & 1 \\ 1 & -1 \end{pmatrix} = \frac{1}{2}\begin{pmatrix} 2 & 0 \\ 0 & 2 \end{pmatrix} = \begin{pmatrix} 1 & 0 \\ 0 & 1 \end{pmatrix} = \mathbf{I}.$$

Let us verify the properties of orthogonal matrices. The sum of the squares of the rows (columns) is equal to one, e.g.

$$a_{1,1}^2 + a_{1,2}^2 = \left(\frac{1}{\sqrt{2}}\right)^2 + \left(\frac{1}{\sqrt{2}}\right)^2 = \frac{1}{2} + \frac{1}{2} = 1.$$

The scalar product of each pair of rows (columns) is zero, e.g.

$$\mathbf{a_1 a_2} = a_{1,1}a_{2,1} + a_{1,2}a_{2,2} = \left(\frac{1}{\sqrt{2}}, \frac{1}{\sqrt{2}}\right)\left(\frac{1}{\sqrt{2}}, -\frac{1}{\sqrt{2}}\right) = \frac{1}{2} - \frac{1}{2} = 0.$$

Example 2.59. The following matrix is orthogonal

$$\mathbf{A} = \frac{1}{3}\begin{pmatrix} 2 & -2 & 1 \\ 1 & 2 & 2 \\ 2 & 1 & -2 \end{pmatrix}.$$

Let us verify that the matrix is orthogonal by calculating $\mathbf{AA^T}$:

$$\mathbf{AA^T} = \frac{1}{3}\begin{pmatrix} 2 & -2 & 1 \\ 1 & 2 & 2 \\ 2 & 1 & -2 \end{pmatrix} \frac{1}{3}\begin{pmatrix} 2 & 1 & 2 \\ -2 & 2 & 1 \\ 1 & 2 & -2 \end{pmatrix} = \frac{1}{9}\begin{pmatrix} 9 & 0 & 0 \\ 0 & 9 & 0 \\ 0 & 0 & 9 \end{pmatrix} = \begin{pmatrix} 1 & 0 & 0 \\ 0 & 1 & 0 \\ 0 & 0 & 1 \end{pmatrix} = \mathbf{I}.$$

Let us verify the properties of orthogonal matrices. The sum of the squares of the rows (columns) is equal to one, e.g.

$$a_{1,1}^2 + a_{1,2}^2 + a_{1,3}^2 = \left(\frac{2}{3}\right)^2 + \left(-\frac{2}{3}\right)^2 + \left(\frac{1}{3}\right)^2 = \frac{4}{9} + \frac{4}{9} + \frac{1}{9} = 1.$$

The scalar product of each pair of rows (columns) is zero, e.g.

$$\mathbf{a_1 a_2} = a_{1,1}a_{2,1} + a_{1,2}a_{2,2} + a_{1,3}a_{2,3} = \left(\frac{2}{3}, -\frac{2}{3}, \frac{1}{3}\right)\left(\frac{1}{3}, \frac{2}{3}, \frac{2}{3}\right) = \frac{2}{9} - \frac{4}{9} + \frac{2}{9} = 0.$$

2.7 Rank of a Matrix

Definition 2.36. Let $\mathbf{A} \in \mathbb{R}_{m,n}$ with $\mathbf{A}$ assumed to be different from the null matrix. The *rank* of the matrix $\mathbf{A}$, indicated as $\rho_\mathbf{A}$, is the highest order of the non-singular submatrix $\mathbf{A}_\rho \subset \mathbf{A}$. If $\mathbf{A}$ is the null matrix then its rank is taken equal to 0.

Example 2.60. The rank of the matrix

$$\begin{pmatrix} 1 & -1 & -2 \\ -1 & 1 & 0 \end{pmatrix}$$

is 2 as the submatrix

$$\begin{pmatrix} -1 & -2 \\ 1 & 0 \end{pmatrix}$$

is non-singular.

Example 2.61. Let us consider the following matrix:

$$\mathbf{A} = \begin{pmatrix} -1 & 2 & -1 \\ 2 & -3 & 2 \\ 1 & -1 & 1 \end{pmatrix}.$$

It can be easily seen that $\det \mathbf{A} = 0$ while there is at least one non-singular square submatrix having order 2. Thus $\rho_\mathbf{A} = 2$.

If a matrix $\mathbf{A} \in \mathbb{R}_{m,n}$ and we indicate with the min a function that detects the lowest element of a (finite) set, the procedure for finding its rank $\rho_\mathbf{A}$ is shown in the pseudocode represented in Algorithm 1.

Algorithm 1 Rank Detection

---

for the matrix $\mathbf{A} \in \mathbb{R}_{m,n}$ find $s = \min\{m,n\}$
while $s \geq 0$ AND $\det \mathbf{A}_s == 0$ **do**
    extract all the order s submatrices $\mathbf{A}_s$
    for all the submatrices $\mathbf{A}_s$ **do**
        calculate $\det(\mathbf{A}_s)$
    end for
    $s = s - 1$
end while
The rank $\rho_\mathbf{A} = s$

Theorem 2.9. *Let $\mathbf{A} \in \mathbb{R}_{n,n}$ and ρ its rank. The matrix $\mathbf{A}$ has ρ linearly independent rows (columns).*

Proof. Let us prove this theorem for the rows. For linearly independent columns the proof would be analogous. Let us apply the procedure in Algorithm 1 to determine the rank of the matrix $\mathbf{A}$. Starting from $s = n$ the determinant is checked, if the determinant is null then the determinants of submatrices of a smaller size are checked. For Proposition 2.13 if $\det \mathbf{A}_s = 0$ then the rows are linearly dependent. When $\det \mathbf{A}_s \neq 0$ the rows of $\mathbf{A}_s$ are linearly independent and the size s of $\mathbf{A}_s$ is the rank ρ of the matrix $\mathbf{A}$. Thus, in $\mathbf{A}$ there are ρ linearly independent rows. $\square$

Example 2.62. Let us consider the following matrix

$$\mathbf{A} = \begin{pmatrix} 1 & 2 & 0 \\ 2 & 1 & 2 \\ 3 & 3 & 2 \end{pmatrix}.$$

We can easily verify that $\det \mathbf{A} = 0$ and that the rank of the matrix is $\rho = 2$. We can observe that the third row is sum of the other two rows:

$$\mathbf{a_3} = \mathbf{a_1} + \mathbf{a_2}$$

that is the rows are linearly dependent. On the other hand, any two rows are linearly independent.

Example 2.63. Let us consider the following matrix

$$\mathbf{A} = \begin{pmatrix} 1 & 2 & 1 \\ 2 & 4 & 2 \\ 0 & 0 & 1 \end{pmatrix}.$$

Also in this case we can verify that $\det \mathbf{A} = 0$ and that the rank of the matrix is $\rho = 2$. Furthermore, the rows are linearly dependent since

$$\mathbf{a_2} = 2\mathbf{a_1} + 0\mathbf{a_3}.$$

This means that also in this case two rows are linearly independent. However, this case is different from the previous one since the linearly independent rows cannot be chosen arbitrarily. More specifically, $\mathbf{a_1}$ and $\mathbf{a_3}$ are linearly independent as well as $\mathbf{a_2}$ and $\mathbf{a_3}$. On the other hand, $\mathbf{a_1}$ and $\mathbf{a_2}$ are linearly dependent ($\mathbf{a_2} = 2\mathbf{a_1}$).

In other words, the rank ρ says how many rows (columns) are linearly independent but does not say that any ρ rows (columns) are linearly independent. Determining which rows (columns) are linearly independent is a separate task which can be not trivial.

Definition 2.37. Let $\mathbf{A} \in \mathbb{R}_{m,n}$ and let us indicate with $\mathbf{M_r}$ a r order square submatrix of $\mathbf{A}$. A $r + 1$ square submatrix of $\mathbf{A}$ that contains also $\mathbf{M_r}$ is here said *edged submatrix*.

Example 2.64. Let us consider a matrix $\mathbf{A} \in \mathbb{R}_{4,3}$.

$$\mathbf{A} = \begin{pmatrix} a_{1,1} & a_{1,2} & a_{1,3} & a_{1,4} \\ a_{2,1} & a_{2,2} & a_{2,3} & a_{2,4} \\ a_{3,1} & a_{3,2} & a_{3,3} & a_{3,4} \end{pmatrix}$$

From **A** let us extract the following submatrix $\mathbf{M_r}$,

$$\mathbf{M_r} = \begin{pmatrix} a_{1,1} & a_{1,3} \\ a_{3,1} & a_{3,3} \end{pmatrix}.$$

Edged submatrices are

$$\begin{pmatrix} a_{1,1} & a_{1,2} & a_{1,3} \\ a_{2,1} & a_{2,2} & a_{2,3} \\ a_{3,1} & a_{3,2} & a_{3,3} \end{pmatrix}$$

and

$$\begin{pmatrix} a_{1,1} & a_{1,3} & a_{1,4} \\ a_{2,1} & a_{2,3} & a_{2,4} \\ a_{3,1} & a_{3,3} & a_{3,4} \end{pmatrix}.$$

Theorem 2.10. Kronecker's Theorem. *Let* $\mathbf{A} \in \mathbb{R}_{m,n}$ *and* $\mathbf{M_r}$ *a r order square submatrix of* $\mathbf{A}$ *with* $1 \leq r < \min\{m,n\}$. *If* $\mathbf{M_r}$ *is non-singular and all the edged submatrices of* $\mathbf{M_r}$ *are singular, then the rank of* $\mathbf{A}$ *is r.*

Example 2.65. Let us consider the following matrix

$$\mathbf{A} = \begin{pmatrix} 1 & 2 & 1 & 3 \\ 1 & 0 & 1 & 1 \\ 1 & 2 & 1 & 3 \end{pmatrix}.$$

The submatrix

$$\begin{pmatrix} 1 & 2 \\ 1 & 0 \end{pmatrix}$$

is non-singular while its edged submatrices

$$\begin{pmatrix} 1 & 2 & 1 \\ 1 & 0 & 1 \\ 1 & 2 & 1 \end{pmatrix}$$

and

$$\begin{pmatrix} 1 & 2 & 3 \\ 1 & 0 & 1 \\ 1 & 2 & 3 \end{pmatrix}$$

are both singular. Kronecker's Theorem states that this is enough to conclude that the rank of the matrix $\mathbf{A}$ is 2. We can verify this fact by considering the other order 3 submatrices, that is

$$\begin{pmatrix} 2 & 1 & 3 \\ 0 & 1 & 1 \\ 2 & 1 & 3 \end{pmatrix}$$

and

$$\begin{pmatrix} 1 & 1 & 3 \\ 1 & 1 & 1 \\ 1 & 1 & 3 \end{pmatrix}.$$

It can be immediately observed that also these two submatrices are singular, hence the rank of $\mathbf{A}$ is indeed 2.

Lemma 2.1. *Let $\mathbf{A} \in \mathbb{R}_{n,n}$ and $\mathbf{B} \in \mathbb{R}_{n,q}$. Let $\mathbf{A}$ be non-singular and $\rho_{\mathbf{B}}$ be the rank of the matrix $\mathbf{B}$. It follows that the rank of the product matrix $\mathbf{AB}$ is $\rho_{\mathbf{B}}$.*

Example 2.66. Let us consider the following matrix

$$\mathbf{A} = \begin{pmatrix} 2 & 1 \\ 1 & 2 \end{pmatrix}$$

and the matrix

$$\mathbf{B} = \begin{pmatrix} 0 & 1 & 1 \\ 1 & 1 & 1 \end{pmatrix}.$$

The matrix $\mathbf{A}$ is non-singular and hence has rank $\rho_{\mathbf{A}} = 2$. The matrix $\mathbf{B}$ has also rank $\rho_{\mathbf{B}} = 2$ since the submatrix

$$\begin{pmatrix} 0 & 1 \\ 1 & 1 \end{pmatrix}$$

is clearly non-singular.

The product matrix

$$\mathbf{AB} = \begin{pmatrix} 1 & 3 & 3 \\ 2 & 3 & 3 \end{pmatrix}$$

has rank $\rho_{\mathbf{AB}} = \rho_{\mathbf{B}} = 2$, as expected from Lemma 2.1.

If we consider a matrix $\mathbf{B}$ having rank equal to 1

$$\mathbf{B} = \begin{pmatrix} 0 & 1 & 1 \\ 0 & 1 & 1 \end{pmatrix}$$

and multiply $\mathbf{A}$ by $\mathbf{B}$ we obtain

$$\mathbf{AB} = \begin{pmatrix} 0 & 3 & 3 \\ 0 & 3 & 3 \end{pmatrix}$$

that has rank $\rho_{\mathbf{AB}} = \rho_{\mathbf{B}} = 1$, in accordance with Lemma 2.1.

Theorem 2.11. Sylvester's Law of Nullity. *Let $\mathbf{A} \in \mathbb{R}_{m,n}$ and $\mathbf{B} \in \mathbb{R}_{n,q}$. Let $\rho_{\mathbf{A}}$ and $\rho_{\mathbf{B}}$ be the ranks of the matrices $\mathbf{A}$ and $\mathbf{B}$, respectively, and $\rho_{\mathbf{AB}}$ be the rank of the product matrix $\mathbf{AB}$. It follows that*

$$\rho_{\mathbf{AB}} \geq \rho_{\mathbf{A}} + \rho_{\mathbf{B}} - n.$$

Example 2.67. In order to check the Sylvester's law of nullity, let us consider the following two matrices:

$$\mathbf{A} = \begin{pmatrix} 0 & 1 \\ 1 & 1 \end{pmatrix}$$

and

$$\mathbf{B} = \begin{pmatrix} 4 & 1 \\ 0 & 0 \end{pmatrix}.$$

Matrix **A** is non-singular, hence has rank $\rho_A = 2$ while **B** is singular and had has rank $\rho_B = 1$.

Let us calculate the product **AB**:

$$\mathbf{AB} = \begin{pmatrix} 0 & 0 \\ 4 & 1 \end{pmatrix}$$

whose rank is $\rho_{AB} = 1$.

We can easily verify that $\rho_A + \rho_B - n = 2 + 1 - 2 = 1$ that is equal to ρ_{AB}.

Example 2.68. Let us consider again the matrix **A** which we know has rank $\rho_A = 2$ and the following matrix ρ_B having $\rho_B = 2$:

$$\mathbf{B} = \begin{pmatrix} 4 & 1 \\ 0 & 1 \end{pmatrix}.$$

In accordance with the Sylvester's law of nullity we expect that the product **AB** has a rank > 1 since $\rho_A + \rho_B - n = 2 + 2 - 2 = 2$. In other words, the Sylvester's law of nullity tells us before calculating it, that **AB** is non-singular.

Let us verify this fact:

$$\mathbf{AB} = \begin{pmatrix} 0 & 1 \\ 4 & 2 \end{pmatrix}.$$

The product matrix is non-singular, i.e. its rank is $\rho_{AB} = 2$, which verifies this theorem.

These examples suggest the following corollary, i.e. the Sylvester's law of nullity for square matrices.

Corollary 2.4. *Let* $\mathbf{A} \in \mathbb{R}_{n,n}$ *and* $\mathbf{B} \in \mathbb{R}_{n,n}$. *Let* ρ_A *and* ρ_B *be the ranks of the matrices* **A** *and* **B**, *respectively, and* ρ_{AB} *be the rank of the product matrix* **AB**. *It follows that*

$$\rho_{AB} \geq \rho_A + \rho_B - n.$$

Example 2.69. Let us consider the following two matrices

$$\mathbf{A} = \begin{pmatrix} 1 & 3 & 1 \\ 1 & 3 & 1 \\ 1 & 3 & 1 \end{pmatrix}$$

and

$$\mathbf{B} = \begin{pmatrix} 2 & 1 & 2 \\ 4 & 2 & 4 \\ 1 & 1 & 1 \end{pmatrix}.$$

The ranks of these two matrices are $\rho_A = 1$ and $\rho_B = 2$ respectively.
The product matrix

$$\mathbf{AB} = \begin{pmatrix} 15 \ 8 \ 15 \\ 15 \ 8 \ 15 \\ 15 \ 8 \ 15 \end{pmatrix}$$

has rank $\rho_{AB} = 1$. Considering that $n = 3$, $\rho_{AB} = 1 \geq \rho_A + \rho_B - n = 1 + 2 - 3 = 0$

Example 2.70. Before entering into the next theorem and its proof, let us introduce the concept of *partitioning* of a matrix. Let us consider the following two matrices:

$$\mathbf{A} = \begin{pmatrix} 2 \ 1 \ 0 \\ 1 \ 3 \ 1 \\ 1 \ 1 \ 0 \end{pmatrix}$$

and

$$\mathbf{B} = \begin{pmatrix} 1 \ 1 \ 1 \\ 2 \ 1 \ 0 \\ 0 \ 1 \ 3 \end{pmatrix}.$$

We can calculate the matrix product

$$\mathbf{AB} = \begin{pmatrix} 4 \ 3 \ 2 \\ 7 \ 5 \ 4 \\ 3 \ 2 \ 1 \end{pmatrix}.$$

Let us now re-write the matrices $\mathbf{A}$ as

$$\mathbf{A} = \begin{pmatrix} \mathbf{A}_{1,1} \ \mathbf{A}_{1,2} \\ \mathbf{A}_{2,1} \ \mathbf{A}_{2,2} \end{pmatrix}$$

where

$$\mathbf{A}_{1,1} = \begin{pmatrix} 2 \ 1 \\ 1 \ 3 \end{pmatrix},$$

$$\mathbf{A}_{1,2} = \begin{pmatrix} 0 \\ 1 \end{pmatrix},$$

$$\mathbf{A}_{2,1} = \begin{pmatrix} 1 \ 1 \end{pmatrix},$$

and

$$\mathbf{A}_{2,2} = \begin{pmatrix} 0 \end{pmatrix}.$$

Analogously, the matrix $\mathbf{B}$ can be written as

$$\mathbf{B} = \begin{pmatrix} \mathbf{B}_{1,1} \\ \mathbf{B}_{2,1} \end{pmatrix}$$

where

$$\mathbf{B}_{1,1} = \begin{pmatrix} 1 & 1 & 1 \\ 2 & 1 & 0 \end{pmatrix}$$

and

$$\mathbf{B}_{2,1} = \begin{pmatrix} 0 & 1 & 3 \end{pmatrix}.$$

This variable replacement is said *partitioning* of a matrix. We can now treat the submatrices as matrix element. This means that in order to calculate $\mathbf{AB}$ we can write

$$\mathbf{AB} = \begin{pmatrix} \mathbf{A}_{1,1} & \mathbf{A}_{1,2} \\ \mathbf{A}_{2,1} & \mathbf{A}_{2,2} \end{pmatrix} \begin{pmatrix} \mathbf{B}_{1,1} \\ \mathbf{B}_{2,1} \end{pmatrix} = \begin{pmatrix} \mathbf{A}_{1,1}\mathbf{B}_{1,1} + \mathbf{A}_{1,2}\mathbf{B}_{2,1} \\ \mathbf{A}_{2,1}\mathbf{B}_{1,1} + \mathbf{A}_{2,2}\mathbf{B}_{2,1} \end{pmatrix}.$$

We can easily verify this fact by calculating the two matrix expressions:

$$\mathbf{A}_{1,1}\mathbf{B}_{1,1} + \mathbf{A}_{1,2}\mathbf{B}_{2,1} = \begin{pmatrix} 2 & 1 \\ 1 & 3 \end{pmatrix} \begin{pmatrix} 1 & 1 & 1 \\ 2 & 1 & 0 \end{pmatrix} + \begin{pmatrix} 0 \\ 1 \end{pmatrix} \begin{pmatrix} 0 & 1 & 3 \end{pmatrix} =$$
$$= \begin{pmatrix} 4 & 3 & 2 \\ 7 & 4 & 1 \end{pmatrix} + \begin{pmatrix} 0 & 0 & 0 \\ 0 & 1 & 3 \end{pmatrix} = \begin{pmatrix} 4 & 3 & 2 \\ 7 & 5 & 4 \end{pmatrix}$$

and

$$\mathbf{A}_{2,1}\mathbf{B}_{1,1} + \mathbf{A}_{2,2}\mathbf{B}_{2,1} = \begin{pmatrix} 1 & 1 \end{pmatrix} \begin{pmatrix} 1 & 1 & 1 \\ 2 & 1 & 0 \end{pmatrix} + \begin{pmatrix} 0 \end{pmatrix} \begin{pmatrix} 0 & 1 & 3 \end{pmatrix} =$$
$$= \begin{pmatrix} 3 & 2 & 1 \end{pmatrix} + \begin{pmatrix} 0 & 0 & 0 \end{pmatrix} = \begin{pmatrix} 3 & 2 & 1 \end{pmatrix}.$$

Hence, we obtain the same $\mathbf{AB}$ matrix as before by operation with the partitions instead of working with the elements:

$$\mathbf{AB} = \begin{pmatrix} \mathbf{A}_{1,1}\mathbf{B}_{1,1} + \mathbf{A}_{1,2}\mathbf{B}_{2,1} \\ \mathbf{A}_{2,1}\mathbf{B}_{1,1} + \mathbf{A}_{2,2}\mathbf{B}_{2,1} \end{pmatrix} = \begin{pmatrix} 4 & 3 & 2 \\ 7 & 5 & 4 \\ 3 & 2 & 1 \end{pmatrix}.$$

Lemma 2.2. *Let* $\mathbf{A} \in \mathbb{R}_{m,n}$ *and* $\rho_\mathbf{A}$ *be its rank. If* $r \leq m$ *rows* ($s \leq n$ *columns*) *are swapped, the rank* $\rho_\mathbf{A}$ *does not change.*

Proof. Every row (column) swaps yields to a change in the determinant sign, see Proposition 2.8. Thus, a row swap cannot make a (square) singular sub-matrix non-singular or vice versa. The rank does not change. □

Example 2.71. The following matrix

$$\mathbf{A} = \begin{pmatrix} 1 & 1 & 3 \\ 0 & 0 & 0 \\ 5 & 3 & 0 \end{pmatrix}$$

is singular and has $\rho_{\mathbf{A}} = 2$ since e.g.

$$\det \begin{pmatrix} 1 & 1 \\ 5 & 3 \end{pmatrix} = -2 \neq 0.$$

If we swap second and third row we obtain

$$\mathbf{A_s} = \begin{pmatrix} 1 & 1 & 3 \\ 5 & 3 & 0 \\ 0 & 0 & 0 \end{pmatrix}$$

the rank $\rho_{\mathbf{A}}$ is still 2.

Thus, without changing the rank of the matrix we can swap the rows (columns) of a matrix to have the non-singular sub-matrix in a specific position of the matrix. We can then partition the matrix to have the non-singular sub-matrix in the first $\rho_{\mathbf{A}}$ rows:

$$\mathbf{A_s} = \begin{pmatrix} \mathbf{A_{1,1}} & \mathbf{A_{1,2}} \\ \mathbf{A_{2,1}} & \mathbf{A_{2,2}} \end{pmatrix}$$

where

$$\mathbf{A_{1,1}} = \begin{pmatrix} 1 & 1 \\ 5 & 3 \end{pmatrix}$$

$$\mathbf{A_{1,2}} = \begin{pmatrix} 3 \\ 0 \end{pmatrix}$$

$$\mathbf{A_{2,1}} = \begin{pmatrix} 0 & 0 \end{pmatrix}$$

$$\mathbf{A_{2,2}} = (0).$$

Lemma 2.3. *Let* $\mathbf{A} \in \mathbb{R}_{m,r}$ *and* $\mathbf{B} \in \mathbb{R}_{r,n}$ *be two compatible matrices and* $\mathbf{C} \in \mathbb{R}_{m,n}$ *the product matrix*

$$\mathbf{AB} = \mathbf{C}.$$

If two arbitrary rows i^{th} *and* j^{th} *of the matrix* $\mathbf{A}$ *are swapped, the product of the resulting matrix* $\mathbf{A_s}$ *by the matrix* $\mathbf{B}$ *is*

$$\mathbf{A_s B} = \mathbf{C_s}$$

where $\mathbf{C_s}$ *is the matrix* $\mathbf{C}$ *after the* i^{th} *and the* j^{th} *row have been swapped.*

Proof. Let us write the matrices **A** and **B** as a vector of row vectors and a vector of column vectors respectively:

$$
\mathbf{A} = \begin{pmatrix} \mathbf{a_1} \\ \mathbf{a_2} \\ \ldots \\ \mathbf{a_i} \\ \ldots \\ \mathbf{a_j} \\ \ldots \\ \mathbf{a_m} \end{pmatrix}
$$

and

$$
\mathbf{B} = \begin{pmatrix} \mathbf{b^1} & \mathbf{b^2} & \ldots & \mathbf{b^n} \end{pmatrix}.
$$

We know that the product matrix **C** is

$$
\mathbf{C} = \begin{pmatrix}
\mathbf{a_1 b^1} & \mathbf{a_1 b^2} & \ldots & \mathbf{a_1 b^n} \\
\mathbf{a_2 b^1} & \mathbf{a_2 b^2} & \ldots & \mathbf{a_2 b^n} \\
\ldots & \ldots & \ldots & \ldots \\
\mathbf{a_i b^1} & \mathbf{a_i b^2} & \ldots & \mathbf{a_i b^n} \\
\ldots & \ldots & \ldots & \ldots \\
\mathbf{a_j b^1} & \mathbf{a_j b^2} & \ldots & \mathbf{a_j b^n} \\
\mathbf{a_m b^1} & \mathbf{a_m b^2} & \ldots & \mathbf{a_m b^n}
\end{pmatrix}.
$$

If we swap the ith and j^{th} rows of **A** we obtain

$$
\mathbf{A_s} = \begin{pmatrix} \mathbf{a_1} \\ \mathbf{a_2} \\ \ldots \\ \mathbf{a_j} \\ \ldots \\ \mathbf{a_i} \\ \ldots \\ \mathbf{a_m} \end{pmatrix}.
$$

Let us now calculate the product $\mathbf{A_s B}$:

$$
\mathbf{A_s B} = \begin{pmatrix}
\mathbf{a_1 b^1} & \mathbf{a_1 b^2} & \ldots & \mathbf{a_1 b^n} \\
\mathbf{a_2 b^1} & \mathbf{a_2 b^2} & \ldots & \mathbf{a_2 b^n} \\
\ldots & \ldots & \ldots & \ldots \\
\mathbf{a_j b^1} & \mathbf{a_j b^2} & \ldots & \mathbf{a_j b^n} \\
\ldots & \ldots & \ldots & \ldots \\
\mathbf{a_i b^1} & \mathbf{a_i b^2} & \ldots & \mathbf{a_i b^n} \\
\mathbf{a_m b^1} & \mathbf{a_m b^2} & \ldots & \mathbf{a_m b^n}
\end{pmatrix} = \mathbf{C_s}. \square
$$

Example 2.72. Let us consider the following two matrices

$$\mathbf{A} = \begin{pmatrix} 1 & 5 & 0 \\ 4 & 2 & 1 \\ 0 & 2 & 3 \end{pmatrix}$$

and

$$\mathbf{B} = \begin{pmatrix} 0 & 2 & 5 \\ 1 & 1 & 0 \\ 5 & 0 & 2 \end{pmatrix}.$$

The product of these two matrices is

$$\mathbf{AB} = \mathbf{C} = \begin{pmatrix} 5 & 7 & 5 \\ 7 & 10 & 22 \\ 17 & 2 & 6 \end{pmatrix}$$

Let us now swap the second and the third rows of the matrix $\mathbf{A}$. The resulting matrix is

$$\mathbf{A_s} = \begin{pmatrix} 1 & 5 & 0 \\ 0 & 2 & 3 \\ 4 & 2 & 1 \end{pmatrix}.$$

Let us now calculate $\mathbf{A_s B}$:

$$\mathbf{A_s B} = \mathbf{C_s} = \begin{pmatrix} 5 & 7 & 5 \\ 17 & 2 & 6 \\ 7 & 10 & 22 \end{pmatrix}.$$

Lemma 2.4. *Let $\mathbf{A} \in \mathbb{R}_{m,r}$ and $\mathbf{B} \in \mathbb{R}_{r,n}$ be two compatible matrices such that*

$$\mathbf{AB} = \mathbf{O}.$$

If two arbitrary rows i^{th} and j^{th} of the matrix $\mathbf{A}$ are swapped, the product of the resulting matrix $\mathbf{A_s}$ by the matrix $\mathbf{B}$ is

$$\mathbf{A_s B} = \mathbf{O}.$$

Proof. For Lemma 2.3 the result of $\mathbf{A_s B}$ is the null matrix with two swapped rows. Since all the rows of the null matrix are identical and composed of only zeros, $\mathbf{A_s B} = \mathbf{O}$. $\square$

Example 2.73. The product of the following matrices

$$\mathbf{A} = \begin{pmatrix} 5 & 1 \\ 0 & 0 \end{pmatrix}$$

and

$$\mathbf{B} = \begin{pmatrix} 0 & -1 \\ 0 & 5 \end{pmatrix}$$

is the null matrix.

Let us now swap first and second rows of the matrix $\mathbf{A}$ and obtain $\mathbf{A_s}$:

$$\mathbf{A_s} = \begin{pmatrix} 0 & 0 \\ 5 & 1 \end{pmatrix}.$$

We can easily verify that $\mathbf{A_s B}$ is also the null matrix.

Lemma 2.5. *Let $\mathbf{A} \in \mathbb{R}_{n,n}$ and $\mathbf{B} \in \mathbb{R}_{n,n}$ be two compatible matrices and $\mathbf{C} \in \mathbb{R}_{n,n}$ the product matrix*

$$\mathbf{AB} = \mathbf{C}.$$

Let us consider that matrix $\mathbf{A_s}$ obtained by swapping two arbitrary columns, the ith and j^{th}, and the matrix $\mathbf{B_s}$ obtained by swapping the ith and j^{th} rows.

It follows that the product of the matrix $\mathbf{A_s}$ by the matrix $\mathbf{B_s}$ is still the matrix $\mathbf{C}$:

$$\mathbf{A_s B_s} = \mathbf{C}.$$

Proof. Let us consider the matrices

$$\mathbf{A} = \begin{pmatrix} a_{1,1} & a_{1,2} & \dots & a_{1,n} \\ a_{2,1} & a_{2,2} & \dots & a_{2,n} \\ \dots & \dots & \dots & \dots \\ a_{n,1} & a_{n,2} & \dots & a_{n,n} \end{pmatrix}$$

and

$$\mathbf{B} = \begin{pmatrix} b_{1,1} & b_{1,2} & \dots & b_{1,n} \\ b_{2,1} & b_{2,2} & \dots & b_{2,n} \\ \dots & \dots & \dots & \dots \\ b_{n,1} & b_{n,2} & \dots & b_{n,n} \end{pmatrix}.$$

Without a loss of generality let us assume $i = 1$ and $j = 2$, i.e. let us swap first and second columns of the matrix $\mathbf{A}$ and first and second rows of the matrix $\mathbf{B}$. The resulting matrices are

$$\mathbf{A_s} = \begin{pmatrix} a_{1,2} & a_{1,1} & \dots & a_{1,n} \\ a_{2,2} & a_{2,1} & \dots & a_{2,n} \\ \dots & \dots & \dots & \dots \\ a_{n,2} & a_{n,1} & \dots & a_{n,n} \end{pmatrix}$$

and

$$\mathbf{B_s} = \begin{pmatrix} b_{2,1} & b_{2,2} & \dots & b_{2,n} \\ b_{1,1} & b_{1,2} & \dots & b_{1,n} \\ \dots & \dots & \dots & \dots \\ b_{n,1} & b_{n,2} & \dots & b_{n,n} \end{pmatrix}.$$

Let us calculate the product $\mathbf{A_s B_s}$. The first scalar product of the first row of $\mathbf{A_s}$ and the first column of $\mathbf{B_s}$ is

$$\mathbf{a_{s1} b_s^1} = a_{1,2} b_{2,1} + a_{1,1} b_{1,1} + \sum_{k=3}^{n} a_{1,k} b_{k,1} = \sum_{k=1}^{n} a_{1,k} b_{k,1} = \mathbf{a_1 b^1}.$$

Analogously, we can verify that $\mathbf{a_{s1}b_s^2} = \mathbf{a_1b^2}$ and, more generally that

$$\forall i, j : \mathbf{a_{si}b_s^j} = \mathbf{a_ib^j}.$$

Thus, $\mathbf{A_sB_s} = \mathbf{AB} = C.$ □

Example 2.74. Let us consider again the matrices

$$\mathbf{A} = \begin{pmatrix} 1 & 5 & 0 \\ 4 & 2 & 1 \\ 0 & 2 & 3 \end{pmatrix}$$

and

$$\mathbf{B} = \begin{pmatrix} 0 & 2 & 5 \\ 1 & 1 & 0 \\ 5 & 0 & 2 \end{pmatrix}.$$

We know that their product is

$$\mathbf{AB} = C = \begin{pmatrix} 5 & 7 & 5 \\ 7 & 10 & 22 \\ 17 & 2 & 6 \end{pmatrix}.$$

Let us swap first and second columns of the matrix $\mathbf{A}$ as well as first and second rows of the matrix $\mathbf{B}$. The resulting matrices are

$$\mathbf{A_s} = \begin{pmatrix} 5 & 1 & 0 \\ 2 & 4 & 1 \\ 2 & 0 & 3 \end{pmatrix}$$

and

$$\mathbf{B} = \begin{pmatrix} 1 & 1 & 0 \\ 0 & 2 & 5 \\ 5 & 0 & 2 \end{pmatrix}.$$

The product matrix is again

$$\mathbf{A_sB_s} = C = \begin{pmatrix} 5 & 7 & 5 \\ 7 & 10 & 22 \\ 17 & 2 & 6 \end{pmatrix}.$$

Lemma 2.6. *Let* $\mathbf{A} \in \mathbb{R}_{n,n}$ *and* $\mathbf{B} \in \mathbb{R}_{n,n}$ *be two compatible matrices such that*

$$\mathbf{AB} = \mathbf{O}.$$

If the ith *and* j^{th} *columns of the matrix* $\mathbf{A}$ *and the* i^{th} *and* j^{th} *rows of the matrix* $\mathbf{B}$ *are swapped, the product of the resulting matrix* $\mathbf{A_s}$ *by the matrix* $\mathbf{B_s}$ *is still null*

$$\mathbf{A_sB_s} = \mathbf{O}.$$

Proof. For Lemma 2.5 the product does not change. Thus it remains the null matrix. □

The Sylvester's Law of Nullity can be also expressed in a weaker form that is often used in applied sciences and engineering.

Theorem 2.12. Weak Sylvester's Law of Nullity. *Let* $\mathbf{A} \in \mathbb{R}_{n,n}$ *and* $\mathbf{B} \in \mathbb{R}_{n,n}$ *such that* $\mathbf{AB} = \mathbf{O}$. *Let* $\rho_\mathbf{A}$ *and* $\rho_\mathbf{B}$ *be the ranks of the matrices* $\mathbf{A}$ *and* $\mathbf{B}$. *It follows that*

$$\rho_\mathbf{A} + \rho_\mathbf{B} \leq n.$$

Proof. Let us swap the rows and the columns of the matrix $\mathbf{A}$ in order to have in the upper left part of the matrix a $\rho_\mathbf{A} \times \rho_\mathbf{A}$ non-singular sub-matrix. For Lemma 2.2, the row and column swap did not alter the rank of the matrix $\mathbf{A}$. Let us indicate with $\mathbf{A_s}$ the matrix resulting from $\mathbf{A}$ after the row and column swaps.

Every time a column swap occurs on $\mathbf{A}$, the ith column is swapped with the jth column, a row swap occurs on $\mathbf{B}$, the ith row is swapped with the jth row. Let us indicate with $\mathbf{B_s}$ the matrix $\mathbf{B}$ after these row swaps.

For Lemmas 2.4 and 2.6 it follows that

$$\mathbf{A_s B_s} = \mathbf{O}.$$

After this rearrangement of rows and columns (after this matrix transformation), let us partition the matrix $\mathbf{A_s}$ in the following way:

$$\mathbf{A_s} = \begin{pmatrix} \mathbf{A}_{1,1} & \mathbf{A}_{1,2} \\ \mathbf{A}_{2,1} & \mathbf{A}_{2,2} \end{pmatrix}$$

where $\mathbf{A}_{1,1}$ has rank $\rho_\mathbf{A}$ and size $\rho_\mathbf{A} \times \rho_\mathbf{A}$.

Analogously, let us perform the partitioning of the matrix $\mathbf{B}$ in order to have $\mathbf{B}_{1,1}$ of size $\rho_\mathbf{A} \times n$.

$$\mathbf{B_s} = \begin{pmatrix} \mathbf{B}_{1,1} \\ \mathbf{B}_{2,1} \end{pmatrix}.$$

For Lemma 2.2 the rank of $\mathbf{B_s}$ is the same as the rank of $\mathbf{B}$:

$$\rho_\mathbf{B} = \rho_{\mathbf{B_s}}.$$

Since $\mathbf{A_s B_s} = \mathbf{O}$, it follows that

$$\mathbf{A}_{1,1}\mathbf{B}_{1,1} + \mathbf{A}_{1,2}\mathbf{B}_{2,1} = \mathbf{O}$$
$$\mathbf{A}_{2,1}\mathbf{B}_{1,1} + \mathbf{A}_{2,2}\mathbf{B}_{2,1} = \mathbf{O}.$$

From the first equation

$$\mathbf{B}_{1,1} = -\mathbf{A}_{1,1}^{-1}\mathbf{A}_{1,2}\mathbf{B}_{2,1}.$$

Let us consider the following non-singular matrix

$$\mathbf{P} = \begin{pmatrix} \mathbf{I} & \mathbf{A}_{1,1}^{-1}\mathbf{A}_{1,2} \\ \mathbf{O} & \mathbf{I} \end{pmatrix}.$$

If we calculate **PB** we obtain

$$\mathbf{PB_s} = \begin{pmatrix} \mathbf{O} \\ \mathbf{B}_{2,1} \end{pmatrix}.$$

Since the matrix **P** is non-singular, for the Lemma 2.1, the matrix $\mathbf{PB_s}$ has rank

$$\rho_{\mathbf{PB}} = \rho_{\mathbf{B}}.$$

It follows that the rank of $\mathbf{B}_{2,1}$ is $\rho_{\mathbf{B}}$:

$$\rho_{\mathbf{B}_{2,1}} = \rho_{\mathbf{B}}.$$

The size of $\mathbf{B}_{2,1}$ is $(n - \rho_{\mathbf{A}}) \times n$. Hence, $\rho_{\mathbf{B}}$ can be at most $(n - \rho_{\mathbf{A}})$. In other words,

$$\rho_{\mathbf{B}} \leq (n - \rho_{\mathbf{A}}) \Rightarrow \rho_{\mathbf{A}} + \rho_{\mathbf{B}} \leq n. \quad \square$$

Example 2.75. Let us consider the following two matrices verifying the hypotheses of the theorem:

$$\mathbf{A} = \begin{pmatrix} 1 & 0 & 0 \\ 5 & 1 & 0 \\ 0 & 0 & 0 \end{pmatrix}$$

and

$$\mathbf{B} = \begin{pmatrix} 0 & 0 & 0 \\ 0 & 0 & 0 \\ 0 & 0 & 1 \end{pmatrix}.$$

The product of the two matrices is $\mathbf{AB} = \mathbf{O}$. Furthermore, $\rho_{\mathbf{A}} = 2$ and $\rho_{\mathbf{B}} = 1$. Clearly $n = 3$ which is $2 + 1$.

Example 2.76. Let us consider now the following matrices:

$$\mathbf{A} = \begin{pmatrix} 5 & 2 & 1 \\ 0 & 1 & -2 \\ 4 & 2 & 0 \end{pmatrix}$$

and

$$\mathbf{B} = \begin{pmatrix} -1 \\ 2 \\ 1 \end{pmatrix}.$$

We can easily verify that $\mathbf{A}$ is singular and its rank is $\rho_{\mathbf{A}} = 2$. Clearly, $\rho_{\mathbf{A}} = 1$. Moreover, we can verify that $\mathbf{AB} = \mathbf{O}$. Considering that $n = 3$, the weak Sylvester's law of nullity is verified since $n = \rho_{\mathbf{A}} + \rho_{\mathbf{B}}$.

It can be observed that if $\mathbf{A}$ is a non-singular matrix of order n and $\mathbf{B}$ is a vector having n rows, the only way to have $\mathbf{AB} = \mathbf{O}$ is if $\mathbf{B}$ is the null vector.

Exercises

2.1. Calculate the product of $\lambda = 3$ by the vector $\mathbf{x} = (2, -3, 0, 5)$

2.2. Calculate the scalar product of $\mathbf{x} = (2, -3, 1, 1)$ by $\mathbf{y} = (3, 3, 4, -1)$

2.3. Multiply the following two matrices:

$$\mathbf{A} = \begin{pmatrix} 2 & 1 & -1 \\ -2 & 3 & 4 \end{pmatrix}$$

$$\mathbf{B} = \begin{pmatrix} 1 & 1 & 2 & 2 \\ 2 & -3 & 1 & 3 \\ -1 & -1 & 5 & 2 \end{pmatrix}.$$

2.4. Multiply the following two matrices:

$$\mathbf{A} = \begin{pmatrix} 7 & -1 & 2 \\ 2 & 3 & 0 \\ 0 & 4 & 1 \end{pmatrix}$$

$$\mathbf{B} = \begin{pmatrix} 0 & -3 & 2 \\ 4 & 1 & -1 \\ 15 & 0 & 1 \end{pmatrix}.$$

2.5. Calculate the determinant the following two matrices:

$$\mathbf{A} = \begin{pmatrix} 2 & 4 & 1 \\ 0 & -1 & 0 \\ 1 & 8 & 0 \end{pmatrix}$$

$$\mathbf{B} = \begin{pmatrix} 0 & 8 & 1 \\ 1 & -1 & 0 \\ 1 & 7 & 1 \end{pmatrix}.$$

2.6. For each of the following matrices, determine the values of k that make the matrix singular.

$$\mathbf{A} = \begin{pmatrix} 2 & -1 & 2 \\ 2 & 5 & 0 \\ k & 4 & k+1 \end{pmatrix}$$

$$\mathbf{B} = \begin{pmatrix} k & -2 & 2-k \\ k+4 & 1 & 0 \\ 3 & 2 & 0 \end{pmatrix}.$$

$$\mathbf{C} = \begin{pmatrix} k & -2 & 2-k \\ k+4 & 1 & 0 \\ 2k+4 & -1 & 2-k \end{pmatrix}.$$

2.7. Compute the adj $(\mathbf{A})$ where $\mathbf{A}$ is the following matrix.

$$\mathbf{A} = \begin{pmatrix} 5 & 0 & 4 \\ 6 & 2 & -1 \\ 12 & 2 & 0 \end{pmatrix}$$

2.8. Invert the following matrices.

$$\mathbf{A} = \begin{pmatrix} 3 & -3 \\ 8 & 2 \end{pmatrix}$$

$$\mathbf{B} = \begin{pmatrix} 2 & 0 & 2 \\ 1 & -2 & -5 \\ 0 & 1 & 1 \end{pmatrix}$$

2.9. Let us consider the following matrix **A**:

$$\mathbf{A} = \begin{pmatrix} 3 & -1 & 2 \\ 0 & 1 & 2 \\ 0 & -2 & 3 \end{pmatrix}.$$

1. Verify that **A** is invertible.
2. Invert the matrix.
3. Verify that $\mathbf{AA}^{-1} = \mathbf{I}$.

2.10. Find the rank of the following matrix

$$\mathbf{A} = \begin{pmatrix} 2 & 1 & 3 \\ 0 & 1 & 1 \\ 1 & 3 & 4 \end{pmatrix}.$$

2.11. Determine the rank of the matrix **A** and invert it, if possible:

$$\mathbf{A} = \begin{pmatrix} 1 & 2 & 1 \\ 2 & 4 & 2 \\ 3 & 6 & 3 \end{pmatrix}$$

Chapter 3
Systems of Linear Equations

3.1 Solution of a System of Linear Equations

Definition 3.1. A *linear equation* in $\mathbb{R}$ in the variables $x_1, x_2, \ldots, x_n$ is an equation of the kind:

$$a_1 x_1 + a_2 x_2 + \ldots + a_n x_n = b$$

where $\forall$ index i, a_i is said *coefficient* of the equation, $a_i x_i$ is said i^{th} *term* of the equation, and b is said *known term*. Coefficients and known term are constant and known numbers in $\mathbb{R}$ while the variables are an unknown set of numbers in $\mathbb{R}$ that satisfy the equality.

Definition 3.2. Let us consider m (with $m > 1$) linear equations in the variables $x_1, x_2, \ldots, x_n$. These equations compose a *system of linear equations* indicated as:

$$\begin{cases} a_{1,1} x_1 + a_{1,2} x_2 + \ldots + a_{1,n} x_n = b_1 \\ a_{2,1} x_1 + a_{2,2} x_2 + \ldots + a_{2,n} x_n = b_2 \\ \ldots \\ a_{n,1} x_1 + a_{n,2} x_2 + \ldots + a_{n,n} x_n = b_n \end{cases}.$$

Every ordered n-tuple of real numbers $y_1, y_2, \ldots y_n$ such that

$$\begin{cases} a_{1,1} y_1 + a_{1,2} y_2 + \ldots + a_{1,n} y_n = b_1 \\ a_{2,1} y_1 + a_{2,2} y_2 + \ldots + a_{2,n} y_n = b_2 \\ \ldots \\ a_{n,1} y_1 + a_{n,2} y_2 + \ldots + a_{n,n} y_n = b_n \end{cases}$$

is said *solution* of the system of linear equations.

© Springer Nature Switzerland AG 2019
F. Neri, *Linear Algebra for Computational Sciences and Engineering*,
https://doi.org/10.1007/978-3-030-21321-3_3

A system can be written as a matrix equation $\mathbf{Ax} = \mathbf{b}$ where

$$\mathbf{A} = \begin{pmatrix} a_{1,1} & a_{1,2} & \ldots & a_{1,n} \\ a_{2,1} & a_{2,2} & \ldots & a_{2,n} \\ \ldots & \ldots & \ldots & \ldots \\ a_{m,1} & a_{m,2} & \ldots & a_{m,n} \end{pmatrix}$$

$$\mathbf{x} = \begin{pmatrix} x_1 \\ x_2 \\ \ldots \\ x_n \end{pmatrix}$$

$$\mathbf{b} = \begin{pmatrix} b_1 \\ b_2 \\ \ldots \\ b_m \end{pmatrix}$$

The coefficient matrix $\mathbf{A}$ is said *incomplete matrix*. The matrix $\mathbf{A^c} \in \mathbb{R}_{m,n+1}$ whose first n columns are those of the matrix $\mathbf{A}$ and the $(n+1)^{th}$ column is the vector $\mathbf{b}$ is said *complete matrix*:

$$\mathbf{A^c} = (\mathbf{A}|\mathbf{b}) = \left(\begin{array}{cccc|c} a_{1,1} & a_{1,2} & \ldots & a_{1,n} & b_1 \\ a_{2,1} & a_{2,2} & \ldots & a_{2,n} & b_2 \\ \ldots & \ldots & \ldots & \ldots & \ldots \\ a_{m,1} & a_{m,2} & \ldots & a_{m,n} & b_m \end{array} \right).$$

Two systems of linear equations are said to be *equivalent* if they have the same solution. It can be observed that if two equations of a system (two rows of the matrix $\mathbf{A^c}$) are swapped, the solution does not change.

Theorem 3.1. Cramer's Theorem. *Let us consider a system of n linear equations in n variables,* $\mathbf{Ax} = \mathbf{b}$. *If* $\mathbf{A}$ *is non-singular, there is only one solution simultaneously satisfying all the equations: if* $\det \mathbf{A} \neq 0$, *then* $\exists! \mathbf{x}$ *such that* $\mathbf{Ax} = \mathbf{b}$.

Proof. Let us consider the system $\mathbf{Ax} = \mathbf{b}$. If $\mathbf{A}$ is non-singular for the Theorem 2.6 the matrix $\mathbf{A}$ is invertible, i.e. a matrix $\mathbf{A^{-1}}$ exists.

Let us multiply $\mathbf{A}^{-1}$ by the equation representing the system:

$$\mathbf{A}^{-1}(\mathbf{A}\mathbf{x}) = \mathbf{A}^{-1}\mathbf{b} \Rightarrow$$
$$\Rightarrow \left(\mathbf{A}^{-1}\mathbf{A}\right)\mathbf{x} = \mathbf{A}^{-1}\mathbf{b} \Rightarrow$$
$$\Rightarrow \mathbf{I}\mathbf{x} = \mathbf{A}^{-1}\mathbf{b} \Rightarrow \mathbf{x} = \mathbf{A}^{-1}\mathbf{b}$$

For Theorem 2.4 the inverse matrix $\mathbf{A}^{-1}$ is unique and thus also the vector $\mathbf{x}$ is unique, i.e. the only one solution solving the system exists. $\quad\square$

We could verify that $\mathbf{A}^{-1}\mathbf{b}$ is the solution of the system by substituting it within the system $\mathbf{A}\mathbf{x} = \mathbf{b}$. Obviously it follows that

$$\mathbf{A}\left(\mathbf{A}^{-1}\mathbf{b}\right) = \left(\mathbf{A}\mathbf{A}^{-1}\right)\mathbf{b} = \mathbf{I}\mathbf{b} = \mathbf{b},$$

i.e. the system is solved.

Thus, on the basis of the Cramer's Theorem, in order to solve a system of linear equations, the inverse matrix of the coefficient matrix should be computed and multiplied by the vector of known terms. In other words, the solution of a system of linear equations $\mathbf{A}\mathbf{x} = \mathbf{b}$ is $\mathbf{x} = \mathbf{A}^{-1}\mathbf{b}$.

Definition 3.3. A system of linear equations that satisfies the hypotheses of the Cramer's Theorem is said a *Cramer system*.

Example 3.1. Solve the following system by inverting the coefficient matrix:

$$\begin{cases} 2x - y + z = 3 \\ x + 2z = 3 \\ x - y = 1 \end{cases}.$$

The system can be re-written as a matrix equation $\mathbf{A}\mathbf{c} = \mathbf{b}$ where

$$\mathbf{A} = \begin{pmatrix} 2 & -1 & 1 \\ 1 & 0 & 2 \\ 1 & -1 & 0 \end{pmatrix}$$

and

$$\mathbf{b} = \begin{pmatrix} 3 \\ 3 \\ 1 \end{pmatrix}$$

In order to verify the non-singularity, let us compute $\det \mathbf{A}$. It can be easily seen that $\det \mathbf{A} = 1$. Thus, the matrix is non-singular and is invertible. The inverse matrix $\mathbf{A}^{-1} = \frac{1}{\det \mathbf{A}} \mathrm{adj}(\mathbf{a})$.

The transpose is:

$$\mathbf{A}^{\mathrm{T}} = \begin{pmatrix} 2 & 1 & 1 \\ -1 & 0 & -1 \\ 1 & 2 & 0 \end{pmatrix}$$

The inverse matrix is:

$$\frac{1}{\det \mathbf{A}} \mathrm{adj}(\mathbf{A}) = \frac{1}{1} \begin{pmatrix} 2 & -1 & -2 \\ 2 & -1 & -3 \\ -1 & 1 & 1 \end{pmatrix}.$$

Thus $\mathbf{x} = \mathbf{A}^{-1}\mathbf{b} = (6-3-2, 6-3-3, -3+3+1) = (1,0,1)$.

Definition 3.4. Let us consider a system of linear equations

$$\mathbf{Ax} = \mathbf{b}$$

where

$$\mathbf{A} = \begin{pmatrix} a_{1,1} & a_{1,2} & \dots & a_{1,n} \\ a_{2,1} & a_{2,2} & \dots & a_{2,n} \\ \dots & \dots & \dots & \dots \\ a_{n,1} & a_{n,2} & \dots & a_{n,n} \end{pmatrix}$$

and

$$\mathbf{b} = \begin{pmatrix} b_1 \\ b_2 \\ \dots b_n \end{pmatrix}.$$

The *hybrid matrix* with respect to the ith column is the matrix $\mathbf{A_i}$ obtained from $\mathbf{A}$ by substituting the ith column with $\mathbf{b}$:

$$\mathbf{A_i} = \begin{pmatrix} a_{1,1} & a_{1,2} & \dots & b_1 & \dots & a_{1,n} \\ a_{2,1} & a_{2,2} & \dots & b_2 & \dots & a_{2,n} \\ \dots & \dots & \dots & \dots & \dots & \dots \\ a_{n,1} & a_{n,2} & \dots & b_n & \dots & a_{n,n} \end{pmatrix}.$$

Equivalently if we write $\mathbf{A}$ as a vector of column vectors:

$$\mathbf{A} = \begin{pmatrix} \mathbf{a}^1 & \mathbf{a}^1 & \dots & \mathbf{a}^{i-1} & \mathbf{a}^i & \mathbf{a}^{i+1} & \dots & \mathbf{a}^n \end{pmatrix}$$

the hybrid matrix $\mathbf{A_i}$ would be

$$\mathbf{A_i} = \begin{pmatrix} \mathbf{a}^1 & \mathbf{a}^1 & \dots & \mathbf{a}^{i-1} & \mathbf{b} & \mathbf{a}^{i+1} & \dots & \mathbf{a}^n \end{pmatrix}.$$

Theorem 3.2. Cramer's Method. *For a given system of linear equations* $\mathbf{Ax} = \mathbf{b}$ *with* $\mathbf{A}$ *non-singular, a generic solution* x_i *element of* $\mathbf{x}$ *can be computed as, see [2]:*

$$x_i = \frac{\det \mathbf{A_i}}{\det \mathbf{A}}$$

where $\mathbf{A_i}$ *is the hybrid matrix with respect to the* i^{th} *column.*

Proof. Let us consider a system of linear equations

$$\begin{cases} a_{1,1}x_1 + a_{1,2}x_2 + \ldots + a_{1,n}x_n = b_1 \\ a_{2,1}x_1 + a_{2,2}x_2 + \ldots + a_{2,n}x_n = b_2 \\ \ldots \\ a_{n,1}x_1 + a_{n,2}x_2 + \ldots + a_{n,n}x_n = b_n \end{cases}.$$

We can compute $\mathbf{x} = \mathbf{A}^{-1}\mathbf{b}$:

$$\begin{pmatrix} x_1 \\ x_2 \\ \ldots \\ x_n \end{pmatrix} = \frac{1}{\det \mathbf{A}} \begin{pmatrix} A_{1,1} & A_{2,1} & \ldots & A_{n,1} \\ A_{1,2} & A_{2,2} & \ldots & A_{n,2} \\ \ldots & \ldots & \ldots & \ldots \\ A_{1,n} & A_{2,n} & \ldots & A_{n,n} \end{pmatrix} \begin{pmatrix} b_1 \\ b_2 \\ \ldots \\ b_n \end{pmatrix}$$

that is

$$\begin{pmatrix} x_1 \\ x_2 \\ \ldots \\ x_n \end{pmatrix} = \frac{1}{\det \mathbf{A}} \begin{pmatrix} A_{1,1}b_1 + A_{2,1}b_2 + \ldots + A_{n,1}b_n \\ A_{1,2}b_1 + A_{2,2}b_2 + \ldots + A_{n,2}b_n \\ \ldots \\ A_{1,n}b_1 + A_{2,n}b_2 + \cdots + A_{n,n}b_n \end{pmatrix}.$$

For the I Laplace Theorem the vector of solutions can be written as:

$$\begin{pmatrix} x_1 \\ x_2 \\ \ldots \\ x_n \end{pmatrix} = \frac{1}{\det \mathbf{A}} \begin{pmatrix} \det \mathbf{A_1} \\ \det \mathbf{A_2} \\ \ldots \\ \det \mathbf{A_n} \end{pmatrix}.$$

□

Example 3.2. Considering the system of the previous example, where $\det \mathbf{A} = 1$,

$$x_1 = \det \begin{pmatrix} 3 & -1 & 1 \\ 3 & 0 & 2 \\ 1 & -1 & 0 \end{pmatrix} = 1$$

$$x_2 = \det \begin{pmatrix} 2 & 3 & 1 \\ 1 & 3 & 2 \\ 1 & 1 & 0 \end{pmatrix} = 0$$

and,

$$x_3 = \det \begin{pmatrix} 2 & -1 & 3 \\ 1 & 0 & 3 \\ 1 & -1 & 1 \end{pmatrix} = 1.$$

Definition 3.5. A system of m linear equations in n variables is said *compatible* if it has at least one solution, *determined* if it has only one solution, *undetermined* if it has infinite solutions, and *incompatible* if it has no solutions.

Theorem 3.3. Rouchè-Capelli Theorem (Kronecker-Capelli Theorem). *A system of m linear equations in n variables* $\mathbf{Ax} = \mathbf{b}$ *is compatible if and only if both the incomplete and complete matrices (*$\mathbf{A}$ *and* $\mathbf{A}^c$ *respectively) are characterised by the same rank* $\rho_{\mathbf{A}} = \rho_{\mathbf{A}^c} = \rho$ *named rank of the system, see [3].*

A proof of the Rouchè-Capelli theorem is given in Appendix B.

The non-singular submatrix having order ρ is said *fundamental submatrix*. The first practical implication of the Rouchè-Capelli Theorem is that when a system of m linear equations in n variables is considered, its compatibility can be verified by computing $\rho_{\mathbf{A}}$ and $\rho_{\mathbf{A}^c}$.

- If $\rho_{\mathbf{A}} < \rho_{\mathbf{A}^c}$ the system is incompatible and thus it has no solutions.
- If $\rho_{\mathbf{A}} = \rho_{\mathbf{A}^c}$ the system is compatible. Under these conditions, three cases can be identified.

  - **Case 1:** If $\rho_{\mathbf{A}} = \rho_{\mathbf{A}^c} = \rho = n = m$, the system is a Cramer's system and can be solved by the Cramer's method.
  - **Case 2:** If $\rho_{\mathbf{A}} = \rho_{\mathbf{A}^c} = \rho = n < m$, ρ equations of the system compose a Cramer's system (and as such has only one solution). The remaining $m - \rho$ equations are a linear combination of the other, these equations are redundant and the system has only one solution.

Algorithm 2 Detection of the General Solution

---

select the ρ rows linearly independent rows of the complete matrix $\mathbf{A}^c$
choose (arbitrarily) $n - \rho$ variables and replace them with parameters
solve the system of linear equations with respect to the remaining variables
express the parametric vector solving the system of linear equations as a sum of vectors where
each parameter appears in only one vector

- **Case 3:** If $\rho_\mathbf{A} = \rho_{\mathbf{A}^c} = \rho \begin{cases} < n \\ \le m \end{cases}$, the system is undetermined and has $\infty^{n-\rho}$ solutions.

In the case of undetermined systems of linear equations the general parametric solution can be found by performing the procedure illustrated in Algorithm 2.

Example 3.3. Let us consider the following system of linear equations:

$$\begin{cases} 3x_1 + 2x_2 + x_3 = 1 \\ x_1 - x_2 = 2 \\ 2x_1 + x_3 = 4 \end{cases} .$$

The incomplete and complete matrices associated with this system are:

$$\mathbf{A} = \begin{pmatrix} 3 & 2 & 1 \\ 1 & -1 & 0 \\ 2 & 0 & 1 \end{pmatrix}$$

and

$$\mathbf{A}^c = \begin{pmatrix} 3 & 2 & 1 & 1 \\ 1 & -1 & 0 & 2 \\ 2 & 0 & 1 & 4 \end{pmatrix}$$

The $\det(\mathbf{A}) = -3$. Hence, the rank $\rho_\mathbf{A} = 3$. It follows that $\rho_{\mathbf{A}^c} = 3$ since a non-singular 3×3 submatrix can be extracted $(\mathbf{A})$ and a 4×4 submatrix cannot be extracted since the size of $\mathbf{A}^c$ is 3×4. Hence, $\rho_\mathbf{A} = \rho_{\mathbf{A}^c} = m = n = 3$ (case 1). The system can be solved by Cramer's Method.

Only one solution exists and is:

$$x_1 = \frac{\det \begin{pmatrix} 1 & 2 & 1 \\ 2 & -1 & 0 \\ 4 & 0 & 1 \end{pmatrix}}{-3} = \frac{1}{3}$$

$$x_2 = \frac{\det \begin{pmatrix} 3 & 1 & 1 \\ 1 & 2 & 0 \\ 2 & 4 & 1 \end{pmatrix}}{-3} = -\frac{5}{3}$$

$$x_3 = \frac{\det \begin{pmatrix} 3 & 2 & 1 \\ 1 & -1 & 2 \\ 2 & 0 & 4 \end{pmatrix}}{-3} = \frac{10}{3}.$$

Example 3.4. Let us now consider the following system of linear equations:

$$\begin{cases} 3x_1 + 2x_2 + x_3 = 1 \\ x_1 - x_2 = 2 \\ 2x_1 + x_3 = 4 \\ 6x_1 + x_2 + 2x_3 = 7 \end{cases}.$$

In this case we have $m = 4$ equations and $n = 3$ variables. From the previous example we already know that $\rho_A = 3$. The complete matrix has also rank $\rho_{A^c} = 3$ because the fourth row is a linear combination of the first three (the fourth row is the sum of the first three rows). Hence $\rho_A = \rho_{A^c} = n = 3 \leq m = 4$. This is a case 2. The system has only one solution, that is that calculated above and the fourth equation is redundant (the same solution satisfies this equation as well).

Example 3.5. Let us now consider the following system of linear equations:

$$\begin{cases} 3x_1 + 2x_2 + 5x_3 = 5 \\ x_1 - x_2 = 0 \\ 2x_1 + 2x_3 = 2 \end{cases}.$$

In this case we have $m = 3$ equations and $n = 3$ variables. The matrix associated with the system is

$$\mathbf{A} = \begin{pmatrix} 3 & 2 & 5 \\ 1 & -1 & 0 \\ 2 & 0 & 2 \end{pmatrix}.$$

It can be observed that the third column is linear combination of the other two. Hence, this matrix is singular. As such, it cannot be solved by Cramer's Method (nor by Cramer's Theorem). The rank of this matrix is 2 as well as the rank of the complete matrix. Hence, we have $\rho_A = \rho_{A^c} = 2 < n = 3$. For the Rouchè-Capelli Theorem the system has $\infty^{n-\rho} = \infty^1$ solutions. This is a case 3.

It can be observed that any solution proportionate to $(1,1,1)$ solves the system above, e.g. $(100, 100, 100)$ would be a solution of the system. We can synthetically write that $(\alpha, \alpha, \alpha) = \alpha(1, 1, 1), \forall \alpha \in \mathbb{R}$.

Example 3.6. Finally let us consider the system of linear equations:

$$\begin{cases} 3x_1 + 2x_2 + x_3 = 1 \\ x_1 - x_2 = 2 \\ 2x_1 + x_3 = 4 \\ 6x_1 + x_2 + x_3 = 6 \end{cases}.$$

We already know from the examples above that $\rho_A = 3$. In this case the $\rho_{A^c} = 4$ because $\det(A^c) = -6 \neq 0$. Hence, $\rho_A \neq \rho_{A^c}$. The system is incompatible., i.e. there is no solution satisfying the system.

Example 3.7. Let us consider the following system of linear equations:

$$\begin{cases} x_1 + x_2 - x_3 = 2 \\ 2x_1 + x_3 = 1 \\ x_2 + 3x_3 = -3 \\ 2x_1 + x_2 + 4x_3 = -2 \\ x_1 + 2x_2 + 2x_3 = -1 \end{cases}.$$

The incomplete and complete matrices associated with this system are:

$$A = \begin{pmatrix} 1 & 1 & -1 \\ 2 & 0 & 1 \\ 0 & 1 & 3 \\ 2 & 1 & 4 \\ 1 & 2 & 2 \end{pmatrix}$$

and

$$A^c = \begin{pmatrix} 1 & 1 & -1 & 2 \\ 2 & 0 & 1 & 1 \\ 0 & 1 & 3 & -3 \\ 2 & 1 & 4 & -2 \\ 1 & 2 & 2 & -1 \end{pmatrix}.$$

It can be verified that the $\rho_A = \rho_{A^c} = 3$. Thus, the system for the Rouchè-Capelli theorem is compatible. Since $\rho = n < m$ we are in case 2. The fourth row of $\mathbf{A}^c$ is a linear combination of the second and third rows (it is the sum of those two rows). The fifth row of $\mathbf{A}^c$ is a linear combination of first and third rows (it is the sum of those two rows). Thus the last two equations are redundant and the solution $(1, 0, -1)$ solves the system of five equations in three variables.

Example 3.8. Let us consider the following system of linear equations:

$$\begin{cases} 5x + y + 6z = 6 \\ 2x - y + z = 1 \\ 3x - 2y + z = 1. \end{cases}$$

The incomplete and complete matrices associated with this system are:

$$\mathbf{A} = \begin{pmatrix} 5 & 1 & 6 \\ 2 & -1 & 1 \\ 3 & -2 & 1 \end{pmatrix}$$

and

$$\mathbf{A}^c = \begin{pmatrix} 5 & 1 & 6 & 6 \\ 2 & -1 & 1 & 1 \\ 3 & -2 & 1 & 1 \end{pmatrix}.$$

The third column of the matrix $\mathbf{A}$ is some of the first two columns, hence it is singular. It can be seen that $\rho_A = 2 = \rho_{A^c}$. The system is therefore compatible. Moreover $\rho_A = \rho_{A^c} = 2 < n = 3$. Hence, the system is undetermined has ∞^1 solutions.

In order to find the general solution, let us cancel the first equation and let us pose $x = \alpha$ with α real parameter. It follows that our problem becomes

$$\begin{cases} 2\alpha - y + z = 1 \\ 3\alpha - 2y + z = 1 \end{cases} \Rightarrow \begin{cases} 2\alpha - y = 1 - z \\ 3\alpha - 2y = 1 - z \end{cases} \Rightarrow 2\alpha - y = 3\alpha - 2y \Rightarrow y = \alpha.$$

By substitution we obtain $z = 1 - \alpha$. Thus, the general solution is

$$(\alpha, \alpha, 1 - \alpha) = (\alpha, \alpha, -\alpha) + (0, 0, 1) = \alpha(1, 1, -1) + (0, 0, 1).$$

Example 3.9. Let us consider the following system of linear equations:

$$\begin{cases} 5x + y + 6z = 6 \\ 2x - y + z = 1 \\ 3x - 2y + z = 0. \end{cases}$$

We know that $\rho_A = 2$. It can be easily seen that $\rho_{A^c} = 3$. Hence $\rho_A = 2 < \rho_{A^c} = 3$, i.e. the system is impossible.

3.2 Homogeneous Systems of Linear Equations

Definition 3.6. A system of linear equations $\mathbf{Ax} = \mathbf{b}$ is said *homogeneous* if the vector of known terms $\mathbf{b}$ is composed of only zeros and is indicated with $\mathbf{O}$:

$$\begin{cases} a_{1,1}x_1 + a_{1,2}x_2 + \ldots + a_{1,n}x_n = 0 \\ a_{2,1}x_1 + a_{2,2}x_2 + \ldots + a_{2,n}x_n = 0 \\ \ldots \\ a_{m,1}x_1 + a_{m,2}x_2 + \ldots + a_{m,n}x_n = 0 \end{cases}.$$

Theorem 3.4. *A homogeneous system of linear equations is always compatible as it always has at least the solution composed of only zeros.*

Proof. From the properties of the matrix product $\mathbf{AO} = \mathbf{O}$, $\forall$ matrix $\mathbf{A} \in \mathbb{R}_{m,n}$ and vector $\mathbf{O} \in \mathbb{R}_{n,1}$. $\square$

Thus, for the Rouchè-Capelli Theorem, if the rank ρ of the system is equal to n, then the system is determined and has only one solution, that is $\mathbf{O}$. If $\rho < n$ the system has $\infty^{n-\rho}$ solutions in addition to $\mathbf{O}$.

Example 3.10. The following homogeneous system of linear equations

$$\begin{cases} x + y = 0 \\ 2x - 3y + 4z = 0 \\ 2y + 5z = 0 \end{cases}$$

is associated with its incomplete matrix

$$\mathbf{A} = \begin{pmatrix} 1 & 1 & 0 \\ 2 & -3 & 4 \\ 0 & 2 & 5 \end{pmatrix}.$$

The determinant of the incomplete matrix is $(-15 - 8 - 10) = -33$. Thus, the matrix is non-singular and consequently $\rho_A = 3$. The system is thus determined. If we apply Cramer's method we can easily find that the only solution of this system is $0, 0, 0$.

Theorem 3.5. *If the n-tuple $\alpha_1, \alpha_2, \ldots, \alpha_n$ is a solution of the homogeneous system $\mathbf{Ax} = \mathbf{O}$, then $\forall \lambda \in \mathbb{R} : \lambda\alpha_1, \lambda\alpha_2, \ldots, \lambda\alpha_n$ is also a solution of the system.*

Proof. Let us consider the generic i^{th} row of the matrix $\mathbf{A}$: $a_{i,1}, a_{i,2}, \ldots, a_{i,n}$. Let us multiply this row by $\lambda \alpha_1, \lambda \alpha_2, \ldots, \lambda \alpha_n$. The result of the multiplication is $a_{i,1} \lambda \alpha_1 + a_{i,2} \lambda \alpha_2 + \ldots + a_{i,n} \lambda \alpha_n = \lambda \left(\alpha_1 a_{i,1} + \alpha_2 a_{i,2} + \ldots + \alpha_n a_{i,n} \right) = \lambda 0 = 0$. This operation can be repeated $\forall i^{th}$ row. Thus $\lambda \alpha_1, \lambda \alpha_2, \ldots, \lambda \alpha_n$ is a solution of the system $\forall \lambda \in \mathbb{R}$. $\square$

Example 3.11. The following homogeneous system of linear equations

$$\begin{cases} 3x + 2y + z = 0 \\ 4x + y + 3z = 0 \\ 3x + 2y + z = 0 \end{cases}$$

has the following incomplete matrix

$$\mathbf{A} = \begin{pmatrix} 3 & 2 & 1 \\ 4 & 1 & 3 \\ 3 & 2 & 1 \end{pmatrix}$$

which is singular. Hence it follows that $\rho_{\mathbf{A}} = \rho_{\mathbf{A}^c} = 2$ and that the system has ∞ solutions. For example $1, -1, -1$ is a solution of the system. Also, $(2, -2, -2)$ is a solution of the system as well as $(5, -5, -5)$ or $(1000, -1000, -1000)$. In general $(\lambda, -\lambda, -\lambda)$ is a solution of this system $\forall \lambda \in \mathbb{R}$.

Theorem 3.6. *Let us consider a homogeneous system* $\mathbf{Ax} = \mathbf{O}$. *If* $(\alpha_1, \alpha_2, \ldots, \alpha_n)$ *and* $(\beta_1, \beta_2, \ldots, \beta_n)$ *are both solutions of the system, then every linear combination of these two n-tuple is solution of the system:* $\forall \lambda, \mu \in \mathbb{R}$, $(\lambda \alpha_1 + \mu \beta_1, \lambda \alpha_2 + \mu \beta_2, \ldots, \lambda \alpha_n + \mu \beta_n)$ *is also a solution.*

Proof. Let us consider the generic i^{th} row of the matrix $\mathbf{A}$: $a_{i,1}, a_{i,2}, \ldots, a_{i,n}$. Let us multiply this row by $\lambda \alpha_1 + \mu \beta_1, \lambda \alpha_2 + \mu \beta_2, \ldots, \lambda \alpha_n + \mu \beta_n$. The result of the multiplication is $a_{i,1} \left(\lambda \alpha_1 + \mu \beta_1 \right) + a_{i,2} \left(\lambda \alpha_2 + \mu \beta_2 \right) + \ldots + a_{i,n} \left(\lambda \alpha_n + \mu \beta_n \right) = \lambda \left(a_{i,1} \alpha_1 + a_{i,2} \alpha_2 + \ldots + a_{i,n} \alpha_n \right) + \mu \left(a_{i,1} \beta_1 + a_{i,2} \beta_2 + \ldots + a_{i,n} \beta_n \right) = 0 + 0 = 0$. $\square$

Example 3.12. Let us consider again the homogeneous system of linear equations

$$\begin{cases} 3x + 2y + z = 0 \\ 4x + y + 3z = 0 \\ 3x + 2y + z = 0. \end{cases}$$

We know that this system has ∞ solutions and that $(1, -1, -1)$ and $(2, -2, -2)$ are two solutions. Let us choose two arbitrary real numbers $\lambda = 4$ and $\mu = 5$. Let us calculate the linear combination of these two solutions by means of the scalars λ and μ:

$$\lambda (1, -1, -1) + \mu (2, -2, -2) = 4 (1, -1, -1) + 5 (2, -2, -2) = (14, -14, -14)$$

that is also a solution of the system.

Theorem 3.7. *Let* $\mathbf{Ax} = \mathbf{O}$ *be a homogeneous system of n equations in n + 1 variables. Let the rank* ρ *associated with the system be n. This system has* ∞^1 *solutions proportionate to the n-tuple* $\left(D_1, -D_2, \ldots, (-1)^{i+1} D_i \ldots (-1)^{n+2} D_{n+1}\right)$ *where* $\forall$ *index i,* D_i *is the determinant of the matrix* $\mathbf{A}$ *after the* i^{th} *column has been cancelled.*

Proof. Let us consider the matrix $\mathbf{A}$:

$$\mathbf{A} = \begin{pmatrix} a_{1,1} & a_{1,2} & \cdots & a_{1,n+1} \\ a_{2,1} & a_{2,2} & \cdots & a_{2,n+1} \\ \cdots & \cdots & \cdots & \cdots \\ a_{n,1} & a_{n,2} & \cdots & a_{n,n+1} \end{pmatrix}.$$

Let us indicate with $\tilde{\mathbf{A}}$ a matrix $\in \mathbb{R}_{n+1,n+1}$ constructed from the matrix $\mathbf{A}$ and adding one row

$$\tilde{\mathbf{A}} = \begin{pmatrix} a_{r,1} & a_{r,2} & \cdots & a_{r,n+1} \\ a_{1,1} & a_{1,2} & \cdots & a_{1,n+1} \\ a_{2,1} & a_{2,2} & \cdots & a_{2,n+1} \\ \cdots & \cdots & \cdots & \cdots \\ a_{n,1} & a_{n,2} & \cdots & a_{n,n+1} \\ a_{n+1,1} & a_{n+1,2} & \cdots & a_{n+1,n+1} \end{pmatrix}.$$

The elements of the n-tuple $D_1, -D_2, \ldots, (-1)^n D_{n+1}$ can be seen as the cofactors $\tilde{A}_{n+1,1}, \tilde{A}_{n+1,2}, \ldots, \tilde{A}_{n+1,n+1}$ related to $(n+1)^{th}$ row of the matrix $\tilde{\mathbf{A}}$. Thus, if we multiply the ith row of the matrix $\mathbf{A}$ by $D_1, -D_2, \ldots, (-1)^n D_{n+1}$ we obtain:

$$a_{i,1} D_1 - a_{i,2} D_2 + \ldots + (-1)^n a_{i,n+1} D_{n+1} =$$
$$= a_{i,1} A_{n+1,1} + a_{i,2} A_{n+1,2} + \ldots + a_{i,n+1} A_{n+1,n+1}.$$

This expression is equal to 0 due to the II Laplace Theorem. Thus $D_1, -D_2, \ldots,$ $(-1)^n D_{n+1}$ is a solution of the system. $\square$

Example 3.13. Let us consider the following homogeneous system of linear equations:

$$\begin{cases} 2x + y + z = 0 \\ x + 0y + z = 0 \end{cases}.$$

The associated matrix is

$$\mathbf{A} = \begin{pmatrix} 2 & 1 & 1 \\ 1 & 0 & 1 \end{pmatrix}.$$

Cancelling first, second, and third column and computing the respective determinants we obtain that the ∞ solutions solving this system are all proportionate to $(1, -1, -1)$.

3.3 Direct Methods

Let us consider a Cramer's system $\mathbf{A}\mathbf{x} = \mathbf{b}$ where $\mathbf{A} \in \mathbb{R}_{n,n}$. The solution of this system can be laborious indeed as, by applying the Cramer's Theorem (matrix inversion in Theorem 3.1), it would require the calculation of one determinant of a n order matrix and n^2 determinants of $n-1$ order matrices. The application of the Cramer's Method (see Theorem 3.2), would require the calculation of one determinant of a n order matrix and n determinants of n order matrices. As shown in Chap. 2, a determinant is the sum of $n!$ terms where each term is the result of a multiplication, see also [4].

Let us identify each term with a mathematical operation and we can conclude that, if we neglect computational simple operations such as the transposition, the solution of a system of linear equations requires at least $n! + n^2 ((n-1)!)$ and $n! + n (n!)$ mathematical operations by matrix inversion and Cramer's Method, respectively. It can be easily verified that the computational cost of the matrix inversion is the same as that of the Cramer's Method. If $n = 6$, the solution of the system requires 5040 mathematical operations by matrix inversion or Cramer's Method. Hence, it can be very laborious to be solved by hand. On the other hand, a modern computer having 2.8 GHz clock frequency can perform 2.8 billions of mathematical operations per second and can quickly solve a system of six linear equations in six variables in a fraction of second. If the system is composed of 50 equations in 50 variables (this would not even be a large problem in many engineering applications) the modern computer will need to perform more than 1.55×10^{66} operations by Cramer's Method, thus requiring over 1.75×10^{46} millennia to be solved. If we consider that the estimated age of the universe approximately 13×10^6 millennia, this waiting time is obviously unacceptable, see [5].

On the basis of this consideration during the last centuries mathematicians investigated methods to solve systems of linear equations by drastically reduction the amount of required calculations. One class of these methods, namely *direct methods* perform a set of matrix transformations to re-write the system of linear equation in a new form that is easy to be solved.

Definition 3.7. Let $\mathbf{A} \in \mathbb{R}_{m,n}$. The matrix $\mathbf{A}$ is said *staircase matrix* (a.k.a. *row echelon form*) if the following properties are verified:

- rows entirely consisting of zeros are placed at the bottom of the matrix
- in each non-null row, the first non-null element cannot be in a column to the right of any non-null element below it: $\forall$ index i, if $a_{i,j} = 0$ then $a_{i+1,1} = 0, a_{i+1,2} = 0, \ldots, a_{i+1,j-1} = 0$

Definition 3.8. Let $\mathbf{A} \in \mathbb{R}_{m,n}$ be a staircase matrix. The first non-null element in each row is said *pivot element* of the row.

Example 3.14. The following matrices are staircase matrices:

$$\begin{pmatrix} 2 & 6 & 1 & 7 \\ 0 & 0 & 1 & 3 \\ 0 & 0 & 2 & 3 \\ 0 & 0 & 0 & 0 \end{pmatrix}$$

$$\begin{pmatrix} 3 & 2 & 1 & 7 \\ 0 & 2 & 1 & 3 \\ 0 & 0 & 2 & 3 \\ 0 & 0 & 0 & 4 \end{pmatrix}.$$

Definition 3.9. Let $\mathbf{A} \in \mathbb{R}_{m,n}$. The following operations on the matrix $\mathbf{A}$ are said *elementary row operations*:

- E1: swap of two rows $\mathbf{a_i}$ and $\mathbf{a_j}$

$$\mathbf{a_i} \leftarrow \mathbf{a_j}$$
$$\mathbf{a_j} \leftarrow \mathbf{a_i}$$

- E2: multiplication of a row $\mathbf{a_i}$ by a scalar $\lambda \in \mathbb{R}$

$$\mathbf{a_i} \leftarrow \lambda \mathbf{a_i}$$

- E3: substitution of a row $\mathbf{a_i}$ by the sum of the row $\mathbf{a_i}$ to another row $\mathbf{a_j}$

$$\mathbf{a_i} \leftarrow \mathbf{a_i} + \mathbf{a_j}$$

By combining E2 and E3, we obtain a transformation consisting of the substitution of the row $\mathbf{a_i}$ by the sum of the row $\mathbf{a_i}$ to another row $\mathbf{a_j}$ multiplied by a scalar λ:

$$\mathbf{a_i} \leftarrow \mathbf{a_i} + \lambda \mathbf{a_j}.$$

It can be easily observed that the elementary row operations do not affect the singularity of the matrix of square matrices or, more generally, the rank of matrices .

Definition 3.10. Equivalent Matrices. Let us consider a matrix $\mathbf{A} \in \mathbb{R}_{m,n}$. If we apply the elementary row operations on $\mathbf{A}$ we obtain a new matrix $\mathbf{C} \in \mathbb{R}_{m,n}$. This matrix is said *equivalent* to $\mathbf{A}$.

Theorem 3.8. *For every matrix $\mathbf{A}$ a staircase matrix equivalent to it exists: $\forall \mathbf{A} \in \mathbb{R}_{m,n}$: exists a staircase matrix $\mathbf{C} \in \mathbb{R}_{m,n}$ equivalent to it.*

Example 3.15. Let us consider the matrix

$$\begin{pmatrix} 0 & 2 & -1 & 2 & 5 \\ 0 & 2 & 0 & 1 & 0 \\ 1 & 1 & 0 & 1 & 2 \\ 1 & 1 & 1 & -1 & 0 \end{pmatrix}.$$

Let us swap first and third row,

$$\begin{pmatrix} 1 & 1 & 0 & 1 & 2 \\ 0 & 2 & 0 & 1 & 0 \\ 0 & 2 & -1 & 2 & 5 \\ 1 & 1 & 1 & -1 & 0 \end{pmatrix},$$

then, let us add to the fourth row the first row multiplied by -1

$$\begin{pmatrix} 1 & 1 & 0 & 1 & 2 \\ 0 & 2 & 0 & 1 & 0 \\ 0 & 2 & -1 & 2 & 5 \\ 0 & 0 & 1 & -2 & -2 \end{pmatrix},$$

then, let us add to third row the second row multiplied by -1

$$\begin{pmatrix} 1 & 1 & 0 & 1 & 2 \\ 0 & 2 & 0 & 1 & 0 \\ 0 & 0 & -1 & 1 & 5 \\ 0 & 0 & 1 & -2 & -2 \end{pmatrix},$$

finally, let us add to the fourth row the third row,

$$\begin{pmatrix} 1 & 1 & 0 & 1 & 2 \\ 0 & 2 & 0 & 1 & 0 \\ 0 & 0 & -1 & 1 & 5 \\ 0 & 0 & 0 & -1 & 3 \end{pmatrix}$$

The obtained matrix is a staircase matrix.

Definition 3.11. Let us consider a system of linear equations $\mathbf{A}\mathbf{x} = \mathbf{b}$. If the complete matrix $\mathbf{A}^c$ is a staircase matrix then the system is said *staircase system*.

Definition 3.12. Equivalent Systems. Let us consider two systems of linear equations in the same variables: $\mathbf{Ax} = \mathbf{b}$ and $\mathbf{Cx} = \mathbf{d}$. These two systems are *equivalent* if they have the same solutions.

Theorem 3.9. *Let us consider a system of m linear equations in n variables* $\mathbf{Ax} = \mathbf{b}$. *Let* $\mathbf{A^c} \in \mathbb{R}_{m,n+1}$ *be the complete matrix associated with this system. If another system of linear equations is associated with a complete matrix* $\tilde{\mathbf{A}}^c \in \mathbb{R}_{m,n+1}$ *equivalent to* $\mathbf{A^c}$, *then the two systems are also equivalent.*

Proof. By following the definition of equivalent matrices, if $\tilde{\mathbf{A}}^c$ is equivalent to $\mathbf{A^c}$, then $\tilde{\mathbf{A}}^c$ can be generated from $\mathbf{A^c}$ by applying the elementary row operations. Each operation of the complete matrix obviously has a meaning in the system of linear equations. Let us analyse the effect of the elementary row operations on the complete matrix.

- When E1 is applied, i.e. the swap of two rows, the equations of the system are swapped. This operation has no effect on the solution of the system. Thus after E1 operation the modified system is equivalent to the original one.
- When E2 is applied, i.e. a row is multiplied by a non-null scalar λ, a scalar is multiplied to all the terms of the equation. In this case the equation $a_{i,1}x_1 + a_{i,2}x_2 + \ldots + a_{i,n}x_n = b_i$ is substituted by

$$\lambda a_{i,1}x_1 + \lambda a_{i,2}x_2 + \ldots + \lambda a_{i,n}x_n = \lambda b_i.$$

The two equations have the same solutions and thus after E2 operation the modified systems is equivalent to the original one.
- When E3 is applied, i.e. a row is added to another row, the equation $a_{i,1}x_1 + a_{i,2}x_2 + \ldots + a_{i,n}x_n = b_i$ is substituted by the equation

$$\left(a_{i,1} + a_{j,1} \right) x_1 + \left(a_{i,2} + a_{j,2} \right) x_2 + \ldots + \left(a_{i,n} + a_{j,n} \right) x_n = b_i + b_j.$$

If the n-tuple $y_1, y_2, \ldots, y_n$ is solution of the original system is obviously solution of $a_{i,1}x_1 + a_{i,2}x_2 + \ldots + a_{i,n}x_n = b_i$ and $a_{j,1}x_1 + a_{j,2}x_2 + \ldots + a_{j,n}x_n = b_j$. Thus, $y_1, y_2, \ldots, y_n$ also verifies $\left(a_{i,1} + a_{j,1} \right) x_1 + \left(a_{i,2} + a_{j,2} \right) x_2 + \ldots + \left(a_{i,n} + a_{j,n} \right) x_n = b_i + b_j$. Thus, after E3 operation the modified system is equivalent to the original one. $\square$

By combining the results of Theorems 3.8 and 3.9, the following Corollary can be easily proved.

Corollary 3.1. *Every system of linear equations is equivalent to a staircase system of linear equations.*

This is the theoretical foundation for the so called direct methods, see [6]. Let us consider a system of n linear equations in n variables $\mathbf{Ax} = \mathbf{b}$ with $\mathbf{A} \in \mathbb{R}_{n,n}$. The complete matrix $\mathbf{A^c}$ can be then manipulated by means of the elementary row operations to generate an equivalent staircase system. The aim of this manipulation is to have a triangular incomplete matrix. The transformed system can then be solved with a modest computational effort. If the matrix $\mathbf{A}$ is triangular the variables are uncoupled: in the case of upper triangular $\mathbf{A}$, the last equation is in only one variable and thus can be independently solved; the last but one equation is in two variables but one of them is known from the last equation thus being in one variable and so on. An upper triangular system is of the kind

$$\begin{cases} a_{1,1}x_1 + a_{1,2}x_2 + \ldots + a_{1,n}x_n = b_1 \\ a_{2,2}x_2 + \ldots + a_{2,n}x_n = b_2 \\ \ldots \\ a_{n,n}x_n = b_n. \end{cases}$$

The system can be solved row by row by sequentially applying

$$\begin{cases} x_n = \frac{b_n}{a_{n,n}} \\ x_{n-1} = \frac{b_{n-1} - a_{n-1,n}x_n}{a_{n-1,n-1}} \\ \ldots \\ x_i = \frac{b_i - \sum_{j=i+1}^{n} a_{i,j}x_j}{a_{i,i}}. \end{cases}$$

In an analogous way, if $\mathbf{A}$ is lower triangular, the first equation is in only one variable and thus can be independently solved; the second equation is in two variables but one of them is known from the first equation thus being in one variable and so on.

3.3.1 Gaussian Elimination

The Gaussian elimination, see [6], is a procedure that transforms any system of linear equations into an equivalent triangular system. This procedure, although named after Carl Friedrich Gauss, was previously presented by Chinese mathematicians in the second century AC. The Gaussian elimination, starting from a system $\mathbf{Ax} = \mathbf{b}$ consists of the following steps.

- Construct the complete matrix $\mathbf{A}^c$
- Apply the elementary row operations to obtain a staircase complete matrix and triangular incomplete matrix
- Write down the new system of linear equations
- Solve the nth equation of the system and use the result to solve the $(n-1)^{th}$
- Continue recursively until the first equation

Example 3.16. Let us solve by Gaussian elimination the following (determined) system of linear equations:

$$\begin{cases} x_1 - x_2 + x_3 = 1 \\ x_1 + x_2 = 4 \\ 2x_1 + 2x_2 + 2x_3 = 9 \end{cases}$$

The associated complete matrix is

$$\mathbf{A}^c = (\mathbf{A}|\mathbf{b}) = \begin{pmatrix} 1 & -1 & 1 & | & 1 \\ 1 & 1 & 0 & | & 4 \\ 2 & 2 & 2 & | & 9 \end{pmatrix}.$$

By applying the elementary row operations we obtain the staircase matrix

$$\tilde{\mathbf{A}}^c = (\mathbf{A}|\mathbf{b}) = \begin{pmatrix} 1 & -1 & 1 & | & 1 \\ 0 & 2 & -1 & | & 3 \\ 0 & 0 & 2 & | & 1 \end{pmatrix}.$$

The matrix corresponds to the system

$$\begin{cases} x_1 - x_2 + x_3 = 1 \\ 2x_2 - x_3 = 3 \\ 2x_3 = 1 \end{cases}.$$

From the last equation, x_3 can be immediately derived: $x_3 = \frac{1}{2}$. Then x_3 is substituted in the second equation and x_2 is detected: $x_2 = \frac{7}{4}$. Finally, after substituting, $x_1 = \frac{9}{4}$.

Let us determine the general transformation formulas of the Gaussian elimination. A system of linear equation $\mathbf{Ax} = \mathbf{b}$ can be re-written as

$$\sum_{j=1}^{n} a_{i,j} x_j = b_i$$

for $i = 1, 2, \ldots, n$. Let us pose $a_{i,j}^{(1)} = a_{i,j}$ and $b_i^{(1)} = b_i$. Hence, the matrix $\mathbf{A}$ at the first step is

$$\mathbf{A}^{(1)} = \begin{pmatrix} a_{1,1}^{(1)} & a_{1,2}^{(1)} & \ldots & a_{1,n}^{(1)} \\ a_{2,1}^{(1)} & a_{2,2}^{(1)} & \ldots & a_{2,n}^{(1)} \\ \ldots & \ldots & \ldots & \ldots \\ a_{n,1}^{(1)} & a_{n,2}^{(1)} & \ldots & a_{n,n}^{(1)} \end{pmatrix}$$

and the system can be written as

$$\begin{cases} a_{1,1}^{(1)} x_1 + \sum_{j=2}^{n} a_{1,j}^{(1)} x_j = b_1^{(1)} \\ a_{i,1}^{(1)} x_1 + \sum_{j=2}^{n} a_{i,j}^{(1)} x_j = b_i^{(1)} \quad i = 2, 3, \ldots, n. \end{cases}$$

Let us consider the first equation of the system. Let us divide this equation by $a_{1,1}^{(1)}$:

$$x_1 + \sum_{j=2}^{n} \frac{a_{1,j}^{(1)}}{a_{1,1}^{(1)}} x_j = b_1^{(1)}.$$

Let us now generate $n-1$ equations by multiplying the latter equation by $-a_{2,1}^{1}, -a_{3,1}^{1}, \ldots, -a_{n,1}^{1}$, respectively:

$$-a_{i,1}^{(1)} x_1 + \sum_{j=2}^{n} \frac{-a_{i,1}^{(1)} a_{1,j}^{(1)}}{a_{1,1}^{(1)}} x_j = \frac{-a_{i,1}^{(1)}}{a_{1,1}^{(1)}} b_1^{(1)} \quad i = 2, 3, \ldots, n.$$

These $n-1$ equations are equal to the first row of the system after a multiplication by a scalar. Thus, if we add the first of these equations to the second of the original system, the second of these equations to the third of the original system, ..., the last of these equations to the nth of the original system, we obtain a new system of linear equations that is equivalent to the original one and is

$$\begin{cases} a_{1,1}^{(1)} x_1 + \sum_{j=2}^{n} a_{1,j}^{(1)} x_j = b_1^{(1)} \\ a_{i,1}^{(1)} x_1 + \sum_{j=2}^{n} a_{i,j}^{(1)} x_j - a_{i,1}^{(1)} x_1 + \sum_{j=2}^{n} \frac{-a_{i,1}^{(1)} a_{1,j}^{(1)}}{a_{1,1}^{(1)}} x_j = b_i^{(1)} - \frac{a_{i,1}^{(1)}}{a_{1,1}^{(1)}} b_1^{(1)} \quad i = 2, 3, \ldots, n \end{cases} \Rightarrow$$

$$\begin{cases} a_{1,1}^{(1)} x_1 + \sum_{j=2}^{n} a_{1,j}^{(1)} x_j = b_1^{(1)} \\ \sum_{j=2}^{n} \left(a_{i,j}^{(1)} - \frac{a_{i,1}^{(1)} a_{1,j}^{(1)}}{a_{1,1}^{(1)}} \right) x_j = b_i^{(1)} - \frac{a_{i,1}^{(1)}}{a_{1,1}^{(1)}} b_1^{(1)} \quad i = 2, 3, \ldots, n \end{cases} \Rightarrow$$

$$\begin{cases} a_{1,1}^{(1)} x_1 + \sum_{j=2}^{n} a_{1,j}^{(1)} x_j = b_1^{(1)} \\ \sum_{j=2}^{n} a_{i,j}^{(2)} x_j = b_i^{(2)} \quad i = 2, 3, \ldots, n \end{cases}$$

where

$$\begin{cases} a_{i,j}^{(2)} = a_{i,j}^{(1)} - \frac{a_{i,1}^{(1)} a_{1,j}^{(1)}}{a_{1,1}^{(1)}} \\ b_i^{(2)} = b_i^{(1)} - \frac{a_{i,1}^{(1)}}{a_{1,1}^{(1)}} b_1^{(1)}. \end{cases}$$

Thus, the matrix $\mathbf{A}$ at the second step is

$$\mathbf{A}^{(2)} = \begin{pmatrix} a_{1,1}^{(1)} & a_{1,2}^{(1)} & \ldots & a_{1,n}^{(1)} \\ 0 & a_{2,2}^{(2)} & \ldots & a_{2,n}^{(2)} \\ \ldots & \ldots & \ldots & \ldots \\ 0 & a_{n,2}^{(2)} & \ldots & a_{n,n}^{(2)} \end{pmatrix}$$

We can now repeat the same steps for the system of $n-1$ equations in $n-1$ unknowns

$$\sum_{j=2}^{n} a_{i,j}^{(2)} x_j = b_i^{(2)}$$

for $i = 2, 3, \ldots, n$. We can re-write this system as

$$\begin{cases} a_{2,2}^{(2)} x_2 + \sum_{j=3}^{n} a_{1,j}^{(2)} x_j = b_2^{(2)} \\ a_{i,2}^{(2)} x_2 + \sum_{j=3}^{n} a_{i,j}^{(2)} x_j = b_i^{(2)} \quad i = 3, 4, \ldots, n. \end{cases}$$

Let us multiply the second equation by $-\dfrac{a_{i,2}^{(2)}}{a_{2,2}^{(2)}}$ for $i = 2, 3, \ldots, n$ thus generating $n - 2$ new equations

$$-a_{i,2}^{(2)} x_2 + \sum_{j=3}^{n} \frac{-a_{i,2}^{(2)} a_{2,j}^{(2)}}{a_{2,2}^{(2)}} x_j = \frac{-a_{i,2}^{(2)}}{a_{2,2}^{(2)}} b_2^{(2)} \quad i = 3, 4, \ldots, n.$$

and sum them, one by one, to the last $n - 2$ equations of the system at the second step

$$\begin{cases} a_{2,2}^{(2)} x_2 + \sum_{j=3}^{n} a_{2,j}^{(2)} x_j = b_2^{(2)} \\ \sum_{j=3}^{n} \left(a_{i,j}^{(2)} - \frac{a_{i,2}^{(2)} a_{2,j}^{(2)}}{a_{2,2}^{(2)}} \right) x_j = b_i^{(2)} - \frac{a_{i,2}^{(2)}}{a_{2,2}^{(2)}} b_2^{(2)} \quad i = 3, 4, \ldots, n \end{cases} \Rightarrow$$

$$\begin{cases} a_{2,2}^{(2)} x_2 + \sum_{j=3}^{n} a_{2,j}^{(2)} x_j = b_2^{(2)} \\ \sum_{j=2}^{n} a_{i,j}^{(3)} x_j = b_i^{(3)} \quad i = 3, 4 \ldots, n \end{cases}$$

where

$$\begin{cases} a_{i,j}^{(3)} = a_{i,j}^{(2)} - \dfrac{a_{i,2}^{(2)} a_{2,j}^{(2)}}{a_{2,2}^{(2)}} \\ b_i^{(3)} = b_i^{(2)} - \dfrac{a_{i,2}^{(2)}}{a_{2,2}^{(2)}} b_2^{(2)}. \end{cases}$$

The matrix associated with the system at the third step becomes

$$\mathbf{A}^{(3)} = \begin{pmatrix} a_{1,1}^{(1)} & a_{1,2}^{(1)} & a_{1,3}^{(1)} & \cdots & a_{1,n}^{(1)} \\ 0 & a_{2,2}^{(2)} & a_{2,3}^{(2)} & \cdots & a_{2,n}^{(2)} \\ 0 & 0 & a_{3,3}^{(3)} & \cdots & a_{3,n}^{(3)} \\ \cdots & \cdots & \cdots & \cdots & \cdots \\ 0 & 0 & a_{n,3}^{(3)} & \cdots & a_{n,n}^{(3)} \end{pmatrix}.$$

By repeating these steps for all the rows we finally reach a triangular system in the form

$$\begin{pmatrix} a_{1,1}^{(1)} & a_{1,2}^{(1)} & a_{1,3}^{(1)} & \cdots & a_{1,n}^{(1)} \\ 0 & a_{2,2}^{(2)} & a_{2,3}^{(2)} & \cdots & a_{2,n}^{(2)} \\ 0 & 0 & a_{3,3}^{(3)} & \cdots & a_{3,n}^{(3)} \\ \cdots & \cdots & \cdots & \cdots & \cdots \\ 0 & 0 & 0 & \cdots & a_{n,n}^{(n)} \end{pmatrix} \begin{pmatrix} x_1 \\ x_2 \\ x_3 \\ \cdots \\ x_n \end{pmatrix} = \begin{pmatrix} b_1^{(1)} \\ b_2^{(2)} \\ b_3^{(3)} \\ \cdots \\ b_n^{(n)} \end{pmatrix}.$$

As shown above, this system is triangular and can be easily solved.

The generic Gaussian transformation formulas at the step k can be thus written as:

$$a_{i,j}^{(k+1)} = a_{i,j}^{(k)} - \frac{a_{i,k}^{(k)}}{a_{k,k}^{(k)}} a_{k,j}^{(k)} \quad i,j = k+1,\ldots,n$$

$$b_i^{(k+1)} = b_i^{(k)} - \frac{a_{i,k}^{(k)}}{a_{k,k}^{(k)}} b_k^{(k)} \quad i = k+1,\ldots,n.$$

3.3.1.1 Row Vector Notation for Gaussian Elimination

Let us know write an equivalent formulation of the Gaussian transformation by using the row vector notation. Let us consider a system of linear equations in a matrix formulation:

$$\mathbf{Ax} = \mathbf{b}$$

and let us write the complete matrix $\mathbf{A^c}$ in terms of its row vectors

$$\mathbf{A^c} = \begin{pmatrix} \mathbf{r_1} \\ \mathbf{r_2} \\ \ldots \\ \mathbf{r_n} \end{pmatrix}$$

and, to emphasize that we are working at the step one we can write the complete matrix as

$$\mathbf{A^{c(1)}} = \begin{pmatrix} \mathbf{r_1^{(1)}} \\ \mathbf{r_2^{(1)}} \\ \ldots \\ \mathbf{r_n^{(1)}} \end{pmatrix}.$$

The Gaussian transformation to obtain the matrix at the step (2) are:

$$\mathbf{r_1^{(2)}} = \mathbf{r_1^{(1)}}$$

$$\mathbf{r_2^{(2)}} = \mathbf{r_2^{(1)}} + \left(\frac{-a_{2,1}^{(1)}}{a_{1,1}^{(1)}} \right) \mathbf{r_1^{(1)}}$$

$$\mathbf{r_3^{(2)}} = \mathbf{r_3^{(1)}} + \left(\frac{-a_{3,1}^{(1)}}{a_{1,1}^{(1)}} \right) \mathbf{r_1^{(1)}}$$

$$\ldots$$

$$\mathbf{r_n^{(2)}} = \mathbf{r_n^{(1)}} + \left(\frac{-a_{n,1}^{(1)}}{a_{1,1}^{(1)}} \right) \mathbf{r_1^{(1)}}.$$

After the application of these steps the complete matrix can be written as

$$\mathbf{A}^{c(2)} = \begin{pmatrix} a_{1,1}^{(2)} & a_{1,2}^{(2)} & \cdots & a_{1,n}^{(2)} & b_1^{(2)} \\ 0 & a_{2,2}^{(2)} & \cdots & a_{2,n}^{(2)} & b_2^{(2)} \\ \cdots & \cdots & \cdots & \cdots & \cdots \\ 0 & a_{n,2}^{(2)} & \cdots & a_{n,n}^{(2)} & b_n^{(2)} \end{pmatrix}.$$

The Gaussian transformation to obtain the matrix at the step (3) are:

$$\mathbf{r}_1^{(3)} = \mathbf{r}_1^{(2)}$$
$$\mathbf{r}_2^{(3)} = \mathbf{r}_2^{(2)}$$
$$\mathbf{r}_3^{(3)} = \mathbf{r}_3^{(2)} + \left(\frac{-a_{3,2}^{(1)}}{a_{2,2}^{(2)}} \right) \mathbf{r}_2^{(2)}$$
$$\cdots$$
$$\mathbf{r}_n^{(3)} = \mathbf{r}_n^{(2)} + \left(\frac{-a_{n,2}^{(2)}}{a_{2,2}^{(2)}} \right) \mathbf{r}_2^{(2)}$$

which leads to the following complete matrix

$$\mathbf{A}^{c(3)} = \begin{pmatrix} a_{1,1}^{(3)} & a_{1,2}^{(3)} & \cdots & a_{1,n}^{(3)} & b_1^{(3)} \\ 0 & a_{2,2}^{(3)} & \cdots & a_{2,n}^{(3)} & b_2^{(3)} \\ 0 & 0 & \cdots & a_{3,n}^{(3)} & b_3^{(3)} \\ \cdots & \cdots & \cdots & \cdots & \cdots \\ 0 & 0 & \cdots & a_{n,n}^{(3)} & b_n^{(3)} \end{pmatrix}.$$

At the generic step $(k+1)$ the Gaussian transformation formulas are

$$\mathbf{r}_1^{(k+1)} = \mathbf{r}_1^{(k)}$$
$$\mathbf{r}_2^{(k+1)} = \mathbf{r}_2^{(k)}$$
$$\cdots$$
$$\mathbf{r}_k^{(k+1)} = \mathbf{r}_k^{(k)}$$
$$\mathbf{r}_{k+1}^{(k+1)} = \mathbf{r}_{k+1}^{(k)} + \left(\frac{-a_{k+1,k}^{(k)}}{a_{k,k}^{(k)}} \right) \mathbf{r}_k^{(k)}$$
$$\mathbf{r}_{k+2}^{(k+1)} = \mathbf{r}_{k+2}^{(k)} + \left(\frac{-a_{k+2,k}^{(k)}}{a_{k,k}^{(k)}} \right) \mathbf{r}_k^{(k)}$$
$$\cdots$$
$$\mathbf{r}_n^{(k+1)} = \mathbf{r}_n^{(k)} + \left(\frac{-a_{n,k}^{(k)}}{a_{k,k}^{(k)}} \right) \mathbf{r}_k^{(k)}$$

which completes the description of the method.

Equivalently we can present the Gaussian elimination as an algorithm that inspects the columns and the rows of the matrix. With reference to a system of n linear equations in n variables and indicating with $\mathbf{r_k}$ the generic row of the complete matrix $\mathbf{A^c}$, the pseudocode of the Gaussian elimination is presented in Algorithm 3.

Algorithm 3 Gaussian Elimination as a Pseudocode

for $k = 1 : n-1$ **do**
   for $j = k+1 : n$ **do**
      $\mathbf{r_j}^{(k+1)} = \mathbf{r_j}^{(k)} + \left(-\frac{a_{jk}}{a_{kk}} \right) \mathbf{r_k}^{(k)}$
   end for
end for

Example 3.17. Let us apply the Gaussian elimination to solve the following system of linear equations:

$$\begin{cases} x_1 - x_2 - x_3 + x_4 = 0 \\ 2x_1 + 2x_3 = 8 \\ -x_2 - 2x_3 = -8 \\ 3x_1 - 3x_2 - 2x_3 + 4x_4 = 7 \end{cases}$$

The associated complete matrix is

$$\mathbf{A}^{c(1)} = (\mathbf{A}|\mathbf{b}) = \begin{pmatrix} 1 & -1 & -1 & 1 & \vert & 0 \\ 2 & 0 & 2 & 0 & \vert & 8 \\ 0 & -1 & -2 & 0 & \vert & -8 \\ 3 & -3 & -2 & 4 & \vert & 7 \end{pmatrix}.$$

Let us apply the Gaussian transformations to move to the step (2)

$$\mathbf{r_1}^{(2)} = \mathbf{r_1}^{(1)}$$

$$\mathbf{r_2}^{(2)} = \mathbf{r_2}^{(1)} + \left(\frac{-a_{2,1}^{(1)}}{a_{1,1}^{(1)}} \right) \mathbf{r_1}^{(1)} = \mathbf{r_2}^{(1)} - 2\mathbf{r_1}^{(1)}$$

$$\mathbf{r_3}^{(2)} = \mathbf{r_3}^{(1)} + \left(\frac{-a_{3,1}^{(1)}}{a_{1,1}^{(1)}} \right) \mathbf{r_1}^{(1)} = \mathbf{r_3}^{(1)} + 0\mathbf{r_1}^{(1)}$$

$$\mathbf{r_4}^{(2)} = \mathbf{r_4}^{(1)} + \left(\frac{-a_{4,1}^{(1)}}{a_{1,1}^{(1)}} \right) \mathbf{r_1}^{(1)} = \mathbf{r_4}^{(1)} - 3\mathbf{r_1}^{(1)}$$

thus obtaining the following complete matrix

$$\mathbf{A}^{c(2)} = (\mathbf{A}|\mathbf{b}) = \begin{pmatrix} 1 & -1 & -1 & 1 & | & 0 \\ 0 & 2 & 4 & -2 & | & 8 \\ 0 & -1 & -2 & 0 & | & -8 \\ 0 & 0 & 1 & 1 & | & 7 \end{pmatrix}.$$

Let us apply the Gaussian transformations to move to the step (2)

$$\mathbf{r}_1^{(3)} = \mathbf{r}_1^{(2)}$$
$$\mathbf{r}_2^{(3)} = \mathbf{r}_2^{(2)}$$
$$\mathbf{r}_3^{(3)} = \mathbf{r}_3^{(2)} + \left(\frac{-a_{3,2}^{(2)}}{a_{2,2}^{(2)}} \right) \mathbf{r}_2^{(2)} = \mathbf{r}_3^{(2)} + \tfrac{1}{2}\mathbf{r}_2^{(2)}$$
$$\mathbf{r}_4^{(2)} = \mathbf{r}_4^{(1)} + \left(\frac{-a_{4,1}^{(2)}}{a_{2,2}^{(2)}} \right) \mathbf{r}_2^{(2)} = \mathbf{r}_4^{(1)} + 0\mathbf{r}_2^{(2)}$$

thus obtaining the following complete matrix

$$\mathbf{A}^{c(2)} = (\mathbf{A}|\mathbf{b}) = \begin{pmatrix} 1 & -1 & -1 & 1 & | & 0 \\ 0 & 2 & 4 & -2 & | & 8 \\ 0 & 0 & 0 & -1 & | & -4 \\ 0 & 0 & 1 & 1 & | & 7 \end{pmatrix}.$$

We would need one more step to obtain a triangular matrix. However, in this case, after two steps the matrix is already triangular. It is enough to swap the third and fourth rows to obtain

$$\mathbf{A}^{c(2)} = (\mathbf{A}|\mathbf{b}) = \begin{pmatrix} 1 & -1 & -1 & 1 & | & 0 \\ 0 & 2 & 4 & -2 & | & 8 \\ 0 & 0 & 1 & 1 & | & 7 \\ 0 & 0 & 0 & -1 & | & -4 \end{pmatrix}.$$

From the system of linear equations associated with this matrix we can easily find that $x_4 = 4$, $x_3 = 3$, $x_2 = 2$, and $x_1 = 1$.

3.3.2 Pivoting Strategies and Computational Cost

From the Gaussian transformation formulas in Sect. 3.3.1, it follows that the Gaussian elimination can be applied only when $a_{k,k}^{(k)} \neq 0$ for $k = 1, 2, \ldots, n$. These elements are said *pivotal* elements of the triangular matrix. The condition that all the

pivotal elements must be non-null is not an actually limiting condition. For the Theorem 3.8, the matrix **A** can be transformed into an equivalent staircase matrix, that
is a triangular matrix in the square case. If the matrix is non-singular, i.e. the system is determined, the product of the diagonal elements, i.e. the determinant, must
be non-null. For this reasons, a non-singular matrix can always be re-arranged as a
triangular matrix displaying non-null diagonal elements.

This means that a system of linear equations might need a preprocessing strategy
prior to the application of the Gaussian elimination. Two simple strategies are here
illustrated. The first one, namely *partial pivoting* consists of swapping the kth row of
the matrix at the step k with that row such that the element in the kth column below
$a_{k,k}^{(k)}$ is the maximum in absolute value. In other words, at first the element $a_{r,k}^{(k)}$

$$\left| a_{r,k}^{(k)} \right| = \max_{k \leq i \leq n} \left| a_{i,k}^{(k)} \right|$$

is found. Then, the rth and k^{th} rows are swapped.

Example 3.18. Let us consider the following system of linear equation

$$\begin{cases} x_2 - x_3 = 4 \\ 2x_1 + 6x_3 = 10 \\ 50x_1 - x_2 - 2x_3 = -8 \end{cases}$$

The associated complete matrix at the step 1 is

$$\mathbf{A}^{e(1)} = (\mathbf{A}|\mathbf{b}) = \begin{pmatrix} 0 & 1 & -1 & 0 \\ 2 & 0 & 6 & 10 \\ 50 & -1 & -2 & -8 \end{pmatrix}.$$

We cannot apply the Gaussian transformation because $a_{1,1}^{(1)} = 0$ and we cannot
perform a division by zero. Nonetheless, we can swap the rows in order to be able
to apply the Gaussian transformation. The partial pivoting at the step 1 consists of
swapping the first and the third rows, that is that row having in the first column the
coefficient having maximum absolute value. The resulting matrix is

$$\mathbf{A}^{e(1)} = (\mathbf{A}|\mathbf{b}) = \begin{pmatrix} 50 & -1 & -2 & -8 \\ 2 & 0 & 6 & 10 \\ 0 & 1 & -1 & 0 \end{pmatrix}.$$

Now we can apply the first step of Gaussian elimination:

$$\mathbf{r}_1^{(2)} = \mathbf{r}_1^{(1)}$$
$$\mathbf{r}_2^{(2)} = \mathbf{r}_2^{(1)} + \left(\frac{-a_{2,1}^{(1)}}{a_{1,1}^{(1)}} \right) \mathbf{r}_1^{(1)}$$
$$\mathbf{r}_3^{(2)} = \mathbf{r}_3^{(1)} + \left(\frac{-a_{3,1}^{(1)}}{a_{1,1}^{(1)}} \right) \mathbf{r}_1^{(1)}.$$

Another option is the *total pivoting*. This strategy at first seeks for the indices r and s such that

$$\left| a_{r,s}^{(k)} \right| = \max_{k \leq i,j \leq n} \left| a_{i,j}^{(k)} \right|$$

and then swaps r^{th} and k^{th} rows as well as s^{th} and k^{th} columns. Total pivoting guarantees that pivotal elements are not small numbers and thus that the multipliers are not big numbers. On the other hand, total pivoting performs a swap of the columns, i.e. a perturbation of the sequence of variables. It follows that if the total pivoting is applied, the elements of the solution vector must be rearranged to obtain the original sequence.

As a conclusion of this overview on Gaussian elimination, the question that could be posed is: "What is the advantage of Gaussian elimination with respect to the Cramer's Method?". It can be proved that Gaussian elimination requires about n^3 arithmetic operations to detect the solution, see [5] and [7]. More specifically, if we neglect the pivotal strategy, in order to pass from the matrix $\mathbf{A}^{(k)}$ to the matrix $\mathbf{A}^{(k+1)}$, $3 (n - k)^2$ arithmetic operations are required while to pass from the vector $\mathbf{b}^{(k)}$ to the vector $\mathbf{b}^{(k+1)}$, $2 (n - k)$ arithmetic operations are required. Hence, in order to determine the matrix $\mathbf{A}^{(n)}$ starting from $\mathbf{A}$ and to determine the vector $\mathbf{b}^{(n)}$ starting from $\mathbf{b}$,

$$3 \sum_{k=1}^{n-1} (n - k)^2 + 2 \sum_{k=1}^{n-1} (n - k)$$

arithmetic operations are totally required. Considering that a triangular system requires n^2 operations, it can be easily proved that the total amount of arithmetic operations necessary to solve a system of linear equations by means of the Gaussian method is

$$2 \left(\frac{n (n - 1) (2n - 1)}{6} \right) + 3 \frac{n (n - 1)}{2} + n^2 = \frac{2}{3} n^3 + \frac{3}{2} n^2 - \frac{7}{6} n.$$

Thus, in the example of a system of 50 linear equations in 50 variables the solution is found after about 8.4×10^4 arithmetic operations. This means that a modern computer can solve this problem in about one thousandth of second.

3.3.3 LU Factorization

Equivalently to the Gaussian elimination, the LU factorization is a direct method that transforms a matrix **A** into a matrix product **LU** where **L** is a lower triangular matrix having the diagonal elements all equal to 1 and **U** is an upper triangular matrix. Thus, if we aim at solving a system of linear equations $\mathbf{Ax} = \mathbf{b}$, we obtain

$$\mathbf{Ax} = \mathbf{b} \Rightarrow$$
$$\Rightarrow \mathbf{LUx} = \mathbf{b}.$$

If we pose $\mathbf{Ux} = \mathbf{y}$, we solve at first the triangular system $\mathbf{Ly} = \mathbf{b}$ and then extract **x** from the triangular system $\mathbf{Ux} = \mathbf{y}$.

The main advantage of factorizing the matrix **A** into **LU** with respect to Gaussian elimination is that the method does not alter the vector of known terms **b**. In applications, such as modelling, where the vector of known terms can vary (e.g. if comes from measurements), a new system of linear equation must be solved. Whereas Gaussian elimination would impose the solution of the entire computational task, LU factorization would require only the last steps to be performed again since the factorization itself would not change.

The theoretical foundation of the LU factorization is given in the following theorem.

Theorem 3.10. *Let* $\mathbf{A} \in \mathbb{R}_{n,n}$ *be a non-singular matrix. Let us indicate with* $\mathbf{A_k}$ *the submatrix having order k composed of the first k rows and k columns of* **A**. *If* $\det \mathbf{A_k} \neq 0$ *for* $k = 1, 2, \ldots, n$ *then* $\exists !$ *lower triangular matrix* **L** *having all the diagonal elements equal to 1 and* $\exists !$ *upper triangular matrix* **U** *such that* $\mathbf{A} = \mathbf{LU}$.

Under the hypotheses of this theorem, every matrix can be decomposed into the two triangular matrices. Before entering into implementation details let consider the LU factorization at the intuitive level throughout the next example.

Example 3.19. If we consider the following system of linear equations

$$\begin{cases} x + 3y + 6z = 17 \\ 2x + 8y + 16z = 42 \\ 5x + 21y + 45z = 91 \end{cases}$$

and the corresponding incomplete matrix **A**

$$\mathbf{A} = \begin{pmatrix} 1 & 3 & 6 \\ 2 & 8 & 16 \\ 5 & 21 & 45 \end{pmatrix},$$

we can impose the factorization $\mathbf{A} = \mathbf{LU}$. This means

$$\mathbf{A} = \begin{pmatrix} 1 & 3 & 6 \\ 2 & 8 & 16 \\ 5 & 21 & 45 \end{pmatrix} = \begin{pmatrix} l_{1,1} & 0 & 0 \\ l_{2,1} & l_{2,2} & 0 \\ l_{3,1} & l_{3,2} & l_{3,3} \end{pmatrix} \begin{pmatrix} u_{1,1} & u_{1,2} & u_{1,3} \\ 0 & u_{2,2} & u_{2,3} \\ 0 & 0 & u_{3,3} \end{pmatrix}.$$

If we perform the multiplication of the two matrices we obtain the following system of 9 equations in 12 variables.

$$\begin{cases} l_{1,1}u_{1,1} = 1 \\ l_{1,1}u_{1,2} = 3 \\ l_{1,1}u_{1,3} = 6 \\ l_{2,1}u_{1,1} = 2 \\ l_{2,1}u_{1,2} + l_{2,2}u_{2,2} = 8 \\ l_{2,1}u_{1,3} + l_{2,2}u_{2,3} = 16 \\ l_{3,1}u_{1,1} = 5 \\ l_{3,1}u_{1,2} + l_{3,2}u_{2,2} = 21 \\ l_{3,1}u_{1,3} + l_{3,2}u_{2,3} + l_{3,3}u_{3,3} = 45. \end{cases}$$

Since this system has infinite solutions we can impose some extra equations. Let us impose that $l_{1,1} = l_{2,2} = l_{3,3} = 1$. By substitution we find that

$$\begin{cases} u_{1,1} = 1 \\ u_{1,2} = 3 \\ u_{1,3} = 6 \\ l_{2,1} = 2 \\ u_{2,2} = 2 \\ u_{2,3} = 4 \\ l_{3,1} = 5 \\ l_{3,2} = 3 \\ u_{3,3} = 3. \end{cases}$$

The $\mathbf{A} = \mathbf{LU}$ factorization is then

$$\begin{pmatrix} 1 & 3 & 6 \\ 2 & 8 & 16 \\ 5 & 21 & 45 \end{pmatrix} = \begin{pmatrix} 1 & 0 & 0 \\ 2 & 1 & 0 \\ 5 & 3 & 1 \end{pmatrix} \begin{pmatrix} 1 & 3 & 6 \\ 0 & 2 & 4 \\ 0 & 0 & 3 \end{pmatrix}.$$

In order to solve the original system of linear equations $\mathbf{Ax} = \mathbf{b}$ we can write

$$\mathbf{Ax} = \mathbf{b} \Rightarrow \mathbf{LUx} = \mathbf{b} \Rightarrow \mathbf{Lw} = \mathbf{b}.$$

where $\mathbf{Ux} = \mathbf{w}$.

Let us solve first $\mathbf{Lw} = \mathbf{b}$, that is

$$\begin{pmatrix} 1 & 0 & 0 \\ 2 & 1 & 0 \\ 5 & 3 & 1 \end{pmatrix} \begin{pmatrix} w_1 \\ w_2 \\ w_3 \end{pmatrix} = \begin{pmatrix} 17 \\ 42 \\ 91 \end{pmatrix}.$$

Since this system is triangular, it can be easily solved by substitution and its solution is $w_1 = 17$, $w_2 = 8$, $w_3 = -18$. With these results, the system $\mathbf{Ux} = \mathbf{w}$ must be solved:

$$\begin{pmatrix} 1 & 3 & 6 \\ 0 & 2 & 4 \\ 0 & 0 & 3 \end{pmatrix} \begin{pmatrix} x \\ y \\ z \end{pmatrix} = \begin{pmatrix} 17 \\ 8 \\ -18 \end{pmatrix}$$

which, by substitution, leads to $z = -6$, $y = 16$, and $x = 5$, that is the solution to the initial system of linear equations by LU factorization.

Let us now derive the general transformation formulas. Let $\mathbf{A}$ be

$$\mathbf{A} = \begin{pmatrix} a_{1,1} & a_{1,2} & \dots & a_{1,n} \\ a_{2,1} & a_{2,2} & \dots & a_{2,n} \\ \dots & \dots & \dots & \dots \\ a_{n,1} & a_{n,2} & \dots & a_{n,n} \end{pmatrix}$$

while $\mathbf{L}$ and $\mathbf{U}$ are respectively

$$\mathbf{L} = \begin{pmatrix} 1 & 0 & \dots & 0 \\ l_{2,1} & 1 & \dots & 0 \\ \dots & \dots & \dots & \dots \\ l_{n,1} & l_{n,2} & \dots & 1 \end{pmatrix}$$

$$\mathbf{U} = \begin{pmatrix} u_{1,1} & u_{1,2} & \dots & u_{1,n} \\ 0 & u_{2,2} & \dots & u_{2,n} \\ \dots & \dots & \dots & \dots \\ 0 & 0 & \dots & u_{n,n} \end{pmatrix}.$$

If we impose $\mathbf{A} = \mathbf{LU}$ we obtain

$$a_{i,j} = \sum_{k=1}^{n} l_{i,k} u_{k,j} = \sum_{k=1}^{\min(i,j)} l_{i,k} u_{k,j}$$

for $i, j = 1, 2, \ldots, n$.

In the case $i \le j$, i.e. in the case of the triangular upper part of the matrix we have

$$a_{i,j} = \sum_{k=1}^{i} l_{i,k} u_{k,j} = \sum_{k=1}^{i-1} l_{i,k} u_{k,j} + l_{i,i} u_{i,j} = \sum_{k=1}^{i-1} l_{i,k} u_{k,j} + u_{i,j}$$

This equation is equivalent to

$$u_{i,j} = a_{i,j} - \sum_{k=1}^{i-1} l_{i,k} u_{k,j}$$

that is the formula to determine the elements of $\mathbf{U}$.

Let us consider the case $j < i$, i.e. the lower triangular part of the matrix

$$a_{i,j} = \sum_{k=1}^{j} l_{i,k} u_{k,j} = \sum_{k=1}^{j-1} l_{i,k} u_{k,j} + l_{i,j} u_{j,j}.$$

This equation is equivalent to

$$l_{i,j} = \frac{1}{u_{j,j}} \left(a_{i,j} - \sum_{k=1}^{j-1} l_{i,k} u_{k,j} \right)$$

that is the formula to determine the elements of $\mathbf{L}$.

In order to construct the matrices $\mathbf{L}$ and $\mathbf{U}$, the formulas to determine their elements should be properly combined. Two procedures (algorithms) are here considered. The first procedure, namely Crout's Algorithm, consists of the steps illustrated in Algorithm 4.

Algorithm 4 Crout's Algorithm

---

compute the first row of $\mathbf{U}$
compute the second row of $\mathbf{L}$
compute the second row of $\mathbf{U}$
compute the third row of $\mathbf{L}$
compute the third row of $\mathbf{U}$
compute the fourth row of $\mathbf{L}$
compute the fourth row of $\mathbf{U}$
. . .

In other words, the Crout's Algorithm computes alternately the rows of the two triangular matrices until the matrices have been filled. Another popular way to full the matrices **L** and **U** is by means the so called Doolittle's Algorithm. This procedure is illustrated in Algorithm 5 and consists of filling the rows of **U** alternately with the columns of **L**.

Algorithm 5 Doolittle's Algorithm
compute the first row of **U**
compute the first column of **L**
compute the second row of **U**
compute the second column of **L**
compute the third row of **U**
compute the third column of **L**
compute the fourth row of **U**
...

Example 3.20. Let us apply the Doolittle's Algorithm to perform **LU** factorization of the following matrix

$$\mathbf{A} = \begin{pmatrix} 1 & -1 & 3 & -4 \\ 2 & -3 & 9 & -9 \\ 3 & 1 & -1 & -10 \\ 1 & 2 & -4 & -1 \end{pmatrix}.$$

At the first step, the first row of the matrix **U** is filled by the formula

$$u_{1,j} = a_{1,j} - \sum_{k=1}^{0} l_{1,k} u_{k,j}$$

for $j = 1, 2, 3, 4$.
    This means

$$u_{1,1} = a_{1,1} = 1$$
$$u_{1,2} = a_{1,2} = -1$$
$$u_{1,3} = a_{1,3} = 3$$
$$u_{1,4} = a_{1,4} = -4.$$

Then, the first column of **L** is filled by the formula

$$l_{i,1} = \frac{1}{u_{1,1}} \left(a_{i,1} - \sum_{k=1}^{0} l_{i,k} u_{k,1} \right)$$

for $i = 2, 3, 4$.
    This means

$$l_{2,1} = \frac{a_{2,1}}{u_{1,1}} = 2$$
$$l_{3,1} = \frac{a_{3,1}}{u_{1,1}} = 3$$
$$l_{4,1} = \frac{a_{4,1}}{u_{1,1}} = 1.$$

Thus, the matrices $\mathbf{L}$ and $\mathbf{U}$ at the moment appear as

$$\mathbf{L} = \begin{pmatrix} 1 & 0 & 0 & 0 \\ 2 & 1 & 0 & 0 \\ 3 & l_{3,2} & 1 & 0 \\ 1 & l_{4,2} & l_{4,3} & 1 \end{pmatrix}.$$

and

$$\mathbf{U} = \begin{pmatrix} 1 & -1 & 3 & -4 \\ 0 & u_{2,2} & u_{2,3} & u_{2,4} \\ 0 & 0 & u_{3,3} & u_{3,4} \\ 0 & 0 & 0 & u_{4,4} \end{pmatrix}$$

Then, the second row of the matrix $\mathbf{U}$ is found by applying

$$u_{2,j} = a_{2,j} - \sum_{k=1}^{1} l_{2,k} u_{k,j} = a_{2,j} - l_{2,1} u_{1,j}$$

for $j = 2,3,4$.

This means

$$u_{2,2} = a_{2,2} - l_{2,1} u_{1,2} = -1$$
$$u_{2,3} = a_{2,3} - l_{2,1} u_{1,3} = 3$$
$$u_{2,4} = a_{2,4} - l_{2,1} u_{1,4} = -1.$$

The second column of $\mathbf{L}$ is given by

$$l_{i,2} = \frac{1}{u_{2,2}} \left(a_{i,2} - \sum_{k=1}^{1} l_{i,k} u_{k,2} \right) = \frac{1}{u_{2,2}} (a_{i,2} - l_{i,1} u_{1,2})$$

for $i = 3,4$.

This means

$$l_{3,2} = \frac{a_{3,2} - l_{3,1} u_{1,2}}{u_{2,2}} = -4$$
$$l_{4,2} = \frac{a_{4,2} - l_{4,1} u_{1,2}}{u_{2,2}} = -3.$$

The third row of the matrix $\mathbf{U}$ is given by

$$u_{3,j} = a_{3,j} - \sum_{k=1}^{2} l_{3,k} u_{k,j} = a_{3,j} - l_{3,1} u_{1,j} - l_{3,2} u_{2,j}$$

for $j = 3,4$.

This means

$$u_{3,3} = 2$$
$$u_{3,4} = -2.$$

In order to complete the matrix $\mathbf{L}$, we compute

$$l_{i,3} = \frac{1}{u_{3,3}} \left(a_{i,1} - \sum_{k=1}^{2} l_{i,k} u_{k,3} \right)$$

for $i = 4$, i.e. $l_{4,3} = 1$.

Finally, the matrix $\mathbf{U}$ is computed by

$$u_{4,j} = a_{4,j} - \sum_{k=1}^{3} l_{4,k} u_{k,j}$$

for $j = 4$, i.e. $u_{4,4} = 2$.

Thus, the $\mathbf{L}$ and $\mathbf{U}$ matrices are

$$\mathbf{L} = \begin{pmatrix} 1 & 0 & 0 & 0 \\ 2 & 1 & 0 & 0 \\ 3 & -4 & 1 & 0 \\ 1 & -3 & 1 & 1 \end{pmatrix}.$$

and

$$\mathbf{U} = \begin{pmatrix} 1 & -1 & 3 & -4 \\ 0 & -1 & 3 & -1 \\ 0 & 0 & 2 & -2 \\ 0 & 0 & 0 & 2 \end{pmatrix}.$$

3.3.4 Equivalence of Gaussian Elimination and LU Factorization

It can be easily shown that Gaussian elimination and LU factorization are essentially two different implementations of the same method. In order to remark this fact it can be shown how a LU factorization can be performed by applying the Gaussian elimination.

Let $\mathbf{Ax} = \mathbf{b}$ be a system of linear equations with

$$\mathbf{A} = \begin{pmatrix} a_{1,1} & a_{1,2} & \dots & a_{1,n} \\ a_{2,1} & a_{2,2} & \dots & a_{2,n} \\ \dots & \dots & \dots & \dots \\ a_{n,1} & a_{n,2} & \dots & a_{n,n} \end{pmatrix}.$$

Let us apply the Gaussian elimination to this system of linear equation. Let us indicate with $\mathbf{G_t}$ the triangular incomplete matrix resulting from the application of the Gaussian elimination:

$$
\mathbf{G_t} = \begin{pmatrix} g_{1,1} & g_{1,2} & \cdots & g_{1,n} \\ 0 & g_{2,2} & \cdots & g_{2,n} \\ \cdots & \cdots & \cdots & \cdots \\ 0 & 0 & \cdots & g_{n,n} \end{pmatrix}.
$$

In order to obtain the matrix $\mathbf{G_t}$ from the matrix $\mathbf{A}$, as shown in Sect. 3.3.1.1, linear combinations of row vectors must be calculated by means of weights

$$
\left(\frac{a_{2,1}^{(1)}}{a_{1,1}^{(1)}} \right), \left(\frac{a_{3,1}^{(1)}}{a_{1,1}^{(1)}} \right), \ldots \left(\frac{a_{3,2}^{(2)}}{a_{2,2}^{(2)}} \right) \ldots
$$

Let us arrange these weights in a matrix in the following way

$$
\mathbf{L_t} = \begin{pmatrix} 1 & 0 & \ldots \ldots & 0 \\ \left(\dfrac{a_{2,1}^{(1)}}{a_{1,1}^{(1)}} \right) & 1 & \ldots \ldots & 0 \\ \left(\dfrac{a_{3,1}^{(1)}}{a_{1,1}^{(1)}} \right) & \left(\dfrac{a_{3,2}^{(2)}}{a_{2,2}^{(2)}} \right) & 1 \ldots & 0 \\ \cdots & \cdots & \ldots \ldots & \\ \left(\dfrac{a_{n,1}^{(1)}}{a_{1,1}^{(1)}} \right) & \left(\dfrac{a_{n,2}^{(2)}}{a_{2,2}^{(2)}} \right) & \ldots \ldots & 1 \end{pmatrix}.
$$

It can be easily shown that

$$
\mathbf{A} = \mathbf{L_t G_t}
$$

and thus Gaussian elimination implicitly performs **LU** factorization where **U** is the triangular Gaussian matrix and **L** is the matrix $\mathbf{L_t}$ of the Gaussian multipliers.

Let us clarify this fact by means of the following example.

Example 3.21. Let us consider again the system of linear equations

$$
\begin{cases} x + 3y + 6z = 17 \\ 2x + 8y + 16z = 42 \\ 5x + 21y + 45z = 91 \end{cases}
$$

and its corresponding incomplete matrix $\mathbf{A}$

$$
\mathbf{A} = \begin{pmatrix} 1 & 3 & 6 \\ 2 & 8 & 16 \\ 5 & 21 & 45 \end{pmatrix},
$$

Let us apply Gaussian elimination to obtain a triangular matrix. At the first step

$$\begin{aligned}
\mathbf{r}_1^{(1)} &= \mathbf{r}_1^{(0)} = (1,3,6) \\
\mathbf{r}_2^{(1)} &= \mathbf{r}_2^{(0)} - 2\mathbf{r}_1^{(1)} = (0,2,4) \\
\mathbf{r}_3^{(1)} &= \mathbf{r}_3^{(0)} - 5\mathbf{r}_1^{(1)} = (0,6,15)
\end{aligned}$$

which leads to

$$\begin{pmatrix} 1 & 3 & 6 \\ 0 & 2 & 4 \\ 0 & 6 & 15 \end{pmatrix}$$

and to a preliminary $\mathbf{L_t}$ matrix

$$\begin{pmatrix} 1 & 0 & 0 \\ 2 & 1 & 0 \\ 5 & \# & 1 \end{pmatrix}.$$

where # simply indicates that there is a nun-null element which has not been calculated yet.

Let us apply the second step:

$$\begin{aligned}
\mathbf{r}_1^{(2)} &= \mathbf{r}_1^{(1)} = (1,3,6) \\
\mathbf{r}_2^{(2)} &= \mathbf{r}_2^{(1)} = (0,2,4) \\
\mathbf{r}_3^{(2)} &= \mathbf{r}_3^{(2)} - 3\mathbf{r}_2^{(2)} = (0,0,3)
\end{aligned}$$

which leads to the following matrices

$$\mathbf{G_t} = \begin{pmatrix} 1 & 3 & 6 \\ 0 & 2 & 4 \\ 0 & 0 & 3 \end{pmatrix}$$

and

$$\mathbf{L_t} = \begin{pmatrix} 1 & 0 & 0 \\ 2 & 1 & 0 \\ 5 & 3 & 1 \end{pmatrix}.$$

It can be easily verified that

$$\mathbf{L_t G_t} = \begin{pmatrix} 1 & 0 & 0 \\ 2 & 1 & 0 \\ 5 & 3 & 1 \end{pmatrix} \begin{pmatrix} 1 & 3 & 6 \\ 0 & 2 & 4 \\ 0 & 0 & 3 \end{pmatrix} = \begin{pmatrix} 1 & 3 & 6 \\ 2 & 8 & 16 \\ 5 & 21 & 45 \end{pmatrix} = \mathbf{A}.$$

3.4 Iterative Methods

These methods, starting from an initial guess $\mathbf{x}^{(0)}$, iteratively apply some formulas to detect the solution of the system. For this reason, these methods are often indicated as *iterative methods*. Unlike direct methods that converge to the theoretical solution in a finite time, iterative methods are *approximate* since they *converge*, under some conditions, to the exact solution of the system of linear equations in infinite steps. Furthermore, it must be remarked that both direct and iterative methods perform subsequent steps to detect the solution of systems of linear equations. However, while direct methods progressively manipulate the matrices (complete or incomplete), iterative methods progressively manipulate a candidate solution.

Definition 3.13. Convergence of an Iterative Method. If we consider a system of linear equations $\mathbf{Ax} = \mathbf{b}$, the starting solution $\mathbf{x}^{(0)}$, and the approximated solution at the kth step $\mathbf{x}^{(k)}$, the solution of the system $\mathbf{c}$, then an approximated method is said to converge to the solution of the system when

$$\lim_{k \to \infty} |\mathbf{x}^{(k)} - \mathbf{c}| = \mathbf{O}$$

On the contrary, if the $\lim_{k \to \infty} |\mathbf{x}^{(k)} - \mathbf{c}|$ grows indefinitely then the method is said to *diverge*.

All Iterative methods are characterized by the same structure. If $\mathbf{Ax} = \mathbf{b}$ is a system of linear equations, it can be expressed as $\mathbf{b} - \mathbf{Ax} = \mathbf{O}$. Let us consider a non-singular matrix $\mathbf{M}$ and write the following equation:

$$\mathbf{b} - \mathbf{Ax} = \mathbf{O} \Rightarrow \mathbf{Mx} + (\mathbf{b} - \mathbf{Ax}) = \mathbf{Mx} \Rightarrow$$
$$\Rightarrow \mathbf{M}^{-1}(\mathbf{Mx} + (\mathbf{b} - \mathbf{Ax})) = \mathbf{M}^{-1}\mathbf{Mx} \Rightarrow$$
$$\Rightarrow \mathbf{x} = \mathbf{M}^{-1}\mathbf{Mx} + \mathbf{M}^{-1}(\mathbf{b} - \mathbf{Ax}) = \mathbf{M}^{-1}\mathbf{Mx} + \mathbf{M}^{-1}\mathbf{b} - \mathbf{M}^{-1}\mathbf{Ax} \Rightarrow$$
$$\Rightarrow \mathbf{x} = \left(\mathbf{I} - \mathbf{M}^{-1}\mathbf{A}\right)\mathbf{x} + \mathbf{M}^{-1}\mathbf{b}.$$

If we replace the variables as $\mathbf{H} = \mathbf{I} - \mathbf{M}^{-1}\mathbf{A}$ and $\mathbf{t} = \mathbf{M}^{-1}\mathbf{b}$, an iterative method is characterized by the update formula

$$\mathbf{x} = \mathbf{Hx} + \mathbf{t}$$

and if we emphasize that this is an update formula

$$\mathbf{x}^{(k+1)} = \mathbf{Hx}^{(k)} + \mathbf{t}.$$

In this formulation the convergence condition can be written easily. An iterative method converges to the solution of a system of linear equations for every initial guess $\mathbf{x}^{(0)}$, if and only if the absolute value of the maximum eigenvalue of matrix $\mathbf{H}$ is < 1. The meaning and calculation procedure for eigenvalues are given in Chap. 10. While a thorough study of iterative methods in not a scope of this book, this section gives some examples of simple iterative methods.

3.4.1 Jacobi's Method

The Jacobi's method is the first and simplest iterative method illustrated in this chapter. The method is named after Carl Gustav Jacob Jacobi (1804–1851). Let us consider a system of linear equations $\mathbf{Ax} = \mathbf{b}$ with $\mathbf{A}$ non-singular, $\mathbf{b} \neq \mathbf{O}$, and $\mathbf{A}$ does not display zeros on its main diagonal. Let us indicate with

$$\mathbf{x}^{(0)} = \begin{pmatrix} x_1^{(0)} \\ x_2^{(0)} \\ \dots \\ x_n^{(0)} \end{pmatrix}$$

the initial guess. At the first step, the system can be written as:

$$\begin{cases} a_{1,1}x_1^{(1)} + a_{1,2}x_2^{(0)} + \dots + a_{1,n}x_n^{(0)} = b_1 \\ a_{2,1}x_1^{(0)} + a_{2,2}x_2^{(1)} + \dots + a_{2,n}x_n^{(0)} = b_2 \\ \dots \\ a_{n,1}x_1^{(0)} + a_{n,2}x_2^{(0)} + \dots + a_{n,n}x_n^{(1)} = b_n \end{cases}.$$

At the generic step k, the system can be written as:

$$\begin{cases} a_{1,1}x_1^{(k+1)} + a_{1,2}x_2^{(k)} + \dots + a_{1,n}x_n^{(k)} = b_1 \\ a_{2,1}x_1^{(k)} + a_{2,2}x_2^{(k+1)} + \dots + a_{2,n}x_n^{(k)} = b_2 \\ \dots \\ a_{n,1}x_1^{(k)} + a_{n,2}x_2^{(k)} + \dots + a_{n,n}x_n^{(k+1)} = b_n \end{cases}.$$

The system of linear equation can be rearranged as:

$$\begin{cases} x_1^{(k+1)} = \left(b_1 - \sum_{j=1, j\neq 1}^{n} a_{1,j}x_j^{(k)} \right) \frac{1}{a_{1,1}} \\ x_2^{(k+1)} = \left(b_2 - \sum_{j=1, j\neq 2}^{n} a_{2,j}x_j^{(k)} \right) \frac{1}{a_{2,2}} \\ \dots \\ x_n^{(k+1)} = \left(b_n - \sum_{j=1, j\neq n}^{n} a_{n,j}x_j^{(k)} \right) \frac{1}{a_{n,n}}. \end{cases}$$

Jacobi's method simply makes use of this manipulation to iteratively detect the solution. The generic update formula for the ith variable is

$$x_i^{(k+1)} = \left(b_i - \sum_{j=1,j\neq i}^{n} a_{i,j} x_j^{(k)} \right) \frac{1}{a_{i,i}}.$$

Jacobi's method is conceptually very simple and also easy to implement. The next example clarifies the implementation of the method in practice.

Example 3.22. The following system of linear equations

$$\begin{cases} 10x + 2y + z = 1 \\ 10y - z = 0 \\ x + y - 10z = 4 \end{cases}$$

is determined as the matrix associated with it is non-singular and has determinant equal to -1002. The solution by the application of Cramer's method is

$$\begin{aligned} x &= \frac{49}{334} \\ y &= -\frac{13}{334} \\ z &= -\frac{65}{167}. \end{aligned}$$

Let us solve now the same system by means of the application of Jacobi's method. Let us write the update formulas at first:

$$\begin{aligned} x^{(k+1)} &= \frac{1}{10} \left(1 - 2y^{(k)} - z^{(k)} \right) \\ y^{(k+1)} &= \frac{1}{10} \left(z^{(k)} \right) \\ z^{(k+1)} &= -\frac{1}{10} \left(4 - x^{(k)} - 10y^{(k)} \right) \end{aligned}$$

and take our initial guess

$$\begin{aligned} x^{(0)} &= 0 \\ y^{(0)} &= 0 \\ z^{(0)} &= 0. \end{aligned}$$

Let us now apply the method

$$\begin{aligned} x^{(1)} &= \frac{1}{10} \left(1 - 2y^{(0)} - z^{(0)} \right) = 0.1 \\ y^{(1)} &= \frac{1}{10} \left(z^{(0)} \right) = 0 \\ z^{(1)} &= -\frac{1}{10} \left(4 - x^{(0)} - 10y^{(0)} \right) = -0.4. \end{aligned}$$

These three values are use to calculate $x^{(2)}, y^{(2)}, z^{(2)}$:

$$x^{(2)} = \tfrac{1}{10}\left(1 - 2y^{(1)} - z^{(1)}\right) = 0.14$$
$$y^{(2)} = \tfrac{1}{10}\left(z^{(1)}\right) = -0.04$$
$$z^{(2)} = -\tfrac{1}{10}\left(4 - x^{(1)} - 10y^{(1)}\right) = -0.39.$$

We can apply Jacobi's method iteratively and obtain

$$x^{(3)} = 0.147$$
$$y^{(3)} = -0.039$$
$$z^{(3)} = -0.39,$$

$$x^{(3)} = 0.1468$$
$$y^{(3)} = -0.039$$
$$z^{(3)} = -0.3892,$$

$$x^{(4)} = 0.14672$$
$$y^{(4)} = -0.03892$$
$$z^{(4)} = -0.389220.$$

The solution at the step (4) approximately solves the system. We can check it by substituting these numbers into the system:

$$10x^{(4)} + 2y^{(4)} + z^{(4)} = 1.0001$$
$$10y^{(4)} - z^{(4)} = 0.00002$$
$$x^{(4)} + y^{(4)} - 10z^{(4)} = 4.$$

This solution already gives an approximation of the exact solution. At the step (10) we have

$$x^{(10)} = 0.146707$$
$$y^{(10)} = -0.038922$$
$$z^{(10)} = -0.389222$$

which is about 10^{-9} distant from the exact solution.

Jacobi's method can be expressed also in a matrix form. If we indicate with

$$\mathbf{E} = \begin{pmatrix} 0 & 0 & \dots & 0 \\ a_{2,1} & 0 & \dots & 0 \\ \dots & \dots & \dots & \dots \\ a_{n,1} & a_{n,2} & \dots & 0 \end{pmatrix},$$

$$F = \begin{pmatrix} 0 & a_{1,2} & \dots & a_{1,n} \\ 0 & 0 & \dots & a_{2,n} \\ \dots & \dots & \dots & \dots \\ 0 & 0 & \dots & 0 \end{pmatrix},$$

and

$$D = \begin{pmatrix} a_{1,1} & 0 & \dots & 0 \\ 0 & a_{2,2} & \dots & 0 \\ \dots & \dots & \dots & \dots \\ 0 & 0 & \dots & a_{n,n} \end{pmatrix}$$

the system of linear equation can be written as

$$\mathbf{E}\mathbf{x}^{(k)} + \mathbf{F}\mathbf{x}^{(k)} + \mathbf{D}\mathbf{x}^{(k+1)} = \mathbf{b},$$

that is

$$\mathbf{x}^{(k+1)} = -\mathbf{D}^{-1}(\mathbf{E}+\mathbf{F})\mathbf{x}^{(k)} + \mathbf{D}^{-1}\mathbf{b}.$$

Considering that $\mathbf{E}+\mathbf{F} = \mathbf{A} - \mathbf{D}$, the equation can be written as

$$\mathbf{x}^{(k+1)} = (\mathbf{I} - \mathbf{D}^{-1}\mathbf{A})\mathbf{x}^{(k)} + \mathbf{D}^{-1}\mathbf{b}.$$

Hence, in the case of Jacobi's method $\mathbf{H} = \mathbf{I} - \mathbf{D}^{-1}\mathbf{A}$ and $\mathbf{t} = \mathbf{D}^{-1}\mathbf{b}$.

Example 3.23. The system of linear equations related to the previous example can be written in a matrix form $\mathbf{A}\mathbf{x} = \mathbf{b}$. Considering that

$$\mathbf{E} = \begin{pmatrix} 0 & 0 & 0 \\ 0 & 0 & 0 \\ 1 & 1 & 0 \end{pmatrix},$$

$$\mathbf{F} = \begin{pmatrix} 0 & 2 & 1 \\ 0 & 0 & -1 \\ 0 & 0 & 0 \end{pmatrix}$$

and

$$\mathbf{D} = \begin{pmatrix} 10 & 0 & 0 \\ 0 & 10 & 0 \\ 0 & 0 & -10 \end{pmatrix},$$

we can calculate the inverse $\mathbf{D}^{-1}$:

$$\mathbf{D}^{-1} = \begin{pmatrix} \frac{1}{10} & 0 & 0 \\ 0 & \frac{1}{10} & 0 \\ 0 & 0 & -\frac{1}{10} \end{pmatrix}.$$

The matrix representation of the Jacobi's method means that the vector of the solution is updated according to the formula

$$\begin{pmatrix} x^{(k+1)} \\ y^{(k+1)} \\ z^{(k+1)} \end{pmatrix} = \mathbf{H} \begin{pmatrix} x^{(k)} \\ y^{(k)} \\ z^{(k)} \end{pmatrix} + \mathbf{t}$$

where

$$\mathbf{H} = \left(\mathbf{I} - \mathbf{D}^{-1}\mathbf{A}\right) =$$

$$= \begin{pmatrix} 1 & 0 & 0 \\ 0 & 1 & 0 \\ 0 & 0 & 1 \end{pmatrix} - \begin{pmatrix} \frac{1}{10} & 0 & 0 \\ 0 & \frac{1}{10} & 0 \\ 0 & 0 & -\frac{1}{10} \end{pmatrix} \begin{pmatrix} 10 & 2 & 1 \\ 0 & 10 & -1 \\ 1 & 1 & -10 \end{pmatrix} =$$

$$= \begin{pmatrix} 0 & -\frac{1}{5} & -\frac{1}{10} \\ 0 & 0 & \frac{1}{10} \\ \frac{1}{10} & \frac{1}{10} & 0 \end{pmatrix}$$

and

$$\mathbf{t} = \mathbf{D}^{-1}\mathbf{b} =$$

$$= \begin{pmatrix} \frac{1}{10} & 0 & 0 \\ 0 & \frac{1}{10} & 0 \\ 0 & 0 & -\frac{1}{10} \end{pmatrix} \begin{pmatrix} 1 \\ 0 \\ 4 \end{pmatrix} = \begin{pmatrix} \frac{1}{10} \\ 0 \\ -\frac{4}{10} \end{pmatrix}.$$

It can be verified that the iterative application of matrix multiplication and sum leads to an approximated solution. Considering our initial guess

$$x^{(0)} = 0$$
$$y^{(0)} = 0$$
$$z^{(0)} = 0,$$

we have

$$\begin{pmatrix} x^{(1)} \\ y^{(1)} \\ z^{(1)} \end{pmatrix} = \begin{pmatrix} 0 & -\frac{1}{5} & -\frac{1}{10} \\ 0 & 0 & \frac{1}{10} \\ \frac{1}{10} & \frac{1}{10} & 0 \end{pmatrix} \begin{pmatrix} 0 \\ 0 \\ 0 \end{pmatrix} + \begin{pmatrix} \frac{1}{10} \\ 0 \\ -\frac{4}{10} \end{pmatrix} = \begin{pmatrix} \frac{1}{10} \\ 0 \\ -\frac{4}{10} \end{pmatrix}.$$

Then,

$$\begin{pmatrix} x^{(2)} \\ y^{(2)} \\ z^{(2)} \end{pmatrix} = \begin{pmatrix} 0 & -\frac{1}{5} & -\frac{1}{10} \\ 0 & 0 & \frac{1}{10} \\ \frac{1}{10} & \frac{1}{10} & 0 \end{pmatrix} \begin{pmatrix} \frac{1}{10} \\ 0 \\ -\frac{4}{10} \end{pmatrix} + \begin{pmatrix} \frac{1}{10} \\ 0 \\ -\frac{4}{10} \end{pmatrix} = \begin{pmatrix} 0.14 \\ -0.04 \\ -0.39 \end{pmatrix}.$$

If we keep on iterating the procedure we obtain the same result at step (10):

$$x^{(10)} = 0.146707$$
$$y^{(10)} = -0.038922$$
$$z^{(10)} = -0.389222.$$

Obviously the two ways of writing Jacobi's method lead to the same results (as they are the same thing).

For the sake of clarity, the pseudocode of the Jacobi's method is shown in Algorithm 6.

Algorithm 6 Jacobi's Method
---
input **A** and **b**
n is the size of **A**
while precision conditions **do**
    for $i = 1 : n$ **do**
      $s = 0$
      for $j = 1 : n$ **do**
        if $j \neq i$ **then**
          $s = s + a_{i,j} x_j$
        end if
      end for
      $y_i = \frac{1}{a_{i,i}} (b_i - s)$
    end for
    x = **y**
end while

3.4.2 Gauss-Seidel's Method

The Gauss-Siedel's method, named after Carl Friedrich Gauss (1777–1855) and Philipp L. Seidel (1821–1896) is a greedy variant of the Jacobi's method. This variant, albeit simplistic, often (but not always) allows a faster convergence to the solution of the system. With the Jacobi's method the update of $\mathbf{x}^{(k+1)}$ occurs only when all the values $x_i^{(k+1)}$ are available. With Gauss-Seidel's method, the x_i values replace the old ones as soon as they have been calculated. Thus, from a system of linear equations the formulas for the Gauss-Siedel's method are:

$$\begin{cases} x_1^{(k+1)} = \left(b_1 - \sum_{j=2}^{n} a_{1,j} x_j^{(k)}\right) \frac{1}{a_{1,1}} \\ x_2^{(k+1)} = \left(b_2 - \sum_{j=3}^{n} a_{2,j} x_j^{(k)} - a_{2,1} x_1^{(k+1)}\right) \frac{1}{a_{2,2}} \\ \dots \\ x_i^{(k+1)} = \left(b_i - \sum_{j=i+1}^{n} a_{i,j} x_j^{(k)} - \sum_{j=1}^{i-1} a_{i,j} x_j^{(k+1)}\right) \frac{1}{a_{i,i}} \\ \dots \\ x_n^{(k+1)} = \left(b_n - \sum_{j=1}^{n-1} a_{i,j} x_i^{(k+1)}\right) \frac{1}{a_{n,n}}. \end{cases}$$

Example 3.24. Let us now solve the same system of linear equations considered above by means of Gauss-Seidel's method. The system is

$$\begin{cases} 10x + 2y + z = 1 \\ 10y - z = 0 \\ x + y - 10z = 4. \end{cases}$$

Let us write the update formulas of Gauss-Seidel's method at first:

$$x^{(k+1)} = \frac{1}{10}\left(1 - 2y^{(k)} - z^{(k)}\right)$$
$$y^{(k+1)} = \frac{1}{10}\left(z^{(k)}\right)$$
$$z^{(k+1)} = -\frac{1}{10}\left(4 - x^{(k+1)} - 10y^{(k+1)}\right).$$

Starting from the initial guess

$$x^{(0)} = 0$$
$$y^{(0)} = 0$$
$$z^{(0)} = 0,$$

we have

$$x^{(1)} = \frac{1}{10}\left(1 - 2y^{(0)} - z^{(0)}\right) = 0.1$$
$$y^{(1)} = \frac{1}{10}\left(z^{(0)}\right) = 0$$
$$z^{(1)} = -\frac{1}{10}\left(4 - x^{(1)} - 10y^{(1)}\right) = -0.39.$$

Iterating the procedure we have

$$x^{(2)} = \frac{1}{10}\left(1 - 2y^{(1)} - z^{(1)}\right) = 0.139$$
$$y^{(2)} = \frac{1}{10}\left(z^{(1)}\right) = -0.039$$
$$z^{(2)} = -\frac{1}{10}\left(4 - x^{(2)} - 10y^{(2)}\right) = -0.39,$$

$$x^{(3)} = \frac{1}{10}\left(1 - 2y^{(2)} - z^{(2)}\right) = 0.1468$$
$$y^{(3)} = \frac{1}{10}\left(z^{(2)}\right) = -0.039$$
$$z^{(3)} = -\frac{1}{10}\left(4 - x^{(3)} - 10y^{(3)}\right) = -0.38922,$$

$$x^{(4)} = \frac{1}{10}\left(1 - 2y^{(3)} - z^{(3)}\right) = 0.146722$$
$$y^{(4)} = \frac{1}{10}\left(z^{(3)}\right) = -0.038922$$
$$z^{(4)} = -\frac{1}{10}\left(4 - x^{(4)} - 10y^{(4)}\right) = -0.389220.$$

At the step (10) the solution is

$$x^{(10)} = 0.146707$$
$$y^{(10)} = -0.038922$$
$$z^{(10)} = -0.389222.$$

which is at most 10^{-12} distant from the exact solution.

Let us re-write the Gauss-Siedel's method in terms of matrix equations. If we pose

$$\mathbf{G} = \begin{pmatrix} a_{1,1} & 0 & \ldots & 0 \\ a_{2,1} & a_{2,2} & \ldots & 0 \\ \ldots & \ldots & \ldots & \ldots \\ a_{n,1} & a_{n,2} & \ldots & a_{n,n} \end{pmatrix}$$

and

$$\mathbf{S} = \begin{pmatrix} 0 & a_{1,2} & \ldots & a_{1,n} \\ 0 & 0 & \ldots & a_{2,n} \\ \ldots & \ldots & \ldots & \ldots \\ 0 & 0 & \ldots & 0 \end{pmatrix}$$

we can write

$$\mathbf{Ax} = \mathbf{b} \Rightarrow \mathbf{Gx}^{(k+1)} + \mathbf{Sx}^{(k)} = \mathbf{b} \Rightarrow$$
$$\Rightarrow \mathbf{x}^{(k+1)} = -\mathbf{G}^{-1}\mathbf{Sx}^{(k)} + \mathbf{G}^{-1}\mathbf{b}.$$

Hence, for the Gauss-Seidel's method the general scheme of the iterative methods can be applied by posing $\mathbf{H} = -\mathbf{G}^{-1}\mathbf{S}$ and $\mathbf{t} = \mathbf{G}^{-1}\mathbf{b}$.

Example 3.25. Let us reach the same result of the system above by means of matrix formulation. The matrices characterizing the method are

$$\mathbf{G} = \begin{pmatrix} 10 & 0 & 0 \\ 0 & 10 & 0 \\ 1 & 1 & -10 \end{pmatrix},$$

$$\mathbf{S} = \begin{pmatrix} 0 & 2 & 1 \\ 0 & 0 & -1 \\ 0 & 0 & 0 \end{pmatrix}.$$

We can calculate the inverse matrix

$$
\mathbf{G}^{-1} = \begin{pmatrix} \frac{1}{10} & 0 & 0 \\ 0 & \frac{1}{10} & 0 \\ \frac{1}{10} & \frac{1}{10} & -\frac{1}{10} \end{pmatrix}
$$

and write the update formula as

$$
\begin{pmatrix} x^{(k+1)} \\ y^{(k+1)} \\ z^{(k+1)} \end{pmatrix} = - \begin{pmatrix} \frac{1}{10} & 0 & 0 \\ 0 & \frac{1}{10} & 0 \\ \frac{1}{10} & \frac{1}{10} & -\frac{1}{10} \end{pmatrix} \begin{pmatrix} 0 & 2 & 1 \\ 0 & 0 & -1 \\ 0 & 0 & 0 \end{pmatrix} \begin{pmatrix} x^{(k} \\ y^{(k} \\ z^{(k} \end{pmatrix} + \begin{pmatrix} \frac{1}{10} & 0 & 0 \\ 0 & \frac{1}{10} & 0 \\ \frac{1}{10} & \frac{1}{10} & -\frac{1}{10} \end{pmatrix} \begin{pmatrix} 1 \\ 0 \\ 4 \end{pmatrix}
$$

that is

$$
\begin{pmatrix} x^{(k+1)} \\ y^{(k+1)} \\ z^{(k+1)} \end{pmatrix} = - \begin{pmatrix} 0 & 0.2 & 0.1 \\ 0 & 0 & -0.1 \\ 0 & 0.02 & 0 \end{pmatrix} \begin{pmatrix} x^{(k)} \\ y^{(k)} \\ z^{(k)} \end{pmatrix} + \begin{pmatrix} 0.1 \\ 0 \\ -0.39 \end{pmatrix}
$$

If we apply iteratively this formula we have the same results above, e.g. at the step (10) we have

$$
\begin{aligned}
x^{(10)} &= 0.146707 \\
y^{(10)} &= -0.038922 \\
z^{(10)} &= -0.389222.
\end{aligned}
$$

For the sake of clarity, the pseudocode of the Gauss-Seidel's method is shown in Algorithm 7.

Algorithm 7 Gauss-Seidel's Method

---

input $\mathbf{A}$ and $\mathbf{b}$
n is the size of $\mathbf{A}$
while precision conditions **do**
  for $i = 1 : n$ **do**
    $s = 0$
    for $j = 1 : n$ **do**
      if $j \neq i$ **then**
        $s = s + a_{i,j} x_j$
      end if
    end for
    $x_i = \frac{1}{a_{i,i}} (b_i - s)$
  end for
end while

3.4.3 The Method of Successive Over Relaxation

The Method of Successive Over Relaxation, briefly indicated as SOR, is a variant of the Gauss-Siedel's method which has been designed to obtain a faster convergence of the original method, see [8]. The SOR method corrects the Gauss-Siedel method by including in the update formula a dependence on the tentative solution at the step k. More specifically, if $\mathbf{x}^{(k)}$ is the solution at the step k and $\mathbf{x_{GS}}^{(k+1)}$ is the update to the step $k+1$ according to the Gauss-Siedel's method, the update formula of the SOR method is:

$$\mathbf{x_{SOR}}^{(k+1)} = \omega \mathbf{x_{GS}}^{(k+1)} + (1-\omega)\mathbf{x}^{(k)}$$

where ω is a parameter to be set. Obviously if $\omega = 1$ the SOR method degenerates into the Gauss-Siedel's method. The explicit update formula of the SOR method can be simply obtained from that of Gauss-Seidel's method by adding the contribution due to $\mathbf{x}^{(k)}$:

$$\begin{cases} x_1^{(k+1)} = \left(b_1 - \sum_{j=2}^{n} a_{1,j}x_j^{(k)}\right)\frac{\omega}{a_{1,1}} + (1-\omega)x_1^{(k)} \\ x_2^{(k+1)} = \left(b_2 - \sum_{j=3}^{n} a_{2,j}x_j^{(k)} - a_{2,1}x_1^{(k+1)}\right)\frac{\omega}{a_{2,2}} + (1-\omega)x_2^{(k)} \\ \dots \\ x_i^{(k+1)} = \left(b_i - \sum_{j=i+1}^{n} a_{i,j}x_j^{(k)} - \sum_{j=1}^{i-1} a_{i,j}x_j^{(k+1)}\right)\frac{\omega}{a_{i,i}} + (1-\omega)x_i^{(k)} \\ \dots \\ x_n^{(k+1)} = \left(b_n - \sum_{j=1}^{n-1} a_{i,j}x_j^{(k+1)}\right)\frac{\omega}{a_{n,n}} + (1-\omega)x_n^{(k)}. \end{cases}$$

Example 3.26. Let us solve again the system of linear equation

$$\begin{cases} 10x + 2y + z = 1 \\ 10y - z = 0 \\ x + y - 10z = 4. \end{cases}$$

This time let us apply SOR method. Let us pose $\omega = 0.9$ and write the update equations

$$x^{(k+1)} = \frac{0.9}{10}\left(1 - 2y^{(k)} - z^{(k)}\right) + 0.1x^{(k)}$$
$$y^{(k+1)} = \frac{0.9}{10}\left(z^{(k)}\right) + 0.1y^{(k)}$$
$$z^{(k+1)} = -\frac{0.9}{10}\left(4 - x^{(k+1)} - 10y^{(k+1)}\right) + 0.1z^{(k)}.$$

Let us start again from the initial guess

$$x^{(0)} = 0$$
$$y^{(0)} = 0$$
$$z^{(0)} = 0,$$

and let us calculate for a few iterations the solution by the SOR method:

$$x^{(1)} = \tfrac{1}{10}\left(1 - 2y^{(0)} - z^{(0)}\right) + 0.1x^{(0)} = 0.09$$
$$y^{(1)} = \tfrac{1}{10}\left(z^{(0)}\right) + 0.1y^{(0)} = 0$$
$$z^{(1)} = -\tfrac{1}{10}\left(4 - x^{(1)} - 10y^{(1)}\right) + +0.1z^{(0)} = -0.3519$$

Iterating the procedure we have

$$x^{(2)} = \tfrac{0.9}{10}\left(1 - 2y^{(1)} - z^{(1)}\right) + 0.1x^{(1)} = 0.130671$$
$$y^{(2)} = \tfrac{0.9}{10}\left(z^{(1)}\right) + 0.1y^{(1)} = -0.031671$$
$$z^{(2)} = -\tfrac{0.9}{10}\left(4 - x^{(2)} - 10y^{(2)}\right) + 0.1z^{(1)} = -0.386280,$$

$$x^{(3)} = \tfrac{1}{10}\left(1 - 2y^{(2)} - z^{(2)}\right) + 0.1x^{(2)} = 0.143533$$
$$y^{(3)} = \tfrac{1}{10}\left(z^{(2)}\right) + 0.1y^{(2)} = -0.037932$$
$$z^{(3)} = -\tfrac{1}{10}\left(4 - x^{(3)} - 10y^{(3)}\right) + 0.1z^{(2)} = -0.389124,$$

$$x^{(4)} = \tfrac{0.9}{10}\left(1 - 2y^{(3)} - z^{(3)}\right) + 0.1x^{(3)} = 0.146202$$
$$y^{(4)} = \tfrac{0.9}{10}\left(z^{(3)}\right) + 0.1y^{(3)} = -0.038814$$
$$z^{(4)} = -\tfrac{0.9}{10}\left(4 - x^{(4)} - 10y^{(4)}\right) + 0.1z^{(3)} = -0.389247.$$

At the step (10) we have

$$x^{(10)} = 0.146707$$
$$y^{(10)} = -0.038922$$
$$z^{(10)} = -0.389222,$$

whose error is at the most in the order of 10^{-8}.

It can be observed that the best results (or better the fastest convergence) were obtained by Gauss-Seidel. The reason behind the use of the SOR method is the parameter ω which allows an easy control of the performance of the method. This topic, as well as the selection of the parameter value is discussed in the following sections.

If, as in the case of Jacobi's method, we indicate with

$$\mathbf{E} = \begin{pmatrix} 0 & 0 & \dots & 0 \\ a_{2,1} & 0 & \dots & 0 \\ \dots & \dots & \dots & \dots \\ a_{n,1} & a_{n,2} & \dots & 0 \end{pmatrix},$$

$$F = \begin{pmatrix} 0 & a_{1,2} & \dots & a_{1,n} \\ 0 & 0 & \dots & a_{2,n} \\ \dots & \dots & \dots & \dots \\ 0 & 0 & \dots & 0 \end{pmatrix},$$

and

$$D = \begin{pmatrix} a_{1,1} & 0 & \dots & 0 \\ 0 & a_{2,2} & \dots & 0 \\ \dots & \dots & \dots & \dots \\ 0 & 0 & \dots & a_{n,n} \end{pmatrix}$$

we can indicate the system $\mathbf{Ax} = \mathbf{b}$ as $\mathbf{Ex} + \mathbf{Fx} + \mathbf{Dx} = \mathbf{b}$. If we consider the update index according to Gauss-Seidel's method, we can write

$$\mathbf{Ex}^{(k+1)} + \mathbf{Fx}^{(k)} + \mathbf{Dx}^{(k+1)} = \mathbf{b} \Rightarrow$$
$$\Rightarrow \mathbf{x}^{(k+1)} = \mathbf{D}^{-1}\left(-\mathbf{Ex}^{(k+1)} - \mathbf{Fx}^{(k)} + \mathbf{b}\right).$$

The SOR method corrects the formula above as

$$\mathbf{x}^{(k+1)} = \omega \mathbf{D}^{-1}\left(-\mathbf{Ex}^{(k+1)} - \mathbf{Fx}^{(k)} + \mathbf{b}\right) + (1 - \omega)\mathbf{x}^{(k)}.$$

Extracting $\mathbf{x}^{(k+1)}$, we obtain

$$\mathbf{Dx}^{(k+1)} = \omega\left(-\mathbf{Ex}^{(k+1)} - \mathbf{Fx}^{(k)} + \mathbf{b}\right) + (1-\omega)\mathbf{Dx}^{(k)} \Rightarrow$$
$$\Rightarrow (\mathbf{D} + \omega\mathbf{E})\mathbf{x}^{(k+1)} = \omega\left(-\mathbf{Fx}^{(k)} + \mathbf{b}\right) + (1-\omega)\mathbf{Dx}^{(k)} = ((1-\omega)\mathbf{D} - \omega\mathbf{F})\mathbf{x}^{(k)} + \omega\mathbf{b} \Rightarrow$$
$$\Rightarrow \mathbf{x}^{(k+1)} = (\mathbf{D} + \omega\mathbf{E})^{-1}\left(((1-\omega)\mathbf{D} - \omega\mathbf{F})\mathbf{x}^{(k)} + \omega\mathbf{b}\right).$$

We can then pose that

$$\mathbf{H} = (\mathbf{D} + \omega\mathbf{E})^{-1}\left((1-\omega)\mathbf{D} - \omega\mathbf{F}\right)$$

and

$$\mathbf{t} = (\mathbf{D} + \omega\mathbf{E})^{-1}\omega\mathbf{b}.$$

Hence, the SOR is expressed in the general form of an iterative method.

Example 3.27. Let us write the matrix update formula of the SOR method for the system of linear equation above. Considering that

$$E = \begin{pmatrix} 0 & 0 & 0 \\ 0 & 0 & 0 \\ 1 & 1 & 0 \end{pmatrix},$$

$$\mathbf{F} = \begin{pmatrix} 0 & 2 & 1 \\ 0 & 0 & -1 \\ 0 & 0 & 0 \end{pmatrix}$$

and

$$\mathbf{D} = \begin{pmatrix} 10 & 0 & 0 \\ 0 & 10 & 0 \\ 0 & 0 & -10 \end{pmatrix},$$

obviously, if $\omega = 0.9$ we have

$$\omega\mathbf{E} = \begin{pmatrix} 0 & 0 & 0 \\ 0 & 0 & 0 \\ 0.9 & 0.9 & 0 \end{pmatrix}$$

and

$$\mathbf{D} + \omega\mathbf{E} = \begin{pmatrix} 10 & 0 & 0 \\ 0 & 10 & 0 \\ 0.9 & 0.9 & -10 \end{pmatrix}.$$

The inverse of this triangular matrix is

$$(\mathbf{D} + \omega\mathbf{E})^{-1} = \begin{pmatrix} 0.1 & 0 & 0 \\ 0 & 0.1 & 0 \\ 0.009 & 0.009 & -0.1 \end{pmatrix}.$$

Let us calculate now

$$((1 - \omega)\mathbf{D} - \omega\mathbf{F}) =$$

$$= \left(0.1 \begin{pmatrix} 10 & 0 & 0 \\ 0 & 10 & 0 \\ 0 & 0 & -10 \end{pmatrix} - 0.9 \begin{pmatrix} 0 & 2 & 1 \\ 0 & 0 & -1 \\ 0 & 0 & 0 \end{pmatrix} \right) =$$

$$= \begin{pmatrix} 1 & -1.8 & -0.9 \\ 0 & 1 & 0.9 \\ 0 & 0 & -1 \end{pmatrix}.$$

Finally, let us calculate $\mathbf{H}$ by multiplying the two matrices:

$$\mathbf{H} = (\mathbf{D} + \omega\mathbf{E})^{-1} ((1 - \omega)\mathbf{D} - \omega\mathbf{F}) =$$

$$= \begin{pmatrix} 0.1 & -0.18 & -0.09 \\ 0 & 0.1 & 0.09 \\ 0.009 & -0.0072 & 0.1 \end{pmatrix}$$

Then to calculate the vector **t** we have

$$\mathbf{t} = (\mathbf{D} + \omega\mathbf{E})^{-1}\omega\mathbf{b} =$$

$$= \begin{pmatrix} 0.1 & 0 & 0 \\ 0 & 0.1 & 0 \\ 0.009 & 0.009 & -0.1 \end{pmatrix} \begin{pmatrix} 0.9 \\ 0 \\ 3.6 \end{pmatrix} =$$

$$= \begin{pmatrix} 0.09 \\ 0 \\ -0.3519 \end{pmatrix}.$$

For the sake of clarity the pseudocode explaining the logical steps of the SOR method is given in Algorithm 8.

Algorithm 8 Method of SOR

---
input **A** and **b**
input ω
n is the size of **A**
while precision conditions **do**
   for $i = 1 : n$ **do**
     $s = 0$
     for $j = 1 : n$ **do**
       if $j \neq i$ **then**
         $s = s + a_{i,j}x_j$
       end if
     end for
     $x_i = \frac{\omega}{a_{i,i}}(b_i - s) + (1 - \omega)x_i$
   end for
end while

3.4.4 Numerical Comparison Among the Methods and Convergence Conditions

In order to understand differences and relative advantages amongst the three iterative methods described in the previous sections, the following example is given.

Example 3.28. Let us consider the following system of linear equations

$$\begin{cases} 5x - 2y + 3z = -1 \\ 3x + 9y + z = 2 \\ 2x + y + 7z = 3. \end{cases}$$

In order to solve it, let us apply at first the Jacobi's method. The system can be re-written as

$$\begin{cases} x = -\frac{1}{5} + \frac{2}{5}y - \frac{3}{5}z \\ y = \frac{2}{9} - \frac{3}{9}x - \frac{1}{9}z \\ z = \frac{3}{7} - \frac{2}{7}x - \frac{1}{7}y \end{cases}$$

while Jacobi's update formulas from step (k) to step $(k+1)$ are

$$x^{(k+1)} = -\frac{1}{5} + \frac{2}{5}y^{(k)} - \frac{3}{5}z^{(k)}$$

$$y^{(k+1)} = \frac{2}{9} - \frac{3}{9}x^{(k)} - \frac{1}{9}z^{(k)}$$

$$z^{(k+1)} = \frac{3}{7} - \frac{2}{7}x^{(k)} - \frac{1}{7}y^{(k)}.$$

If we choose $\mathbf{x}^{(0)} = (0,0,0)$ as an initial guess, we obtain $\mathbf{x}^{(1)} = \left(-\frac{1}{5}, \frac{2}{9}, \frac{3}{7}\right)$. If we substitute iteratively the guess solutions we obtain

k	x	y	z
0	0	0	0
1	−0.20000	0.2222222	0.4285714
2	−0.3682540	0.2412698	0.4539683
3	−0.3758730	0.2945326	0.4993197
4	−0.3817788	0.2920333	0.4938876
5	−0.3795193	0.2946054	0.4959320
6	−0.3797171	0.2936251	0.4949190
7	−0.3795014	0.2938036	0.4951156
8	−0.3795479	0.2937098	0.4950285

After eight iterations the Jacobi's method returns a solution $\mathbf{x}^{(8)}$ such that

$$|\mathbf{A}\mathbf{x}^{(8)} - \mathbf{b}| = \begin{pmatrix} 0.0000739 \\ 0.0002267 \\ 0.0001868 \end{pmatrix}.$$

Let us now apply the Gauss-Seidel's method to the same system of linear equations. The Gauss-Seidel's update formulas from step (k) to step $(k+1)$ are

$$x^{(k+1)} = -\frac{1}{5} + \frac{2}{5}y^{(k)} - \frac{3}{5}z^{(k)}$$

$$y^{(k+1)} = \frac{2}{9} - \frac{3}{9}x^{(k+1)} - \frac{1}{9}z^{(k)}$$

$$z^{(k+1)} = \frac{3}{7} - \frac{2}{7}x^{(k+1)} - \frac{1}{7}y^{(k+1)}.$$

If the initial guess is $\mathbf{x}^{(0)} = (0,0,0)$, at first $x^{(1)} = -0.200$, and then $y^1 = \frac{2}{9} - \frac{3}{9}x^{(1)} - \frac{1}{9}z^{(0)} = 0.288$ and $z^{(1)} = \frac{3}{7} - \frac{2}{7}x^{(1)} - \frac{1}{7}y^{(1)} = 0.444$. The application of the Gauss-Seidel's method leads to the following results:

k	x	y	z
0	0	0	0
1	−0.20000	0.2888889	0.4444444
2	−0.3511111	0.2898765	0.4874780
3	−0.3765362	0.2935701	0.4942146
4	−0.3791007	0.2936764	0.4949322
5	−0.3794887	0.293726	0.4950359
6	−0.3795312	0.2937286	0.4950477
7	−0.3795372	0.2937293	0.4950493
8	−0.3795378	0.2937294	0.4950495

It can be observed that Jacobi's and Gauss-Seidel's methods converge to very similar solutions. However, for the same amount of iterations, Gauss-Siedel is more accurate since

$$|\mathbf{A}\mathbf{x}^{(8)} - \mathbf{b}| = \begin{pmatrix} 0.0000005 \\ 0.0000002 \\ 0 \end{pmatrix}.$$

Finally, let us solve the system above by applying the SOR method. The SOR update formulas from step (k) to step $(k+1)$ are

$$x^{(k+1)} = \omega\left(-\frac{1}{5} + \frac{2}{5}y^{(k)} - \frac{3}{5}z^{(k)}\right) + (1-\omega)x^{(k)}$$

$$y^{(k+1)} = \omega\left(\frac{2}{9} - \frac{3}{9}x^{(k+1)} - \frac{1}{9}z^{(k)}\right) + (1-\omega)y^{(k)}$$

$$z^{(k+1)} = \omega\left(\frac{3}{7} - \frac{2}{7}x^{(k+1)} - \frac{1}{7}y^{(k+1)}\right) + (1-\omega)z^{(k)}.$$

Let us set $\omega = 0.9$. If the initial guess is $\mathbf{x}^{(0)} = (0,0,0)$, at first $x^{(1)} = 0.9\,(-0.200) + 0.1\,(0) = -0.18$, and then $y^1 = 0.9\left(\frac{2}{9} - \frac{3}{9}x^{(1)} - \frac{1}{9}z^{(0)}\right) + 0.1y^{(0)} = 0.254$ and $z^{(1)} = 0.9\left(\frac{3}{7} - \frac{2}{7}x^{(1)} - \frac{1}{7}y^{(1)}\right) + 0.5z^{(0)} = 0.399$.

k	x	y	z
0	0	0	0
1	−0.18000	0.254000	0.3993429
2	−0.3222051	0.2821273	0.4722278
3	−0.3656577	0.2906873	0.4895893
4	−0.3762966	0.2929988	0.4937639
5	−0.3787826	0.2935583	0.4947487
6	−0.3793616	0.2936894	0.4949792
7	−0.3794967	0.2937200	0.4950331
8	−0.3795283	0.2937272	0.4950457

Again, the SOR method detected a very similar solution with respect to that found by Jacobi's and Gauss-Seidel's methods. After eight steps the solution $\mathbf{x}^{(8)}$ is such that

$$|\mathbf{A}\mathbf{x}^{(8)} - \mathbf{b}| = \begin{pmatrix} 0.0000410 \\ 0.0000054 \\ 0.0000098 \end{pmatrix}$$

that is slightly worse than that detected by Gauss-Siedel. On the other hand, the SOR method has the advantage that the convergence of the method can be explicitly controlled by tuning ω. This tuning can be a difficult task but may lead, in some cases, to a higher performance of the method. A wrong choice can make the method diverge away from the solution of the system. For example, given the previous system of linear equations, if ω is chosen equal to 8, after eight steps $\mathbf{x}^{(8)} = (1.7332907e10, -5.9691761e10, 3.7905479e10)$. The error related to this solution would be in the order of 10^{11} and would grow over the subsequent iterations. The relation between ω and convergence of SOR is given by the following theorem.

As a further remark, although Jacobi's method appears to be the least powerful method out of the three under examination, it hides a precious advantage in the computational era. At each iteration, the calculation of each row occurs independently on the other rows. Hence, an iteration of Jacobi's method can be easily parallelized by distributing the calculations related to each row to a different CPU. A parallelization would not be so easy for Gauss-Seidel's method since each row requires a value calculated in the previous row. Obviously, the natural way Jacobi's method can be distributed make the method appealing when large linear systems (when the order n of the system is a large number) must be solved and a cluster is available.

Theorem 3.11. *Let us consider a system of n linear equations in n variables* $\mathbf{A}\mathbf{x} = \mathbf{b}$. *If the SOR method converges to the solution of the system for the given matrix and known terms, the parameter* ω *is such that*

$$|\omega - 1| < 1.$$

The selection of ω is not the only issue relevant to the convergence of the method. The following example better clarifies this point.

Example 3.29. Let us consider the following system of linear equations:

$$\begin{cases} 5x - 2y = 4 \\ 9x + 3y + z = 2 \\ 8x + y + z = 2. \end{cases}$$

The application Jacobi's and Gauss-Seidel's methods do not converge to the solution of the system. More specifically, after 100 iterations, we obtain

$$|\mathbf{Ax}^{(100)} - \mathbf{b}| = \begin{pmatrix} 93044.372 \\ 116511.88 \\ 95058.989 \end{pmatrix}$$

for Jacobi's method and

$$|\mathbf{Ax}^{(100)} - \mathbf{b}| = 10^{14} \begin{pmatrix} 22.988105 \\ 7.2843257 \\ 0 \end{pmatrix}$$

for Gauss-Seidel's method, respectively. In other words, the system above cannot be tackled by means of Jacobi's nor Gauss-Seidel's methods. On the other hand, a tuning of ω can lead to the detection of a good approximation of the solution. For example, if ω is set equal to 0.5, we obtain

$$|\mathbf{Ax}^{(100)} - \mathbf{b}| = \begin{pmatrix} 0 \\ 0 \\ 8.882e - 16 \end{pmatrix}.$$

This example suggests that not all the systems of linear equations can be tackled by an iterative method. The following definition and theorem clarify the reason.

Definition 3.14. Let $\mathbf{A} \in \mathbb{R}_{n,n}$ be a square matrix. The matrix $\mathbf{A}$ is said to be **strictly diagonally dominant** if the absolute value of each entry on the main diagonal is greater than the sum of the absolute values of the other entries in the same row, that is,

$$|a_{1,1}| > |a_{1,2}| + |a_{1,3}| + \ldots + |a_{1,n}|$$
$$|a_{2,2}| > |a_{2,1}| + |a_{1,3}| + \ldots + |a_{2,n}|$$
$$\ldots$$
$$|a_{i,i}| > |a_{i,1}| + |a_{i,2}| + \ldots + |a_{i,n}|$$
$$\ldots$$
$$|a_{n,n}| > |a_{n,1}| + |a_{n,2}| + \ldots + |a_{n,n-1}|$$

Theorem 3.12. *Let* $\mathbf{Ax} = \mathbf{b}$ *be a system of n linear equations in n variables. If* $\mathbf{A}$ *is strictly diagonally dominant, then this system of linear equations has a unique solution to which the Jacobi's and Gauss-Seidel's methods will converge for any initial approximation* $\mathbf{x}^{(0)}$.

It must be noted that this theorem states that the strict diagonal dominance guarantees the convergence of the Jacobi's and Gauss-Seidel's methods. The reverse implication is not true, i.e. some systems of linear equations can be solved by Jacobi's and Gauss-Seidel methods even though the associated matrix $\mathbf{A}$ is not strictly diagonal dominant.

Moreover, since a row swap (elementary row operation E1) leads to an equivalent system of linear equations, if the rows of the complete matrix $\mathbf{A}^c$ can be rearranged so that $\mathbf{A}$ can be transformed into a strictly diagonally dominant matrix $\mathbf{C}$, then $\mathbf{Ax} = \mathbf{b}$ can still be solved by Jacobi's and Gauss-Seidel's methods.

Example 3.30. The system of linear equations

$$\begin{cases} x - 10y + 2z = -4 \\ 7x + y + 2z = 3 \\ x + y + 8z = -6. \end{cases}$$

is associated with the matrix

$$\mathbf{A} = \begin{pmatrix} 1 & -10 & 2 \\ 7 & 1 & 2 \\ 1 & 1 & 8 \end{pmatrix},$$

that is not strictly diagonally dominant. Nonetheless, if we swap the first and second equations, we obtain an equivalent system whose associated incomplete matrix is

$$\mathbf{C} = \begin{pmatrix} 7 & 1 & 2 \\ 1 & -10 & 2 \\ 1 & 1 & 8 \end{pmatrix}.$$

Obviously this matrix is strictly diagonally dominant. Hence, Jacobi's and Gauss-Seidel's methods will converge to the solution.

A synoptic scheme of the methods for solving systems of linear equation is displayed in Table 3.1 where the symbol $\mathscr{O}$ represents the computational complexity of the method, see Chap. 11.

Table 3.1 Synopsis of the methods for solving linear systems

	Cramer (Rouchè-Capelli)	Direct	Iterative
Operational feature	Determinant	Manipulate matrix	Manipulate guess solution
Outcome	Exact solution	Exact solution	Approximate solution
Computational cost	Unacceptably high ($\mathscr{O}(n!)$, see Chap. 11)	High $\mathscr{O}(n^3)$, see Chap. 11	∞ to the exact solution but it can be stopped after k steps with $k \cdot \mathscr{O}(n^2)$, see Chap. 11
Practical usability	Very small matrices (up to approx. 10×10)	Medium matrices (up to approx 1000×1000)	Large matrices
Hypothesis	No hypothesis	$a_{kk}^{(k)} \neq 0$ (solvable by pivoting)	Conditions on the eigenvalues of the matrix

Exercises

3.1. Solve, if possible, the following homogeneous system of linear equations by applying matrix theory of Cramer and Rouchè-Capelli:

$$\begin{cases} x - 2y + z = 2 \\ x + 5y = 1 \\ -3y + z = 1 \end{cases}.$$

3.2. Determine for what values of the parameter k, the system is determined, undetermined, and incompatible.

$$\begin{cases} (k+2)x + (k-1)y - z = k-2 \\ kx - ky = 2 \\ 4x - y = 1 \end{cases}.$$

3.3. Solve, if possible, the following homogeneous system of linear equations by applying matrix theory of Cramer and Rouchè-Capelli

$$\begin{cases} x + y - z = 0 \\ y - z = 0 \\ x + 2y - 2z = 0 \end{cases}.$$

Find, if possible, the unique solution or the general solution of the system.

3.4. Solve, if possible, the following system of linear equations by applying matrix theory of Cramer and Rouchè-Capelli

$$\begin{cases} x + 2y + 3z = 1 \\ 4x + 4y + 8z = 2 \\ 3x - y + 2z = 1 \end{cases}.$$

Find, if possible, the unique solution or the general solution of the system.

3.5. Solve, if possible, the following system of linear equations by applying matrix theory of Cramer and Rouchè-Capelli

$$\begin{cases} x + 2y + 3z = 1 \\ 2x + 4y + 6z = 2 \\ 3x + 6y + 9z = 3 \end{cases}.$$

Find, if possible, the unique solution or the general solution of the system.

3.6. Apply the Gaussian elimination to the following system of linear equations to find the equivalent triangular matrix/system (the solution of the equivalent system is not required).

$$\begin{cases} x - y + z = 1 \\ x + y = 4 \\ 2x + 2y + 2z = 9 \end{cases}.$$

3.7. Perform the **LU** factorization of the following matrix **A**:

$$\mathbf{A} = \begin{pmatrix} 5 & 0 & 5 \\ 10 & 1 & 13 \\ 15 & 2 & 23 \end{pmatrix}$$

3.8. For the following system of linear equation

$$\begin{cases} x + 2y = 0 \\ 2x - y + 6z = 2 \\ 4y + z = 8 \end{cases}$$

starting from $x^{(0)} = 0, y^{(0)} = 0, z^{(0)} = 0$,

1. apply the first step of Jacobi's Method to obtain $x^{(1)}, y^{(1)}, z^{(1)}$;
2. apply the first step of Gauss-Siedel's Method to obtain $x^{(1)}, y^{(1)}, z^{(1)}$.

Chapter 4
Geometric Vectors

4.1 Basic Concepts

It can be proved that $\mathbb{R}$ is a continuous set. As such, it can be graphically represented as an infinite continuous line, see [1].

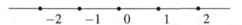

$$-2 \quad -1 \quad 0 \quad 1 \quad 2$$

The set $\mathbb{R}^2 = \mathbb{R} \times \mathbb{R}$ is also continuous and infinite. As such, it can be graphically represented by a plane [1]. Each element of $\mathbb{R}^2$ can be thus seen as a point $P = (x_1, x_2)$ belonging to this plane. Without a generality loss, let us fix a Cartesian reference system within this plane, see [9]. Within a Cartesian reference system, the horizontal reference axis is referred to as *abscissa's axis* while the vertical reference axis is referred to as *ordinate's axis*.

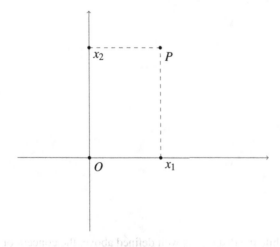

© Springer Nature Switzerland AG 2019
F. Neri, *Linear Algebra for Computational Sciences and Engineering*,
https://doi.org/10.1007/978-3-030-21321-3_4

In summary, since there is a bijection between lines and $\mathbb{R}$ and between $\mathbb{R}^2$ and planes, we can identify the concept of line (and plane) with the concepts of continuous set in one and two dimensions, respectively.

Definition 4.1. Two lines belonging to the same plane are said *parallel* if they have no common points.

Definition 4.2. A *direction* of a line is another line parallel to it and passing through the origin of the Cartesian system.

Let us consider an arbitrary pair of points $\mathbf{P} = (x_1, x_2)$ and $\mathbf{Q} = (y_1, y_2)$ belonging to the same plane. By means of simple considerations of Euclidean geometry, only one line passes through $\mathbf{P}$ and $\mathbf{Q}$. This line identifies univocally a direction. This direction is *oriented* in two ways on the basis of the starting and final point. The first is following the line from $\mathbf{P}$ to $\mathbf{Q}$ and the second from $\mathbf{Q}$ to $\mathbf{P}$. Along this line, the points between $\mathbf{P}$ and $\mathbf{Q}$ are a *segment*, characterised by the Euclidean distance $d(PQ) = \sqrt{(y_1 - x_1)^2 + (y_2 - x_2)^2}$, see [9].

Example 4.1. Let us consider the following two points of the plane $\mathbf{P} = (2, 1)$ and $\mathbf{Q} = (2, 2)$. The Euclidean distance

$$d(PQ) = \sqrt{(2-2)^2 + (2-1)^2} = 1.$$

Definition 4.3. Let $\mathbf{P}$ and $\mathbf{Q}$ be two points in the plane. A *geometric vector in the plane* $\vec{v}$ with starting point $\mathbf{P}$ and final point $\mathbf{Q}$ is a mathematical entity characterized by:

1. its oriented direction, identified by $\mathbf{P}$ and $\mathbf{Q}$
2. its module $||\vec{v}||$, that is the distance between $\mathbf{P}$ and $\mathbf{Q}$

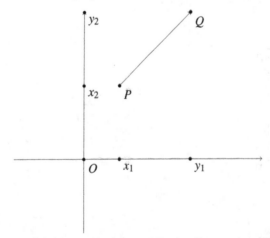

It must be remarked that while the distance is well defined above, the concept of direction is not formally defined at this stage. A more clear explanation of direction will be clear at the end of this chapter and will be formalized in Chap. 6.

The concept of geometric vector is the same as the number vector defined in Chap. 2 from a slightly different perspective. It can be observed that if the starting point is the origin $\mathbf{O}$, the geometric vector is the restriction to $\mathbb{R}^2$ of the number vector. Considering that the origin of the reference system is arbitrary, the two concepts coincide. More formally, between the set of vectors in the plane and the set of points in the plane (i.e. $\mathbb{R}^2$) there is a bijective relation.

Example 4.2. A vector of the plane can be $(1,2)$, $(6,5)$, or $(7,3)$.

The definition above can be extended to vectors in the space. In the latter case, two points belonging to $\mathbb{R}^3$ are represented as $\mathbf{P} = (x_1, x_2, x_3)$ and $\mathbf{Q} = (y_1, y_2, y_3)$. The segment from $\mathbf{P}$ to $\mathbf{Q}$ is identified by the distance of the two points and the direction of the line passing though $\mathbf{P}$ and $\mathbf{Q}$.

Definition 4.4. Let $\mathbf{P}$ and $\mathbf{Q}$ be two points in the space. A *geometric vector in the space* $\vec{v}$ with starting point $\mathbf{P}$ and final point $\mathbf{Q}$ is a mathematical entity characterized by:

1. its oriented direction, identified by $\mathbf{P}$ and $\mathbf{Q}$
2. its module $||\vec{v}||$, that is the distance between $\mathbf{P}$ and $\mathbf{Q}$

A point in the plane belongs to infinite lines. Analogously, a vector in the (3D) space belongs to infinite planes. If two vectors in the space do not have the same direction, they determine a unique plane which they both belong to.

Definition 4.5. When three vectors in the space all belong to the same plane are said *coplanar*.

Definition 4.6. Sum of Two Vectors. Let $\vec{u}$ and $\vec{v}$ be two vectors. Let $\vec{u} = \overrightarrow{\mathbf{AB}} = (\mathbf{B} - \mathbf{A})$ and $\vec{v} = \overrightarrow{\mathbf{BC}} = (\mathbf{C} - \mathbf{B})$. The *sum* $\vec{w} = \vec{u} + \vec{v} = (\mathbf{B} - \mathbf{A}) + (\mathbf{C} - \mathbf{B}) = \mathbf{C} - \mathbf{A} = \overrightarrow{\mathbf{AC}}$, where $\vec{w} \in \mathbb{V}_3$ $\vec{w}$ belongs to the same plane where $\vec{u}$ and $\vec{v}$ lie.

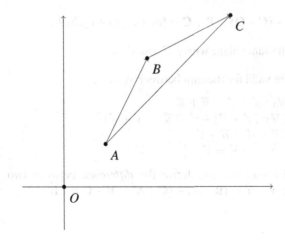

Definition 4.7. A special vector in the space is the vector with starting and final point in the origin **O** of the reference system. This vector, namely *null vector* is indicated with $\vec{o}$.

The null vector $\vec{o}$ of the space coincides with $\mathbf{O} = (0,0,0)$

It can be observed that there is a bijective relation also between the set of vectors in the space and points in the space.

In this chapter we will focus on those vectors of the type $\overrightarrow{\mathbf{PO}}$, i.e. those vectors whose starting point is the origin **O**.

Definition 4.8. Let us indicate with $\mathbb{V}_3$ the set containing all the possible geometric vectors of the space whose starting point is the origin **O**.

Example 4.3. A vector $\vec{v} \in \mathbb{V}_3$ is, for example, $(2,5,8)$ where one of the points identifying the direction is the origin **O**.

Let us define within this $\mathbb{V}_3$ the sum of vectors as a special case of the sum of vectors in Definition 4.6.

Definition 4.9. Sum of Two Vectors with Starting Point in O. Let **B** and **C** be the following two points of the space

$$\mathbf{B} = (b_1, b_2, b_3)$$
$$\mathbf{C} = (c_1, c_2, c_3).$$

Let $\vec{u}$ and $\vec{v}$ be the following two vectors:

$$\vec{u} = \overrightarrow{\mathbf{OB}} = (\mathbf{B} - \mathbf{O})$$
$$\vec{v} = \overrightarrow{\mathbf{OC}} = (\mathbf{C} - \mathbf{O})$$

The *sum* of vectors of $\mathbb{V}_3$ is

$$\vec{w} = \vec{u} + \vec{v} = (\mathbf{B} - \mathbf{O}) + (\mathbf{C} - \mathbf{O}) = \mathbf{B} + \mathbf{C} = (b_1 + c_1, b_2 + c_2, b_3 + c_3)$$

where $\vec{w} \in \mathbb{V}_3$ $\vec{w}$ belongs to the same plane where $\vec{u}$ and $\vec{v}$ lie.

The following properties are valid for the sum between vectors.

- commutativity: $\forall \vec{u}, \vec{v} \in \mathbb{V}_3 : \vec{u} + \vec{v} = \vec{v} + \vec{u}$
- associativity: $\forall \vec{u}, \vec{v}, \vec{w} \in \mathbb{V}_3 : (\vec{u} + \vec{v}) + \vec{w} = \vec{u} + (\vec{v} + \vec{w})$
- neutral element: $\forall \vec{v} : \exists! \vec{o} \mid \vec{v} + \vec{o} = \vec{o} + \vec{v} = \vec{v}$
- opposite element: $\forall \vec{v} : \exists! \overrightarrow{-v} \mid \overrightarrow{-v} + \vec{v} = \vec{v} + (\overrightarrow{-v}) = \vec{o}$

From the properties of the sum we can derive the *difference* between two vectors $\vec{u} = \overrightarrow{\mathbf{AB}}$ and $\vec{w} = \overrightarrow{\mathbf{AC}}$: $\vec{u} - \vec{w} = (\mathbf{B} - \mathbf{A}) - (\mathbf{C} - \mathbf{A}) = \mathbf{B} - \mathbf{C} = \overrightarrow{\mathbf{CB}}$

Example 4.4. Let us consider two points $\mathbf{P}, \mathbf{Q} \in \mathbb{R}^3$, where $\mathbf{P} = (1,3,4)$ and $\mathbf{Q} = (2,5,6)$. The vectors $\overrightarrow{\mathbf{OP}}$ and $\overrightarrow{\mathbf{OQ}}$ are, respectively,

$$\overrightarrow{\mathbf{OP}} = (1,3,4)$$
$$\overrightarrow{\mathbf{OQ}} = (2,5,6).$$

The sum of these two vectors is

$$\overrightarrow{\mathbf{OP}} + \overrightarrow{\mathbf{OQ}} = (3,8,10).$$

Example 4.5. Let us consider the following points $\in \mathbb{R}^2$:

$$\mathbf{A} = (1,1)$$
$$\mathbf{B} = (2,3)$$
$$\mathbf{C} = (4,4)$$

and the following vectors

$$\overrightarrow{\mathbf{AB}} = \mathbf{B} - \mathbf{A} = (1,2)$$
$$\overrightarrow{\mathbf{BC}} = \mathbf{C} - \mathbf{B} = (2,1).$$

The sum of these two vectors is

$$\overrightarrow{\mathbf{AC}} = \overrightarrow{\mathbf{AB}} + \overrightarrow{\mathbf{BC}} = \mathbf{B} - \mathbf{A} + \mathbf{C} - \mathbf{B} = \mathbf{C} - \mathbf{A} = (3,3). \qquad (4.1)$$

It can be observed that the vector $\overrightarrow{\mathbf{AC}}$ would be the coordinates of the point $\mathbf{C}$ if the origin of the reference system is $\mathbf{A}$.

Definition 4.10. Product of a Scalar by a Vector. Let $\vec{v}$ be a vector $\in V_3$ and λ a scalar $\in \mathbb{R}$. The *product* of a scalar λ by a vector $\vec{v}$ is a new vector $\in V_3$, $\lambda \vec{v}$ having the same direction of $\vec{v}$, module $||\overrightarrow{\lambda v}|| = |\lambda| \, ||\vec{v}||$, and orientation is the same of $\vec{v}$ if λ is positive and the opposite if λ is negative.

The following properties are valid for the product of a scalar by a vector.

- commutativity: $\forall \lambda \in \mathbb{R}$ and $\forall \vec{v} \in V_3 : \lambda \vec{v} = \vec{v} \lambda$
- associativity: $\forall \lambda, \mu \in \mathbb{R}$ and $\forall \vec{v} \in V_3 : \lambda (\mu \vec{v}) = (\lambda \vec{v}) \mu$
- distributivity 1: $\forall \lambda \in \mathbb{R}$ and $\forall \vec{u}, \vec{v} \in V_3 : \lambda (\vec{u} + \vec{v}) = \lambda \vec{u} + \lambda \vec{v}$
- distributivity 2: $\forall \lambda, \mu \in \mathbb{R}$ and $\forall \vec{v} \in V_3 : (\lambda + \mu) \vec{v} = \lambda \vec{v} + \mu \vec{v}$
- neutral element: $\forall \vec{v} : 1\vec{v} = \vec{v}$

Example 4.6. If $\lambda = 2$ and $\vec{v} = (1,1,1)$, the vector $\lambda \vec{v} = (2,2,2)$.

Proposition 4.1. *Let $\lambda \in \mathbb{R}$ and $\vec{v} \in V_3$. If either $\lambda = 0$ or $\vec{v} = \vec{o}$, then $\lambda \vec{v}$ is equal to the null vector $\vec{o}$*

Proof. Let us prove that if $\lambda = 0$ then $\lambda \vec{v} = \vec{o}$. From the properties of the sum of vectors we know that

$$\vec{o} = \lambda \vec{v} + (-\lambda \vec{v}).$$

Since $\lambda = 0$ then, from basic arithmetic, $\lambda = 0 + 0$. By substituting we obtain

$$\vec{o} = 0\vec{v} + 0\vec{v} + (-0\vec{v}) = 0\vec{v} + \vec{o} = 0\vec{v}.\square$$

Let us prove that if $\vec{v} = \vec{o}$ then $\lambda \vec{v} = \vec{o}$. Considering that $\vec{o} = \vec{o} + \vec{o}$, then

$$\lambda \vec{o} = \lambda (\vec{o} + \vec{o}) = \lambda \vec{o} + \lambda \vec{o}.$$

If we sum to both the terms $-\lambda \vec{o}$, it follows that

$$\lambda \vec{o} + (-\lambda \vec{o}) = (\lambda \vec{o} + \lambda \vec{o}) + (-\lambda \vec{o}).$$

From the properties of the sum of vectors, it follows that $\vec{o} = \lambda \vec{o}$. $\square$

4.2 Linear Dependence and Linear Independence

Definition 4.11. Let $\lambda_1, \lambda_2, \ldots, \lambda_n$ be n scalars $\in \mathbb{R}$ and $\vec{v_1}, \vec{v_2}, \ldots, \vec{v_n}$ be n vectors $\in \mathbb{V}_3$. The *linear combination* of the n vectors by means of the n scalars is the vector

$$\vec{w} = \lambda_1 \vec{v_1} + \lambda_2 \vec{v_2} + \ldots, \lambda_n \vec{v_n}.$$

Definition 4.12. Let $\vec{v_1}, \vec{v_2}, \ldots, \vec{v_n}$ be n vectors $\in \mathbb{V}_3$. These vectors are said *linearly dependent* if the null vector can be expressed as their linear combination by means of and n-tuple of non-null coefficients:

$$\exists \lambda_1, \lambda_2, \ldots, \lambda_n \in \mathbb{R} \ni `$$
$$\vec{o} = \lambda_1 \vec{v_1} + \lambda_2 \vec{v_2} + \ldots + \lambda_n \vec{v_n}$$

with $\lambda_1, \lambda_2, \ldots, \lambda_n \neq 0, 0, \ldots, 0$.

Definition 4.13. Let $\vec{v_1}, \vec{v_2}, \ldots, \vec{v_n}$ be n vectors $\in \mathbb{V}_3$. These vectors are said *linearly independent* if the null vector can be expressed as their linear combination only by means of null coefficients:

$$\nexists \lambda_1, \lambda_2, \ldots, \lambda_n \in \mathbb{R} \ni `$$
$$\vec{o} = \lambda_1 \vec{v_1} + \lambda_2 \vec{v_2} + \ldots + \lambda_n \vec{v_n}$$

with $\lambda_1, \lambda_2, \ldots, \lambda_n \neq 0, 0, \ldots, 0$.

Example 4.7. Let us consider three vectors $\vec{v_1}$, $\vec{v_2}$, and $\vec{v_3} \in \mathbb{V}_3$. If at least one tuple $\lambda_1, \lambda_2, \lambda_3 \in \mathbb{R}$ and $\neq 0, 0, 0$ such that $\vec{o} = \lambda_1 \vec{v_1} + \lambda_2 \vec{v_2} + \lambda_3 \vec{v_3}$ can be found, the vectors are linearly dependent. For example if the tuple $-4, 5, 0$ is such that $\vec{o} = -4\vec{v_1} + 5\vec{v_2} + 0\vec{v_3}$ then the tree vectors are linearly dependent.

Obviously, if $\lambda_1, \lambda_2, \lambda_3 = 0, 0, 0$ the equation $\vec{o} = \lambda_1 \vec{v_1} + \lambda_2 \vec{v_2} + \lambda_3 \vec{v_3}$ is always verified for the Proposition 4.1 for both, linearly dependent and linearly independent vectors. If the only way to obtain a null linear combination is by means of coefficients all null, then the vectors are linearly independent.

Example 4.8. Let us consider the following vectors $\in \mathbb{V}_3$:

$$\vec{v_1} = (1, 2, 1)$$
$$\vec{v_2} = (1, 1, 1)$$
$$\vec{v_3} = (2, 4, 2).$$

Let us check whether or not these vectors are linearly dependent. From its definition, these vectors are linearly independent if there exists a triple $\lambda_1, \lambda_2, \lambda_3 \neq 0, 0, 0$ such that

$$\vec{o} = \lambda_1 \vec{v_1} + \lambda_2 \vec{v_2} + \lambda_3 \vec{v_3}.$$

The latter equation, in our case, is

$$(0, 0, 0) = \lambda_1 (1, 2, 1) + \lambda_2 (1, 1, 1) + \lambda_3 (2, 4, 2),$$

which can be written as

$$\begin{cases} \lambda_1 + \lambda_2 + 2\lambda_3 = 0 \\ 2\lambda_1 + \lambda_2 + 4\lambda_3 = 0 \\ \lambda_1 + \lambda_2 + 2\lambda_3 = 0. \end{cases}$$

This is a homogeneous system of linear equations. It can be observed that the matrix associated with the system is singular and has rank $\rho = 2$. Thus the system has ∞^1 solutions. Hence, not only $\lambda_1, \lambda_2, \lambda_3 = 0, 0, 0$ solves the system. For example, $\lambda_1, \lambda_2, \lambda_3 = 2, 0, -1$ is a solution of the system. Thus, the vectors $\vec{v_1}, \vec{v_2}, \vec{v_3}$ are linearly dependent.

Example 4.9. Let us check the linear dependence for the following vectors

$$\vec{v_1} = (1, 0, 0)$$
$$\vec{v_2} = (0, 1, 0)$$
$$\vec{v_3} = (0, 0, 2).$$

This means that we have to find those scalars $\lambda_1, \lambda_2, \lambda_3$

$$(0, 0, 0) = \lambda_1 (1, 0, 0) + \lambda_2 (0, 1, 0) + \lambda_3 (0, 0, 2),$$

which leads to

$$\begin{cases} \lambda_1 = 0 \\ \lambda_2 = 0 \\ 2\lambda_3 = 0 \end{cases}$$

whose only solution is $\lambda_1, \lambda_2, \lambda_3 = 0, 0, 0$. Thus the vectors $\vec{v_1}, \vec{v_2}, \vec{v_3}$ are linearly independent.

Example 4.10. Let us check the linear dependence for the following vectors

$$\vec{v_1} = (1,2,0)$$
$$\vec{v_2} = (3,1,0)$$
$$\vec{v_3} = (4,0,1).$$

This means that we have to find those scalars $\lambda_1, \lambda_2, \lambda_3$

$$(0,0,0) = \lambda_1 (1,2,0) + \lambda_2 (3,1,0) + \lambda_3 (4,0,1),$$

which leads to

$$\begin{cases} \lambda_1 + 3\lambda_2 + 4\lambda_3 = 0 \\ 2\lambda_1 + \lambda_2 = 0 \\ \lambda_3 = 0. \end{cases}$$

This system is determined and its only solution is $\lambda_1, \lambda_2, \lambda_3 = 0,0,0$. The vectors are linearly independent.

Theorem 4.1. *Let $\vec{v_1}, \vec{v_2}, \ldots, \vec{v_n}$ be n vectors $\in \mathbb{V}_3$. These vectors are linearly dependent if and only if at least one of them can be expressed as a linear combination of the others.*

Proof. If the vectors are linearly dependent, the

$$\exists \lambda_1, \lambda_2, \ldots, \lambda_n \neq 0,0,\ldots,0 \ni '$$
$$\vec{o} = \lambda_1 \vec{v_1} + \lambda_2 \vec{v_2} + \ldots + \lambda_n \vec{v_n}.$$

Without a loss of generality, let us suppose that $\lambda_1 \neq 0$. Thus,

$$\vec{o} = \lambda_1 \vec{v_1} + \lambda_2 \vec{v_2} + \ldots + \lambda_n \vec{v_n} \Rightarrow$$
$$\Rightarrow -\lambda_1 \vec{v_1} = \lambda_2 \vec{v_2} + \ldots + \lambda_n \vec{v_n} \Rightarrow$$
$$\Rightarrow \vec{v_1} = \frac{\lambda_2}{-\lambda_1} \vec{v_2} + \ldots + \frac{\lambda_n}{-\lambda_1} \vec{v_n}$$

One vector has been expressed as linear combination of the others.   □

If one vector can be expressed as a linear combination of the others then we can write

$$\vec{v_n} = \lambda_1 \vec{v_1} + \lambda_2 \vec{v_2} + \ldots + \lambda_{n-1} \vec{v_{n-1}}.$$

Thus,

$$\vec{v_n} = \lambda_1 \vec{v_1} + \lambda_2 \vec{v_2} + \ldots + \lambda_{n-1} \vec{v_{n-1}} \Rightarrow$$
$$\Rightarrow \vec{o} = \lambda_1 \vec{v_1} + \lambda_2 \vec{v_2} + \ldots + \lambda_{n-1} \vec{v_{n-1}} - \vec{v_n}.$$

The null vector $\vec{o}$ has been expressed as a linear combination of the n vectors by means of the coefficients $\lambda_1, \lambda_2, \ldots, \lambda_{n-1}, -1 \neq 0,0,\ldots,0$. The vectors are linearly dependent.   □

Example 4.11. We know that the vectors

$$\vec{v_1} = (1,2,1)$$
$$\vec{v_2} = (1,1,1)$$
$$\vec{v_3} = (2,4,2)$$

are linearly dependent. We can express one of them as linear combination of the other two:

$$\vec{v_3} = \mu_1 \vec{v_1} + \mu_2 \vec{v_2}$$

with $\mu_1, \mu_2 = 2, 0$.

Let us try to express $\vec{v_2}$ as a linear combination of $\vec{v_1}$ and $\vec{v_3}$:

$$\vec{v_2} = v_2 \vec{v_1} + v_3 \vec{v_3}.$$

In order to find v_2 and $v3$ let us write

$$(1,1,1) = v_2 (1,2,1) + v_3 (2,4,2)$$

which leads to

$$\begin{cases} v_2 + 2v_3 = 1 \\ 2v_2 + 4v_3 = 1 \\ v_2 + 2v_3 = 1 \end{cases}$$

which is an impossible system of linear equations. Thus $\vec{v_2}$ cannot be expressed as a linear combination of $\vec{v_1}$ and $\vec{v_3}$.

The latter example has been reported to remark that Theorem 4.1 states that in a list of linearly dependent vectors at least one can be expressed as a linear combination of the other. Thus, not necessarily all of them can be expressed as a linear combination of the others.

Proposition 4.2. *Let $\vec{v_1}, \vec{v_2}, \ldots, \vec{v_n}$ be n vectors $\in \mathbb{V}_3$. Let $h \in \mathbb{N}$ with $0 < h < n$. If h vectors are linearly dependent, then all the n vectors are linearly dependent.*

Proof. If h vectors a linearly dependent, then

$$\vec{o} = \lambda_1 \vec{v_1} + \lambda_2 \vec{v_2} + \ldots + \lambda_h \vec{v_h}$$

with $\lambda_1, \lambda_2, \ldots, \lambda_h \neq 0, 0, \ldots, 0$.

Even if we assume that all the other λ values $\lambda_{h+1}, \ldots, \lambda_n = 0, \ldots, 0$, then we can write

$$\vec{o} = \lambda_1 \vec{v_1} + \lambda_2 \vec{v_2} + \ldots + \lambda_h \vec{v_h} + \lambda_{h+1} \vec{v_{h+1}} + \ldots + \lambda_n \vec{v_n}$$

where $\lambda_1, \lambda_2, \ldots, \lambda_h, \lambda_{h+1}, \ldots, \lambda_n \neq 0, 0, \ldots, 0$. Thus, the n vectors are linearly dependent. $\square$

Example 4.12. Let us consider the following vectors

$$\vec{v_1} = (2,2,1)$$
$$\vec{v_2} = (1,1,1)$$
$$\vec{v_3} = (3,3,2)$$
$$\vec{v_4} = (5,1,2).$$

We can easily notice that $\vec{v_3} = \vec{v_1} + \vec{v_2}$, hence $\vec{v_1}, \vec{v_2}, \vec{v_3}$ are linearly dependent. Let us check that all four are linearly dependent. We have to find those scalars $\lambda_1, \lambda_2, \lambda_3, \lambda_4$ such that

$$(0,0,0,0) = \lambda_1 (2,2,1) + \lambda_2 (1,1,1) + \lambda_3 (3,3,2) + \lambda_4 (5,1,2),$$

which leads to

$$\begin{cases} 2\lambda_1 + \lambda_2 + 3\lambda_3 + 5\lambda_4 = 0 \\ 2\lambda_1 + \lambda_2 + 3\lambda_3 + \lambda_4 = 0 \\ \lambda_1 + \lambda_2 + 2\lambda_3 + 2\lambda_4 = 0. \end{cases}$$

This system has ∞^1 solutions. For example, $\lambda_1, \lambda_2, \lambda_3, \lambda_4 = 1,1,-1,0$ is a solution of the system. Thus, $\vec{v_1}, \vec{v_2}, \vec{v_3}, \vec{v_4}$ are linearly dependent.

Proposition 4.3. *Let $\vec{v_1}, \vec{v_2}, \ldots, \vec{v_n}$ be n vectors $\in \mathbb{V}_3$. If one of these vectors is null $\vec{o}$, then the n vectors are linearly dependent.*

Proof. We know from hypothesis that one vector is null. Without a loss of generality let us assume that $\vec{v_1} = \vec{o}$. Thus, if we consider the linear combination of these vectors:

$$\lambda_1 \vec{o} + \lambda_2 \vec{v_2} + \ldots + \lambda_n \vec{v_n}$$

If λ_1 is chosen equal to a real scalar $k \neq 0$, then the linear combination will be equal to the null vector $\vec{o}$ for

$$\lambda_1, \lambda_2, \ldots, \lambda_n = k, 0, \ldots, 0 \neq 0, 0, \ldots, 0.$$

Thus, the vectors are linearly dependent. $\square$

Example 4.13. Let us consider the following vectors

$$\vec{v_1} = (0,0,0)$$
$$\vec{v_2} = (1,5,1)$$
$$\vec{v_3} = (8,3,2).$$

Let us check the linear dependence

$$(0,0,0) = \lambda_1 (0,0,0) + \lambda_2 (1,5,1) + \lambda_3 (8,3,2).$$

For example $\lambda_1, \lambda_2, \lambda_3 = 50, 0, 0$ satisfies the equation. Thus, the vectors are linearly dependent.

Definition 4.14. Let $\vec{u}$ and $\vec{v} \in \mathbb{V}_3$. The two vectors are *parallel* if they have the same direction.

It can be observed that for the parallelism relation the following properties are valid:

- reflexivity: a vector (line) is parallel to itself
- symmetry: if $\vec{u}$ is parallel to $\vec{v}$ then $\vec{v}$ is parallel to $\vec{u}$
- transitivity: if $\vec{u}$ is parallel to $\vec{v}$ and $\vec{v}$ is parallel to $\vec{w}$, then $\vec{u}$ is parallel to $\vec{w}$

Since these three properties are simultaneously verified the parallelism is an equivalence relation. Two vectors having same module, direction, and orientation are said *equipollent*. It can be proved that equipollence is an equivalence relation. This means that a vector having a first point **A** is equivalent to one with same module, direction, and orientation having a different starting point, e.g. **B**.

In addition, it can be observed that every vector is parallel to the null vector $\vec{o}$.

Lemma 4.1. *Let $\vec{u}$ and $\vec{v} \in \mathbb{V}_3$. If the two vectors are parallel, then they could be expressed as*

$$\vec{u} = \lambda \vec{v}$$

with $\lambda \in \mathbb{R}$.

Theorem 4.2. *Let $\vec{u}$ and $\vec{v} \in \mathbb{V}_3$. The two vectors are linearly dependent if and only if they are parallel.*

Proof. If the two vectors are linearly dependent then the null vector can be expressed as linear combination by means of non-null coefficients:

$$\exists \lambda, \mu \in \mathbb{R}$$

with $\lambda, \mu \neq 0, 0$ such that

$$\vec{o} = \lambda \vec{u} + \mu \vec{v}.$$

Let us suppose that $\lambda \neq 0$. Thus,

$$\vec{o} = \lambda \vec{u} + \mu \vec{v} \Rightarrow$$
$$\Rightarrow \lambda \vec{u} = -\mu \vec{v} \Rightarrow$$
$$\Rightarrow \vec{u} = -\frac{\mu}{\lambda} \vec{v}.$$

The two vectors are parallel. $\square$

If the two vectors are parallel then

$$\exists \lambda \in \mathbb{R} \ni '$$
$$\vec{u} = \lambda \vec{v}.$$

Thus,

$$\vec{u} = \lambda \vec{v} \Rightarrow \vec{o} = \vec{u} - \lambda \vec{v}.$$

The null vector has been expressed as the linear combination of the two vector by means of the coefficients $1, -\lambda \neq 0, 0$. Hence, the vectors are linearly dependent. □

Example 4.14. The vectors

$$\vec{v_1} = (1, 1, 1)$$
$$\vec{v_2} = (2, 2, 2).$$

These two vectors are parallel since $\vec{v_2} = 2\vec{v_1}$. We can easily check that these vectors are linearly dependent since

$$\vec{o} = \lambda_1 \vec{v_2} + \lambda_2 \vec{v_1}$$

with $\lambda_1, \lambda_2 = 1, -2$.

Let us indicate with $\mathbb{V}_1$ the set of vectors belonging to one line, i.e. the unidimensional vectors.

Theorem 4.3. *Let $\vec{u}$ and $\vec{v}$ be two parallel vectors $\in \mathbb{V}_3$ with $\vec{u} \neq \vec{o}$ and $\vec{v} \neq \vec{o}$. The vector $\vec{u}$ can be expressed in only one way as the product of a scalar λ and the vectors $\vec{v}$:*

$$\exists! \lambda \ni \text{'} \vec{u} = \lambda \vec{v}.$$

Proof. By contradiction let us assume that $\exists \lambda, \mu \ni$ '

$$\vec{u} = \lambda \vec{v}$$

and

$$\vec{u} = \mu \vec{v}$$

with $\lambda \neq \mu$.
Thus,

$$\vec{o} = (\lambda - \mu) \vec{v} \Rightarrow$$

since for the hypothesis $\vec{v} \neq \vec{o}$,

$$\Rightarrow \lambda - \mu = 0 \Rightarrow \lambda = \mu$$

Since $\lambda \neq \mu$, by contradiction hypothesis, then we reached a contradiction   □

Example 4.15. Let us consider the following vector $\vec{v} \in \mathbb{V}_3$:

$$\vec{v} = (6, 6, 6)$$

The previous theorem simply says that for a vector $\vec{u} \in \mathbb{V}_3$ parallel to $\vec{v}$, there exists only one scalar λ such that

$$\vec{v} = \lambda \vec{u}.$$

If $\vec{u} = (2, 2, 2)$, the only λ value such that $\vec{v} = \lambda \vec{u}$ is $\lambda = 3$.

Theorem 4.4. *Let $\vec{u}$, $\vec{v}$, and $\vec{w} \in \mathbb{V}_3$ and $\neq \vec{o}$. The three vectors are coplanar if and only if $\vec{u}$, $\vec{v}$, and $\vec{w}$ are linearly dependent.*

Proof. If $\vec{u}$, $\vec{v}$, and $\vec{w}$ are coplanar, they all belong to the same plane. Let us consider three coplanar vectors having all an arbitrary starting point **O** and final points, **A, b**, and **c**, respectively, where $\vec{u} = \overrightarrow{OA}$. Let us indicated with **B**, and **C** the projection of the point **A** over the directions determined by the segments $\overrightarrow{Ob}$ and $\overrightarrow{Oc}$ (or by the vectors $\vec{v}$ and $\vec{w}$) respectively. A parallelogram is determined by the vertices **ABCD**.

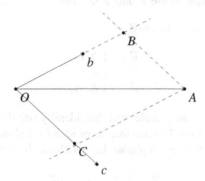

On the basis of the geometrical construction,

$$\overrightarrow{OA} = \overrightarrow{OB} + \overrightarrow{OC}$$

where $\overrightarrow{OA} = \vec{u}$, $\overrightarrow{OB} = \lambda\vec{v}$, and $\overrightarrow{OC} = \mu\vec{w}$. Thus, $\vec{u} = \lambda\vec{v} + \mu\vec{w}$. For the Theorem 4.1, the vectors are linearly dependent. □

If the vectors are linearly dependent they can be expressed as $\vec{u} = \lambda\vec{v} + \mu\vec{w}$.

- If $\vec{v}$ and $\vec{w}$ are parallel then $\vec{u}$ is parallel to both of them. Thus, the three vectors determine a unique direction. Since infinite planes include a line (direction), there is at least one plane that contains the three vectors. The three vectors would be coplanar.
- If $\vec{v}$ and $\vec{w}$ are not parallel, they determine one unique plane containing both of them. Within this plane the sum $\vec{u} = \lambda\vec{v} + \mu\vec{w}$ is verified. The vector $\vec{u}$ belongs to a plane where $\lambda\vec{v}$ and $\mu\vec{w}$ belong to. This is true because the sum of two vectors in a plane is a vector belonging to the same plane (the sum is a closed operator, see e.g. [10]). Obviously $\lambda\vec{v}$ is parallel to $\vec{v}$ and $\mu\vec{w}$ is parallel to $\vec{w}$. Thus $\vec{u}$, $\vec{v}$, and $\vec{w}$ belong to the same plane. □

Example 4.16. Let us consider the following three vectors $\in \mathbb{V}_3$:

$$\vec{u} = (1, 2, 1)$$
$$\vec{v} = (2, 2, 4)$$
$$\vec{w} = (4, 6, 6).$$

We can easily observe that

$$(4,6,6) = 2(1,2,1) + (2,2,4)$$

that is

$$\vec{w} = \lambda\,\vec{u} + \mu\,\vec{v}.$$

This means that the vectors $\vec{u}, \vec{v}, \vec{w}$ are linearly dependent. The fact that one vector is linear combination of the other two has a second meaning. The vectors $\vec{u}$ and $\vec{v}$ identify a plane. The vector $\vec{w}$ is the weighted sum of the other two and thus belongs to the same plane where $\vec{u}$ and $\vec{v}$ lie. Thus $\vec{u}, \vec{v}, \vec{w}$ are also coplanar.

Example 4.17. The three vectors $in\mathbb{V}_3$

$$\vec{u} = (1,2,1)$$
$$\vec{v} = (2,4,2)$$
$$\vec{w} = (4,6,6).$$

In this case $\vec{u}$ and $\vec{v}$ are parallel and thus identify one direction. The vector $\vec{w}$ identifies another direction. The two directions identify a plane containing the three vectors. Thus, the vectors are coplanar. Let us check the linear dependence. The equation

$$\vec{o} = \lambda\,\vec{u} + \mu\,\vec{v} + v\vec{w}$$

is verified for infinite values of λ, μ, v, e.g. $1, -\frac{1}{2}, 0$.

Theorem 4.5. *Let $\vec{u}$, $\vec{v}$, and $\vec{w} \in \mathbb{V}_3$ and $\neq \vec{o}$. If these three vectors are coplanar and two of them ($\vec{v}$ and $\vec{w}$) are linearly independent ($\vec{v}$ and $\vec{w}$ are not parallel to each other), the third ($\vec{u}$) can be expressed as the linear combination of the other two in only one way (i.e. by means of only one tuple of scalars λ, μ):*

$$\exists!\lambda, \mu \neq 0, 0 \ni `$$
$$\vec{u} = \lambda\,\vec{v} + \mu\,\vec{w}.$$

Proof. Let us assume by contradiction that $\exists \lambda, \mu \in \mathbb{R}$ such that $\vec{u} = \lambda\mathbf{v} + \mu\vec{w}$, and also $\exists \lambda', \mu' \in \mathbb{R}$ such that $\vec{u} = \lambda'\vec{v} + \mu'\vec{w}$ with $\lambda, \mu \neq \lambda', \mu'$. Thus,

$$\vec{o} = (\lambda - \lambda')\,\vec{v} + (\mu - \mu')\,\vec{w}.$$

Since the vectors are linearly independent $\lambda - \lambda' = 0$ and $\mu - \mu' = 0$. Thus $\lambda = \lambda'$ and $\mu = \mu'$. $\square$

Example 4.18. If we consider again

$$\vec{u} = (1,2,1)$$
$$\vec{v} = (2,2,4)$$
$$\vec{w} = (4,6,6)$$

we know that

$$\vec{w} = \lambda\,\vec{u} + \mu\,\vec{v}.$$

with $\lambda, \mu = 2, 1$.

The theorem above says that the couple $\lambda, \mu = 2, 1$ is unique. Let us verify it:

$$(4,6,6) = \lambda\,(1,2,1) + \mu\,(2,2,4),$$

which leads to

$$\begin{cases} \lambda + 2\mu = 4 \\ 2\lambda + 2\mu = 6 \\ \lambda + 4\mu = 6. \end{cases}$$

The system is determined and the only solution is $\lambda, \mu = 2, 1$.

Theorem 4.6. *Let $\vec{u}$, $\vec{v}$, $\vec{w}$, and $\vec{t} \in \mathbb{V}_3$. These vectors are always linearly dependent: $\forall\ \vec{u}$, $\vec{v}$, $\vec{w}$, and $\vec{t} \in \mathbb{V}_3$:*

$$\exists \lambda, \mu, \nu \neq 0,0,0 \ni \text{`}$$
$$\vec{t} = \lambda\,\vec{u} + \mu\,\vec{v} + \nu\vec{w}.$$

Proof. If one vector is the null vector $\vec{o}$ or two vectors are parallel or three vectors are coplanar, the vectors are linearly dependent for Proposition 4.3 and Theorems 4.2 and 4.4.

Let us consider the case of four vectors such that each triple is not coplanar. Without a loss of generality let us consider all the vectors have the same arbitrary starting point **O**.

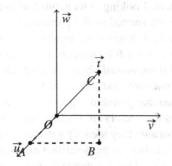

It can be observed that

$$\overrightarrow{OC} = \overrightarrow{OA} + \overrightarrow{AB} + \overrightarrow{BC}.$$

This equation can be written as

$$\vec{t} = \lambda\,\vec{u} + \mu\,\vec{v} + \nu\vec{w} \Rightarrow$$
$$\Rightarrow \vec{o} = \lambda\,\vec{u} + \mu\,\vec{v} + \nu\vec{w} - \vec{t}.$$

The null vector has been expressed as linear combination of the four vectors by means of non-null coefficients as $\lambda, \mu, \nu, -1 \neq 0,0,0,0$. The vectors are linearly dependent. $\square$

Example 4.19. The theorem above tells us that three arbitrary vectors $\in \mathbb{V}_3$

$$\vec{u} = (0,1,1)$$
$$\vec{v} = (0,1,0)$$
$$\vec{w} = (1,0,0)$$
$$\vec{t} = (4,6,5)$$

are necessarily linearly dependent. If we write

$$\vec{t} = \lambda \vec{u} + \mu \vec{v} + \nu \vec{w}$$

that is

$$(4,6,5) = \lambda (0,1,1) + \mu (0,1,0) + \nu (1,0,0)$$

which leads to the following system of linear equations:

$$\begin{cases} \nu = 4 \\ \lambda + \mu = 6 \\ \lambda = 5. \end{cases}$$

The system is determined and $\lambda, \mu, \nu = 5, 1, 4$ is its solution.

In summary, two parallel vectors, three vectors in a plane, or four vectors in the space are linearly dependent. Looking at its geometric implications, the concept of linear dependence can be interpreted as the presence of redundant pieces of information within a mathematical description. Intuitively, we can easily see that in a unidimensional space one number fully describes an object. There is no need of two numbers to represent a unidimensional object. In a similar way, two parallel vectors in the space identify only one direction and are de facto describing a unidimensional problem. The three dimensional problem degenerates in a unidimensional problem. When this situation occurs the vectors are linearly dependent.

If two vectors are not parallel they identify a plane and describe a point within it. This object requires two numbers to be described and a third coordinate would be unnecessary. In a similar way, three (or more) coplanar vectors in the space are de facto a two-dimensional problems. Also in this case redundant pieces information are present in tha mathematical description and the vectors are linearly dependent.

Finally, three vectors are needed to describe an object in the space. Any fourth vector would give redundant pieces of information. This fact is mathematically translated into the statement that four (or more) vectors in the space are surely linearly dependent.

Each point of the space can then be detected by three linearly independent vectors. Equivalently, each vector of $\mathbb{V}_3$ can be univocally expressed as the linear combination of three linearly independent vectors. Thus, when three linearly independent vectors $\vec{u}$, $\vec{v}$, and $\vec{w}$ are fixed, every vector $\vec{t}$ in $\mathbb{V}_3$ is univocally identified by those three coefficients λ, μ, ν such that

$$\vec{t} = \lambda \vec{u} + \mu \vec{v} + \nu \vec{w}.$$

Let us formally prove this statement

Theorem 4.7. *Let $\vec{u}$, $\vec{v}$, and $\vec{w}$ be three linearly independent vectors $\in \mathbb{V}_3$ and $\lambda, \mu, \nu \neq 0,0,0$ be a triple of scalars. Let a vector $\vec{t}$ be another vector $\in \mathbb{V}_3$. It follows that the vector $\vec{t}$ can be expressed as the linear combination of the other three in only one way (i.e. by means of only one tuple of scalars λ, μ, ν):*

$$\exists! \lambda, \mu, \nu \neq 0,0,0 \ni `$$
$$\vec{t} = \lambda \vec{u} + \mu \vec{v} + \nu \vec{w}$$

Proof. If, by contradiction, there exists another triple $\lambda', \mu', \nu' \neq 0,0,0$ such that $\vec{t} = \lambda' \vec{u} + \mu' \vec{v} + \nu' \vec{w}$, it follows that

$$\vec{t} = \lambda \vec{u} + \mu \vec{v} + \nu \vec{w} = \lambda' \vec{u} + \mu' \vec{v} + \nu' \vec{w} \Rightarrow$$
$$\Rightarrow (\lambda - \lambda') \vec{u} + (\mu - \mu') \vec{v} + (\nu - \nu') \vec{w} = \vec{o}.$$

Since, for hypothesis, $\vec{u}$, $\vec{v}$, and $\vec{w}$ are linearly independent, it follows that

$$(\lambda - \lambda') = 0 \Rightarrow \lambda = \lambda'$$
$$(\mu - \mu') = 0 \Rightarrow \mu = \mu'$$
$$(\nu - \nu') = 0 \Rightarrow \nu = \nu'. \square$$

Example 4.20. Let us consider again

$$\vec{u} = (0,1,1)$$
$$\vec{v} = (0,1,0)$$
$$\vec{w} = (1,0,0)$$
$$\vec{t} = (4,6,5).$$

We know that

$$\vec{t} = \lambda \vec{u} + \mu \vec{v} + \nu \vec{w}$$

with $\lambda, \mu, nu = 5,1,4$.

We can easily verify that if λ, μ, nu and $\vec{u}, \vec{v}, \vec{w}$ are fixed, then $\vec{t}$ is implicitly identified.

Also, if $\vec{t}, \vec{u}, \vec{v}, \vec{w}$ are fixed, the scalars λ, μ, ν are univocally determined (the resulting system is determined).

4.3 Matrices of Vectors

Proposition 4.4. *Let* $\vec{u}, \vec{v} \in \mathbb{V}_3$ *where*

$$\vec{u} = (u_1, u_2, u_3)$$
$$\vec{v} = (v_1, v_2, v_3)$$

and $\mathbf{A}$ *be a* 2×3 *matrix whose elements are the components of* $\vec{u}$ *and* $\vec{v}$

$$\mathbf{A} = \begin{pmatrix} u_1 & u_2 & u_3 \\ v_1 & v_2 & v_3 \end{pmatrix}.$$

These two vectors are parallel (and thus linearly dependent) if and only if the rank of the matrix $\mathbf{A}$ *associated with the corresponding components is* < 2: $\rho_{\mathbf{A}} < 2$.

Proof. If $\vec{u}$ and $\vec{v}$ are parallel they could be expressed as $\vec{u} = \lambda \vec{v}$ with $\lambda \in \mathbb{R}$. Thus,

$$\vec{u} = \lambda \vec{v} \Rightarrow$$
$$\Rightarrow u_1 \vec{e_1} + u_2 \vec{e_2} + u_3 \vec{e_3} = \lambda \left(v_1 \vec{e_1} + v_2 \vec{e_2} + v_3 \vec{e_3} \right).$$

Since two vectors are the equal if and only if they have the same components,

$$u_1 = \lambda v_1$$
$$u_2 = \lambda v_2$$
$$u_3 = \lambda v_3.$$

Since the two rows are proportional, there is no non-singular order 2 submatrix. Thus $\rho_{\mathbf{A}} < 2$. $\square$

If $\rho_{\mathbf{A}} < 2$, every two submatrix has null determinant. This can happen in the following cases.

- A row is composed of zeros. This means that one vector is the null vector $\vec{o}$, e.g.

$$\mathbf{A} = \begin{pmatrix} 0 & 0 & 0 \\ v_1 & v_2 & v_3 \end{pmatrix}.$$

Since every vector is parallel to $\vec{o}$, the vectors are parallel.
- Two columns are composed of zeros. The vectors are of the kind

$$\vec{u} = (u_1, 0, 0)$$
$$\vec{v} = (v_1, 0, 0).$$

These vectors can always be expressed as

$$\vec{u} = \lambda \vec{v}.$$

Thus, the vectors are parallel.

- Two rows are proportional.

$$\vec{u} = (u_1, u_2, u_3)$$
$$\vec{v} = (\lambda u_1, \lambda u_2, \lambda u_3).$$

The vectors are parallel.

- Each pair of columns is proportional. The vectors are of the kind

$$\vec{u} = (u_1, \lambda u_1, \mu u_1)$$
$$\vec{v} = (v_1, \lambda v_1, \mu v_1).$$

If we pose

$$\frac{v_1}{u_1} = k$$

then

$$\vec{v} = (ku_1, \lambda ku_1, \mu ku_1) \Rightarrow \vec{v} = (ku_1, ku_2, ku_3) = k\vec{u}.$$

The vectors are parallel.

$\square$

Example 4.21. The following vectors

$$\vec{u} = (1, 3, 6)$$
$$\vec{v} = (2, 6, 12)$$

are parallel since $\vec{v} = 2\vec{u}$. Obviously the matrix

$$\begin{pmatrix} 1 & 3 & 6 \\ 2 & 6 & 12 \end{pmatrix}$$

has rank equal to 1, that is < 2.

Example 4.22. The following vectors

$$\vec{u} = (2, 4, 6)$$
$$\vec{v} = (3, 6, 9)$$

are associated with a matrix

$$\begin{pmatrix} 2 & 4 & 6 \\ 3 & 6 & 9 \end{pmatrix} = \begin{pmatrix} 2 & 2\lambda & 2\mu \\ 3 & 3\lambda & 3\mu \end{pmatrix}$$

with $\lambda, \mu = 2, 3$. Ever pair of columns is proportional. Thus, this matrix has rank $\rho < 2$.

If we pose $k = \frac{v_1}{u_1} = 1.5$ we can write the two vectors as

$$\vec{u} = (2, 4, 6)$$
$$\vec{v} = k(2, 4, 6).$$

Thus, these two vectors are parallel.

Example 4.23. Let us determine the values of h that make $\vec{u}$ parallel to $\vec{v}$.

$$\vec{u} = (h-1)\,\vec{e_1} + 2h\vec{e_2} + \vec{e_3}$$
$$\vec{v} = \vec{e_1} + 4\vec{e_2} + \vec{e_3}$$

These two vectors are parallel if and only if the matrix $\mathbf{A}$ has a rank $\rho_{\mathbf{A}} < 2$:

$$\mathbf{A} = \begin{pmatrix} h-1 & 2h & 1 \\ 1 & 4 & 1 \end{pmatrix}.$$

Let us compute $\det \begin{pmatrix} h-1 & 2h \\ 1 & 4 \end{pmatrix} = 4h - 4 - 2h = 2h - 4$. The vectors are parallel if $2h - 4 = 0 \Rightarrow h = 2$. In addition, we have to impose that $\det \begin{pmatrix} 2h & 1 \\ 4 & 1 \end{pmatrix} = 0 \Rightarrow h = 2$ and $\det \begin{pmatrix} h-1 & 1 \\ 1 & 1 \end{pmatrix} = h - 1 - 1 = 0$. Thus, the vectors are parallel if $h = 2$.

Proposition 4.5. *Let $\vec{u}$, $\vec{v}$, and $\vec{w} \in \mathbb{V}_3$ be*

$$\vec{u} = (u_1, u_2, u_3)$$
$$\vec{v} = (v_1, v_2, v_3)$$
$$\vec{w} = (w_1, w_2, w_3)$$

and $\mathbf{A}$ be the matrix whose elements are the components of $\vec{u}$, $\vec{v}$, and $\vec{w}$:

$$\mathbf{A} = \begin{pmatrix} u_1 & u_2 & u_3 \\ v_1 & v_2 & v_3 \\ w_1 & w_2 & w_3 \end{pmatrix}.$$

The three vectors are coplanar (and thus linearly dependent) if and only if the determinant of the matrix $\mathbf{A}$ is equal to 0:

$$\det(\mathbf{A}) = 0.$$

Proof. If the vectors are coplanar then they are linearly dependent for Theorem 4.4. For Theorem 4.1 one of them can be expressed as linear combination of the others:

$$\vec{u} = \lambda \vec{v} + \mu \vec{w} \Rightarrow$$
$$(u_1, u_2, u_3) = \lambda (v_1, v_2, v_3) + \mu (w_1, w_2, w_3) \Rightarrow$$
$$(u_1, u_2, u_3) = (\lambda v_1 + \mu w_1, \lambda v_2 + \mu w_2, \lambda v_3 + \mu w_3).$$

The first row of the matrix $\mathbf{A}$ has been expressed as linear combination of the other two rows:

$$\mathbf{A} = \begin{pmatrix} \lambda v_1 + \mu w_1 & \lambda v_2 + \mu w_2 & \lambda v_3 + \mu w_3 \\ v_1 & v_2 & v_3 \\ w_1 & w_2 & w_3 \end{pmatrix}.$$

Thus

$$\det(\mathbf{A}) = 0. \Box$$

If $\det(\mathbf{A}) = 0$ the following conditions may occur.

- A row is null, e.g.:

$$\mathbf{A} = \begin{pmatrix} 0 & 0 & 0 \\ v_1 & v_2 & v_3 \\ w_1 & w_2 & w_3 \end{pmatrix}.$$

This means that one vector is null and two vectors determine a plane. Thus, the three vectors are coplanar.

- A column is null, e.g.

$$\mathbf{A} = \begin{pmatrix} u_1 & u_2 & 0 \\ v_1 & v_2 & 0 \\ w_1 & w_2 & 0 \end{pmatrix}.$$

One component is null for all the vectors and thus the vectors are in $\mathbb{V}_2$, i.e. the three vectors are in the (same) plane.

- One column is linear combination of the other two columns. This means that the three vectors can be expressed as

$$\vec{u} = (u_1, u_2, \lambda u_1 + \mu u_2)$$
$$\vec{v} = (v_1, v_2, \lambda v_1 + \mu v_2)$$
$$\vec{w} = (w_1, w_2, \lambda w_1 + \mu w_2)$$

that is

$$\mathbf{A} = \begin{pmatrix} u_1 & u_2 & \lambda u_1 + \mu u_2 \\ v_1 & v_2 & \lambda v_1 + \mu v_2 \\ w_1 & w_2 & \lambda w_1 + \mu w_2 \end{pmatrix}$$

with the scalars $\lambda, \mu \in \mathbb{R}$.

Since one component is not independent, the vectors are in $\mathbb{V}_2$, i.e. the three vectors are in the (same) plane. In other words, since the vectors are linearly dependent are coplanar.

- One row is linear combination of the other two rows, e.g.

$$\mathbf{A} = \begin{pmatrix} \lambda v_1 + \mu w_1 & \lambda v_2 + \mu w_2 & \lambda v_3 + \mu w_3 \\ v_1 & v_2 & v_3 \\ w_1 & w_2 & w_3 \end{pmatrix}.$$

The vectors are linearly dependent and thus coplanar.
$\Box$

Propositions 4.4 and 4.5 clarify the meaning of determinant and rank of a matrix by offering an immediate geometric interpretation. The determinant of a 3×3 matrix can be interpreted as a volume generated by three vectors. If the vectors are coplanar the volume is zero as well as the associated determinant. This situation occurs when a redundancy appears in the mathematical description. In a similar way, if we consider only two vectors in the space we can geometrically interpret the concept of rank of a matrix. If the vectors are not parallel, they identify a plane and the rank of the associated matrix is 2. If the two vectors are parallel, the problem is practically unidimensional and the rank is 1. In other words, the rank of a matrix can be geometrically interpreted as the actual dimensionality of a mathematical description.

Example 4.24. Let us verify whether or not the following three vectors are coplanar:

$$\vec{u} = (5, 3, 12)$$
$$\vec{v} = (2, 8, 4)$$
$$\vec{w} = (1, -13, 4)$$

The det $\begin{pmatrix} 5 & 3 & 12 \\ 2 & 8 & 4 \\ 1 & -13 & 4 \end{pmatrix} = 160 + 12 - 312 - 96 - 24 + 260 = 0$. The vectors are coplanar. It can be observed that $\vec{w} = \vec{u} - 2\vec{v}$.

4.4 Bases of Vectors

Definition 4.15. A *vector basis* in $\mathbb{V}_3$ is a triple of linearly independent vectors and is indicated with

$$B = \{\vec{e_1}, \vec{e_2}, \vec{e_3}\}.$$

For Theorem 4.7 every vector belonging to $\mathbb{V}_3$ can be univocally expressed as the linear combination of the vectors composing the basis: $\forall \vec{v} \in \mathbb{V}_3$:

$$\vec{v} = v_1 \vec{e_1} + v_2 \vec{e_2} + v_3 \vec{e_3}$$

where $v_1, v_2, v_3 \in \mathbb{R}$ and can be indicated as

$$\vec{v} = (v_1, v_2, v_3).$$

Each v_i, $\forall i$ is named *component* of the vector $\vec{v}$. In this case the vector $\vec{v}$ is represented in the basis $B = \{\vec{e_1}, \vec{e_2}, \vec{e_3}\}$. In general, for a fixed basis $B = \{\vec{e_1}, \vec{e_2}, \vec{e_3}\}$, each vector belonging to $\mathbb{V}_3$ is identified by its component and, thus, two vectors are equal if they have the same components in the same basis.

Every time in the previous sections of this chapter a vector has been indicated e.g. $\vec{v} = (x, y, z)$, we implicitly meant that

$$\vec{v} = x\vec{e_1} + y\vec{e_2} + z\vec{e_3}.$$

where $\vec{e_1}, \vec{e_2}, \vec{e_3}$ is a basis of vector having module equal to 1 and direction of the axes of the reference system, i.e.

$$\vec{e_1} = (1,0,0)$$
$$\vec{e_2} = (0,1,0)$$
$$\vec{e_3} = (0,0,1).$$

Example 4.25. We can easily verify that the vectors

$$\vec{e_1} = (0,1,2)$$
$$\vec{e_2} = (0,1,0)$$
$$\vec{e_3} = (4,0,0)$$

are linearly independent. As such, these vectors are a basis $B = \{\vec{e_1}, \vec{e_2}, \vec{e_3}\}$. For Theorem 4.7 any vector $\in \mathbb{V}_3$ can be univocally expressed as a linear combination of the vectors of this basis.

Let us express an arbitrary vector $\vec{t} = (4,8,6)$:

$$\vec{t} = \lambda \vec{e_1} + \mu \vec{e_2} + \nu \vec{e_3}$$

with $\lambda, \mu, \nu = 3,5,1$.

Thus, when we write $\vec{t} = (3,5,1)$ in the basis $B = \{\vec{e_1}, \vec{e_2}, \vec{e_3}\}$.

A basis of $\mathbb{V}_3$ can be seen as a set of vectors able to generate, by linear combination, any arbitrary vector $\in \mathbb{V}_3$. The following corollary gives a formalization of this statement.

Corollary 4.1. *Every vector $\vec{t} \in \mathbb{V}_3$ can be represented as a linear combination of the vectors $\vec{e_1}, \vec{e_2}, \vec{e_3}$ composing a basis of $\mathbb{V}_3$.*

Proof. Let us consider a generic vector $\vec{t} \in \mathbb{V}_3$. For Theorem 4.6 $\vec{t}, \vec{e_1}, \vec{e_2}, \vec{e_3}$ are linearly dependent. For Theorem 4.7 it follows that

$$\vec{t} = \lambda \vec{e_1} + \mu \vec{e_2} + \nu \vec{e_3}$$

by means of a unique triple λ, μ, ν. Thus, in the basis the vector $\vec{t}$ is univocally determined by the triple (λ, μ, ν).

If we considered another vector $\vec{t'} \in \mathbb{V}_3$ with $\vec{t} \neq \vec{t'}$ it would follow that another triple $\lambda', \mu', \nu' \neq \lambda, \mu, \nu$ would be univocally associated with $\vec{t'}$. We can indefinitely reiterate this operation and discover that any arbitrary vector $\in \mathbb{V}_3$ can be represented as linear combination of $\vec{e_1}, \vec{e_2}, \vec{e_3}$ by means of a unique triple of scalars. $\square$

The following corollary revisits the previous result and formally states a concept previously introduced in an intuitive way: there is an equivalence between the vectors of the space and the points of the space.

Corollary 4.2. *If an arbitrary basis $B = \{\vec{e_1}, \vec{e_2}, \vec{e_3}\}$ is fixed, there exists a bijection between the set of vectors in the space $\mathbb{V}_3$ and the set of points $\mathbb{R}^3$.*

Proof. For a fixed basis $B = \{\vec{e_1}, \vec{e_2}, \vec{e_3}\}$ let us consider the mapping $\phi : \mathbb{V}_3 \to \mathbb{R}^3$ defined as

$$\vec{t} = (\lambda, \mu, \nu),$$

where

$$\vec{t} = \lambda \vec{e_1} + \mu \vec{e_2} + \nu \vec{e_3}$$

This mapping is injective since for $\vec{t} \neq \vec{t'}$ with

$$\vec{t} = (\lambda, \mu, \nu)$$
$$\vec{t'} = (\lambda', \mu', \nu')$$

for Theorem 4.6, it follows that the triple $\lambda', \mu', \nu' \neq \lambda, \mu, \nu$.

This mapping is surjective since, for Theorem 4.7, a vector is always representable as a linear combination of $\vec{e_1}, \vec{e_2}, \vec{e_3}$ and thus, always associated with a triple λ, μ, ν.

Thus, the mapping is bijective. $\square$

Definition 4.16. Two vectors (or two lines) in $\mathbb{V}_3$ are said to be *perpendicular* when their direction compose four angles of $90°$.

Definition 4.17. Three vectors in $\mathbb{V}_3$ are said to be *orthogonal* if each of them is perpendicular to the other two.

It must be remarked that the notion of perpendicularity does not coincide with the notion of orthogonality. However, the two concepts are closely related. While perpendicularity refers to lines (or vectors) in the plane and means that these two objects generate a $90°$ angle, orthogonality is a more general term and refers to multidimensional objects (such as planes in the space). This concept is better explained in Chap. 8. Intuitively we may think that two multi-dimensional objects are orthogonal when all the angles generated by the intersection of these two objects are $90°$. For example, three vectors in $\mathbb{V}_3$ can be interpreted as an object in the space. The definition above states that this solid is orthogonal when all the angles involved are $90°$. We may conclude that perpendicularity is orthogonality among lines of the plane.

Definition 4.18. If a basis of vectors in $\mathbb{V}_3$ is composed of three orthogonal vectors $\{\vec{i}, \vec{j}, \vec{k}\}$ the basis is said *orthonormal*.

Definition 4.19. When the vectors composing an orthonormal basis have modulus equal to 1, the vectors composing this basis are said *versors*.

In $\mathbb{V}_3$, an orthonormal basis composed of vectors is

$$\vec{i} = (1, 0, 0)$$
$$\vec{j} = (0, 1, 0)$$
$$\vec{k} = (0, 0, 1).$$

Vectors are in general represented in an orthonormal basis of versors but can be rewritten in another basis by simply imposing the equivalence of each component. The following example clarifies this fact.

Example 4.26. Let us consider the following vectors in an orthonormal basis of versors $\{\vec{i}, \vec{j}, \vec{k}\}$.

$$\vec{u} = (2, 0, -1)$$
$$\vec{v} = (1, 2, 1)$$
$$\vec{w} = (1, 0, 3)$$
$$\vec{t} = (2, -1, -1)$$

Let us verify that $\vec{u}$, $\vec{v}$, and $\vec{w}$ are linearly independent:

$$\det \begin{pmatrix} 2 & 0 & -1 \\ 1 & 2 & 1 \\ 1 & 0 & 3 \end{pmatrix} = 12 + 2 = 14 \neq 0.$$

The vectors $\vec{u}$, $\vec{v}$, and $\vec{w}$ are linearly independent. Now, let us determine the components of $\vec{t}$ in the new basis $\{\vec{u}, \vec{v}, \vec{w}\}$.

$$\vec{t} = \lambda \vec{u} + \mu \vec{v} + v \vec{w} \Rightarrow$$
$$\Rightarrow (2, -1, -1) = \lambda (2, 0, -1) + \mu (1, 2, 1) + v (1, 0, 3) \Rightarrow$$
$$\Rightarrow (2, -1, -1) = (2\lambda + \mu + v, 2\mu, -\lambda + \mu + 3v).$$

In order to detect the values of λ, μ, v the following linear system must be solved:

$$\begin{cases} 2\lambda + \mu + v = 2 \\ 2\mu = -1 \\ -\lambda + \mu + 3v = -1 \end{cases}$$

It can be easily observed that the matrix associated with this linear system is the transpose of the matrix A associated with the three vectors. The system is determined. The solution of the linear system is $\lambda = \frac{8}{7}$, $\mu = -\frac{1}{2}$, and $v = \frac{3}{14}$. Thus, $\vec{t} = \frac{8}{7}\vec{u} - \frac{1}{2}\vec{v} + \frac{3}{14}\vec{w}$.

Example 4.27. Let us consider the following vectors in an orthonormal basis of versors $\{\vec{i}, \vec{j}, \vec{k}\}$.

$$\vec{u} = (2, 0, -1)$$
$$\vec{v} = (1, 2, 1)$$
$$\vec{w} = (3, 2, 0)$$
$$\vec{t} = (2, -1, -1)$$

It can be easily verified that $\vec{u}$, $\vec{v}$, and $\vec{w}$ are linearly dependent since the matrix

$$\det \begin{pmatrix} 2 & 0 & -1 \\ 1 & 2 & 1 \\ 3 & 2 & 0 \end{pmatrix} = 0.$$

If we try to express anyway $\vec{t}$ as a linear combination of $\vec{u}$, $\vec{v}$, and $\vec{w}$ we obtain

$$\vec{t} = \lambda \vec{u} + \mu \vec{v} + v \vec{w} \Rightarrow$$
$$\Rightarrow (2, -1, -1) = \lambda (2, 0, -1) + \mu (1, 2, 1) + v (3, 2, 0) \Rightarrow$$
$$\Rightarrow (2, -1, -1) = (2\lambda + \mu + 3v, 2\mu + 2v, -\lambda + \mu).$$

In order to detect the values of λ, μ, v the following linear system must be solved:

$$\begin{cases} 2\lambda + \mu + 3v = 2 \\ 2\mu + 2v = -1 \\ -\lambda + \mu = -1 \end{cases}.$$

This system is impossible because the rank of the incomplete matrix is 2 whilst the rank of the complete matrix is 3. The system has no solutions. This fact can be geometrically seen with the following sentence: it is impossible to generate a vector in the space starting from three coplanar vectors. The three coplanar vectors are effectively in two dimensions and their combination cannot generate a three-dimensional object.

Example 4.28. Let us consider the following vectors in an orthonormal basis of versors $\{\vec{i}, \vec{j}, \vec{k}\}$.

$$\vec{u} = (2, 0, -1)$$
$$\vec{v} = (1, 2, 1)$$
$$\vec{w} = (3, 2, 0)$$
$$\vec{t} = (2, 4, 2)$$

We already know that $\vec{u}$, $\vec{v}$, and $\vec{w}$ are coplanar. If we try to express $\vec{t}$ as linear combination of them, we obtain the following system of linear equations:

$$\begin{cases} 2\lambda + \mu + 3v = 2 \\ 2\mu + 2v = 4 \\ -\lambda + \mu = 2 \end{cases}.$$

The rank of the incomplete matrix is 2 as well as the rank of the complete matrix. The system in undetermined and, hence, has ∞ solutions. It can be observed that $\vec{t}$ and $\vec{v}$ are parallel. Hence $\vec{t}$, $\vec{u}$, $\vec{v}$, and $\vec{w}$ are coplanar. In the same plane one vector can be expressed as the linear combination of three other vectors in an infinite number of ways.

Example 4.29. Let us consider the following vectors in an orthonormal basis of versors $\{\vec{i}, \vec{j}, \vec{k}\}$.

$$\vec{u} = (2,0,-1)$$
$$\vec{v} = (1,2,1)$$
$$\vec{w} = (1,0,3)$$
$$\vec{t} = (0,0,0)$$

We know that $\vec{u}$, $\vec{v}$, and $\vec{w}$ are linearly independent and thus are a basis. Let us express $\vec{t}$ in the basis of $\vec{u}$, $\vec{v}$, and $\vec{w}$:

$$\vec{t} = \lambda \vec{u} + \mu \vec{v} + v \vec{w} \Rightarrow$$
$$\Rightarrow (0,0,0) = \lambda (2,0,-1) + \mu (1,2,1) + v (1,0,3) \Rightarrow$$
$$\Rightarrow (0,0,0) = (2\lambda + \mu + v, 2\mu, -\lambda + \mu + 3v).$$

In order to detect the values of λ, μ, v the following linear system must be solved:

$$\begin{cases} 2\lambda + \mu + v = 0 \\ 2\mu = 0 \\ -\lambda + \mu + 3v = 0 \end{cases}.$$

This is a homogeneous system of linear equations. The system is determined as the associated incomplete matrix is non-singular. The only solution of the system is $\lambda, \mu, v = 0, 0, 0$. It can be observed that $\vec{t}$ is the null vector. Hence, we found out that the only linear combination of $\vec{u}$, $\vec{v}$, and $\vec{w}$ that return the null vector $\vec{o}$ is by means of the tuple $\lambda, \mu, v = 0, 0, 0$. We have verified the linear independence of the vectors $\vec{u}$, $\vec{v}$, and $\vec{w}$.

Example 4.30. If we consider three linearly dependent vectors , such as

$$\vec{u} = (2,0,-1)$$
$$\vec{v} = (1,2,1)$$
$$\vec{w} = (3,2,0)$$

and we try to express the vector $\vec{t} = (0,0,0)$ as a linear combination of $\vec{u}$, $\vec{v}$, and $\vec{w}$ by means of three scalars λ, μ, ν we have the following homogeneous system of linear equations:

$$\begin{cases} 2\lambda + \mu + 3\nu = 0 \\ 2\mu + 2\nu = 0 \\ -\lambda + \mu = 0 \end{cases}.$$

The rank of the incomplete matrix is 2 as well as the rank of the complete matrix. The system in undetermined and, hence, has ∞ solutions besides $0,0,0$. This means that at least one solution $\neq 0,0,0$ such that $\vec{o} = \lambda\vec{u} + \mu\vec{v} + \nu\vec{w}$ exists. This is another way to express the linear dependence of the vectors.

These examples are extremely important as they link systems of linear equations and vectors highlighting how they correspond to different formulations of the same concepts.

4.5 Products of Vectors

Definition 4.20. Let $\vec{u}, \vec{v} \in \mathbb{V}_3$ having both an arbitrary starting point **O**. The convex angle determined by the vectors is said *angle of the vectors*. This angle can be between 0 and π.

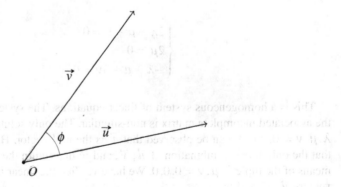

Definition 4.21. Scalar Product (Dot Product). Let $\vec{u}, \vec{v} \in \mathbb{V}_3$ having both an arbitrary starting point **O** and let us indicate with ϕ their angle. The *scalar product* is an operator that associates a scalar to two vectors ($\mathbb{V}_3 \times \mathbb{V}_3 \rightarrow \mathbb{R}$) according to the following formula:

$$\vec{u}\,\vec{v} = \|\vec{u}\|\|\vec{v}\|\cos\phi.$$

Proposition 4.6. *Let $\vec{u}, \vec{v} \in \mathbb{V}_3$. The scalar product of these two vectors is equal to 0 ($\vec{u}\,\vec{v} = 0$) if and only if they are perpendicular.*

Proof. If $\vec{u}\,\vec{v} = 0 \Rightarrow \|\vec{u}\|\|\vec{v}\|\cos\phi = 0$. This equality is verified either if at least one of the modules in 0 or if $\cos\phi = 0$. If a module is 0, the one vector is the null vector $\vec{o}$. Since a null vector has an undetermined direction, every vector is perpendicular to the null vector, the vectors are perpendicular. If $\cos\phi = 0 \Rightarrow \phi = \frac{\pi}{2}$ ($+k\pi$ with $k \in \mathbb{N}$). Thus the vectors are perpendicular. $\square$

If the vectors are perpendicular, their angle $\phi = \frac{\pi}{2}$. Thus, $\cos\phi = 0$ and the scalar product is 0. $\square$

The following properties are valid for the scalar product with $\vec{u}$, $\vec{v}$, and $\vec{w} \in \mathbb{V}_3$ and $\lambda \in \mathbb{R}$.

- commutativity: $\vec{u}\,\vec{v} = \vec{v}\,\vec{u}$
- homogeneity: $\lambda\,(\vec{u}\,\vec{v}) = (\lambda\,\vec{u})\,\vec{v} = \vec{u}\,(\lambda\,\vec{v})$
- associativity: $\vec{w}\,(\vec{u}\,\vec{v}) = (\vec{w}\,\vec{u})\,\vec{v}$
- distributivity with respect to the sum of vectors: $\vec{w}\,(\vec{u} + \vec{v}) = \vec{w}\,\vec{u} + \vec{w}\,\vec{v}$

This is the second time in this book that the term "scalar product is used". A natural question would be "How does the latter definition relate to that given in Chap. 2?". The following proposition addresses this question.

Proposition 4.7. *Let $\vec{u}, \vec{v} \in \mathbb{V}_3$, with $\vec{u} = u_1\,\vec{i} + u_2\,\vec{j} + u_3\,\vec{k}$ and $\vec{v} = v_1\,\vec{i} + v_2\,\vec{j} + v_3\,\vec{k}$. The scalar product is equal to:*

$$\vec{u}\,\vec{v} = \|\vec{u}\|\|\vec{v}\|\cos\phi = u_1 v_1 + u_2 v_2 + u_3 v_3.$$

Proof. The scalar product can be expressed in the following way.

$$\vec{u}\,\vec{v} = \left(u_1\,\vec{i} + u_2\,\vec{j} + u_3\,\vec{k}\right)\left(v_1\,\vec{i} + v_2\,\vec{j} + v_3\,\vec{k}\right) =$$
$$= (u_1 v_1)\,\vec{i}\,\vec{i} + (u_1 v_2)\,\vec{i}\,\vec{j} + (u_1 v_3)\,\vec{i}\,\vec{k} +$$
$$+ (u_2 v_1)\,\vec{j}\,\vec{i} + (u_2 v_2)\,\vec{j}\,\vec{j} + (u_2 v_3)\,\vec{j}\,\vec{k} +$$
$$(u_3 v_1)\,\vec{k}\,\vec{i} + (u_3 v_2)\,\vec{k}\,\vec{j} + (u_3 v_3)\,\vec{k}\,\vec{k}$$

Considering that $\{\vec{i}, \vec{j}, \vec{k}\}$ is taken orthonormal, the scalar product of a basis vector by itself, e.g. $\vec{i}\,\vec{i}$, is equal to 1 because the vectors are parallel ($\phi = 0 \Rightarrow \cos\phi = 1$) and the module is unitary, while the scalar product of a basis vector by another basis vector, e.g. $\vec{i}\,\vec{j}$, is equal to 0 because the vectors are perpendicular.

Thus,
$$\vec{u}\,\vec{v} = u_1v_1 + u_2v_2 + u_3v_3.$$
$\square$

In other words, the two scalar products defined in this book are homonyms because they are the same concept from different perspectives.

Example 4.31. The two vectors
$$\vec{u} = (1,5,-3)$$
$$\vec{v} = (0,6,1)$$

have scalar product
$$\vec{u}\,\vec{v} = 1(0) + 5(6) - 3(1) = 27.$$

These two vectors are not perpendicular.

Example 4.32. The two vectors
$$\vec{u} = (2,5,-3)$$
$$\vec{v} = (1,2,4)$$

have scalar product
$$\vec{u}\,\vec{v} = 2(1) + 5(2) - 3(4) = 0.$$

These two vectors are perpendicular.

Definition 4.22. Vector Product (Cross Product). Let $\vec{u}, \vec{v} \in \mathbb{V}_3$ having both an arbitrary starting point **O** and let us indicate with ϕ their angle. The *vector product* is an operator that associates a vector to two vectors ($\mathbb{V}_3 \times \mathbb{V}_3 \to \mathbb{V}_3$) and is indicated with $\vec{u} \otimes \vec{v}$. The resulting vector has module according to the following formula:

$$\|\vec{u} \otimes \vec{v}\| = \|\vec{u}\|\|\vec{v}\|\sin\phi.$$

The direction of $\vec{u} \otimes \vec{v}$ is perpendicular to that of $\vec{u}$ and $\vec{v}$. The orientation of $\vec{u} \otimes \vec{v}$ is given by the so-called right-hand-rule, graphically represented below.

Proposition 4.8. *Let $\vec{u}, \vec{v} \in \mathbb{V}_3$ and ϕ their angle. The vector product of the vectors $\vec{u}, \vec{v}$ is equal to $\vec{o}$ if and only if the two vectors are parallel.*

Proof. The vector product is equal to the null vector $\vec{o}$ either if one of the two vectors is the null vector or if the $\sin\phi = 0$. If one of the vector is the null vector $\vec{o}$ then the vectors are parallel. If $\sin\phi = 0 \Rightarrow \phi = 0$ and the vectors are parallel. $\square$

If the vectors are parallel $\phi = 0 \Rightarrow \sin\phi = 0$. Thus the vector product is equal to $\vec{o}$. $\square$

The following properties for the vector product are valid.

- anticommutativity: $\vec{u} \otimes \vec{v} = -\vec{v} \otimes \vec{u}$
- homogeneity: $(\lambda\,\vec{u}) \otimes \vec{v} = \vec{u} \otimes (\lambda\,\vec{v})$
- distributivity with respect to the sum of vectors: $\vec{w} \otimes (\vec{u} + \vec{v}) = \vec{w} \otimes \vec{u} + \vec{w} \otimes \vec{v}$

It can be observed that the associativity is not valid for the vector product, i.e.

$$(\vec{u} \otimes \vec{v}) \otimes \vec{w} \neq \vec{u} \otimes (\vec{v} \otimes \vec{w}).$$

Proposition 4.9. *Let $\vec{u}, \vec{v} \in \mathbb{V}_3$ and ϕ their angle. The vector product : $\vec{u} \otimes \vec{v}$ is equal to the (symbolic) determinant of the matrix $\mathbf{A}$ where,*

$$\mathbf{A} = \begin{pmatrix} \vec{i} & \vec{j} & \vec{k} \\ u_1 & u_2 & u_3 \\ v_1 & v_2 & v_3 \end{pmatrix}.$$

Proof. The vector product of $\vec{u}$ by $\vec{v}$ is calculated in the following way.

$$\vec{u} \otimes \vec{v} = \left(u_1\,\vec{i} + u_2\,\vec{j} + u_3\,\vec{k}\right) \otimes \left(v_1\,\vec{i} + v_2\,\vec{j} + v_3\,\vec{k}\right) =$$
$$= (u_1 v_1)\,\vec{i} \otimes \vec{i} + (u_1 v_2)\,\vec{i} \otimes \vec{j} + (u_1 v_3)\,\vec{i} \otimes \vec{k} +$$
$$+ (u_2 v_1)\,\vec{j} \otimes \vec{i} + (u_2 v_2)\,\vec{j} \otimes \vec{j} + (u_2 v_3)\,\vec{j} \otimes \vec{k} +$$
$$(u_3 v_1)\,\vec{k} \otimes \vec{i} + (u_3 v_2)\,\vec{k} \otimes \vec{j} + (u_3 v_3)\,\vec{k} \otimes \vec{k}$$

The vector product of a basis vector by itself, e.g. $\vec{i} \otimes \vec{i}$, is equal to $\vec{o}$ since the vectors are parallel. Since $\{\vec{i}, \vec{j}, \vec{k}\}$ is orthonormal, the vector product of two vectors composing the basis is the third vector. Hence, we obtain:

$$\vec{i} \otimes \vec{j} = \vec{k}$$
$$\vec{j} \otimes \vec{k} = \vec{i}$$
$$\vec{k} \otimes \vec{i} = \vec{j}$$

and, for anticommutativity

$$\vec{j} \otimes \vec{i} = -\vec{k}$$
$$\vec{k} \otimes \vec{j} = -\vec{i}$$
$$\vec{i} \otimes \vec{k} = -\vec{j}.$$

Thus, the equation becomes:

$$\vec{u} \otimes \vec{v} = (u_1 v_2)\,\vec{k} - (u_1 v_3)\,\vec{j} - (u_2 v_1)\,\vec{k} +$$
$$+ (u_2 v_3)\,\vec{i} + (u_3 v_1)\,\vec{j} - (u_3 v_2)\,\vec{i} =$$
$$= (u_2 v_3)\,\vec{i} + (u_3 v_1)\,\vec{j} + (u_1 v_2)\,\vec{k} +$$
$$- (u_3 v_2)\,\vec{i} - (u_1 v_3)\,\vec{j} - (u_2 v_1)\,\vec{k} =$$
$$= \det(\mathbf{A}).$$

□

The $\det(\mathbf{A})$ is said symbolic because it is based on a matrix composed of heterogeneous elements (instead of being composed of only numbers).

Example 4.33. Let us consider the following two vectors

$$\vec{u} = (2, 5, 1)$$
$$\vec{v} = (4, 10, 2).$$

These vectors are parallel since $\vec{v} = 2\vec{u}$. Let us check their parallelism by cross product

$$\vec{v} \otimes \vec{u} = \det \begin{pmatrix} \vec{i} & \vec{j} & \vec{k} \\ 2 & 5 & 1 \\ 4 & 10 & 2 \end{pmatrix} = 10\vec{i} + 4\vec{j} + 20\vec{k} - 20\vec{k} - 4\vec{j} - 10\vec{i} = \vec{o}.$$

Example 4.34. Let us consider the following two vectors:

$$\vec{u} = (4, 1, -2)$$
$$\vec{v} = (1, 0, 2).$$

The scalar product of these two vectors is

$$4(1) + 1(0) + 2(-2) = 0.$$

The vectors are perpendicular. If we calculate the vector product we obtain

$$\det \begin{pmatrix} \vec{i} & \vec{j} & \vec{k} \\ 4 & 1 & -2 \\ 2 & 0 & 2 \end{pmatrix} = 2\vec{i} - 12\vec{j} - 2\vec{k}.$$

Definition 4.23. Mixed Product (Triple Product). Let $\vec{u}, \vec{v}, \vec{w} \in \mathbb{V}_3$ having all an arbitrary starting point **O**. The *mixed product* is an operator that associates a scalar to three vectors ($\mathbb{V}_3 \times \mathbb{V}_3 \times \mathbb{V}_3 \to \mathbb{R}$) and is defined as the scalar product of one of the three vectors by the vector product of the other two vectors:

$$(\vec{u} \otimes \vec{v})\,\vec{w}$$

Proposition 4.10. *Let* $\vec{u}, \vec{v}, \vec{w} \in \mathbb{V}_3$ *having all an arbitrary starting point* **O**. *The three vectors are coplanar if and only if their mixed product* $(\vec{u} \otimes \vec{v}) \vec{w}$ *is equal to* 0.

Proof. If the tree vectors are coplanar, $\vec{u}$ and $\vec{v}$ are also coplanar. Thus, the vector product $\vec{u} \otimes \vec{v}$ is a vector perpendicular to both and to the plane that contains them. Since $\vec{w}$ belongs to the same plane, $\vec{w}$ and $\vec{u} \otimes \vec{v}$ are perpendicular. Thus the scalar product between them is equal to 0, i.e. the mixed product is equal to 0. □

If the mixed product $(\vec{u} \otimes \vec{v}) \vec{w}$ is equal to 0, $\vec{w}$ and $\vec{u} \otimes \vec{v}$ are perpendicular. By definition of vector product, $\vec{u} \otimes \vec{v}$ is perpendicular to both $\vec{u}$ and $\vec{v}$ and to the plane determined by them. Since $\vec{u} \otimes \vec{v}$ is perpendicular to the three vectors $\vec{u}, \vec{v}, \vec{w}$, there is only one plane that contains all of them. Thus, $\vec{u}, \vec{v}, \vec{w}$ are coplanar. □

Proposition 4.11. *Let* $B = \{\vec{i}, \vec{j}, \vec{k}\}$ *be an orthonormal basis. Let* $\vec{u}, \vec{v}, \vec{w} \in \mathbb{V}_3$ *having components* $\vec{u} = (u_1, u_2, u_3)$, $\vec{v} = (v_1, v_2, v_3)$, *and* $\vec{w} = (w_1, w_2, w_3)$, *respectively, in the basis B. The mixed product* $(\vec{u} \otimes \vec{v}) \vec{w} = \det(\mathbf{A})$ *where the matrix* $\mathbf{A}$ *is:*

$$\mathbf{A} = \begin{pmatrix} u_1 & u_2 & u_3 \\ v_1 & v_2 & v_3 \\ w_1 & w_2 & w_3 \end{pmatrix}.$$

Proof. The vector product is equal to

$$\vec{u} \otimes \vec{v} = \det \begin{pmatrix} \vec{i} & \vec{j} & \vec{k} \\ u_1 & u_2 & u_3 \\ v_1 & v_2 & v_3 \end{pmatrix} =$$

$$= \left(\det \begin{pmatrix} u_2 & u_3 \\ v_2 & v_3 \end{pmatrix} \vec{i} - \det \begin{pmatrix} u_1 & u_3 \\ v_1 & v_3 \end{pmatrix} \vec{j} + \det \begin{pmatrix} u_1 & u_2 \\ v_1 & v_2 \end{pmatrix} \vec{k} \right)$$

because of the I Laplace Theorem.

The mixed product is then obtained by calculating the scalar product between $\vec{w}$ and $\vec{u} \otimes \vec{v}$:

$$(\vec{u} \otimes \vec{v}) \vec{w} = \left(\det \begin{pmatrix} u_2 & u_3 \\ v_2 & v_3 \end{pmatrix} w_1 - \det \begin{pmatrix} u_1 & u_3 \\ v_1 & v_3 \end{pmatrix} w_2 + \det \begin{pmatrix} u_1 & u_2 \\ v_1 & v_2 \end{pmatrix} w_3 \right) = \det(\mathbf{A})$$

for the I Laplace Theorem. □

Example 4.35. The following three vectors are coplanar since $\vec{w} = 2\vec{u} + 3\vec{v}$:

$$(1, 2, 1)$$
$$(0, 4, 2)$$
$$(4, 16, 8).$$

Let us check that the vectors are coplanar by verifying that the matrix associated with these vectors is singular

$$\det \begin{pmatrix} 1 & 2 & 1 \\ 0 & 4 & 2 \\ 4 & 16 & 8 \end{pmatrix} = 0.$$

This result could have seen immediately by considering that the third row is a linear combination of the first two.

Example 4.36. The following three vectors are coplanar since two of these vectors are parallel (hence only two directions are under consideration):

$$(2,2,1)$$
$$(6,6,3)$$
$$(5,1,2).$$

Let us check that the vectors are coplanar by verifying that the matrix associated with these vectors is singular

$$\det \begin{pmatrix} 2 & 2 & 1 \\ 6 & 6 & 3 \\ 5 & 1 & 2 \end{pmatrix} = 0.$$

This result could have seen immediately by considering that the second row is the first two multiplied by 3.

Exercises

4.1. Let us consider the two following vectors$\in \mathbb{V}_3$:

$$\vec{u} = 2\vec{e_1} + 1\vec{e_2} - 2\vec{e_3}$$
$$\vec{v} = -8\vec{e_1} - 4\vec{e_2} + 8\vec{e_3}$$

expressed in the orthonormal basis.

Determine whether or not the two vectors are parallel.

4.2. Let us consider the three following vectors in an orthonormal basis of versors

$$\vec{v} = (1,0,1-k)$$
$$\vec{w} = (2,0,1)$$

1. Determine the value of k that makes the two vectors perpendicular;
2. Determine the value of k that makes the two vectors parallel.

4.3. Let us consider the following vectors

$$\vec{u} = (2, -3, 2)$$
$$\vec{v} = (3, 0, -1)$$
$$\vec{w} = (1, 0, 2).$$

Determine whether or not the vectors are linearly independent.

4.4. Let us consider the three following vectors in an orthonormal basis of versors

$$\vec{u} = (6, 2, 3)$$
$$\vec{v} = (1, 0, 1)$$
$$\vec{w} = (0, 0, 1)$$

1. Check whether or not the vectors are linearly independent;
2. State whether or not $\vec{u}$, $\vec{v}$, $\vec{w}$ are a basis (and justify the answer) and express, if possible, the vector $\vec{t} = (1, 1, 1)$ in the basis $\vec{u}, \vec{v}, \vec{w}$.

4.5. Let us consider the following vectors $\in \mathbb{V}_3$ expressed in the orthonormal basis:

$$\vec{u} = (1, 0, 1)$$
$$\vec{v} = (2, 1, 1)$$
$$\vec{w} = (3, 1, 2)$$
$$\vec{t} = (1, 1, 1).$$

1. Verify whether or not the vectors $\vec{u}$, $\vec{v}$, $\vec{w}$ are a basis of $\mathbb{V}_3$;
2. Express, if possible, the vector $\vec{t}$ in the basis identified by $\vec{u}$, $\vec{v}$, $\vec{w}$.

4.6. Check whether or not the vectors $\mathbf{u} = (4, 2, 12)$, $\mathbf{v} = (1, 1, 4)$, and $\mathbf{w} = (2, 2, 8)$ are linearly dependent and coplanar.

4.7. Determine, if they exist, the values of h that make $\vec{u}$ parallel to $\vec{v}$.

$$\vec{u} = (3h - 5)\vec{e_1} + (2h - 1)\vec{e_2} + 3\vec{e_3}$$
$$\vec{v} = \vec{e_1} - \vec{e_2} + 3\vec{e_3}$$

4.8. Determine, if they exists, the values of h that make $\vec{u}$, $\vec{v}$, and $\vec{w}$ coplanar.

$$\vec{u} = 2\vec{e_1} - \vec{e_2} + 3\vec{e_3}$$
$$\vec{v} = \vec{e_1} + \vec{e_2} - 2\vec{e_3}$$
$$\vec{w} = h\vec{e_1} - \vec{e_2} + (h - 1)\vec{e_3}$$

Chapter 5
Complex Numbers and Polynomials

5.1 Complex Numbers

As mentioned in Chap. 1, for a given set and an operator applied to its elements, if the result of the operation is still an element of the set regardless of the input of the operator, then the set is said closed with respect to that operator. For example it is easy to verify that $\mathbb{R}$ is closed with respect to the sum as the sum of two real numbers is certainly a real number. On the other hand, $\mathbb{R}$ is not closed with respect to the square root operation. More specifically, if a square root of a negative number has to be calculated the result is not determined and is not a real number. In order to represent these numbers Gerolamo Cardano in the sixteenth century introduced the concept of *Imaginary numbers*, see [11], by defining the imaginary unit j as the square root of -1: $j = \sqrt{-1}$. This means that the square roots of negative numbers can be represented.

Example 5.1. $\sqrt{-9} = j3$.

Imaginary numbers compose a set of numbers represented by the symbol $\mathbb{I}$. The basic arithmetic operations can be applied to imaginary numbers.

- sum: $ja + jb = j(a+b)$
- difference: $ja - jb = j(a-b)$
- product: $ja\,jb = -ab$
- division: $\frac{ja}{jb} = \frac{a}{b}$

Example 5.2. Let us consider the imaginary numbers $j2$ and $j5$. It follows that

$$j2 + j5 = j7$$
$$j2 - j5 = -j3$$
$$j2\,j5 = -10$$
$$\frac{j2}{j5} = \frac{2}{5}.$$

© Springer Nature Switzerland AG 2019

F. Neri, *Linear Algebra for Computational Sciences and Engineering*,
https://doi.org/10.1007/978-3-030-21321-3_5

It can be observed that while the set $\Im$ is closed with respect to sum and difference operations, it is not closed with respect to product and division. For example, the product of two imaginary numbers is a real number.

In addition, the zero has an interesting role. Since $0j = j0 = 0$, the zero is both a real and imaginary number, i.e. it can be seen as the intersection of the two sets.

Definition 5.1. A *complex number* is a number that can be expressed as $z = a + jb$ where a and b are real numbers and j is the imaginary unit. Furthermore, a is the *real part* of the complex number while jb is its imaginary part. The set of complex numbers is indicated with $\mathbb{C}$.

Example 5.3. The number $a + jb = 3 + j2$ is a complex number.

Complex numbers can be graphically represented as points in the so called *Gaussian plane* where real and imaginary parts are, respectively, the projections of the point on the real and imaginary axes.

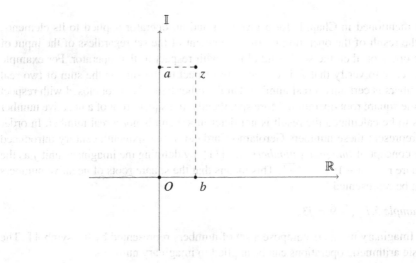

The representation of a complex number in a Gaussian plan must not be confused with the representation of a point in $\mathbb{R}^2$. Although in both cases there is a bijection between the set and the points of the plane, while the set $\mathbb{R}^2$ is the Cartesian product $\mathbb{R} \times \mathbb{R}$, the set of complex numbers $\mathbb{C}$ contains numbers that are the sum of real and imaginary parts.

If $z_1 = a + jb$ and $z_2 = c + jd$, the basic arithmetic operations can be applied to complex numbers.

- sum: $z_1 + z_2 = a + c + j(c + d)$
- product: $z_1 z_2 = (a + jb)(c + jd) = ac + jad + jbc - bd = ac - bd + j(ad + bc)$
- division: $\frac{z_1}{z_2} = \frac{a+jb}{c+jd} = \frac{(a+jb)(c-jd)}{(c+jd)(c-jd)} = \frac{ac+bd+j(bc-ad)}{c^2+d^2}$

Example 5.4. Let us consider the complex numbers $z_1 = 4 + j3$ and $z_2 = 2 - j5$. Let us compute their sum, product, and division.

$$z_1 + z_2 = 6 - j2$$
$$z_1 z_2 = 8 - j20 + j6 + 15 = 23 - j14$$
$$\frac{z_1}{z_2} = \frac{4+j3}{2-j5} = \frac{(4+j3)(2+j5)}{(2-j5)(2+j5)} = \frac{-7+j26}{29}.$$

From the division of complex numbers, the inverse of a complex number $z = a + jb$ can be easily verified as

$$\frac{1}{z} = \frac{1}{a+jb} = \frac{a-jb}{(a+jb)(a-jb)} = \frac{a-jb}{a^2+b^2}.$$

Example 5.5.

$$\frac{1}{2+j2} = \frac{2-j2}{8} = \frac{1}{8} - j\frac{1}{8}.$$

Definition 5.2. Let $z = a + jb$ be a complex number. The complex number $a - jb$ is said *conjugate* of z and is indicated with $\dot{z}$.

The following basic arithmetic operations can be defined for a complex number and its conjugate.

- sum: $z + \dot{z} = a + jb + a - jb = 2a$
- difference: $z - \dot{z} = a + jb - a + jb = j2b$
- product: $z\dot{z} = (a + jb)(a - jb) = a^2 - jab + jab - j^2b^2 = a^2 + b^2$

Example 5.6. Let us consider the following conjugate complex numbers $z = 3 + j2$ and $\dot{z} = 3 - j2$. It follows that

$$z + \dot{z} = 6$$
$$z - \dot{z} = j4$$
$$z\dot{z} = 9 + 4 = 13.$$

From the first basic arithmetic operations we can extract that if $z = a + jb$,

- $a = \frac{z+\dot{z}}{2}$
- $b = \frac{z-\dot{z}}{2}$.

Proposition 5.1. *Let z_1 and z_2 be two complex numbers ($z_1, z_2 \in \mathbb{C}$) and $\dot{z}_1$ and $\dot{z}_2$ be the corresponding conjugate numbers. It follows that*

$$\dot{z_1 z_2} = \dot{z}_1 \dot{z}_2.$$

Proof. Let z_1, z_2 be two complex numbers

$$z_1 = a + jb$$
$$z_2 = c + jd$$

and the corresponding complex numbers

$$\bar{z}_1 = a - jb$$
$$\bar{z}_2 = c - jd.$$

Let us calculate

$$z_1 z_2 = ac - bd + j(ad + bc)$$

and

$$\bar{z}_1 \bar{z}_2 = ac - bd - j(ad + bc).$$

Let us calculate now

$$\bar{z}_1 \bar{z}_2 = (a - jb)(c - jd) = ac - jad - jbc - bd = ac - bd - j(ad + bc). \square$$

Example 5.7. Let us consider the following two complex numbers

$$z_1 = 1 + j2$$
$$z_2 = 4 + j3.$$

Let us calculate

$$z_1 z_2 = 4 - 6 + j(3 + 8) = -2 + j11$$

and then

$$\overline{z_1 z_2} = -2 - j11.$$

Let us calculate now

$$\bar{z}_1 \bar{z}_2 = (1 - j2)(4 - j3) = -2 - j11.$$

An important characterization of complex numbers can be done on the basis of the definitions of ordered set and field given in Chap. 1. We know that the field of real number is the set $\mathbb{R}$ with its sum and product operations. In addition, we have defined the operations of sum and product over the set of complex numbers $\mathbb{C}$. It can easily be verified that the field properties are valid for sum and product over complex numbers. In a similar way we can define a field of imaginary numbers.

The real field $\mathbb{R}$ is *totally ordered*, i.e. the following properties are valid.

- $\forall x_1, x_2 \in \mathbb{R}$ with $x_1 \neq x_2$: either $x_1 \leq x_2$ or $x_2 \leq x_1$
- $\forall x_1, x_2 \in \mathbb{R}$: if $x_1 \leq x_2$ and $x_2 \leq x_1$ then $x_1 = x_2$
- $\forall x_1, x_2, x_3 \in \mathbb{R}$: if $x_1 \leq x_2$ and $x_2 \leq x_3$ then $x_1 \leq x_3$
- $\forall x_1, x_2, c \in \mathbb{R}$ with $c > 0$: if $x_1 \leq x_2$ then $x_1 + c \leq x_2 + c$
- $\forall x_1, x_2 \in \mathbb{R}$ with $x_1 > 0$ and $x_2 > 0$: $x_1 x_2 > 0$

Proposition 5.2. *The imaginary field $\mathbb{I}$ is not totally ordered.*

Proof. Let us prove that the property:

$$\forall x_1, x_2 \in \mathbb{I} \quad \text{with } x_1 > 0 \text{ and } x_2 > 0 : x_1 x_2 > 0$$

is not valid in the imaginary field.

Let us consider $x_1, x_2 \in \mathbb{I}$. Let $x_1 = jb$ and $x_2 = jd$. Let $b > 0$. and $d > 0$. Then, $x_1 x_2 = j^2 bd$ with $bd > 0$. Thus, $x_1 x_2 = -bd < 0$. Since one of the total order requirement is not respected the imaginary field is not totally ordered. $\square$

It follows that the complex field is not totally ordered. As an intuitive explanation of this fact, two complex numbers cannot be in general sorted in the same way as there is no explicit criterion to sort two points in a plane.

The representation of a complex number as $z = a + jb$ is said in *rectangular coordinates*. A complex number can have an equivalent representation using a system of *polar coordinates*. More specifically, from a complex number $z = a + jb$, we can represent the same number in terms of radius (or module) ρ and phase θ, and indicate as $(\rho; \angle \theta)$, where

$$\rho = \sqrt{a^2 + b^2}$$

$$\theta = \begin{cases} \arctan\left(\frac{b}{a}\right) & \text{if } a > 0 \\ \arctan\left(\frac{b}{a}\right) + \pi & \text{if } a < 0. \end{cases}$$

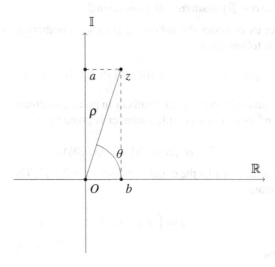

Example 5.8. Let us consider the two following complex numbers: $z_1 = 2 + j3$ and $z_2 = -4 + j8$. Let us represent these two numbers in polar coordinates. As for z_1, the radius $\rho_1 = \sqrt{2^2 + 3^2} = \sqrt{13}$ and the phase $\theta_1 = \arctan\left(\frac{3}{2}\right) = 56,3°$. As for z_2, the radius $\rho_2 = \sqrt{-4^2 + 8^2} = \sqrt{80}$ and the phase $\theta_1 = \arctan\left(\frac{8}{-4} + 180°\right) = 116,6°$.

Let us now compute sum and product of these two complex numbers. The sum is $z_1 + z_2 = -2 + j11$. The product can be computed from the polar representation of the numbers: $z_1 z_2 = \left(\left(\sqrt{13}\sqrt{80}\right); \angle 172,9°\right)$.

From simple trigonometric considerations on the geometric representation of a complex number we can derive that

$$a = \rho \cos \theta$$
$$b = \rho \sin \theta.$$

Thus, we can represent a complex number as

$$z = a + jb = \rho \left(\cos \theta + j \sin \theta \right).$$

Let us consider two complex numbers $z_1 = \rho_1 \left(\cos \theta_1 + j \sin \theta_1 \right)$, $z_2 = \rho_2 \left(\cos \theta_2 + j \sin \theta_2 \right)$ and compute their product.

$$z_1 z_2 = \rho_1 \left(\cos \theta_1 + j \sin \theta_1 \right) \rho_2 \left(\cos \theta_2 + j \sin \theta_2 \right) =$$
$$= \rho_1 \rho_2 \left(\cos \theta_1 + j \sin \theta_1 \right) \left(\cos \theta_2 + j \sin \theta_2 \right) =$$
$$= \rho 1 \rho_2 \left(\cos \theta_1 \cos \theta_2 + j \cos \theta_1 \sin \theta_2 + j \sin \theta_1 \cos \theta_2 - \sin \theta_1 \sin \theta_2 \right) =$$
$$= \rho_1 \rho_2 \left(\cos \left(\theta_1 + \theta_2 \right) + j \sin \left(\theta_1 + \theta_2 \right) \right).$$

This means that if two complex numbers are represented in polar coordinates, in order to computer their product, it is enough to calculate the product of their modules and to sum their phases. It is here reminded that $\cos \left(\alpha - \beta \right) = \cos \alpha \cos \beta + sin \alpha sin \beta$ and $\sin \left(\alpha + \beta \right) = \sin \alpha cos \beta + \cos \alpha \sin \beta$.

Example 5.9. Let us consider the following complex numbers $z_1 = \left(5; \angle 30° \right)$ and $z_2 = \left(2; \angle 45° \right)$. It follows that

$$z_1 z_2 = \left(10; \angle 75° \right) = 10 \left(\cos \left(75° \right) + j \sin \left(75° \right) \right).$$

From the product of two complex numbers in polar coordinates , it immediately follows that the nth power of a complex number is given by

$$z^n = \rho^n \left(\cos \left(n\theta \right) + j \sin \left(n\theta \right) \right).$$

Example 5.10. Let us consider the complex number $z = 2 + j2$. The complex number z in polar coordinates is

$$z = \left(\sqrt{8}; \angle 45° \right).$$

Let us calculate

$$z^4 = 64 \left(\cos \left(180° \right) + j \sin \left(180° \right) \right) = -64.$$

From this formula, the nth root can be derived. More specifically, let us suppose that $\left(z_1 \right)^n = z_2$. If $z_1 = \rho_1 \left(\cos \theta_1 + j \sin \theta_1 \right)$ and $z_2 = \rho_2 \left(\cos \theta_2 + j \sin \theta_2 \right)$ then

$$z_2 = \left(\rho_1 \left(\cos \theta_1 + j \sin \theta_1 \right) \right)^n \Rightarrow$$
$$\Rightarrow \rho_2 \left(\cos \theta_2 + j \sin \theta_2 \right) = \rho_1^n \left(\cos \left(n\theta_1 \right) + j \sin \left(n\theta_1 \right) \right).$$

From these formulas we can write

$$\begin{cases} \rho_2 = \rho_1^n \\ \cos\theta_2 = \cos(n\theta_1) \\ \sin\theta_2 = \sin(n\theta_1) \end{cases} \Rightarrow$$

$$\Rightarrow \begin{cases} \rho_1 = \sqrt[n]{\rho_2} \\ \theta_1 = \frac{\theta_2 + 2k\pi}{n} \\ \theta_1 = \frac{\theta_2 + 2k\pi}{n}. \end{cases}$$

where $k \in \mathbb{N}$. Thus, the formula for the nth root is

$$\sqrt[n]{z} = \sqrt[n]{\rho_2}\left(\cos\left(\frac{\theta_2 + 2k\pi}{n} \right) + j\sin\left(\frac{\theta_2 + 2k\pi}{n} \right) \right).$$

In a more compact way and neglecting $2k\pi$, if $z = \rho\left(\cos\theta + j\sin\theta\right)$, then

$$\sqrt[n]{z} = \sqrt[n]{\rho}\left(\cos\left(\frac{\theta}{n} \right) + j\sin\left(\frac{\theta}{n} \right) \right).$$

Example 5.11. Let us consider the complex number $z = (8; \angle 45°)$ and calculate

$$\sqrt[3]{z} = (2; \angle 15°) = 2\left(\cos(15°) + j\sin(15°)\right).$$

An alternative (and equivalent) formulation of the nth power of a complex number is the so-called De Moivre formula.

Theorem 5.1. De Moivre's Formula. *For every real number $\theta \in \mathbb{R}$ and integer $n \in \mathbb{N}$,*

$$\left(\cos\theta + j\sin\theta\right)^n = \cos(n\theta) + j\sin(n\theta).$$

Finally, the following Theorem broadens the interpretation of the concept of complex numbers.

Theorem 5.2. Euler's Formula. *For every real number $\theta \in \mathbb{R}$,*

$$e^{j\theta} = \cos\theta + j\sin\theta,$$

where e *is the Euler's number 2.71828, base of natural logarithm.*

The Euler formula is an important result that allows to connect exponential functions to sinusoidal functions to complex numbers by means of their polar representation, see [12]. A proof of the Euler's Formula is reported in Appendix B.

Example 5.12. For $\theta = 45° = \frac{\pi}{4}$,

$$e^{j\frac{\pi}{4}} = \cos 45° + j\sin 45° = \frac{\sqrt{2}}{2} + j\frac{\sqrt{2}}{2}.$$

As a remark, the De Moivre's formula is anterior with respect to the Euler's formula, and thus this is not the original proof. Nonetheless, it can be seen as an extension of the Euler's formula and, thus, may appear as its logical consequence, see [13].

Proposition 5.3. *Let* $z = \rho e^{j\theta} = \rho\left(\cos\theta + j\sin\theta\right)$ *be a complex number. It follows that*

$$jz = \rho\left(\cos\left(\theta + 90°\right) + j\sin\left(\theta + 90°\right)\right) = \rho e^{j(\theta + 90°)}.$$

Proof. If we multiply z by j we obtain:

$$
\begin{aligned}
jz &= j\rho e^{j\theta} = j\rho\left(\cos\theta + j\sin\theta\right) = \\
&= \rho\left(j\cos\theta - \sin\theta\right) = \rho\left(-\sin\theta + j\cos\theta\right) = \\
&= \rho\left(\sin\left(-\theta\right) + j\cos\left(\theta\right)\right) = \rho\left(\cos\left(\theta + 90°\right) + j\sin\left(\theta + 90°\right)\right) = \\
&= \rho e^{j(\theta + 90°)}. \square
\end{aligned}
$$

This means that the multiplication of a complex number by the imaginary unit j can be interpreted as a $90°$ rotation of the complex number within the Gaussian plane.

Example 5.13. Let us consider the complex number $z = 2 + j2$ and multiply is by j:

$$jz = j2 - 2 = -2 + j2.$$

In the Gaussian plane this means

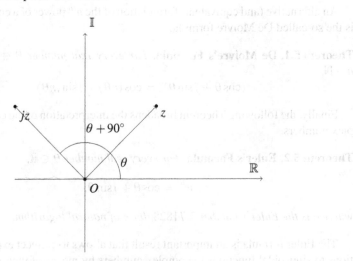

Example 5.14. Let us consider the complex number $z = \rho\angle 15°$, which can be written as $5\left(\cos 15° + j\sin 15°\right)$. Let us multiply this number by j:

$$
\begin{aligned}
jz &= j5\left(\cos 15° + j\sin 15°\right) = 5\left(-\sin 15° + j\cos 15°\right) = \\
&= 5\left(\sin -15° + j\cos 15°\right) = 5\left(\cos 105° + j\sin 105°\right) = \\
&\quad 5\left(\cos\left(90° + 15°\right) + j\sin\left(90° + 15°\right)\right) = 5e^{j(15° + 90°)}
\end{aligned}
$$

Finally, if we calculate Euler's formula for $\theta = \pi$ we obtain:

$$e^{j\pi} + 1 = 0,$$

that is the so called **Euler's identity**. This equation is historically considered as an example of mathematical beauty as it contains the basic numbers of mathematics, i.e. $0, 1, e, \pi, j$, as well as the basic operations, i.e. sum, multiplication, and exponentiation, all these elements appearing only once in the equation.

5.2 Complex Vectors, Matrices and Systems of Linear Equation

All the theory and examples in the previous chapters refer to real numbers. However, it is worth mentioning that the algebra analysed up to this point can be straightforwardly extended to the complex case. This section elaborates on this statement by showing several examples.

Definition 5.3. Let $\mathbb{C}$ be the complex set and $\mathbb{C}^n = \mathbb{C} \times \mathbb{C} \times \ldots \times \mathbb{C}$ the Cartesian product obtained by the composition of the complex set calculated n times.

A generic element $\mathbf{u} \in \mathbb{C}^n$ is named *complex vector* and is an n-tuple of the type

$$\mathbf{u} = (a_1 + jb_1, a_2 + jb_2, \ldots, a_n + jb_n)$$

where each component $a_k + jb_k$ is a complex number.

Example 5.15. The following is a complex vector of $\mathbb{C}^3$:

$$\mathbf{u} = (3 - j2, 4, 1 + j7).$$

By using the operation of sum and product of complex numbers we can define the scalar product of complex vectors.

Definition 5.4. Let $\mathbf{u} = (u_1, u_2, \ldots, u_n)$ and $\mathbf{v} = (v_1, v_2, \ldots, v_n)$ be two vectors of $\mathbb{C}^n$. The *scalar product* $\mathbf{uv}$ is

$$\mathbf{uv} = \sum_{j=1}^{n} u_j v_j. \tag{5.1}$$

Example 5.16. Let us consider the following complex vectors of $\mathbb{C}^3$:

$$\mathbf{u} = (1, 1 + j2, 0)$$
$$\mathbf{v} = (3 - j, j5, 6 - j2).$$

The scalar product is

$$\mathbf{uv} = 3 - j + -10 + j5 + 0 = -7 + j4.$$

Similarly, we may think about a matrix whose elements are complex numbers.

Definition 5.5. A *complex matrix* $\mathbf{C}$ composed of m rows and n columns is a table of the type

$$\mathbf{C} = \begin{pmatrix} c_{1,1} & c_{1,2} & \cdots & c_{1,n} \\ c_{2,1} & c_{2,2} & \cdots & c_{2,n} \\ \cdots & \cdots & \cdots & \cdots \\ c_{m,1} & c_{m,2} & \cdots & c_{m,n} \end{pmatrix}$$

where $c_{i,j}$ is a complex number:

$$c_{i,j} = a_{i,j} + jb_{i,j}.$$

The set of all the possible $m \times n$ complex matrices is indicated with $\mathbb{C}_{m,n}$.

Example 5.17. The follow matrix is a complex matrix

$$\mathbf{C} = \begin{pmatrix} 3 & -1+j3 & j5 \\ 1+j8 & -9 & 0 \\ 0 & 2 & j12 \end{pmatrix}.$$

For complex matrices can be summed like matrices composed of real numbers by adding the complex elements one by one.

Example 5.18. Let us consider the following matrices:

$$\mathbf{C} = \begin{pmatrix} 3 & -1+j3 & j5 \\ 1+j8 & -9 & 0 \\ 0 & 2 & j12 \end{pmatrix}.$$

and

$$\mathbf{D} = \begin{pmatrix} 1+j2 & 0 & -j1 \\ 4-j8 & 4 & 0 \\ 0 & -2-j3 \end{pmatrix}.$$

The sum of these matrices is

$$\mathbf{C} + \mathbf{D} = \begin{pmatrix} 4+j2 & -1+j3 & j4 \\ 5 & -5 & 0 \\ 0 & 0 & -j9 \end{pmatrix}.$$

From the definition of scalar product we can easily extend also the matrix product to complex matrices.

Definition 5.6. Let us consider the complex matrices $\mathbf{C} \in \mathbb{C}_{m,r}$ and $\mathbf{D} \in \mathbb{C}_{r,n}$. Let us represent $\mathbf{C}$ as a vector of row vectors and $\mathbf{D}$ as a vector of column vectors:

$$\mathbf{C} = \begin{pmatrix} \mathbf{c_1} \\ \mathbf{c_2} \\ \cdots \\ \mathbf{c_m} \end{pmatrix}$$

and

$$D = \left(\mathbf{d}^1 \ \mathbf{d}^2 \ \ldots \ \mathbf{d}^n \right)$$

The product matrix $\mathbf{CD}$ is

$$\mathbf{CD} = \begin{pmatrix} \mathbf{c}_1\mathbf{d}^1 & \mathbf{c}_1\mathbf{d}^2 & \ldots\mathbf{c}_1\mathbf{d}^n \\ \mathbf{c}_2\mathbf{d}^1 & \mathbf{c}_2\mathbf{d}^2 & \ldots\mathbf{c}_2\mathbf{d}^n \\ \ldots & \ldots & \ldots \quad \ldots \\ \mathbf{c}_m\mathbf{d}^1 & \mathbf{c}_m\mathbf{d}^2 & \ldots\mathbf{c}_m\mathbf{d}^n \end{pmatrix}$$

where $\mathbf{c}_i\mathbf{d}^j$ is the scalar product of two complex vectors, i.e. $\mathbf{c}_i$ and $\mathbf{d}^j$.

Example 5.19. Let us consider the following two complex matrices:

$$\mathbf{C} = \begin{pmatrix} 2+j & 0 & 5-j3 \\ 1 & 1 & 0 \\ j5 & 3+j2 & 0 \end{pmatrix}$$

and

$$\mathbf{D} = \begin{pmatrix} 1 & 0 & 1 \\ 0 & 1 & 0 \\ j5 & 3+j2 & 0 \end{pmatrix}$$

The product matrix is

$$\mathbf{CD} = \begin{pmatrix} 17+j26 & 21+j & 2+j \\ 1 & 1 & 1 \\ j5 & 3+j2 & j5 \end{pmatrix}.$$

All the properties valid for sum and product of matrices of real numbers can be straightforwardly extended to complex numbers. Similarly we can extend to complex numbers the definition of determinant according to the description in Sect. 2.4 and inverse matrix according to the theory outlined in Sect. 2.5.

Example 5.20. Let us consider again the complex matrix

$$\mathbf{C} = \begin{pmatrix} 2+j & 0 & 5-j3 \\ 1 & 1 & 0 \\ j5 & 3+j2 & 0 \end{pmatrix}.$$

The determinant of the matrix is $\det(\mathbf{C}) = 6 - j24 \neq 0$. Thus the matrix $\mathbf{C}$ is invertible. By applying the formula to invert the matrix

$$\mathbf{C}^{-1} = \frac{1}{\det(\mathbf{C})}\text{adj}(\mathbf{C})$$

we can calculate the inverse of a complex matrix:

$$\mathbf{C}^{-1} = \frac{1}{6 - j24} \begin{pmatrix} 0 & 21 + j & -5 + j3 \\ 0 & -15 - j25 & 5 - j3 \\ 3 - j3 & -4 - j7 & 2 + j \end{pmatrix}.$$

Example 5.21. The complex matrix

$$\mathbf{C} = \begin{pmatrix} 2 + j & 0 & 2 + j \\ 1 & 1 & 2 \\ j5 & 3 + j2 & 3 + j7 \end{pmatrix}$$

is singular. The third column is sum of the other two. The matrix is not invertible.

We can extend to complex numbers also the theory of systems of linear equations.

Definition 5.7. Let us consider m (with $m > 1$) linear equations in the variables $x_1, x_2, \ldots, x_n$. These equations compose a *system of complex linear equations* indicated as:

$$\begin{cases} c_{1,1}x_1 + c_{1,2}x_2 + \ldots + c_{1,n}x_n = d_1 \\ c_{2,1}x_1 + c_{2,2}x_2 + \ldots + c_{2,n}x_n = d_2 \\ \ldots \\ c_{n,1}x_1 + c_{n,2}x_2 + \ldots + c_{n,n}x_n = d_n \end{cases}$$

where $\forall i, j \; c_{i,j} \in \mathbb{C}$ and $d_i \in \mathbb{C}$.

For systems of complex linear equations, the same theorems valid for real linear equations are still valid. For examples, Cramer's and Rouchè-Capelli theory is valid as well as Gaussian elimination, LU factorisation and iterative methods. The following examples clarify this fact.

Example 5.22. Let us solve by applying Cramer's method the following system of linear equations.

$$\begin{cases} 2x - 4jy = 0 \\ (1 + j)x + y = 2. \end{cases}$$

The determinant of the associated incomplete matrix is $-2 + j4$. Let us apply Cramer's method. The solution is

$$x = \frac{\det \begin{pmatrix} 0 & -j4 \\ 2 & 1 \end{pmatrix}}{-2 + j4} = 1.6 - j0.8$$

and

$$y = \frac{\det \begin{pmatrix} 2 & 0 \\ (1+j) & 2 \end{pmatrix}}{-2+j4} = -0.4 - j0.8$$

respectively.

Example 5.23. Let us consider the following system of linear equations

$$\begin{cases} 4x + 3jy + z = 0 \\ (2+j)x - 2y + z = 0 \\ (6+j)x + (-2+j3)y + 2z = 0. \end{cases}$$

We can easily see that the third equation is the sum of the first two. Since the system is homogeneous, by applying the Rouchè-Capelli Theorem, the rank $\rho_A = \rho_{A^c} = 2 < n = 3$. Thus the system has ∞^1 solutions.

Let us find the general solution to the system by placing $x = \alpha$. It follows that

$$z = -4\alpha - j3y$$

and

$$(2+j)\alpha - 2y - 4\alpha - j3y = 0$$

which yields to

$$y = \frac{(-2+j)}{(2+j3)}\alpha = \left(-\frac{1}{13} + j\frac{8}{13}\right)\alpha$$

and

$$z = -4\alpha - j3\frac{(-2+j)}{(2+j3)}\alpha = \left(-\frac{28}{13} + j\frac{3}{13}\right)\alpha.$$

Example 5.24. Let us solve by Gaussian elimination the following system of linear equations

$$\begin{cases} x - jy + 2z = 1 \\ 4x - 10y - z = 2 \\ 2x + 2y + 10z = 4 \end{cases}$$

Starting from the complete matrix

$$\mathbf{A}^c = \begin{pmatrix} 1 & -j & 2 & 1 \\ 4 & -10 & -1 & 2 \\ 2 & 2 & 10 & 4 \end{pmatrix}$$

Let us apply the row transformations

$$\mathbf{r}_2 = \mathbf{r}_2 - 4\mathbf{r}_1$$
$$\mathbf{r}_3 = \mathbf{r}_3 - 2\mathbf{r}_1$$

and obtain

$$\mathbf{A}^c = \begin{pmatrix} 1 & -j & 2 & 1 \\ 0 & (-10+j4) & -9 & -2 \\ 0 & (2+j2) & 6 & 2 \end{pmatrix}.$$

We now need to apply the following row transformation

$$\mathbf{r}_3 = \mathbf{r}_3 - \frac{2+j2}{-10+j4}\mathbf{r}_2 = \mathbf{r}_3 - \left(-\frac{3}{29} - j\frac{7}{29}\right)\mathbf{r}_2 = \mathbf{r}_3 + \left(\frac{3}{29} + j\frac{7}{29}\right)\mathbf{r}_2$$

which leads to

$$\mathbf{A}^c = \begin{pmatrix} 1 & -j & 2 & 1 \\ 0 & (-10+j4) & -9 & -2 \\ 0 & 0 & (5.068 - j2.172) & (1.793 - j0.483) \end{pmatrix}.$$

The solution of this triangular system of linear equation is

$$x = 0.336 - j0.006$$
$$y = -0.101 + j0.002$$
$$z = 0.333 + j0.048$$

From the final $\mathbf{A}^c$ we can extract the matrix $\mathbf{U}$ while the matrix $\mathbf{L}$ is

$$\mathbf{L} = \begin{pmatrix} 1 & 0 & 0 \\ 4 & 1 & 0 \\ 2 & \left(-\frac{3}{29} - j\frac{7}{29}\right) & 1 \end{pmatrix}$$

Example 5.25. Let us consider the following matrix

$$\mathbf{A} = \begin{pmatrix} 1 & 3j & (4+j) \\ j2 & (-6+j5) & (-1+j8) \\ (1+j) & (-3+j3) & (4+j5) \end{pmatrix}.$$

The matrix $\mathbf{A}$ can be factorized as $\mathbf{LU}$ where

$$\mathbf{L} = \begin{pmatrix} 1 & 0 & 0 \\ j2 & 1 & 0 \\ (1+j) & 0 & 1 \end{pmatrix}$$

and

$$\mathbf{U} = \begin{pmatrix} 1 & j3 & (4+j) \\ 0 & j5 & 1 \\ 0 & 0 & 1 \end{pmatrix}.$$

5.3 Complex Polynomials

5.3.1 Operations of Polynomials

Definition 5.8. Let $n \in \mathbb{N}$ and $a_0, a_1, \ldots, a_n \in \mathbb{C}$. The function $p(z)$ in the complex variable $z \in \mathbb{C}$ defined as

$$p(z) = a_0 + a_1 z + a_2 z^2 + \ldots + \ldots a_n z^n = \sum_{k=0}^{n} a_k z^k$$

is said *complex polynomial* in the coefficients a_k and complex variable z. The order n of the polynomial is the maximum value of k corresponding to a non-null coefficient a_k.

Example 5.26. The following function

$$p(z) = 4z^4 - 5z^3 + z^2 - 6$$

is a polynomial.

Definition 5.9. Let $p(z) = \sum_{k=0}^{n} a_k z^k$ be a polynomial. If $\forall k \in \mathbb{N}$ with $k \leq n: a_k = 0$, the polynomial is said *null polynomial*.

Definition 5.10. Let $p(z) = \sum_{k=0}^{n} a_k z^k$ be a polynomial. If $\forall k \in \mathbb{N}$ with $0 < k \leq n:$ $a_k = 0$ and $a_0 \neq 0$, the polynomial is said *constant polynomial*.

Definition 5.11. Identity Principle. Let $p_1(z) = \sum_{k=0}^{n} a_k z^k$ and $p_2(z) = \sum_{k=0}^{n} b_k z^k$ be two complex polynomials. The two polynomials are said *identical* $p_1(z) = p_2(z)$ if and only if the following two conditions are both satisfied:

* the order n of the two polynomials is the same
* $\forall k \in \mathbb{N}$ with $k \leq n: a_k = b_k$.

Example 5.27. Let $p_1(z) = \sum_{k=0}^{n} a_k z^k$ and $p_2(z) = \sum_{k=0}^{m} b_k z^k$ be two complex polynomials with $m < n$. The two polynomials are identical if and only if

* $\forall k \in \mathbb{N}$ with $k \leq m: a_k = b_k$
* $\forall k \in \mathbb{N}$ with $m < k \leq n: a_k = 0$

Definition 5.12. Let $p_1(z) = \sum_{k=0}^{n} a_k z^k$ and $p_2(z) = \sum_{k=0}^{m} b_k z^k$ be two polynomials of orders n and m, respectively. The *sum polynomial* is the polynomial $p_3(z) = \sum_{k=0}^{n} a_k z^k + \sum_{k=0}^{m} b_k z^k$.

Example 5.28. Let us consider the polynomials

$$p_1(z) = z^3 - 2z$$
$$p_2(z) = 2z^3 + 4z^2 + 2z + 2.$$

The sum polynomial is

$$p_3(z) = 3z^3 + 4z^2 + 2.$$

Proposition 5.4. *Let $p_1(z) = \sum_{k=0}^{n} a_k z^k$ and $p_2(z) = \sum_{k=0}^{m} b_k z^k$ be two polynomials of orders n and m, respectively.*

• *If $m \neq n$, the order of the sum polynomial $p_3(z) = p_1(z) + p_2(z)$ is the greatest among n and m.*

• *If $m = n$, the order of the sum polynomial $p_3(z) = p_1(z) + p_2(z)$ is $\leq n$.*

Example 5.29. To clarify the meaning of this proposition let us consider the following polynomials: $p_1(z) = 5z^3 + 3z - 2$ and $p_2(z) = 4z^2 + z + 8$. It is obvious that the sum polynomial is of the same order of the greatest among the orders of the two polynomials, i.e. 3 and 2. Hence the sum polynomial is of order 3.

On the other hand, if we consider two polynomials of the same order such as $p_1(z) = 5z^3 + 3z - 2$ and $p_2(z) = -5z^3 + z + 8$, their sum results into a polynomial of the first order. The sum polynomial could have a lower order with respect to the starting ones.

Definition 5.13. Let $p_1(z) = \sum_{k=0}^{n} a_k z^k$ and $p_2(z) = \sum_{k=0}^{m} b_k z^k$ be two polynomials of orders n and m, respectively. The *product polynomial* is a polynomial $p_3 = \left(\sum_{k=0}^{n} a_k z^k \right) \left(\sum_{k=0}^{m} b_k z^k \right)$.

Proposition 5.5. *Let $p_1(z) = \sum_{k=0}^{n} a_k z^k$ and $p_2(z) = \sum_{k=0}^{m} b_k z^k$ be two polynomials of orders n and m, respectively. The order of the product polynomial $p_3(z) = p_1(z) p_2(z)$ is $n + m$.*

Example 5.30. The following two polynomials

$$p_1(z) = z^2 - 2z$$
$$p_2(z) = 2z + 2.$$

are of order 2 and 1 respectively.

The product polynomial

$$p_3(z) = 2z^3 - 4z^2 + 2z^2 - 4z = 2z^3 - 2z^2 - 4z$$

is of order $2 + 1 = 3$.

Theorem 5.3. Euclidean Division. *Let $p_1(z) = \sum_{k=0}^{n} a_k z^k$ and $p_2(z) = \sum_{k=0}^{m} b_k z^k$ be two polynomials of orders n and m, respectively and $p_2(z) \neq 0$. The division of polynomials $p_1(z)$ (dividend) by $p_2(z)$ (divisor) results into a polynomial*

$$p_3(z) = \frac{p_1(z)}{p_2(z)} = q(z) + d(z)$$

which can be rewritten as

$$p_1(z) = p_2(z) q(z) + r(z)$$

where $q(z)$ is said polynomial quotient *and $r(z)$ is said polynomial remainder. The order r or the polynomial remainder is strictly less than the order m of the divisor $p_2(z)$:*

$$r < m.$$

Example 5.31. Let us consider the following polynomials

$$p_1(z) = z^2 - z + 5$$
$$p_2(z) = z - 4.$$

It follows that

$$p_1(z) = p_2(z)q(z) + r(z)$$

where

$$q(z) = z + 3$$
$$r(z) = 17.$$

We have that the order n of $p_1(z)$ is 2, the order m of $p_2(z)$ is 1 and the order r of the remainder is $< m$, i.e. it is zero.

The following theorem shows that polynomial quotient and remainder are unique for a given pair of polynomials $p_1(z)$ and $p_2(z)$.

Theorem 5.4. Uniqueness of Polynomial Quotient and Remainder. *Let $p_1(z) = \sum_{k=0}^{n} a_k z^k$ and $p_2(z) = \sum_{k=0}^{m} b_k z^k$ be two complex polynomials with $m < n$. $\exists!$ complex polynomial $q(z)$ and $\exists!$ complex polynomial $r(z)$ having order $r < m | p_1(z) = p_2(z)q(z) + r(z)$.*

Proof. By contradiction, let us assume that two pairs of complex polynomials $q(z)$, $r(z)$ and $q_0(z)$, $r_0(z)$ exist such that

$$p_1(z) = p_2(z)q(z) + r(z)$$
$$p_1(z) = p_2(z)q_0(z) + r_0(z)$$

where the order of $r(z)$, r, and the order of $r_0(z)$, r_0, are both $< m$.
Thus, the following equality is verified:

$$0 = p_2(z)(q(z) - q_0(z)) + (r(z) - r_0(z)) \Rightarrow$$
$$\Rightarrow r_0(z) - r(z) = p_2(z)(q(z) - q_0(z)).$$

From the hypothesis we know that the order of $p_2(z)$ is m. Let us name l the order of $(q(z) - q_0(z))$. The order of $p_2(z)(q(z) - q_0(z))$ is $m + l \geq m$. Since the order of $r_0(z) - r(z)$ can be at most $m - 1$, the equation above violates the identity principle of two polynomials. Thus, we reached a contradiction as the polynomial quotient and remainder must be unique. $\square$

Example 5.32. Let us consider a special case of the Theorem 5.4 where the order of $p_1(z)$ is n while the order of $p_2(z)$ is 1. More specifically, $p_2(z) = (z - \alpha)$.

From Theorem 5.4 we know that $\exists! q(z)$ and $\exists! r(z)$ such that $p_1(z) = (z - \alpha) q(z) + r(z)$.

The order r of the polynomial $r(z) <$ than the order of $p_2(z)$, that is 1. Thus the order of the polynomial $r(z)$ is 0, i.e. the polynomial $r(z)$ is either the constant or the null polynomial.

Definition 5.14. Let $p_1(z) = \sum_{k=0}^{n} a_k z^k$ and $p_2(z) = \sum_{k=0}^{m} b_k z^k$ be two complex polynomials. The polynomial $p_1(z)$ is said to be *divisible by* $p_2(z)$ if $\exists!$ polynomial $q(z)$ such that $p_1(z) = p_2(z) q(z)$ (with $r(z) = 0 \ \forall z$).

In the case $p_2(z) = (z - \alpha)$, a polynomial $p_1(z)$ is divisible by $p_2(z)$ if $\exists!$ polynomial $q(z)$ such that $p_1(z) = (z - \alpha) q(z)$ (with $r(z) = 0 \ \forall z$).

It must be observed that the null polynomial is divisible by all polynomials while all polynomials are divisible by a constant polynomial.

Theorem 5.5. Polynomial Remainder Theorem or Little Bézout's Theorem. *Let* $p(z) = \sum_{k=0}^{n} a_k z^k$ *be a complex polynomial having order* $n \geq 1$. *The polynomial remainder of the division of* $p(z)$ *by* $(z - \alpha)$ *is* $r(z) = p(\alpha)$.

Proof. From the Euclidean division in Theorem 5.3 we know that

$$p(z) = (z - \alpha) q(z) + r(z)$$

with the order of $r(z)$ less than the order of $(z - \alpha)$. Hence, the polynomial remainder $r(z)$ has order 0, i.e. the polynomial remainder $r(z)$ is a constant. To highlight that the polynomial remainder is a constant, let us indicate it with r. Hence, the Euclidean division is

$$p(z) = (z - \alpha) q(z) + r.$$

Let us calculate the polynomial $p(z)$ in α

$$p(\alpha) = (\alpha - \alpha) q(\alpha) + r = r.$$

Hence, $r = p(\alpha)$. $\quad\square$

5.3.2 Roots of Polynomials

Definition 5.15. Let $p(z)$ be a polynomial. The values of z such that $p(z) = 0$ are said *roots* or *solutions* of the polynomial.

Corollary 5.1. Ruffini's Theorem. *Let* $p(z) = \sum_{k=0}^{n} a_k z^k$ *be a complex polynomial having order* $n \geq 1$. *The polynomial* $p(z)$ *is divisible by* $(z - \alpha)$ *if and only if* $p(\alpha) = 0$ (α *is a root of the polynomial*).

Proof. If $p(z)$ is divisible by $(z - \alpha)$ then we may write

$$p(z) = (z - \alpha) q(z).$$

Thus, for $z = \alpha$ we have

$$p(\alpha) = (\alpha - \alpha)q(\alpha) = 0.\ \square$$

If α is a root of the polynomial, then $p(\alpha) = 0$. Considering that

$$p(z) = (z - \alpha)q(z) + r(z)$$

and for the little Bézout's Theorem $p(\alpha) = r$, it follows that $r = 0$ and that

$$p(z) = (z - \alpha)q(z)$$

that is $p(z)$ is divisible by $(z - \alpha)$. $\square$

Example 5.33. Let us consider the division of polynomials

$$\frac{(-z^4 + 3z^2 - 5)}{(z + 2)}.$$

It can be easily verified that the polynomial reminder of this division is

$$r = p(-2) = -9.$$

On the contrary, in the case of the division of

$$\frac{(-z^4 + 3z^2 + 4)}{(z + 2)}$$

we obtain

$$r = p(-2) = 0.$$

In the latter case the two polynomials are divisible.

A practical implication of Polynomial Reminder and Ruffini's Theorems is the so called **Ruffini's rule** that is an algorithm for dividing a polynomial $p(z) = \sum_{k=0}^{n} a_k z^k$ by a first order polynomial $(z - \alpha)$. Obviously, for the Euclidean division and Polynomial Remainder Theorem it results that

$$p(z) = (z - \alpha)q(z) + r$$

where r is a constant and $q(z) = \sum_{k=0}^{n-1} b_k z^k$.

The algorithm consists of the following steps. At the beginning the coefficients are arranged as

$$\begin{array}{c|c|c|c|c||c} & a_n & a_{n-1} & \dots & a_1 & a_0 \\ \hline \alpha & & & & & \end{array}$$

and the coefficient corresponding to the maximum power in the polynomial a_n is initialized in the second row. Let us rename it as b_{n-1} as it is the coefficient of the maximum power of $q(z)$:

$$
\begin{array}{c|c|ccc||c}
 & a_n & a_{n-1} & \ldots & a_1 & a_0 \\
\alpha & & & & & \\
\hline
 & b_{n-1} = a_n & & & & \\
\end{array}
$$

From this point, each coefficient b_k of $q(z)$ can be recursively calculated as $b_k = a_{k+1} + b_{k+1}\alpha$ for $k = n-1, n-1, \ldots, 0$:

$$
\begin{array}{c|c|c|c|c||c}
 & a_n & a_{n-1} & \ldots & a_1 & a_0 \\
\alpha & & & & & \\
\hline
 & b_{n-1} = a_n & b_{n-2} = a_{n-1} + b_{n-1}\alpha & \ldots & b_0 = a_1 + b_1\alpha & \\
\end{array}
$$

Finally, the remainder r is $r = a_0 + b_0\alpha$.

Example 5.34. Let us consider the division of the polynomial $\left(-z^4 + 3z^2 - 5\right)$ by $(z+2)$. By applying Ruffini's rule we obtain

$$
\begin{array}{c|c|c|c|c||c}
 & -1 & 0 & 3 & 0 & -5 \\
-2 & & & & & \\
\hline
 & -1 & 2 & -1 & 2 & \\
\end{array}
$$

Hence the polynomial quotient is $\left(-z^3 + 2z^2 - z + 2\right)$ and the polynomial reminder as expected from the Polynomial Remainder Theorem is $r = a_0 + b_0\alpha = -9$.

Theorem 5.6. Fundamental Theorem of Algebra. *If $p(z) = \sum_{k=0}^{n} a_k z^k$ is a complex polynomial having order $n \geq 1$, then this polynomial has at least one root.*

Obviously, for the Ruffini's Theorem if $\alpha \in \mathbb{C}$ is the root of the polynomial, then $p(z)$ is divisible by $(z - \alpha)$. A proof of this theorem is given in Appendix B.

As a first observation of the Fundamental Theorem of Algebra, since real numbers are special case of complex numbers, i.e. complex numbers with null imaginary part, the theorem is valid for real polynomials too. This means that a real polynomial always has at least one root. This root is not necessarily a real number but it could be a complex number as in the case of $x^2 + 1$.

Furthermore, let us give a second interpretation of the Fundamental Theorem of Algebra. In order to do this, let us consider the set of natural numbers $\mathbb{N}$. Let us consider the following natural polynomial (a polynomial where all the coefficients are natural numbers):

$$8 - x = 9.$$

Although all the coefficients of this polynomial are natural numbers, the root of this polynomial is not a natural number. For this reason we have the need of "expanding" the set of natural numbers to the set of relative numbers $\mathbb{Z}$. This statement can be written as: "the set of natural numbers is not closed with respect to the subtraction operation".

Now, let us consider the following relative polynomial (a polynomial where all the coefficients are relative numbers):

$$-5x = 3.$$

In a similar way, although all the coefficients of the polynomial are relative numbers the root of this polynomial is not a relative number. To find the root, we need a set expansion. Thus, we introduce the set of relative numbers $\mathbb{Q}$ and we conclude that " the set of relative number is not closed with respect to the division operation".

Now, let us consider the following rational polynomial (a polynomial where all the coefficients are relative numbers):

$$x^2 = 2.$$

The roots of this polynomial are not relative numbers. Thus we need a further set expansion and we need to introduce the set of real numbers $\mathbb{R}$. We can conclude that "the set of rational numbers is not closed with respect to the nth root operation".

Finally, let us consider the real polynomial

$$x^2 = -1.$$

The roots of this polynomial are not real numbers. In order to solve this equation we had to introduce the set of complex numbers. Hence, we can conclude that also "the set of real numbers is not closed with respect to the nth root operation". More specifically, it is not closed when the square root of a negative number is taken.

Now, the Fundamental Theorem of Algebra guarantees that if we consider a complex polynomial, it will sure have at least one complex root. This means that we can conclude that "the set of complex numbers is closed with respect to the operations of sum (subtraction), multiplication (division), exponentiation (n^{th} root)."

Remembering that a field is essentially a set with its operations, the latter statement can be re-written according to the equivalent and alternative formulation of the Fundamental Theorem of Algebra.

Theorem 5.7. Fundamental Theorem of Algebra (Alternative Formulation). *The field of complex numbers is algebraically closed.*

An algebraically closed field is a field that contains the roots for every non-constant polynomial.

Definition 5.16. Let $p(z)$ be a complex polynomial and α its root. The root is said *single* or *simple* is $p(z)$ is divisible by $(z - \alpha)$ but not $(z - \alpha)^2$.

Definition 5.17. If a polynomial can be expressed as

$$p(z) = h(z - \alpha_1)(z - \alpha_2)\ldots(z - \alpha_n)$$

with h constant and $\alpha_1 \neq \alpha_2 \neq \ldots \neq \alpha_n$, the polynomial is said to have n *distinct roots* $\alpha_1, \alpha_2, \ldots, \alpha_n$.

Theorem 5.8. Theorem on the Distinct Roots of a Polynomial. *If* $p(z) = \sum_{k=0}^{n} a_k z^k$ *is a complex polynomial having order* $n \geq 1$, *then this polynomial has at most n distinct solutions.*

Proof. Let us assume, by contradiction, that the polynomial has $n+1$ distinct roots $\alpha_1, \alpha_2, \ldots, \alpha_{n+1}$. Hence, $p(z)$ is divisible by $(z - \alpha_1), (z - \alpha_2), \ldots$, and $(z - \alpha_{n+1})$. Since $p(z)$ is divisible by $(z - \alpha_1)$, then we can write

$$p(z) = (z - \alpha_1) q(z).$$

Let us consider

$$p(\alpha_2) = (\alpha_2 - \alpha_1) q(\alpha_2)$$

with $\alpha_1 \neq \alpha_2$. Since α_2 is a root of the polynomial $p(\alpha_2) = 0$. It follows that also $q(\alpha_2) = 0$. If α_2 is a root of $q(\alpha_2)$, then $q(z)$ is divisible by $(z - \alpha_2)$, i.e. we can write

$$q(z) = (z - \alpha_2) q_1(z).$$

Hence,

$$p(z) = (z - \alpha_1)(z - \alpha_2) q_1(z).$$

Let us consider the root $\alpha_3 \neq \alpha_2 \neq \alpha_1$. Then, $p(\alpha_3) = 0$. It follows that $q_1(\alpha_3) = 0$ and that we can write

$$q_1(z) = (z - \alpha_3) q_2(z).$$

Hence,

$$p(z) = (z - \alpha_1)(z - \alpha_2)(z - \alpha_3) q_2(z).$$

If we iterate this procedure we obtain

$$p(z) = (z - \alpha_1)(z - \alpha_2) \ldots (z - \alpha_{n+1}) q_n$$

with q_n constant.

We have written an equality between a n order polynomial ($p(z)$ has order n for hypothesis) and a $n+1$ order polynomial, against the identity principle. We have reached a contradiction. $\square$

Corollary 5.2. *If two complex polynomials* $p_1(z)$ *and* $p_2(z)$ *of order* $n \geq 1$ *take the same value in* $n+1$ *points, then the two polynomials are identical.*

Example 5.35. Let us consider the following order 2 polynomial:

$$z^2 + 5z + 4.$$

For the theorem on the distinct roots of a polynomial, this polynomial cannot have more than two distinct roots. In particular, the roots of this polynomial are -1 and -4 and can be written as

$$(z^2 + 5z + 4) = (z + 1)(z + 4).$$

Example 5.36. The polynomial

$$z^3 + 2z^2 - 11z - 12$$

cannot have more than three distinct roots. The roots are -1, -4, and 3 and the polynomial can be written as

$$\left(z^3 + 2z^2 - 11z - 12\right) = (z+1)(z+4)(z-3).$$

Example 5.37. The polynomial

$$z^2 + 2z + 5$$

cannot have more than two distinct roots. In this case the roots are not simple real roots but two distinct complex roots, i.e. $-1 + j2$ and $-1 - j2$. The polynomial can be written as

$$\left(z^2 + 2z + 5\right) = (z+1-j2)(z+1+2j).$$

Obviously, a polynomial can have both real and complex roots as in the case of $\left(z^3 + z^2 + 3z - 5\right) = (z+1-j2)(z+1+2j)(z-1)$.

Example 5.38. Finally, the polynomial

$$z^4 - z^3 - 17z^2 + 21z + 36$$

is of order 4 and cannot have more than four distinct roots. It can be verified that this polynomial has three roots, that is -1, -4, and 3 as the polynomial above but the root -3 is repeated twice. The polynomial can be written as

$$\left(z^4 - z^3 - 17z^2 + 21z + 36\right) = (z+1)(z+4)(z-3)(z-3).$$

This situation is explained in the following definition.

Definition 5.18. Let $p(z)$ be a complex polynomial in the variable z. A solution is said *multiple with algebraic multiplicity* $k \in \mathbb{N}$ and $k > 1$ if $p(z)$ is divisible by $(z-\alpha)^k$ but not divisible by $(z-\alpha)^{k+1}$.

Example 5.39. In the example above, 3 is a solution (or a root) of multiplicity 2 because the polynomial $\left(z^4 - z^3 - 17z^2 + 21z + 36\right)$ is divisible by $(z-3)^2$ and not by $(z-3)^3$.

Theorem 5.9. Let $p(z) = \sum_{k=0}^{n} a_k z^k$ be a complex polynomial of order $n > 1$ in the variable z. If $\alpha_1, \alpha_2, \ldots, \alpha_s$ are its roots having algebraic multiplicity $h_1, h_2, \ldots, h_s$ then

$$h_1 + h_2 + \cdots + h_s = n$$

and

$$p(z) = a_n (z-\alpha_1)^{h_1} (z-\alpha_2)^{h_2} \ldots (z-\alpha_s)^{h_s}.$$

Proof. Since the polynomial $p(z)$ has roots $\alpha_1, \alpha_2, \ldots, \alpha_n$ with the respective multiplicity values $h_1, h_2, \ldots, h_s$ then we can write that $\exists! q(z)$ such that

$$p(z) = q(z)(z - \alpha_1)^{h_1}(z - \alpha_2)^{h_2} \ldots (z - \alpha_s)^{h_s}.$$

At first, we need to prove that $q(z)$ is a constant. Let us assume, by contradiction, that $q(z)$ is of order ≥ 1. For the Fundamental Theorem of Algebra, this polynomial has at least one root $\alpha \in \mathbb{C}$. Thus, $\exists q_1(z) | q(z) = (z - \alpha) q_1(z)$. If we substitute in the $p(z)$ expression we obtain

$$p(z) = (z - \alpha) q_1(z)(z - \alpha_1)^{h_1}(z - \alpha_2)^{h_2} \ldots (z - \alpha_s)^{h_s}.$$

This means that α is also a root of $p(z)$. Since for hypothesis the roots of $p(z)$ are $\alpha_1, \alpha_2, \ldots, \alpha_s$, α must be equal to one of them. Let us consider a generic index i such that $\alpha = \alpha_i$. In this case $p(z)$ must be divisible by $(z - \alpha_i)^{h_i+1}$. Since this is against the definition of multiplicity of a root, we reached a contradiction. It follows that $q(z)$ is a constant q and the polynomial is

$$p(z) = q(z - \alpha_1)^{h_1}(z - \alpha_2)^{h_2} \ldots (z - \alpha_s)^{h_s}.$$

Let us re-write the polynomial $p(z)$ as

$$p(z) = a_n z^n + a_{n-1} z^{n-1} + \ldots a_2 z^2 + a_1 z + a_0.$$

It follows that the addend of order n is

$$a_n z^n = q z^{h_1 + h_2 + \ldots + h_s}.$$

For the identity principle

$$h_1 + h_2 + \cdots + h_s = n$$

$$a_n = q.$$

Thus, the polynomial can be written as

$$p(z) = a_n (z - \alpha_1)^{h_1}(z - \alpha_2)^{h_2} \ldots (z - \alpha_s)^{h_s}. \quad \square$$

Example 5.40. The polynomial $(z^4 - z^3 - 17z^2 + 21z + 36)$ in the previous example has three roots, two of them having multiplicity 1 and one having multiplicity 2: $h_1 = 1, h_2 = 1, h_3 = 2$. It follows that $h_1 + h_2 + h_3 = 4$ that is the order of the polynomial. As shown above, the polynomial can be written as

$$(z + 1)(z + 4)(z - 3)^2.$$

With reference to Theorem 5.9, $a_n = 1$.

Example 5.41. Let us consider the polynomial

$$2z^7 - 20z^6 + 70z^5 - 80z^4 - 90z^3 + 252z^2 - 30z + 200$$

which can be written as

$$2(z+1)^2(z-4)(z^2 - 4z + 5)^2.$$

The roots are -1 with multiplicity 1, 4 with multiplicity 2, $2 - j$ with multiplicity 2 and $2 + j$ with multiplicity 2. Hence we have that the sum of the multiplicity values is $1 + 2 + 2 + 2 = 7$ that is the order of the polynomial and $a_n = 2$.

Definition 5.19. Let $p(z) = \sum_{k=0}^{n} a_k z^k$ be a complex polynomial of order $n \geq 1$. A *conjugate complex polynomial* $\dot{p}(z)$ is a polynomial whose coefficients are conjugate of the coefficients of $p(z)$: $\dot{p}(z) = \sum_{k=0}^{n} \dot{a}_k z^k$.

Proposition 5.6. *Let $p(z)$ be a complex polynomial of order $n \geq 1$. If $\alpha_1, \alpha_2, \ldots, \alpha_s$ are its roots having algebraic multiplicity $h_1, h_2, \ldots, h_s$ then $\dot{\alpha}_1, \dot{\alpha}_2, \ldots, \dot{\alpha}_s$ with algebraic multiplicity $h_1, h_2, \ldots, h_s$ are roots of $\dot{p}(z)$.*

Proposition 5.7. *Let $p(z) = \sum_{k=0}^{n} a_k z^k$ be a complex polynomial of order $n \geq 1$. If $\alpha_1, \alpha_2, \ldots, \alpha_n$ are its roots it follows that*

- $\alpha_1 + \alpha_2 + \ldots + \alpha_s = -\frac{a_{n-1}}{a_n}$
- $\alpha_1 \alpha_2 + \alpha_2 \alpha_3 \ldots + \alpha_{s-1} \alpha_s = \frac{a_{n-2}}{a_n}$
- $\alpha_1 \alpha_2 \ldots \alpha_s = (-1)^n \frac{a_0}{a_n}$.

5.3.2.1 How to Determine the Roots of a Polynomial

The previous sections explain what a root is, what type of roots exist, and how many roots are in a polynomial. It has not been explained yet how to determine these roots. In order to pursue this aim let us consider polynomials of increasing orders.

The detection of the root of a polynomial of order 1, $az - b$ is a trivial problem:

$$az - b \Rightarrow \alpha = \frac{b}{a}.$$

The roots of a polynomial of order 2, $az^2 + bz + c$, can be found analytically by the popular formula developed by antique Indian, Babylonian and Chinese mathematicians:

$$\alpha_1 = \frac{-b}{2a} + \frac{\sqrt{b^2 - 4ac}}{2a}$$
$$\alpha_2 = \frac{-b}{2a} - \frac{\sqrt{b^2 - 4ac}}{2a}$$

Let us prove this formula.

Proof. The roots are the solution of the equation

$$az^2 + bz + c = 0 \Rightarrow az^2 + bz = -c.$$

Let us multiply both members of the equation by $4a$:

$$4a^2 z^2 + 4abz = -4ac \Rightarrow (2az)^2 + 2(2az)b = -4ac.$$

Let us add b^2 to both members

$$(2az)^2 + 2(2az)b + b^2 = -4ac + b^2 \Rightarrow ((2az) + b)^2 = b^2 - 4ac \Rightarrow ((2az) + b) = \pm\sqrt{b^2 - 4ac}.$$

From this equation we obtain

$$\alpha_1 = \frac{-b}{2a} + \frac{\sqrt{b^2 - 4ac}}{2a} \quad \square$$
$$\alpha_2 = \frac{-b}{2a} - \frac{\sqrt{b^2 - 4ac}}{2a}$$

The method for the calculation of the roots of a polynomial of order 3, $az^3 + bz^2 + cz + d$ has been introduced in the sixteenth century thanks to the studies of Girolamo Cardano and Niccoló Tartaglia. The solution of $az^3 + bz^2 + cz + d = 0$ can be calculated by posing $x = y - \frac{b}{3a}$ and thus obtaining a new equation:

$$y^3 + py + q = 0.$$

where

$$p = \frac{c}{a} - \frac{b^2}{3a^2}$$
$$q = \frac{d}{a} - \frac{bc}{3a^2} + \frac{2b^3}{27a^3}.$$

The solutions of this equation are given by $y = u + v$ where

$$u = \sqrt[3]{-\frac{q}{2} + \sqrt{\frac{q^2}{4} + \frac{p^3}{27}}}$$
$$v = \sqrt[3]{-\frac{q}{2} - \sqrt{\frac{q^2}{4} + \frac{p^3}{27}}}$$

and two solutions whose values depend of $\Delta = \frac{q^2}{4} + \frac{p^3}{27}$. If $\Delta > 0$ the roots are

$$\alpha_1 = u + v$$
$$\alpha_2 = u\left(-\frac{1}{2} + j\frac{\sqrt{3}}{2}\right) + v\left(-\frac{1}{2} - j\frac{\sqrt{3}}{2}\right)$$
$$\alpha_3 = u\left(-\frac{1}{2} - j\frac{\sqrt{3}}{2}\right) + v\left(-\frac{1}{2} + j\frac{\sqrt{3}}{2}\right).$$

If $\Delta < 0$, in order to find the roots, the complex number $-\frac{q}{2} + j\sqrt{-\Delta}$ must be expressed in polar coordinates as $(\rho\angle\theta)$. The roots are:

$$\alpha_1 = 2\sqrt{-\frac{p}{3}} + \cos\left(\frac{\theta}{3}\right)$$
$$\alpha_2 = 2\sqrt{-\frac{p}{3}} + \cos\left(\frac{\theta+2\pi}{3}\right)$$
$$\alpha_3 = 2\sqrt{-\frac{p}{3}} + \cos\left(\frac{\theta+4\pi}{3}\right).$$

If $\Delta = 0$ the roots are

$$\alpha_1 = -2\sqrt[3]{-\frac{q}{2}}$$
$$\alpha_2 = \alpha_3 = \sqrt[3]{-\frac{q}{2}}.$$

The proof of the solving formulas are not reported as they are outside the scopes of this book.

If the polynomial is of order 4, i.e. in the form $ax^4 + bx^3 + cx^2 + dx + e$, the detection of the roots have been investigated by Lodovico Ferrari and Girolamo Cardano in the sixteenth century. A representation of the solving method is the following.

$$\alpha_1 = -\frac{b}{4a} - S + \frac{1}{2}\sqrt{-4S^2 - 2p + \frac{q}{s}}$$
$$\alpha_2 = -\frac{b}{4a} - S - \frac{1}{2}\sqrt{-4S^2 - 2p + \frac{q}{s}}$$
$$\alpha_3 = -\frac{b}{4a} + S + \frac{1}{2}\sqrt{-4S^2 - 2p + \frac{q}{s}}$$
$$\alpha_4 = -\frac{b}{4a} + S - \frac{1}{2}\sqrt{-4S^2 - 2p + \frac{q}{s}}$$

where

$$p = \frac{8ac - 3b^2}{8a^2}$$
$$q = \frac{b^3 - 4abc + 8a^2 d}{8a^3}.$$

The value of S is given by

$$S = \frac{1}{2}\sqrt{-\frac{2}{3}p + \frac{1}{3a}\left(Q + \frac{\Delta_0}{Q}\right)}$$

where

$$Q = \sqrt[3]{\frac{\Delta_1 + \sqrt{\Delta_1^2 - \Delta_0^3}}{2}}$$
$$\Delta_0 = c^2 - 3bd + 12ae$$
$$\Delta_1 = 2c^3 - 9bcd + 27b^2 e + 27ad^2 - 72ace.$$

The proof of the solving method is also not reported in this book. However, it is clear that the detection of the roots of a polynomial is in general a difficult task. More drastically, the detection of the roots of a polynomial having order 5 or higher is an impossible task. This fact is proved in the so called Abel-Ruffini's Theorem.

Theorem 5.10. Abel-Ruffini's Theorem. *There is no general algebraic solution to polynomial equations of degree five or higher with arbitrary coefficients.*

This means that if the calculation of the roots of a polynomial of order ≥ 5 must be calculated a numerical method must be implemented to find an approximated solution as the problem has no analytic solution. A description of the numerical methods does not fall within the scopes of this book. However, some examples of numerical methods for finding the roots of a high order polynomial are the bisection and secant methods, see e.g. [14].

5.4 Partial Fractions

Definition 5.20. Let $p_1(z) = \sum_{k=0}^{m} a_k z^k$ and $p_2(z) = \sum_{k=0}^{n} b_k z^k$ be two complex polynomials. The function $Q(z)$ obtained by dividing $p_1(z)$ by $p_2(z)$,

$$Q(z) = \frac{p_1(z)}{p_2(z)}$$

is said *rational fraction* in the variable z.

Let $Q(z) = \frac{p_1(z)}{p_2(z)}$ be a rational fraction in the variable z. The *partial fraction decomposition* (or *partial fraction expansion*) is a mathematical procedure that consists of expressing the fraction as a sum of rational fractions where the denominators are of lower order that that of $p_2(z)$:

$$Q(z) = \frac{p_1(z)}{p_2(z)} = \sum_{k=1}^{n} \frac{f_i(z)}{g_i(z)}$$

This decomposition can be of great help to break a complex problem into many simple problems. For example, the integration term by term can be much easier if the fraction has been decomposed, see Chap. 13.

Let us consider the case of a *proper fraction*, i.e. the order of $p_1(z) \leq$ than the order of $p_2(z)$ $(m \leq n)$. Let us indicate with the term *zeros* the values $\alpha_1, \alpha_2, \ldots, \alpha_m$ such that $p_1(\alpha_k) = 0$ $\forall k \in \mathbb{N}$ with $1 \leq k \leq n$, and the term *poles* the values $\beta_1, \beta_2, \ldots, \beta_n$ such that $p_2(\beta_k) = 0$ $\forall k \in \mathbb{N}$ with $1 \leq k \leq n$.

Let us distinguish three cases:

- rational fractions with distinct/single real or complex poles
- rational fractions with multiple real or complex poles
- rational fractions with conjugate complex poles

Rational fractions with only distinct poles are characterized by a denominator of the kind

$$p_2(z) = (z - \beta_1)(z - \beta_2) \ldots (z - \beta_n)$$

i.e. the poles have null imaginary parts (the poles are real numbers).

In this first case the rational fraction can be written as

$$Q(z) = \frac{p_1(z)}{(z-\beta_1)(z-\beta_2)\ldots(z-\beta_n)} = \frac{A_1}{(z-\beta_1)} + \frac{A_2}{(z-\beta_1)} + \ldots + \frac{A_n}{(z-\beta_n)}$$

where $A_1, A_2, \ldots, A_n$ are constant coefficients.

If the rational fraction contains multiple poles, each multiple pole in the denominator appears as

$$p_2(z) = (z-\beta_k)^h$$

i.e. some poles are real numbers with multiplicity > 1.

In this second case the rational fraction can be written as

$$Q(z) = \frac{p_1(z)}{(z-\beta_k)^h} = \frac{A_k^1}{(z-\beta_k)} + \frac{A_k^2}{(z-\beta_k)^2} + \ldots + \frac{A_k^h}{(z-\beta_k)^h}$$

where $A_k^1, A_k^2, \ldots, A_k^h$ are constant coefficients.

rational fractions with quadratic terms are characterized by a denominator of the kind

$$p_2(z) = (z^2 + \xi z + \zeta)$$

i.e. some poles are conjugate imaginary or conjugate complex numbers.

In this third case the rational fraction can be written as

$$Q(z) = \frac{p_1(z)}{(z-\beta_2)\ldots(z-\beta_n)} = \frac{Bz+C}{(z^2+\xi z+\zeta)}$$

where B, C are constant coefficients.

Obviously, the polynomial can contain single and multiple poles as well as real and complex poles. In the case of multiple complex poles, the corresponding constant coefficients are indicated with B_k^j and C_k^j.

In order to find these coefficients, from the equation

$$p_1(z) = \sum_{k=1}^{n} \frac{f_k(z)\, p_2(z)}{g_k(z)}$$

the coefficients a_k of the polynomial $p_1(z)$ are imposed to be equal to the corresponding ones on the left hand side of the equation. This operation leads to a system of linear equations in the variables A_k, A_k^j, B_k, and B_k^j whose solution completes the partial fraction decomposition.

Example 5.42. Let us consider the following rational fraction

$$\frac{8z-42}{z^2+3z-18}.$$

This rational fraction has two single poles and can be written as

$$\frac{8z-42}{(z+6)(z-3)} = \frac{A_1}{z+6} + \frac{A_2}{z-3}.$$

Thus, we can write the numerator as

$$8z-42 = A_1(z-3) + A_2(z+6) = A_1z - A_13 + A_2z + A_26 = (A_1+A_2)z - 3A_1 + 6A_2.$$

We can now set the following system of linear equations in the variables A_1 and A_2

$$\begin{cases} A_1 + A_2 = 8 \\ -3A_1 + 6A_2 = -42 \end{cases}$$

whose solution is $A_1 = 10$ and $A_2 = -2$. Hence the partial fraction decomposition is

$$\frac{8z-42}{z^2+3z-18} = \frac{10}{z+6} - \frac{2}{z-3}.$$

Example 5.43. Let us consider the following rational fraction

$$\frac{4z^2}{z^3 - 5z^2 + 8z - 4}.$$

This rational fraction has one single pole and one double pole. The fraction can be written as

$$\frac{4z^2}{(z-1)(z-2)^2} = \frac{A_1}{z-1} + \frac{A_2^1}{z-2} + \frac{A_2^2}{(z-2)^2}.$$

The numerator can be written as

$$\begin{aligned} 4z^2 &= A_1(z-2)^2 + A_2^1(z-2)(z-1) + A_2^2(z-1) = \\ &= z^2A_1 + 4A_1 - 4zA_1 + z^2A_2^1 - 3zA_2^1 + 2A_2^1 + zA_2^2 - A_2^2 = \\ &= z^2\left(A_1 + A_2^1\right) + z\left(A_2^2 - 3A_2^1 - 4A_1\right) + 4A_1 + 2A_2^1 - A_2^2 \end{aligned}$$

and the following system of linear equations can be set

$$\begin{cases} A_1 + A_2^1 = 4 \\ -4A_1 - 3A_2^1 + A_2^2 = 0 \\ 4A_1 + 2A_2^1 - A_2^2 = 0, \end{cases}$$

whose solution is $A_1 = 4$, $A_2^1 = 0$, and $A_2^2 = 16$. The partial fraction decomposition is

$$\frac{4z^2}{z^3 - 5z^2 + 8z - 4} = \frac{4}{z-1} + \frac{16}{(z-2)^2}.$$

Example 5.44. Let us consider the following rational fraction

$$\frac{8z^2 - 12}{z^3 + 2z^2 - 6z}.$$

This rational fraction has one pole in the origin and two conjugate complex poles. The fraction can be written as

$$\frac{8z^2 - 12}{z(z^2 + 2z - 6)} = \frac{A_1}{z} + \frac{B_1 z + C_1}{z^2 + 2z - 6}.$$

The numerator can be written as

$$8z^2 - 12 = A_1 \left(z^2 + 2z - 6\right) + (B_1 z + C_1) z =$$
$$= z^2 A_1 + 2z A_1 - 6A_1 + z^2 B_1 + z C_1 = z^2 (A_1 + B_1) + z (2A_1 + C_1) - 6A_1$$

and the following system of linear equations can be set

$$\begin{cases} A_1 + B_1 = 8 \\ 2A_1 + C_1 = 0 \\ -6A_1 = -12, \end{cases}$$

whose solution is $A_1 = 2$, $B_1 = 6$, and $C_1 = -4$. The partial fraction decomposition is

$$\frac{8z^2 - 12}{z^3 + 2z^2 - 6z} = \frac{2}{z} + \frac{6z - 4}{z^2 + 2z - 6}.$$

Let us now consider a rational fraction

$$Q(z) = \frac{p_1(z)}{p_2(z)}$$

where the order m of $p_1(z)$ is $>$ than the order n of $p_2(z)$. This rational fraction is called *improper fraction*.

A partial fraction expansion can be performed also in this case but some considerations must be done. By applying the Theorem 5.4, we know that every polynomial $p_1(z)$ can be expressed as

$$p_1(z) = p_2(z) q(z) + r(z)$$

by means of unique $q(z)$ and $r(z)$ polynomials. We know also that the order of $p_1(z)$ is equal to the sum of the orders of $p_2(z)$ and $q(z)$. Thus, we can express the improper fraction as

$$\frac{p_1(z)}{p_2(z)} = q(z) + \frac{r(z)}{p_2(z)}.$$

The polynomial $q(z)$ is of order $m - n$ and can be expressed as :

$$q(z) = E_0 + E_1 z + E_2 z^2 + \ldots + E_{m-n} z^{m-n}$$

and the improper fraction can be expanded as

$$\frac{p_1(z)}{p_2(z)} = E_0 + E_1 z + E_2 z^2 + \ldots + E_{m-n} z^{m-n} + \frac{r(z)}{p_2(z)}$$

and apply the partial fraction expansion for $\frac{r(z)}{p_2(z)}$ that is certainly proper as the order of $r(z)$ is always $<$ than the order of $p_2(z)$ for the Theorem 5.4. The coefficients $E_0, E_1, \ldots, E_{m-n}$ can be determined at the same time of the coefficients resulting from the expansion of $\frac{r(z)}{p_2(z)}$ by posing the identity of the coefficients and solving the resulting system of linear equations.

Example 5.45. Let us consider the following improper fraction

$$\frac{4z^3 + 10z + 4}{2z^2 + z}.$$

The rational fraction can be written as

$$\frac{4z^3 + 10z + 4}{2z^2 + z} = \frac{4z^3 + 10z + 4}{z(2z+1)} = zE_1 + E_0 + \frac{A_1}{z} + \frac{A_2}{2z+1}.$$

The numerator can be expressed as

$$4z^3 + 10z + 4 = z^2(2z+1)E_1 + z(2z+1)E_0 + (2z+1)A_1 + zA_2 =$$
$$= 2z^3 E_1 + z^2 E_1 + 2z^2 E_0 + zE_0 + 2zA_1 + zA_2 + A_1 = z^3 2E_1 + z^2(E_1 + 2E_0) + z(2A_1 + A_2 + E_0) + A_1.$$

We can then pose the system of linear equations

$$\begin{cases} 2E_1 = 4 \\ E_1 + 2E_0 = 0 \\ 2A_1 + A_2 + E_0 = 10 \\ A_1 = 4 \end{cases}$$

whose solution is $A_1 = 4$, $A_2 = 3$, $E_0 = -1$, and $E_1 = 2$.

Thus, the partial fraction expansion

$$\frac{4z^3 + 10z + 4}{2z^2 + z} = 2z - 1 + \frac{4}{z} + \frac{3}{2z + 1}.$$

Exercises

5.1. Verify that if $z = a + jb$, then

$$\frac{1}{z} = \frac{a - jb}{a^2 + b^2}.$$

5.2. Express the complex number $z = 1 - j$ in polar coordinates.

5.3. Express the complex number $z = 4; \angle 90°$ in rectangular coordinates.

5.4. Calculate $\sqrt[3]{5 + j5}$

5.5. Apply Ruffini's theorem to check whether or not $z^3 - 3z^2 - 13z + 15$ is divisible by $(z - 1)$.

5.6. Invert, if possible, the matrix $\mathbf{A}$

$$\mathbf{A} = \begin{pmatrix} 1 & 6 & 1 \\ 2 & j2 & 2 \\ 3 & 6 + 2j & 3 \end{pmatrix}$$

5.7. Calculate the remainder of the division

$$\frac{z^3 + 2z^2 + 4z - 8}{z - 2j}$$

where z is a complex variable.

5.8. Expand in partial fractions the following rational fraction

$$\frac{-9z + 9}{2z^2 + 7z - 4}.$$

5.9. Expand in partial fractions the following rational fraction

$$\frac{3z + 1}{(z - 1)^2 (z + 2)}.$$

5.10. Expand in partial fractions the following rational fraction

$$\frac{5z}{(z^3 - 3z^2 - 3z - 2)}.$$

Chapter 6
An Introduction to Geometric Algebra and Conics

6.1 Basic Concepts: Lines in the Plane

This chapter introduces the conics and characterizes them from an algebraic perspective. While in depth geometrical aspects of the conics lie outside the scopes of this chapter, this chapter is an opportunity to revisit concepts studied in other chapters such as matrix and determinant and assign a new geometric characterization to them.

In order to achieve this aim, let us start with considering the three-dimensional space. Intuitively, we may think that, within this space, points, lines, and planes exist.

We have previously introduced, in Chap. 4, the concept of line as representation of the set $\mathbb{R}$. If $\mathbb{R}^2$ can be represented as the plane, a line is an infinite subset of $\mathbb{R}^2$. We have also introduced in Chap. 4 the concepts of point, distance between two points, segment, and direction of a line. From the algebra of the vectors we also know that the direction of a line is identified by the components of a vector having the same direction, i.e. a line can be characterized by two numbers which we will indicate here as (l, m).

Definition 6.1. Let $\mathbf{P}$ and $\mathbf{Q}$ be two points of the plane and $d_{\overline{\mathbf{PQ}}}$ be the distance between two points. The point $\mathbf{M}$ of the segment $\overline{\mathbf{PQ}}$ such that $d_{\overline{\mathbf{PM}}} = d_{\overline{\mathbf{MQ}}}$ is said *middle point*.

6.1.1 Equations of the Line

Let $\vec{v} \neq \vec{o}$ be a vector of the plane having components (l, m) and $\mathbf{P_0}(x_0, y_0)$ be a point of the plane. Let us think about the line passing through $\mathbf{P_0}$ and having direction (l, m).

© Springer Nature Switzerland AG 2019

F. Neri, *Linear Algebra for Computational Sciences and Engineering*,
https://doi.org/10.1007/978-3-030-21321-3_6

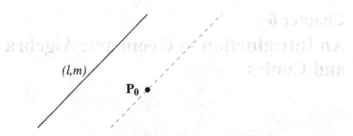

Now let us consider a point $\mathbf{P}(x,y)$ of the plane. The segment $\overline{\mathbf{P_0P}}$ can be interpreted as a vector having components $(x-x_0, y-y_0)$. For Proposition 4.4 vectors $\vec{v} = (l,m)$ and $\overline{\mathbf{P_0P}} = (x-x_0, y-y_0)$ are parallel if and only if

$$\det \left(\begin{array}{cc} (x-x_0) & (y-y_0) \\ l & m \end{array} \right) = 0.$$

This situation occurs when

$$(x-x_0)m - (y-y_0)l = 0 \Rightarrow$$
$$\Rightarrow mx - ly - mx_0 + ly_0 = 0 \Rightarrow$$
$$ax + by + c = 0$$

where

$$a = m$$
$$b = -l$$
$$c = -mx_0 + ly_0.$$

Example 6.1. Let is consider the point $P_0 = (1,1)$ and the vector $\vec{v} = (3,4)$. Let us determine the equation of the line passing through P_0 and having the direction of $\vec{v}$. Let us impose the parallelism between $\vec{v}$ and $\overline{\mathbf{P_0P}}$:

$$\det \left(\begin{array}{cc} (x-1) & (y-1) \\ 3 & 4 \end{array} \right) = 4(x-1) - 3(y-1) = 4x - 4 - 3y + 3 = 4x - 3y - 1 = 0.$$

Definition 6.2. A line in a plane is a set of points $\mathbf{P}(x,y) \in \mathbb{R}^2$ such that $ax+by+c = 0$ where a, b, and c are three coefficients $\in \mathbb{R}$.

The equation $ax+by+c = 0$ is said *analytic representation of the line* or *analytic equation of the line in its explicit form*. The coefficients a and b are non-null because $(l,m) \neq (0,0)$.

By applying some simple arithmetic operations

$$ax + by + c = 0 \Rightarrow by = -ax - c \Rightarrow$$
$$\Rightarrow y = -\frac{a}{b}x - \frac{c}{b} \Rightarrow y = kx + q$$

which is known as *analytic equation of the line in its explicit form*.

It must be observed that a line having direction (l,m) has equation $ax+by+c = 0$ where $a = m$ and $b = -l$. Analogously a line having equation $ax+by+c = 0$ has direction $(-b,a) = (b,-a)$.

Example 6.2. The line having equation

$$5x + 4y - 2 = 0$$

has direction $(-4, 5) = (4, -5)$.

Definition 6.3. The components of a vectors parallel to a line are said *direction numbers of the line*.

Let us now calculate the scalar product $((x - x_0), (y - y_0)) (a, b)$ and let us impose that it is null:

$$(x - x_0), (y - y_0) (a, b) = a(x - x_0) + b(y - y_0) =$$
$$= ax + by - ax_0 - by_0 = ax + by + c = 0.$$

This equation means that a line having equation $ax + by + c = 0$ is perpendicular to the direction (a, b). In other words, the direction identified by the coefficients (a, b) of the line equation in its explicit form is perpendicular to the line.

Let us write an alternative representation of the line. The equation above $(x - x_0) m - (y - y_0) l = 0$ can be re-written as

$$\frac{(x - x_0)}{l} = \frac{(y - y_0)}{m}.$$

This equation yields to the following system of linear equations

$$\begin{cases} x - x_0 = lt \\ y - y_0 = mt \end{cases} \Rightarrow \begin{cases} x(t) = lt + x_0 \\ y(t) = mt + y_0 \end{cases}$$

where t is a parameter. While t varies the line is identified. The equations of the system are said *parametric equations of the line*.

Example 6.3. Let $5x - 4y - 1 = 0$ be the equation of a line in the plane. We know that the direction numbers $(l, m) = (4, 5)$. Hence, we can write the parametric equations as

$$\begin{cases} x(t) = 4t + x_0 \\ y(t) = 5t + y_0. \end{cases}$$

In order to find x_0 and y_0, let us choose an arbitrary value for x_0 and let us use the equation of the line to find the corresponding y_0 value. For example if we choose $x_0 = 1$ we have $y_0 = 1$. The parametric equations are then

$$\begin{cases} x(t) = 4t + 1 \\ y(t) = 5t + 1. \end{cases}$$

Let us now write the equation of the line in a slightly different way. Let $\mathbf{P_1}(x_1, y_1)$ and $\mathbf{P_2}(x_2, y_2)$ be two points of the plane. We would like to write the equation of the line between the two points. In order to do it, let us impose that for a generic point $P(x,y)$, the segment $\overline{P_1P_2} = (x_1 - x_2, y_1 - y_2)$ is parallel to the segment $\overline{PP_2} = (x - x_2, y - y_2)$. If the two segments have the same direction are thus aligned and belong to the same line.

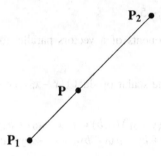

The parallelism is given by

$$\det \begin{pmatrix} (x - x_2) & (y - y_2) \\ (x_1 - x_2) & (y_1 - y_2) \end{pmatrix} = 0 \Rightarrow \frac{(x - x_2)}{(y - y_2)} = \frac{(x_1 - x_2)}{(y_1 - y_2)}$$

that is the *equation of a line between two points*.

Example 6.4. A line passing through the points $P_1(1,5)$ and $P_2(-2,8)$ has equation

$$\frac{x+2}{y-8} = \frac{1+2}{5-8}$$

which can equivalently be written as

$$(5-8)(x+2) - (1+2)(x+2).$$

6.1.2 Intersecting Lines

Let l_1 and l_2 be two lines of the plane having equation, respectively,

$$l_1 : a_1x + b_1y + c_1 = 0$$
$$l_2 : a_2x + b_2y + c_2 = 0.$$

We aim at studying the position of these two line in the plane. If these two line intersect in a point P_0 it follows that the point P_0 belongs to both the line. Equivalently we may state that the coordinates (x_0, y_0) of this point P_0 simultaneously satisfy the equations of the lines l_1 and l_2.

In other words, (x_0, y_0) is the solution of the following system of linear equations:

$$\begin{cases} a_1x + b_1y + c_1 = 0 \\ a_2x + b_2y + c_2 = 0. \end{cases}$$

At first, we may observe that a new characterization of the concept of system of linear equations is given. A system of linear equation can be seen as a set of lines and its solution, when it exists, is the intersection of these lines. In this chapter we study lines in the plane. Thus, the system has two linear equations in two variables. A system having size 3×3 can be seen as the equation of three lines in the space. By extension an $n \times n$ system of linear equation represents lines in a n-dimensional space. In general, even when not all the equations are equations of the line, the solutions of a system of equations can be interpreted as the intersection of objects.

Let us focus on the case of two lines in the plane. The system above is associated with the following incomplete and complete matrices, respectively,

$$\mathbf{A} = \begin{pmatrix} a_1 & b_1 \\ a_2 & b_2 \end{pmatrix}$$

and

$$\mathbf{A^c} = \begin{pmatrix} a_1 & b_1 & -c_1 \\ a_2 & b_2 & -c_2 \end{pmatrix}.$$

Indicating with $\rho_{\mathbf{A}}$ and $\rho_{\mathbf{A^c}}$ the ranks of the matrices $\mathbf{A}$ and $\mathbf{A^c}$, respectively, and by applying the Rouchè-Capelli Theorem the following cases are distinguished.

- **Case 1:** If $\rho_{\mathbf{A}} = 2$ (which yields that also $\rho_{\mathbf{A^c}} = 2$), the system is determined and has only one solution. Geometrically, this means that the lines intersect in a single point.
- **Case 2:** If $\rho_{\mathbf{A}} = 1$ and $\rho_{\mathbf{A^c}} = 2$, the system is incompatible and has no solutions. Geometrically, this means that the two lines are parallel, thus the system is of the kind

$$\begin{cases} ax + by + c_1 = 0 \\ \lambda ax + \lambda by + c_2 = 0 \end{cases}$$

  with $\lambda \in \mathbb{R}$.
- **Case 3:** If $\rho_{\mathbf{A}} = 1$ and $\rho_{\mathbf{A^c}} = 1$, the system is undetermined and has ∞ solutions. Geometrically this means that the two lines are overlapped and the system is of the kind

$$\begin{cases} ax + by + c = 0 \\ \lambda ax + \lambda by + \lambda c = 0 \end{cases}$$

  with $\lambda \in \mathbb{R}$.

It can be observed that if $\det(\mathbf{A}) \neq 0$ it follows that $\rho_{\mathbf{A}} = 2$ and that also $\rho_{\mathbf{A^c}} = 2$. If $\det(\mathbf{A}) = 0$ it follows that $\rho_{\mathbf{A}} = 1$ ($\rho_{\mathbf{A}} = 0$ would mean that there are no lines on the plane).

Example 6.5. Let us find, if possible, the intersection point of the lines having equation $2x + y - 1 = 0$ and $4x - y + 2 = 0$. This means that the following system of linear equations must be set

$$\begin{cases} 2x + y - 1 = 0 \\ 4x - y + 2 = 0 \end{cases}.$$

The incomplete matrix is non-singular as it has determinant -6. Thus, the lines are intersecting. The solution of the system is

$$x = \frac{\det \begin{pmatrix} 1 & 1 \\ -2 & -1 \end{pmatrix}}{-6} = -\frac{1}{6}$$

and

$$y = \frac{\det \begin{pmatrix} 2 & 1 \\ 4 & -2 \end{pmatrix}}{-6} = \frac{4}{3}.$$

The intersection point is $\left(-\frac{1}{6}, \frac{4}{3}\right)$.

Example 6.6. Let us find, if possible, the intersection point of the lines having equation $2x + y - 1 = 0$ and $4x + 2y + 2 = 0$. The associated incomplete matrix

$$\begin{pmatrix} 2 & 1 \\ 4 & 2 \end{pmatrix}$$

is singular and thus $\rho_A = 1$ while the rank of the complete matrix

$$\begin{pmatrix} 2 & 1 & 1 \\ 4 & 2 & -2 \end{pmatrix}$$

is 2. The system is incompatible and thus the two lines are parallel.

Example 6.7. Let us find, if possible, the intersection point of the lines having equation $2x + y - 1 = 0$ and $4x + 2y - 2 = 0$. It can be easily seen that the second equations is the first one multiplied by 2. Thus, the two equations represent the same line. The two lines are overlapped and have infinite points in common.

6.1.3 Families of Straight Lines

Definition 6.4. A *family of intersecting straight lines* is the set of infinite lines of the plane that contain a common point. The intersection point of all the infinite lines is said *center of the family*.

Definition 6.5. A *family of parallel straight lines* is the set of infinite lines of the plane having the same direction. The lines can be parallel or overlapped.

Theorem 6.1. *Let l_1, l_2, and l_3 be three lines of the plane having equations*

$$l_1 : a_1x + b_1y + c_1 = 0$$
$$l_2 : a_2x + b_2y + c_2 = 0$$
$$l_3 : a_3x + b_3y + c_3 = 0.$$

The lines l_1, l_2, and l_3 belong to the same family if and only if

$$\det(\mathbf{A}) = \begin{pmatrix} a_1 \ b_1 \ c_1 \\ a_2 \ b_2 \ c_2 \\ a_3 \ b_3 \ c_3 \end{pmatrix} = 0.$$

Proof. If l_1, l_2, and l_3 belong to the same family of intersecting lines, they are simultaneously verified for the center of the family. The associated system of linear equations

$$\begin{cases} a_1x + b_1y + c_1 = 0 \\ a_2x + b_2y + c_2 = 0 \\ a_3x + b_3y + c_3 = 0 \end{cases}$$

composed of three equations in two variables, has only one solution. Thus, for Rouchè-Capelli Theorem, the rank of the matrix $\mathbf{A}$ is 2. This means that $\det(\mathbf{A}) = 0$.

If l_1, l_2, and l_3 belong to the same family of parallel lines, the rank of the matrix $\mathbf{A}$ is 1. This means that if an arbitrary 2×2 submatrix is extracted it is singular. For the I Laplace Theorem $\det(\mathbf{A}) = 0$.

In summary, if the lines belong to the same family of either intersecting or parallel lines then $\det(\mathbf{A}) = 0$. $\square$

If $\det(\mathbf{A}) = 0$ the rank of the matrix $\mathbf{A}$ is < 3.

If the rank is 2 the system is compatible and determined. The solution is then the intersection point of the family. Thus, the three lines belong to the same family of intersecting lines.

If the rank is 1 each pair of lines is of the kind

$$ax + by + c = 0$$
$$\lambda ax + \lambda b + \lambda c = 0.$$

This means that all the lines are parallel (have the same direction). Thus, the three lines belong to the same family of parallel lines. Hence if $\det(\mathbf{A}) = 0$ the three lines belong to the same family of either intersecting or parallel lines. $\square$

If $\det(\mathbf{A}) = 0$ then at least a row of $\mathbf{A}$ is linear combination of the other two. This means that $\exists$ a pair of real scalar $(\lambda, \mu) \neq (0,0)$ such that

$$a_3 = \lambda a_1 + \mu a_2$$
$$b_3 = \lambda b_1 + \mu b_2$$
$$c_3 = \lambda c_1 + \mu c_2.$$

If we substitute these values in the third equation of the system of linear equations above we obtain

$$(\lambda a_1 + \mu a_2)x + (\lambda b_1 + \mu b_2)y + (\lambda c_1 + \mu c_2) = 0$$

which can be re-written as

$$\lambda (a_1 x + b_1 y + c_1) + \mu (a_2 x + b_2 y + c_2) = 0$$

which is said *equation of the family of lines*. When the parameters (λ, μ) vary a line of the family is identified.

We know that $(\lambda, \mu) \neq (0,0)$. Without a loss of generality let assume that $\lambda \neq 0$. It follows that

$$\lambda (a_1 x + b_1 y + c_1) + \mu (a_2 x + b_2 y + c_2) = 0 \Rightarrow$$
$$\Rightarrow (a_1 x + b_1 y + c_1) + k (a_2 x + b_2 y + c_2) = 0$$

with $k = \frac{\mu}{\lambda}$.

Each value of k identifies a line. If $k = 0$ the line l_1 is obtained, if $k = \infty$ the line l_2 is identified.

If the lines l_1 and l_2 are parallel then can be written as

$$ax + by + c_1 = 0$$
$$vax + vb + c_2 = 0$$

and the family of lines becomes

$$(ax + by + c_1) + k(vax + vb + c_2) = 0 \Rightarrow$$
$$\Rightarrow a(1 + vk)x + b(1 + vk)y + c_1 + c_2 k = 0 \Rightarrow$$
$$\Rightarrow ax + by + h = 0$$

with

$$h = \frac{c_1 + c_2 k}{1 + vk}.$$

Thus, if the lines are parallel, all the lines have the same direction.

Example 6.8. A family of straight lines is for example

$$(5x + 3y - 1)\lambda + (4x - 2y + 6)\mu = 0.$$

Example 6.9. The center of the family

$$(2x + 2y + 4)l + (2x - 4y + 8)m = 0.$$

is the solution of the system of linear equations

$$\begin{cases} 2x + 2y + 4 = 0 \\ 2x - 4y + 8 = 0 \end{cases}$$

that is $x = -\frac{8}{3}$ and $y = \frac{2}{3}$.

6.2 An Intuitive Introduction to the Conics

Definition 6.6. The *orthogonal projection of a point* **P** *of the space on a plane* is a point of this plane obtained by connecting **P** with the plane by means of a line orthogonal to the plane (orthogonal to all the lines contained in the plane).

Following the definition of orthogonal projection of a point on a plane, we can perform the orthogonal projection for all the infinite points contained in a line. In this way, we can perform *the orthogonal projection of a line on a plane*.

Definition 6.7. An *angle between a line and a plane* is the angle $\leq 90°$ between that line and its orthogonal projection on the plane.

Now, let us consider two lines that are chosen to be not parallel. Let us indicate one of the two lines with z and let us refer to it as "axis" while the other one is simply indicated as "line". the angle between the line and the axis is indicated as θ. In Fig. 6.1 the axis is denoted with an arrow. Let us imagine that the line performs a full rotation (360°) around the axis. This rotation generates a solid object which we will refer to as "cone".

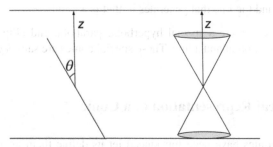

Fig. 6.1 Cone as rotational solid

If we take into consideration a plane in the space, it can intersect the cone in three ways according to the angle ϕ between this plane and the axis z. More specifically the following cases can occur:

- if $0 \leq \phi < \theta$ the intersection is an open figure, namely *hyperbola*
- if $\phi = \theta$ the intersection is an open figure, namely *parabola*

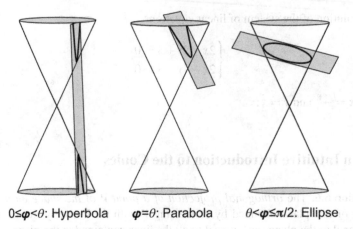

0≤**φ**<θ: Hyperbola   **φ**=θ: Parabola   θ<**φ**≤π/2: Ellipse

Fig. 6.2 Conics as conical sections

- if $\theta < \phi \leq \frac{\pi}{2}$ the intersection is a closed figure, namely *ellipse*

The special ellipse corresponding to $\phi = \frac{\pi}{2}$ is named *circumference*. These figures, coming from the intersection of a cone with a plane are said *conics*. A graphical representation of the conics is given in Fig. 6.2.

Besides the cases listed above, three more cases can be distinguished:

- a plane intersecting the cone with $0 \leq \phi < \theta$ and passing for the intersection between the axis of the cone and the line that generates it, that is two intersecting lines
- a plane tangent to the cone with $\phi = \theta$, that is a line
- a plane intersecting the cone with $0 \leq \phi < \theta$ in the intersection between the axis of the cone and the line that generates it, that is a point

These three cases are a special hyperbola, parabola, and ellipse, respectively, corresponding to a point and a line. These special conics are said *degenerate conics*.

6.3 Analytical Representation of a Conic

Now that the conics have been introduced let us define them again in a different way.

Definition 6.8. A set of points whose location satisfies or is determined by one or more specified conditions is said *locus of points*.

The easiest example of locus of points is a line where the condition is given by its equation $ax + by + c = 0$.

If we consider a point $\mathbf{Q}\left(x_\mathbf{Q}, y_\mathbf{Q}\right)$ not belonging to this line, from basic geometry we know that the distance between $\mathbf{Q}$ and the line is given by, see e.g. [15]:

$$\frac{ax_\mathbf{Q} + by_\mathbf{Q} + c}{\sqrt{a^2 + b^2}}.$$

Definition 6.9. Conic as a Locus of Points. Let $\mathbf{F}$ be a point of a plane, namely *focus* and d be a line in the same plane, namely *directrix*. Let us consider the generic point $\mathbf{P}$ of the plane. Let us indicate with $d_{\mathbf{PF}}$ the distance from $\mathbf{P}$ to $\mathbf{F}$ and with $d_{\mathbf{P}d}$ the distance from $\mathbf{P}$ to d. A *conic* $\mathscr{C}$ is the locus of points $\mathbf{P} \in \mathbb{R}^2$ of a plane such that the ratio $\frac{d_{\mathbf{PF}}}{d_{\mathbf{P}d}}$ is constant.

$$\mathscr{C} = \left\{ \mathbf{P} \,\middle|\, \frac{d_{\mathbf{PF}}}{d_{\mathbf{P}d}} = e \right\}$$

where e is a constant namely *eccentricity* of the conic.

If the coordinates of the focus $\mathbf{F}$ are (α, β), and the equation of the directrix is indicated with $ax + by + c = 0$ where x and y are variable while a, b, and c are coefficients, we can re-write $\frac{d_{\mathbf{PF}}}{d_{\mathbf{P}d}} = e$ in the following way:

$$\frac{d_{\mathbf{PF}}}{d_{\mathbf{P}d}} = e \Rightarrow d_{\mathbf{PF}} = e d_{\mathbf{P}d} \Rightarrow d_{\mathbf{PF}}^2 = e^2 d_{\mathbf{P}d}^2 \Rightarrow$$
$$\Rightarrow (x - \alpha)^2 + (y - \beta)^2 = e^2 \frac{(ax + by + c)^2}{a^2 + b^2}. \tag{6.1}$$

which is called *analytical representation of a conic*. This is a second order (quadratic) algebraic equation in two variables, x and y respectively.

6.4 Simplified Representation of Conics

This section works out the analytical representation of a conic given in Eq. (6.1) in the case of a reference system that allows a simplification of the calculations. This simplified representation allows a better understanding of the conic equations and a straightforward graphic representation by means of conic diagrams.

6.4.1 Simplified Representation of Degenerate Conics

Let us consider, at first, the special case where the focus belongs to the directrix, i.e. $\mathbf{F} \in d$. Without a loss of generality, let us choose a reference system having its origin in $\mathbf{F}$, that is $\alpha = \beta = 0$, and the directrix coinciding with the ordinate axis, that is the equation $ax + by + c = 0$ becomes $x = 0$ (equivalently $a = 1, b = 0, c = 0$).

In this specific case the analytical representation of the conic can be re-written as:

$$d_{PF}^2 = e^2 d_{Pd}^2 \Rightarrow (x - \alpha)^2 + (y - \beta)^2 = e^2 \frac{(ax+by+c)^2}{a^2+b^2} \Rightarrow$$
$$\Rightarrow (x)^2 + (y)^2 = e^2 \frac{(x)^2}{1} \Rightarrow x^2 + y^2 = e^2 x^2 \Rightarrow \left(1 - e^2\right) x^2 + y^2 = 0.$$

From this equation and considering that the eccentricity for its definition can take only non-negative values, we can distinguish three cases.

- **The conic is two intersecting lines.** $1 - e^2 < 0 \Rightarrow e > 1$: the equation of the conic becomes $-kx^2 + y^2 = 0$, with $k = -\left(1 - e^2\right) > 0$. If we solve this equation we obtain, $y = \pm\sqrt{k}x$, that is the equations of two intersecting lines.
- **The conic is two overlapped lines.** $1 - e^2 = 0 \Rightarrow e = 1$: the equation of the conic becomes $0x^2 + y^2 = 0 \Rightarrow y^2 = 0$. This is the equation of the abscissa's axis ($y = 0$) counted twice. The geometrical meaning of this equation is two overlapped lines.
- **The conic is one point.** $1 - e^2 > 0 \Rightarrow 0 \leq e < 1$: the equation of the conic becomes $kx^2 + y^2 = 0$, with $k = \left(1 - e^2\right) > 0$. This equation has only one solution in $\mathbb{R}^2$, that is $(0, 0)$. This means that the conic is only one point, that is the focus and the origin of the axes.

These three situations correspond to the degenerate conics, hyperbola, parabola, and ellipse, respectively.

6.4.2 Simplified Representation of Non-degenerate Conics

Let us now consider the general case $\mathbf{F} \notin d$. Without a loss of generality, let us choose a reference system such that a directrix has equation $x - h = 0$ with h constant $\in \mathbb{R}$ and $\mathbf{F}$ has coordinates $(F, 0)$. Under these conditions, the analytical representation of the conic can be re-written as:

$$d_{PF}^2 = e^2 d_{Pd}^2 \Rightarrow (x - \alpha)^2 + (y - \beta)^2 = e^2 \frac{(ax+by+c)^2}{a^2+b^2} \Rightarrow$$
$$\Rightarrow (x - F)^2 + y^2 = e^2 (x - h)^2 \Rightarrow$$
$$\Rightarrow x^2 + F^2 - 2Fx + y^2 = e^2 \left(x^2 + h^2 - 2hx\right) \tag{6.2}$$
$$\Rightarrow x^2 + F^2 - 2Fx + y^2 - e^2 x^2 - e^2 h^2 - e^2 2hx = 0 \Rightarrow$$
$$\Rightarrow \left(1 - e^2\right) x^2 + y^2 - 2\left(F - he^2\right) x + F^2 - e^2 h^2 = 0$$

Let us now consider the intersection of the conic with the abscissa's axis, $y = 0$.

$$\left(1 - e^2\right) x^2 - 2\left(F - he^2\right) x + F^2 - e^2 h^2 = 0$$

This is a second order polynomial in the real variable x. This polynomial has at most two real (distinct) solutions for the Theorem 5.8.

If we fix the reference axis so that its origin is between the intersections of the conic with the line $y = 0$, the two intersection points can be written as $(a, 0)$ and $(-a, 0)$, respectively. For the Proposition 5.7 if $p(z) = \sum_{k=0}^{n} a_k z^k$ is a complex (real is a special case of complex) polynomial of order $n \geq 1$ and $\alpha_1, \alpha_2, \ldots, \alpha_n$ are its roots it follows that $\alpha_1 + \alpha_2 + \ldots + \alpha_s = -\frac{a_{n-1}}{a_n}$. Thus, in this case the sum of the roots $a - a$ is equal to

$$\frac{2(F - he^2)}{(1 - e^2)} = a - a = 0.$$

From this equation we obtain

$$(F - he^2) = 0 \Rightarrow h = \frac{F}{e^2} \tag{6.3}$$

given that we suppose $1 - e^2 \neq 0 \Rightarrow e \neq 1$.

Furthermore, for the Proposition 5.7 it also occurs that $\alpha_1 \alpha_2 \ldots \alpha_s = (-1)^n \frac{a_0}{a_n}$ which, in this case, means

$$\frac{F^2 - e^2 h^2}{(1 - e^2)} = -a^2.$$

From this equation we obtain

$$F^2 - e^2 h^2 = a^2 (e^2 - 1). \tag{6.4}$$

Substituting equation (6.3) into (6.4) we have

$$F^2 - e^2 \left(\frac{F}{e^2}\right)^2 = a^2 (e^2 - 1) \Rightarrow c^2 - \frac{F^2}{e^2} = a^2 (e^2 - 1) \Rightarrow$$
$$\Rightarrow F^2 (e^2 - 1) = e^2 a^2 (e^2 - 1) \Rightarrow F^2 = e^2 a^2 \Rightarrow e^2 = \frac{F^2}{a^2}. \tag{6.5}$$

Substituting equation (6.5) into (6.3) we obtain

$$h = \frac{a^2}{F} \tag{6.6}$$

and substituting equations (6.5) and (6.6) into the general equation of a conic in Eq. (6.2) we have

$$\left(1 - \frac{F^2}{a^2}\right) x^2 + y^2 - 2 \left(c - \frac{a^2}{F} \frac{F^2}{a^2}\right) x + c^2 - \frac{F^2}{a^2} \left(\frac{a^2}{F}\right)^2 = 0 \Rightarrow$$
$$\Rightarrow \left(1 - \frac{F^2}{a^2}\right) x^2 + y^2 - 2(F - F)x + F^2 - a^2 = 0 \Rightarrow \tag{6.7}$$
$$\Rightarrow (a^2 - F^2) x^2 + a^2 y^2 + a^2 (F^2 - a^2) = 0.$$

This formulation of a conic has been done under the hypothesis that $e \neq 1$. For Eq. (6.5) it occurs that $a^2 \neq F^2$. This means that Eq. (6.7) can be considered in only two cases: $a^2 < F^2$ ($e > 1$) and $a^2 > F^2$ ($e < 1$), respectively.

Equation of the Hyperbola

If $a^2 < F^2$, it follows that $e > 1$. Let us pose $b^2 = (F^2 - a^2) > 0$ and let us substitute into the equation of the conic in Eq. (6.7).

$$-b^2 x^2 + a^2 y^2 + a^2 b^2 = 0 \Rightarrow -\frac{x^2}{a^2} + \frac{y^2}{b^2} + 1 = 0 \Rightarrow \frac{x^2}{a^2} - \frac{y^2}{b^2} - 1 = 0.$$

This is the *analytical equation of the hyperbola*. From this simplified equation we can easily see that the fraction $\frac{x^2}{a^2}$ is always non-negative. From this statement it follows that

$$\frac{y^2}{b^2} + 1 > 0$$

regardless of the value of y. This means that a hyperbola has y values that can be in the $]-\infty, +\infty[$ interval.

In a similar way, since $\frac{y^2}{b^2}$ is always non-negative, it follows that

$$\frac{x^2}{a^2} - 1 \geq 0.$$

This happens when $x \geq a$ and $x \leq -a$. In other words, a graphic of an hyperbola (in its simplified equation form) can be only in the area marked in figure.

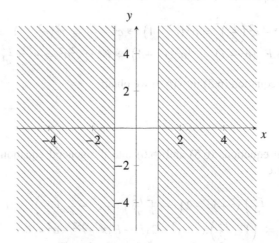

The position of the foci can be easily identified by considering that in an hyperbola $F^2 > a^2$. It follows that $F > a$ or $F < -a$, i.e. the foci are in the marked area of the figure. In order to calculate the position of the directrix le us consider only the right hand side of the graphic. The foci have coordinates $\left(\sqrt{a^2+b^2},0\right)$ and $\left(-\sqrt{a^2+b^2},0\right)$, respectively.

We know that $F > a$. Thus, $Fa > a^2 \Rightarrow a > \frac{a^2}{F}$. Taking into consideration that, as shown in Eq. (6.6), the equation of the directrix is $x = \frac{a^2}{F}$, it follows that the directrix of the right hand side part of the graphic falls outside and at the left of the marked area. It can be easily verified that there is a symmetric directrix associated with the left hand side of the graphic.

If we now consider from the equations of the line and from basic analytical geometry, see e.g. [16], that the equation of a line passing for the origin the Cartesian system of coordinates can be written in its *implicit form* as $y = mx$ with x variable and m coefficient, namely *angular coefficient*, we may think about the infinite lines of the plane passing through the origin. Each line is univocally identified by its angular coefficient m. Obviously, this equation is equivalent to the equation $ax + by + c = 0$.

In order to narrow down the areas where the graphic of the hyperbola (of the simplified equation) can be plotted, let us calculate those values of m that result into an intersection of the line with the hyperbola. This means that we want to identify the values of m that satisfy the following system of equations:

$$\begin{cases} \frac{x^2}{a^2} - \frac{y^2}{b^2} - 1 = 0 \\ y = mx \end{cases}.$$

Substituting the second equation into the first one we obtain

$$\frac{x^2}{a^2} - \frac{m^2 x^2}{b^2} = 1 \Rightarrow \left(\frac{1}{a^2} - \frac{m^2}{b^2}\right) x^2 = 1.$$

Since x^2 must be non-negative (and it is actually positive because the multiplication must be 1) it follows that the values of m satisfying the inequality

$$\left(\frac{1}{a^2} - \frac{m^2}{b^2}\right) > 0$$

are those values that identify the region of the plane where the hyperbola can be plotted. By solving the inequality in the variable m we have

$$\frac{1}{a^2} > \frac{m^2}{b^2} \Rightarrow m^2 < \frac{b^2}{a^2} \Rightarrow -\frac{b}{a} < m < \frac{b}{a}.$$

This means that the graphic of the (simplified) hyperbola delimited by the lines having angular coefficient $-\frac{b}{a}$ and $\frac{b}{a}$ respectively. The corresponding two lines having equation $y = -\frac{b}{a}x$ and $y = \frac{b}{a}x$, respectively, are named *asymptotes of the hyperbola*.

The figure below highlights the areas where the hyperbola can be plotted.

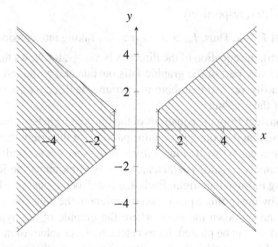

By using the simplified analytical equation of the hyperbola we can now plot the conic, as shown in the following figure.

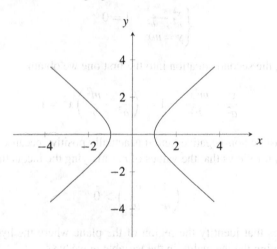

Example 6.10. Let us consider the equation

$$9x^2 - 16y^2 - 144 = 0$$

which can be re-written as

$$\frac{x^2}{a^2} - \frac{y^2}{b^2} = \frac{x^2}{4^2} - \frac{y^2}{3^2} = 1.$$

This is the equation of an hyperbola in the simplified conditions mentioned above. The coordinates of the foci are $(F,0)$ and $(-F,0)$ where

$$F = \sqrt{a^2 + b^2} = 5.$$

The directrices have equations $x = \frac{a^2}{F}$ and $x = -\frac{a^2}{F}$, respectively, where

$$\frac{a^2}{F} = \frac{4^2}{5} = \frac{16}{5}.$$

The asymptotes have equations $y = \frac{b}{a}x$ and $y = -\frac{b}{a}x$, respectively, where

$$\frac{b}{a} = \frac{3}{4}.$$

Let us work the equation of the hyperbola out to obtain an alternative definition.

Theorem 6.2. *Let* **F** *and* **F'** *be two foci of an hyperbola. An hyperbola is the locus of points* **P** *such that* $d_{\textbf{PF'}} - d_{\textbf{PF}}$ *is constant and equal to* $2a$.

Proof. The distance of a generic point of the conic from the focus is given by the equations

$$d_{\textbf{PF}} = \sqrt{(x - F)^2 + y^2}$$
$$d_{\textbf{PF'}} = \sqrt{(x - F)^2 + y^2}.$$

For the simplified equation of the hyperbola it happens that

$$y^2 = b^2 \left(\frac{x^2}{a^2} - 1\right),$$

and that

$$b^2 = \left(F^2 - a^2\right) \Rightarrow F = \sqrt{a^2 + b^2}.$$

We can write now the distance of a point from the focus as

$$d_{\textbf{PF}} = \sqrt{\left(x - \sqrt{a^2 + b^2}\right)^2 + b^2 \left(\frac{x^2}{a^2} - 1\right)} =$$
$$= \sqrt{\left(x^2 + a^2 + b^2 - 2\sqrt{a^2 + b^2}x\right) - b^2 + b^2 \frac{x^2}{a^2}} =$$
$$= \sqrt{x^2 + a^2 - 2\sqrt{a^2 + b^2}x + b^2 \frac{x^2}{a^2}} = \sqrt{a^2 - 2\sqrt{a^2 + b^2}x + \frac{(a^2 + b^2)x^2}{a^2}} =$$
$$= \sqrt{\left(a - \frac{\sqrt{a^2 + b^2}}{a}\right)^2} = |a - \frac{Fx}{a}|$$

and analogously

$$d_{\mathbf{PF'}} = \sqrt{\left(x+\sqrt{a^2+b^2}\right)^2 + b^2\left(\tfrac{x^2}{a^2}-1\right)} = \left|a+\tfrac{Fx}{a}\right|.$$

The absolute values are to highlight that the expressions above have a geometric meaning, i.e. they are distances. In order to be always positive $d_{\mathbf{PF}} = a - \tfrac{Fx}{a}$ when $a - \tfrac{Fx}{a}$ is positive and $d_{\mathbf{PF}} = -a + \tfrac{Fx}{a}$ when $a - \tfrac{Fx}{a}$ is negative. Similarly, $d_{\mathbf{PF'}} = a + \tfrac{Fx}{a}$ when $a + \tfrac{Fx}{a}$ is positive and $d_{\mathbf{PF'}} = -a - \tfrac{Fx}{a}$ when $a + \tfrac{Fx}{a}$ is negative. If we solve the inequality $a - \tfrac{Fx}{a} \geq 0$, we find that the inequality is verified when $x < \tfrac{a^2}{F}$. Considering that $x = \tfrac{a^2}{F}$ is the equation of the directrix which, as shown above, is on the left of the right hand side of the hyperbola, the inequality is verified only for the left hand side of the hyperbola. A symmetric consideration can be done for $a + \tfrac{Fx}{a} > 0 \Rightarrow x > -\tfrac{a^2}{F}$ (right hand side of the hyperbola).

In summary we have two possible scenarios:

- Right hand side: $d_{\mathbf{PF}} = -a + \tfrac{Fx}{a}$ and $d_{\mathbf{PF'}} = a + \tfrac{Fx}{a}$

- Left hand side: $d_{\mathbf{PF}} = a - \tfrac{Fx}{a}$ and $d_{\mathbf{PF'}} = -a - \tfrac{Fx}{a}$.

  In both cases we obtain

$$d_{\mathbf{PF}} - d_{\mathbf{PF'}} = |2a|.\ \square \tag{6.8}$$

Equation (6.8) gives an alternative characterization of an hyperbola.

Equation of the Parabola

If $e = 1$ the Eq. (6.7) cannot be used. However, if we substitute $e = 1$ into Eq. (6.2) we obtain

$$y^2 - 2(F - h)x + F^2 - h^2 = 0.$$

Without a loss of generality let us fix the reference system so that the conic passes through its origin. Under these conditions, the term $F^2 - h^2 = 0$. This means that $F^2 = h^2 \Rightarrow h = \pm F$. However, $h = F$ is impossible because we have supposed $\mathbf{F} \notin d$ with $\mathbf{F} : (F,0)$ and $d : x - h = 0$. Thus the statement $h = 0$ would be equivalent to say $\mathbf{F} \in d$. Thus, the only possible value of h is $-F$. If we substitute it into the equation above we have

$$y^2 - 4Fx = 0$$

that is the *analytical equation of the parabola*.

Let us try to replicate the same procedure done for the hyperbola in order to narrow down the area where a parabola can be plotted. At first let us observe that this equation, written as $y^2 = 4Fx$ imposes that the focus coordinate F and the variable x are either both positive or both negative. This means that the graphic of a (simplified) parabola is plotted either all in the positive semi-plane $x \geq 0$ or in the negative semi-plane $x \leq 0$. Let us consider the case when both $F \geq 0$ and $x \geq 0$.

Furthermore, the fact that $\forall x$ there are two values of y (one positive and one negative respectively) make the graphic of this parabola symmetric with respect to the abscissa's axis. Finally, if we look for the value of m such that the lines $y = mx$ intersect the parabola, we can easily verify that there are no impossible m values. This means that the parabola has no asymptotes.

A graphic of the parabola is shown in the figure below.

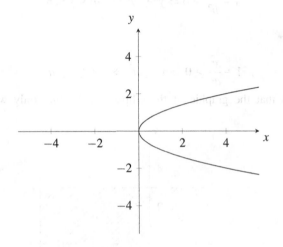

Example 6.11. Let us consider the equation

$$2y^2 - 16x = 0.$$

This is the equation of a parabola and can be re-written as

$$y^2 - 4Fx = y^2 - 8x = 0.$$

The coordinates of the focus are simply $(2,0)$ and the directrix has equation $x = -2$.

Theorem 6.3. *Let* **F** *be the focus and d be the directrix of a parabola, respectively. A parabola is the locus of points* **P** *such that* $d_{\mathbf{PF}} = d_{\mathbf{P}d}$.

Proof. Considering that $d_{\mathbf{PF}} = ed_{\mathbf{P}d}$ with $e = 1$, it occurs that $d_{\mathbf{PF}} = d_{\mathbf{P}d}$, that is an alternative definition of the parabola.

Equation of the Ellipse

If $a^2 > F^2$, it follows that $e < 1$. Let us pose $b^2 = (a^2 - F^2) > 0$ and let us substitute this piece of information into the equation of the conic in Eq. (6.7)

$$b^2x^2 + a^2y^2 - a^2b^2 = 0 \Rightarrow \frac{x^2}{a^2} + \frac{y^2}{b^2} - 1 = 0.$$

This is the *analytical equation of the ellipse*. Analogous to what done for the hyperbola, this equation is defined when $\frac{x^2}{a^2} \geq 0$ and when $\frac{y^2}{b^2} \geq 0$. This occurs when the following two inequality are verified:

$$1 - \frac{y^2}{b^2} \geq 0 \Rightarrow y^2 \leq b^2 \Rightarrow -b \leq y \leq b$$

and

$$1 - \frac{x^2}{a^2} \geq 0 \Rightarrow x^2 \leq a^2 \Rightarrow -a \leq x \leq a.$$

This means that the graphic of the ellipse can be plot only within a $a \times b$ rectangle.

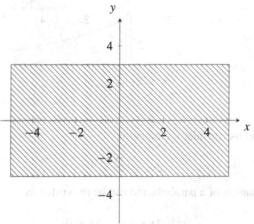

Since a is the horizontal (semi-)length of this rectangle and for an ellipse $a^2 - F^2 > 0 \Rightarrow -a < F < a$, i.e. the focus is within the rectangle. The foci have coordinates $(F, 0)$ and $(-F, 0)$, respectively where $F = \sqrt{a^2 - b^2}$.

Since $a > F$, it follows that $a^2 > aF \Rightarrow \frac{a^2}{F} > a$. Since, as shown in Eq. (6.6), the directrix has equation $x = \frac{a^2}{F}$ it follows that the directrix is outside the rectangle.

It can be easily checked that the ellipse, like the hyperbola, is symmetric with respect to both abscissa's and ordinate's axes ($y = 0$ and $x = 0$). Hence the ellipse has two foci having coordinates $(F, 0)$ and $(-F, 0)$, respectively, and two directrices having equation $x = \frac{a^2}{F}$ and $x = -\frac{a^2}{F}$. Each pair focus-directrix is associated with half graphic below (right and left hand side respectively).

The graphic of the ellipse is plotted in the figure below.

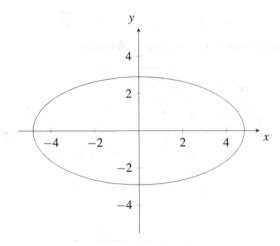

Example 6.12. Let us consider the following equation

$$4x^2 + 9y^2 - 36 = 0.$$

This equation can be re-written as

$$\frac{x^2}{a^2} + \frac{y^2}{b^2} = \frac{x^2}{3^2} + \frac{y^2}{2^2} = 1$$

that is the simplified equation of an ellipse where $a = 3$ and $b = 2$. The foci have co-ordinates $(F, 0)$ and $(-F, 0)$, respectively, where $F = \sqrt{a^2 - b^2} = \sqrt{5}$. The directrix has equation $x = \frac{a^2}{F} = \frac{9}{\sqrt{5}}$

Let us now work the equations out to give the alternative definition also for the ellipse.

Theorem 6.4. *Let* **F** *and* **F'** *be the two foci of an ellipse. An ellipse is the locus of points* **P** *such that* $d_{PF'} + d_{PF}$ *is constant and equal to 2a.*

Proof. Considering that

$$d_{PF} = \sqrt{(x - F)^2 + y^2}$$
$$d_{PF'} = \sqrt{(x - F)^2 + y^2},$$

that for the simplified equation of the ellipse it happens that

$$y^2 = b^2 \left(1 - \frac{x^2}{a^2}\right),$$

and that

$$b^2 = (a^2 - F^2) \Rightarrow F = \sqrt{a^2 - b^2},$$

we can write the distance of a point from the focus as

$$
\begin{aligned}
d_{\mathbf{PF}} &= \sqrt{\left(x - \sqrt{a^2 - b^2}\right)^2 + b^2 \left(1 - \frac{x^2}{a^2}\right)} = \\
&= \sqrt{\left(x^2 + a^2 - b^2 - 2x\sqrt{a^2 - b^2}\right) + b^2 - \left(\frac{b^2 x^2}{a^2}\right)} = \\
&= \sqrt{a^2 - 2x\sqrt{a^2 - b^2} + \frac{(a^2 - b^2)x^2}{a^2}} = \sqrt{\left(a - \frac{\sqrt{a^2 - b^2}}{a}x\right)^2} = \\
&= \left(a - \frac{\sqrt{a^2 - b^2}}{a}x\right) = a - \frac{Fx}{a}
\end{aligned}
$$

and analogously

$$d_{\mathbf{PF'}} = \sqrt{\left(x + \sqrt{a^2 - b^2}\right)^2 + b^2 \left(1 - \frac{x^2}{a^2}\right)} = a + \frac{Fx}{a}.$$

If we now sum these two distances we obtain

$$d_{\mathbf{PF}} + d_{\mathbf{PF'}} = 2a. \tag{6.9}$$

In other words, the sum of these two distances does not depend on any variable, i.e. it is a constant equal to $2a$. $\quad\square$

Equation (6.9) represents an alternative definition of the ellipse.

6.5 Matrix Representation of a Conic

In general, a conic is shifted and rotated with respect to the reference axes. In the following sections a characterization of a conic in its generic conditions is given.

Let us consider again the analytical representation of a conic. If we work out the expression in Eq. (6.1) we have:

$$
\begin{aligned}
\left(x^2 + \alpha^2 - 2\alpha x + y^2 + \beta^2 - 2\beta y\right)\left(a^2 + b^2\right) &= a^2 x^2 + b^2 y^2 + c^2 + 2abxy + 2acx + 2bcy \Rightarrow \\
\Rightarrow b^2 x^2 + a^2 y^2 - 2abxy - 2\left(\alpha a^2 + \alpha b^2 + ac\right)x - 2\left(\beta a^2 + \beta b^2 + bc\right)y + \\
+ \left(a^2\alpha^2 + a^2\beta^2 + b^2\alpha^2 + b^2\beta^2 - c^2\right).
\end{aligned}
$$

If we now perform the following replacements:

$$a_{1,1} = b^2$$
$$a_{1,2} = -ab$$
$$a_{1,3} = -\left(\alpha a^2 + \alpha b^2 + ac\right)$$
$$a_{2,2} = a^2$$
$$a_{2,3} = -\left(\beta a^2 + \beta b^2 + bc\right)$$
$$a_{3,3} = \left(a^2\alpha^2 + a^2\beta^2 + b^2\alpha^2 + b^2\beta^2 - c^2\right)$$

we can write the analytical representation of a conic as

$$a_{1,1}x^2 + 2a_{1,2}xy + 2a_{1,3}x + a_{2,2}y^2 + 2a_{2,3}y + a_{3,3} = 0 \qquad (6.10)$$

which will be referred to as *matrix representation of a conic*.

Example 6.13. The equation

$$5x^2 + 26xy + 14x + 8y^2 + 5y + 9 = 0$$

represents a conic. However, from the knowledge we have at this point, we cannot assess which kind of conic this equation represents and we cannot identify its other features, such as the position of foci and directrices. In the following pages, a complete and general characterization of the conics is presented.

6.5.1 Intersection with a Line

Let us consider two points of the plane, **T** and **R**, respectively, whose coordinates in a Cartesian reference system are (x_t, y_t) and (x_r, y_r). From Sect. 6.1.1 we know that the equation of a line passing from **T** and **R** is

$$\frac{x - x_t}{x_r - x_t} = \frac{y - y_t}{y_r - y_t}$$

where x and y are variables.

Since $x_r - x_t$ and $y_r - y_t$ are constants, let us pose $l = x_r - x_t$ and $m = y_r - y_t$. Hence, the equation of the line becomes

$$\frac{x - x_t}{l} = \frac{y - y_t}{m} \qquad (6.11)$$

where l and m are supposed $\neq 0$.

Let us pose now

$$\frac{y - y_t}{m} = t$$

where t is a parameter. It follows that

$$\begin{cases} \frac{y-y_t}{m} = t \\ \frac{x-x_t}{l} = t \end{cases} \Rightarrow \begin{cases} y = mt + y_t \\ x = lt + x_t \end{cases}$$

Let us now consider again the matrix representation of a conic in Eq. (6.10). We would like to find the intersections between the line and the conic. The search of an intersection point is the search for a point that simultaneously belongs to two objects. This means that an intersection point can be interpreted as the solution that simultaneously satisfies multiple equations. In our case the intersection point of the line with the conic is given by

$$\begin{cases} y = mt + y_t \\ x = lt + x_t \\ a_{1,1}x^2 + 2a_{1,2}xy + 2a_{1,3}x + a_{2,2}y^2 + 2a_{2,3}y + a_{3,3} = 0 \end{cases}$$

which leads to the equation

$$a_{1,1}(lt+x_t)^2 + 2a_{1,2}(lt+x_t)(mt+y_t) + 2a_{1,3}(lt+x_t) + a_{2,2}(mt+y_t)^2 + 2a_{2,3}(mt+y_t) + a_{3,3} = 0 \Rightarrow$$
$$\Rightarrow (a_{1,1}l^2 + 2a_{1,2}lm + a_{2,2}m^2)t^2 +$$
$$+ 2((a_{1,1}x_t + a_{1,2}y_t + a_{1,3})l + (a_{1,2}x_t + a_{2,2}y_t + a_{2,3})m) +$$
$$+ (a_{1,1}x_t^2 + a_{2,2}y_t^2 + 2a_{1,2}x_ty_t + 2a_{1,2}x_ty_t + 2a_{1,3}x_t + 2a_{2,3}y_t + a_{3,3}) = 0$$

that can be re-written as

$$\alpha t^2 + 2\beta t + \gamma = 0 \qquad\qquad (6.12)$$

where

$$\alpha = (a_{1,1}l^2 + 2a_{1,2}lm + a_{2,2}m^2)$$
$$\beta = ((a_{1,1}x_t + a_{1,2}y_t + a_{1,3})l + (a_{1,2}x_t + a_{2,2}y_t + a_{2,3})m)$$
$$\gamma = (a_{1,1}x_t^2 + a_{2,2}y_t^2 + 2a_{1,2}x_ty_t + 2a_{1,2}x_ty_t + 2a_{1,3}x_t + 2a_{2,3}y_t + a_{3,3}).$$

It must be observed that γ is the Eq. (6.10) calculated in the point $\mathbf{T}$. Let us now consider Eq. (6.12). It is a second order polynomial in the variable t. As such we can distinguish the following three cases:

- if the equation has two real solutions then the line crosses the conic, i.e. the line intersects the conic in two distinct points of the plane (the line is said *secant to the conic*)
- if the equation has two coinciding solution then the line is tangent to the conic
- if the equation has two complex solutions then the line does not intersect the conic (the line is said *external to the conic*)

6.5.2 Line Tangent to a Conic

Let us focus on the case when Eq. (6.12) has two coinciding solutions, i.e. when the line is tangent to the conic, and let us write down the equation of a line tangent to a conic.

The solution of Eq. (6.12) corresponds to a point $\mathbf{T}$ of the plane belonging to both the line and the conic. Since this point belongs to the conic, its coordinates satisfy Eq. (6.10) that, as observed above, can be written as $\gamma = 0$. Since $\gamma = 0$ Eq. (6.12) can be written as

$$\alpha t^2 + \beta t = 0.$$

Obviously this equation can be re-written as

$$t(\alpha t + \beta) = 0$$

which has $t = 0$ as a solution. Since the solutions are coinciding with $t = 0$ it follows that

$$\beta = 0 \Rightarrow ((a_{1,1}x_t + a_{1,2}y_t + a_{1,3})l + (a_{1,2}x_t + a_{2,2}y_t + a_{2,3})m) = 0.$$

This is an equation in two variables, l and m respectively. Obviously, this equation has ∞ solutions. The solution $(0,0)$, albeit satisfying the equation, is unacceptable because it has no geometrical meaning, see Eq. (6.11). Since ∞ solutions satisfy the equation if we find one solution, all the others are proportionate to it. A solution that satisfies the equation is the following:

$$l = (a_{1,2}x_t + a_{2,2}y_t + a_{2,3})$$
$$m = -(a_{1,1}x_t + a_{1,2}y_t + a_{1,3}).$$

If we substitute the values of l and m into Eq. (6.11) we obtain

$$(a_{1,1}x_t + a_{1,2}y_t + a_{1,3})(x - x_t) + (a_{1,2}x_t + a_{2,2}y_t + a_{2,3})(y - y_t) = 0 \Rightarrow$$
$$\Rightarrow a_{1,1}x_t x + a_{1,2}y_t x + a_{1,3}x - a_{1,1}x_t^2 - a_{1,2}y_t x_t - a_{1,3}x_t +$$
$$+ a_{1,2}x_t y + a_{2,2}y_t y + a_{2,3}y - a_{1,2}x_t y_t - a_{2,2}y_t^2 - a_{2,3}y_t = 0 \Rightarrow \qquad (6.13)$$
$$\Rightarrow a_{1,1}x_t x + a_{1,2}y_t x + a_{1,3}x + a_{1,2}x_t y + a_{2,2}y_t y + a_{2,3}y +$$
$$- a_{1,1}x_t^2 - 2a_{1,2}x_t y_t - a_{1,3}x_t - a_{2,2}y_t^2 - a_{2,3}y_t = 0.$$

Let us consider again Eq. (6.10) and let us consider that $\mathbf{T}$ is a point of the conic. Hence, the coordinates of $\mathbf{T}$ verify the equation of the conic:

$$a_{1,1}x_t^2 + 2a_{1,2}x_t y_t + 2a_{1,3}x_t + a_{2,2}y_t^2 + 2a_{2,3}y_t + a_{3,3} = 0 \Rightarrow$$
$$\Rightarrow a_{1,3}x_t + a_{2,3}y_t + a_{3,3} = -a_{1,1}x_t^2 - a_{2,2}y_t^2 - 2a_{1,2}x_t y_t - a_{2,3}y_t - a_{1,3}x_t.$$

Substituting this result into Eq. (6.13) we obtain

$$a_{1,1}x_t x + a_{1,2}y_t x + a_{1,3}x + a_{1,2}x_t y + a_{2,2}y_t y + a_{2,3}y + a_{1,3}x_t + a_{2,3}y_t + a_{3,3} = 0 \Rightarrow$$
$$\Rightarrow (a_{1,1}x_t + a_{1,2}y_t + a_{1,3})x + (a_{1,2}x_t + a_{2,2}y_t + a_{2,3})y + (a_{1,3}x_t + a_{2,3}y_t + a_{3,3}) = 0$$
$$(6.14)$$

that is the *equation of a line tangent to a conic* in the point **T** with coordinates (x_t, y_t).

6.5.3 Degenerate and Non-degenerate Conics: A Conic as a Matrix

The equation of a line tangent to a conic is extremely important to study and understand conics. In order to do so, let us consider the very special case when Eq. (6.14) is verified regardless of the values of x and y or, more formally $\forall x, y$ the line is tangent to the conic. This situation occurs when the coefficients of x and y are null as well as the constant coefficient. In other words, this situation occurs when

$$\begin{cases} a_{1,1}x_t + a_{1,2}y_t + a_{1,3} = 0 \\ a_{1,2}x_t + a_{2,2}y_t + a_{2,3} = 0 \\ a_{1,3}x_t + a_{2,3}y_t + a_{3,3} = 0. \end{cases}$$

Algebraically, this is a system of three linear equations in two variables. From Rouchè-Capelli Theorem we know that, since the rank of the incomplete matrix is at most 2, if the determinant of the complete matrix is null then the system is certainly incompatible. In our case, if

$$\det \mathbf{A^c} = \det \begin{pmatrix} a_{1,1} & a_{1,2} & a_{1,3} \\ a_{1,2} & a_{2,2} & a_{2,3} \\ a_{1,3} & a_{2,3} & a_{3,3} \end{pmatrix} \neq 0$$

then the system is surely incompatible, i.e. it is impossible to have a line tangent to the conic regardless of x and y. On the contrary, if $\det \mathbf{A^c} = 0$ the system has at least one solution, i.e. the special situation when $\forall x, y$ the line is tangent to the conic is verified.

Geometrically, a line can be tangent to a conic regardless of its x and y values only when the conic is a line itself or a point of the line. These situations correspond to a degenerate conic. From the considerations above, given a generic conic having equation

$$a_{1,1}x^2 + 2a_{1,2}xy + 2a_{1,3}x + a_{2,2}y^2 + 2a_{2,3}y + a_{3,3} = 0,$$

we can associate to it the following matrix

$$\mathbf{A^c} = \begin{pmatrix} a_{1,1} & a_{1,2} & a_{1,3} \\ a_{1,2} & a_{2,2} & a_{2,3} \\ a_{1,3} & a_{2,3} & a_{3,3} \end{pmatrix}.$$

The determinant of this matrix tells us whether or not this conic is degenerate. More specifically the following cases can occur:

- $\det \mathbf{A^c} \neq 0$: the conic is non-degenerate
- $\det \mathbf{A^c} = 0$: the conic is degenerate.

This finding allows us to see how a conic can also be considered as a matrix. Furthermore, the understanding of the determinant can also be revisited: a line tangent to a conic means that two figures in the plane have one point in common or that the intersection of two sets is a single point. If this intersection point is the only point of the conic or if the intersection is the entire line we can conclude that the set of the points of the conic is contained within the set of the line. This can be seen as an initial problem involving two sets is collapsed into a problem involving only one set. Yet again, a null determinant can be seen as the presence of redundant pieces of information.

The figure below depicts the two situations. On the left a line tangent to a non-degenerate conic is shown. The tangent point **T** is also indicated. On the right, the line is tangent to the conic in all its points, i.e. the conic is part of the line or coincides with it. In figure, the overlap of the conic with the line is graphically represented with a thick line.

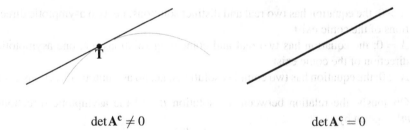

$$\det \mathbf{A^c} \neq 0 \qquad\qquad\qquad \det \mathbf{A^c} = 0$$

Example 6.14. The conic having equation

$$5x^2 + 2xy + 14x + 8y^2 + 5y + 9 = 0$$

is associated with the matrix

$$\begin{pmatrix} 5 & 1 & 7 \\ 1 & 8 & 3 \\ 7 & 3 & 9 \end{pmatrix}$$

whose determinant is -44. The conic is non-degenerate.

6.5.4 Classification of a Conic: Asymptotic Directions of a Conic

Let us consider again Eq. (6.12) and let us analyse again its meaning. It represents the intersection of a line with a generic conic. It is a second order polynomial because the conic has a second order equation. If $\alpha = 0$ the polynomial becomes of the first order.

Algebraically, the condition $\alpha = 0$ can be written as

$$\alpha = \left(a_{1,1}l^2 + 2a_{1,2}lm + a_{2,2}m^2\right) = 0.$$

This equation would be verified for $(l,m) = (0,0)$. However, this solution cannot be considered for what written in Eq. (6.11). Hence, we need to find a solution $(l,m) \neq (0,0)$. A pair of these numbers, when existent, are said *asymptotic direction of the conic*.

Let us divide the equation $\alpha = 0$ by l^2. If we pose $\mu = \frac{m}{l}$ the equation becomes

$$\left(a_{2,2}\mu^2 + 2a_{1,2}\mu + a_{1,1}\right) = 0$$

namely, *equation of the asymptotic directions*.

Let us solve the equation above as a second order polynomial in the variable μ. In order to do it we need to discuss the sign of

$$\Delta = a_{1,2}^2 - a_{1,1}a_{2,2}.$$

More specifically the following cases can be distinguished:

- $\Delta > 0$: the equation has two real and distinct solutions, i.e. two asymptotic directions of the conic exist
- $\Delta = 0$: the equation has two real and coinciding solutions, i.e. one asymptotic direction of the conic exist
- $\Delta < 0$: the equation has two complex solutions, i.e. no asymptotic directions exist

Obviously, the relation between the solution μ and the asymptotic directions (l,m) is given by

$$(1,\mu) = \left(1, \frac{m}{l}\right).$$

The amount of asymptotic directions is a very important feature of conics. The following theorem describes this fact.

Theorem 6.5. *An hyperbola has two asymptotic directions, a parabola has one asymptotic direction, an ellipse has no asymptotic directions.*

Furthermore, since

$$\Delta = a_{1,2}^2 - a_{1,1}a_{2,2} = -\left(a_{1,2}^2 - a_{1,1}a_{2,2}\right) = -\det\begin{pmatrix} a_{1,1} & a_{1,2} \\ a_{1,2} & a_{2,2} \end{pmatrix}$$

we can study a conic from directly its matrix representation in Eq. (6.10).

Theorem 6.6. Classification Theorem. *Given the a conic having equation*

$$a_{1,1}x^2 + 2a_{1,2}xy + a_{2,2}y^2 + a_{1,3}x + a_{2,3}y + a_{3,3} = 0$$

and associated matrix

$$\begin{pmatrix} a_{1,1} & a_{1,2} & a_{1,3} \\ a_{1,2} & a_{2,2} & a_{2,3} \\ a_{1,3} & a_{2,3} & a_{3,3} \end{pmatrix},$$

this conic is classified by the determinant of its submatrix

$$\mathbf{I}_{3,3} = \begin{pmatrix} a_{1,1} & a_{1,2} \\ a_{1,2} & a_{2,2} \end{pmatrix}.$$

It occurs that

- *if* $\det \mathbf{I}_{33} < 0$ *the conic is an hyperbola*
- *if* $\det \mathbf{I}_{33} = 0$ *the conic is a parabola*
- *if* $\det \mathbf{I}_{33} > 0$ *the conic is an ellipse*

Example 6.15. Let us consider again the conic having equation

$$5x^2 + 2xy + 14x + 8y^2 + 5y + 9 = 0.$$

We already know that this conic is non-degenerate. Let us classify this conic by calculating the determinant of $\mathbf{I}_{33}$:

$$\det \mathbf{I}_{3,3} = \det \begin{pmatrix} 5 & 1 \\ 1 & 8 \end{pmatrix} = 38 > 0.$$

The conic is an ellipse.

Example 6.16. We know that the conic having equation

$$\frac{x^2}{25} + \frac{y^2}{16} = 1$$

is an ellipse in its simplified form. Let us verify it by applying the Classification Theorem. Let us verify at first that this conic is non-degenerate. In order to do it, let us write the equation in its matrix representation:

$$16x^2 + 25y^2 - 400 = 0.$$

Let us write its associated matrix:

$$\mathbf{A}^c = \begin{pmatrix} 16 & 0 & 0 \\ 0 & 25 & 0 \\ 0 & 0 & -400 \end{pmatrix}.$$

It can be immediately observed that this matrix is diagonal. A such the matrix is non-singular, hence the conic is non-degenerate. Furthermore, since it happens that

$$\det \mathbf{I}_{3,3} = \det \begin{pmatrix} 16 & 0 \\ 0 & 25 \end{pmatrix} = 400 > 0,$$

the conic is an ellipse.

It can be shown that all the simplified equations of ellipses and hyperbolas correspond to diagonal matrices. More rigorously, the simplified equations are written by choosing a reference system that leads to a diagonal matrix. Let us verify this statement by considering the simplified equation of an hyperbola

$$\frac{x^2}{25} - \frac{y^2}{16} = 1.$$

The matrix representation of this conic is

$$16x^2 - 25y^2 - 400 = 0$$

with associated matrix

$$\mathbf{A}^c = \begin{pmatrix} 16 & 0 & 0 \\ 0 & -25 & 0 \\ 0 & 0 & -400 \end{pmatrix}.$$

Since this matrix is non-singular it is non-degenerate. By applying the Classification Theorem we observe that since

$$\det \mathbf{I}_{3,3} = \det \begin{pmatrix} 16 & 0 \\ 0 & -25 \end{pmatrix} = -400 < 0,$$

the conic is an hyperbola.

Example 6.17. Let us consider the following simplified equation of a parabola:

$$2y^2 - 2x = 0.$$

The associated matrix is

$$\mathbf{A}^c = \begin{pmatrix} 0 & 0 & 1 \\ 0 & 2 & 0 \\ 1 & 0 & 0 \end{pmatrix},$$

that is non-singular. Hence, the conic is non-degenerate. If we apply the Classification Theorem we obtain

$$\det \mathbf{I}_{3,3} = \det \begin{pmatrix} 0 & 0 \\ 0 & 2 \end{pmatrix} = 0,$$

i.e. the conic is a parabola.

It can be observed that the simplified equation of a parabola corresponds to a matrix where only its elements on the secondary diagonal are non-null.

Example 6.18. Let us now write the equation of a generic conic and let us classify it:

$$6y^2 - 2x + 12xy + 12y + 1 = 0.$$

At first, the associated matrix is

$$\mathbf{A}^c = \begin{pmatrix} 0 & 6 & -1 \\ 6 & 6 & 6 \\ -1 & 6 & 0 \end{pmatrix},$$

whose determinant is $-36 - 36 - 6 = -78 \neq 0$. The conic is non-degenerate. Let us classify it

$$\det \mathbf{I}_{3,3} = \det \begin{pmatrix} 0 & 6 \\ 6 & 6 \end{pmatrix} = -36 < 0.$$

The conic is an hyperbola.

Proposition 6.1. *The asymptotic directions of an hyperbola are perpendicular if and only if the trace of the associated submatrix* $\mathbf{I}_{33}$ *is equal to 0.*

Proof. For an hyperbola the equation of the asymptotic directions

$$\left(a_{2,2}\mu^2 + 2a_{1,2}\mu + a_{1,1} \right) = 0$$

has two distinct real roots μ_1 and μ_2. The associated asymptotic directions are $(1, \mu_1)$ and $(1, \mu_2)$, respectively. These two directions can be seen as two vectors in the plane.

If these two directions are perpendicular for Proposition 4.6, their scalar product is null:

$$1 + \mu_1\mu_2 = 0.$$

For Proposition 5.7 $\mu_1\mu_2 = \frac{a_{1,1}}{a_{2,2}}$. Hence,

$$1 + \frac{a_{1,1}}{a_{2,2}} = 0 \Rightarrow a_{11} = -a_{2,2}.$$

The trace of $\mathbf{I}_{33}$ is $a_{1,1} + a_{2,2} = 0$ $\square$

If $\mathrm{tr}(\mathbf{I}_{33}) = 0$, then $a_{11} = -a_{2,2}$, that is $\frac{a_{1,1}}{a_{2,2}} = -1$. For Proposition 5.7 $\mu_1\mu_2 = -1 \Rightarrow (1, \mu_1)(1, \mu_2) = 0$, i.e. the two directions are perpendicular. $\square$

Example 6.19. The hyperbola having equation

$$-2x^2 + 2y^2 - x + 3xy + 5y + 1 = 0$$

has perpendicular asymptotic directions since $a_{11} + a_{2,2} = 2 - 2 = 0$.

Let us find the asymptotic directions. In order to do it let us solve the equation of the asymptotic directions:

$$\left(a_{2,2}\mu^2 + 2a_{1,2}\mu + a_{1,1}\right) = 2\mu^2 + 3\mu - 2 = 0,$$

whose solutions are $\mu_1 = -2$ and $\mu_2 = \frac{1}{2}$. The corresponding asymptotic directions are $(1,-2)$ and $\left(1,\frac{1}{2}\right)$.

Definition 6.10. An hyperbola whose asymptotic directions are perpendicular is said to be *equilateral*.

Let us see a few degenerate examples.

Example 6.20. Let us consider the following conic

$$2x^2 + 4y^2 + 6xy = 0$$

and its associated matrix

$$\mathbf{A}^c = \begin{pmatrix} 2 & 3 & 0 \\ 3 & 4 & 0 \\ 0 & 0 & 0 \end{pmatrix}.$$

The determinant of the matrix is null. Hence, the conic is degenerate. By applying the classification theorem we find out that since

$$\det \mathbf{I}_{3,3} = \det \begin{pmatrix} 2 & 3 \\ 3 & 4 \end{pmatrix} = -1 < 0$$

the conic is an hyperbola. More specifically the conic is a degenerate hyperbola, i.e. a pair of intersecting lines. By working out the original equation we obtain

$$2x^2 + 4y^2 + 6xy = 0 \Rightarrow 2\left(x^2 + 2y^2 + 3xy\right) = 0 \Rightarrow 2\left(x+y\right)\left(x+2y\right) = 0.$$

Hence the conic is the following pair of lines:

$$y = -x$$
$$y = -\frac{x}{2}.$$

Example 6.21. If we consider the conic having equation

$$2x^2 + 4y^2 + 2xy = 0,$$

its associated matrix is

$$\mathbf{A}^c = \begin{pmatrix} 2 & 1 & 0 \\ 1 & 4 & 0 \\ 0 & 0 & 0 \end{pmatrix}$$

is singular and its classification leads to

$$\det \mathbf{I}_{3,3} = \det \begin{pmatrix} 2 & 1 \\ 1 & 4 \end{pmatrix} = 7 > 0,$$

i.e. the conic is an ellipse. More specifically, the only real point that satisfies the equation of this conic (and thus the only point with geometrical meaning) is $(0,0)$. The conic is a point.

Example 6.22. Let us consider the conic having equation

$$y^2 + x^2 - 2xy - 9 = 0.$$

The matrix associated with this conic is

$$\mathbf{A^c} = \begin{pmatrix} 1 & -1 & 0 \\ -1 & 1 & 0 \\ 0 & 0 & -9 \end{pmatrix}$$

whose determinant is null. The conic is degenerate and, more specifically, a degenerate parabola since

$$\det \mathbf{I_{3,3}} = \det \begin{pmatrix} 1 & -1 \\ -1 & 1 \end{pmatrix} = 0.$$

This equation of the conic can be written as $(x - y + 3)(x - y - 3) = 0$ that is the equation of two parallel lines:

$$y = x + 3$$
$$y = x - 3.$$

Example 6.23. Let us consider the conic having equation

$$x^2 + 4y^2 + 4xy = 0.$$

Since the matrix associated with this conic is

$$\mathbf{A^c} = \begin{pmatrix} 1 & 2 & 0 \\ 2 & 4 & 0 \\ 0 & 0 & 0 \end{pmatrix},$$

the conic is degenerate. Considering that

$$\det \mathbf{I_{3,3}} = \det \begin{pmatrix} 1 & 2 \\ 2 & 4 \end{pmatrix} = 0,$$

it follows that the conic is a parabola. The conic can written as $(x + 2y)^2 = 0$ that is the equation of two coinciding lines.

The last two examples show that a degenerate parabola can be of two kinds. In Example 6.21 the parabola breaks into two parallel lines. On the contrary, in Example 6.23 the two parallel lines are also overlapping. In the first case the rank of the matrix $\mathbf{A^c}$ is 2 while in the second one the rank of $\mathbf{A^c}$ is 1. Degenerate conics belonging to the latter group are said *twice degenerate*.

6.5.5 Diameters, Centres, Asymptotes, and Axes of Conics

Definition 6.11. A *chord of a conic* is a segment connecting two arbitrary points of a conic.

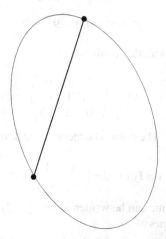

Obviously, since infinite points belong to a conic, a conic has infinite chords.

Definition 6.12. Let (l, m) be an arbitrary direction. A *diameter of a conic conjugate to the direction* (l, m) (indicated with *diam*) is the locus of the middle points of all the possible chords of a conic that have direction (l, m). The direction (l, m) is said to be *direction conjugate to the diameter* of the conic.

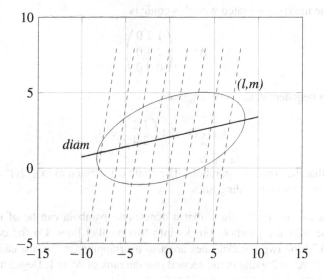

It must be noted that in this context, the term conjugated is referred with respect to the conic. In other words the directions of line and diameter are conjugated to each other with respect to the conic.

Definition 6.13. Two diameters are said to be conjugate if the direction of one diameter is the conjugate direction of the other and vice-versa.

The conjugate diameters *diam* and *diam'* are depicted in figure below.

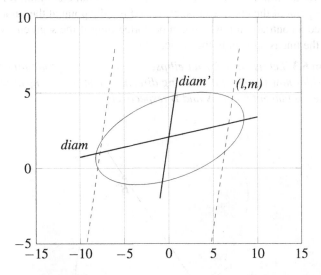

Proposition 6.2. *Let diam be the diameter of a conic conjugated to a direction* (l, m). *If this diameter intersects the conic in a point* **P** *the line containing* **P** *and parallel to the direction* (l, m) *is tangent to the conic in the point* **P**.

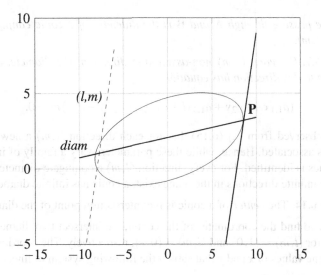

Proof. Since **P** belongs to both the conic and the line, it follows that the line cannot be external to the conic. It must be either tangent or secant.

Let us assume by contradiction, that the line is secant to the conic and thus intersects it in two points. We know that one point is **P**. Let us name the second intersection point **Q**. It follows that the chord $\overline{\mathbf{PQ}}$ is parallel to (l,m). The middle point of this chord, for definition of diameter, is a point of the diameter *diam*. It follows that **P** is middle point of the chord $\overline{\mathbf{PQ}}$ (starting and middle point at the same time). We have reached a contradiction which can be sorted only if the segment is one point, i.e. only if the line is tangent to the conic. □

Proposition 6.3. *Let us consider an ellipse and a non-asymptotic direction* (l,m). *Let us consider now the two lines having direction* (l,m) *and tangent to the conic. Let us name the tangent points* **A** *and* **B** *respectively.*

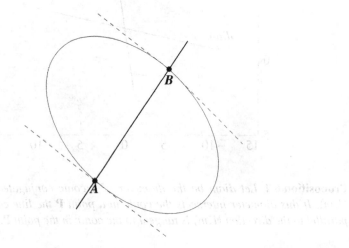

The line passing through **A** *and* **B** *is the diameter of a conic conjugate to the direction* (l,m).

Theorem 6.7. *For each* (l,m) *non-asymptotic direction, the diameter of a conic conjugated to this direction has equation*

$$(a_{1,1}x + a_{1,2}y + a_{1,3})l + (a_{1,2}x + a_{2,2}y + a_{2,3})m = 0. \qquad (6.15)$$

It can be observed from Eq. (6.15) that for each direction (l,m) a new conjugate diameter is associated. Hence, while these parameters vary a family of intersecting straight lines is identified. For every direction (l,m) a conjugate diameter is identified. Since infinite directions in the plane exist, a conic has infinite diameters.

Definition 6.14. The *center* of a conic is the intersection point of the diameters.

In order to find the coordinates of the center, we intersect two diameters that is Eq. (6.15) for $(l,m) = (1,0)$ and $(l,m) = (0,1)$, respectively. The coordinates of the center are the values of x and y that satisfy the following system of linear equations:

$$\begin{cases} a_{1,1}x + a_{1,2}y + a_{1,3} = 0 \\ a_{1,2}x + a_{2,2}y + a_{2,3} = 0. \end{cases} \qquad (6.16)$$

The system is determined when

$$\det \begin{pmatrix} a_{1,1} & a_{1,2} \\ a_{1,2} & a_{2,2} \end{pmatrix} \neq 0.$$

In this case the conic is an either an hyperbola or an ellipse. These conics have a center and infinite diameters passing through it (family of intersecting straight lines). On the contrary, if

$$\det \begin{pmatrix} a_{1,1} & a_{1,2} \\ a_{1,2} & a_{2,2} \end{pmatrix} = 0$$

the conic is a parabola and two cases can be distinguished:

- $\frac{a_{1,1}}{a_{1,2}} = \frac{a_{1,2}}{a_{2,2}} \neq \frac{a_{1,3}}{a_{2,3}}$: the system is incompatible, the conic has no center within the plane;
- $\frac{a_{1,1}}{a_{1,2}} = \frac{a_{1,2}}{a_{2,2}} = \frac{a_{1,3}}{a_{2,3}}$: the system is undetermined and has infinite centres.

In the first case, the two equations of system (6.16) are represented by two parallel lines (hence with no intersections). Since there are no x and y values satisfying the system also Eq. (6.15) is never satisfied. Hence, the conic has infinite parallel diameters (a family of parallel straight lines).

In the second case, the parabola is degenerate. The rows of the equations of the system are proportionate, i.e. $a_{1,2} = \lambda a_{1,1}$, $a_{2,2} = \lambda a_{1,2}$, and $a_{2,3} = \lambda a_{1,3}$ with $\lambda \in \mathbb{R}$. The two equations of system (6.16) are represented by two coinciding lines.

It follows that

$$(a_{1,1}x + a_{1,2}y + a_{1,3})l + (a_{1,2}x + a_{2,2}y + a_{2,3})m = 0 \Rightarrow$$
$$\Rightarrow (a_{1,1}x + a_{1,2}y + a_{1,3})(l + \lambda m) = 0 \Rightarrow (a_{1,1}x + a_{1,2}y + a_{1,3}) = 0.$$

In this degenerate case, the parabola has thus only one diameter (all the diameters are overlapped on the same line) having equation $(a_{1,1}x + a_{1,2}y + a_{1,3}) = 0$ and where each of its points is center.

Example 6.24. If we consider again the hyperbola having equation

$$6y^2 - 2x + 12xy + 12y + 1 = 0,$$

we can write the equation of the diameters

$$(6y - 1)l + (6x + 6y + 6)m = 0.$$

In order to find two diameters let us write the equations in the specific case $(l,m) = (0,1)$ and $(l,m) = (1,0)$. The two corresponding diameters are

$$6y - 1 = 0$$
$$6x + 6y + 6 = 0.$$

Let us now simultaneously solve the equations to obtain the coordinates of the center. The coordinates are $\left(-\frac{7}{6}, \frac{1}{6}\right)$.

Example 6.25. Let us consider the conic having equation

$$4y^2 + 2x + 2y - 4 = 0.$$

The associated matrix

$$\mathbf{A}^c = \begin{pmatrix} 0 & 0 & 1 \\ 0 & 4 & 1 \\ 1 & 1 & -4 \end{pmatrix}$$

has determinant -4. The conic is non-degenerate and, more specifically, a parabola since

$$\det \mathbf{I}_{3,3} = \det \begin{pmatrix} 0 & 0 \\ 0 & 4 \end{pmatrix} = 0.$$

This parabola has no center. If we tried to look for one we would need to solve the following system of linear equations

$$\begin{cases} 1 = 0 \\ 4y + 1 = 0 \end{cases}$$

which is incompatible. The parabola has infinite diameters parallel to the line having equation $4y + 1 = 0 \Rightarrow y = -\frac{1}{4}$.

Example 6.26. Let us consider again the conic having equation

$$x^2 + 4y^2 + 4xy = 0.$$

We know that this conic is a degenerate parabola consisting of two coinciding lines. Let us write the system of linear equations for finding the coordinates of the center

$$\begin{cases} x + 2y = 0 \\ 2x + 4y = 0. \end{cases}$$

This system is undetermined and has ∞ solutions. The conic has ∞ centres belonging to the diameter. The latter is the line having equation $x + 2y = 0$.

Proposition 6.4. *Let the conic $\mathscr{C}$ be an ellipse or an hyperbola. The diameters of this conic are a family of intersecting straight lines whose intersection point is the center of symmetry of the conic $\mathscr{C}$.*

Proof. Let (l, m) be a non-asymptotic direction and *diam* the diameter conjugated to (l, m). Let us indicate with (l', m') the direction of the diameter *diam*. The diameter *diam* intersects the conic in two points $\mathbf{A}$ and $\mathbf{B}$, respectively. Thus, the segment $\overline{\mathbf{AB}}$ is a chord. We may think of the infinite chords parallel to $\overline{\mathbf{AB}}$ and thus to a diameter *diam'* conjugated to the direction (l, m). The diameter *diam'* intersects the conic in the points $\mathbf{C}$ and $\mathbf{D}$, respectively, and the segment $\overline{\mathbf{AB}}$ in its middle point $\mathbf{M}$ (for definition of diameter). The point $\mathbf{M}$ is also middle point for the segment $\overline{\mathbf{CD}}$.

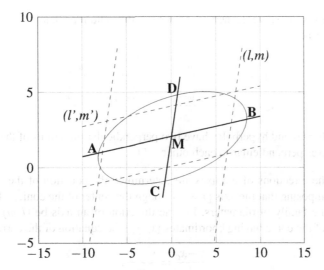

We can make the same consideration for any arbitrary direction (l,m), i.e. every chord is intersected in its middle point by a chord having direction conjugate to (l,m). Hence, the diameters are a family of straight lines intersecting in **M** and **M** is the center of symmetry. □

Definition 6.15. An *axis of a conic* is a diameter when it is perpendicular to its conjugate direction (equivalently: an axis is a diameter whose direction is conjugated to its perpendicular direction).

In order to calculate the equation of an axis, let us prove the following theorem.

Theorem 6.8. *An ellipse or an hyperbola has two axes, perpendicular to each other.*

Proof. Let us consider a direction (l,m) and Eq. (6.15):

$$(a_{1,1}x + a_{1,2}y + a_{1,3})\,l + (a_{1,2}x + a_{2,2}y + a_{2,3})\,m = 0 \Rightarrow$$
$$\Rightarrow (a_{1,1}lx + a_{1,2}ly + a_{1,3}) + (a_{1,2}mx + a_{2,2}my + a_{2,3}m) = 0 \Rightarrow$$
$$\Rightarrow (a_{1,1}l + a_{1,2}m)\,x + (a_{1,2}l + a_{2,2}m)\,y + (a_{1,3}l + a_{2,3}m) = 0.$$

The direction of the diameter is $\left(-\,(a_{1,2}l + a_{2,2}m)\,,(a_{1,1}l + a_{1,2}m)\right)$. The diameter is an axis if its direction is perpendicular to (l,m) and hence if their scalar product is null:

$$(a_{1,2}l + a_{2,2}m)\,l - (a_{1,1}l + a_{1,2}m)\,m = 0 \Rightarrow$$
$$a_{1,2}l^2 + a_{2,2}lm - a_{1,1}lm - a_{1,2}m^2 = 0 \Rightarrow$$
$$a_{1,2}l^2 + (a_{2,2} - a_{1,1})\,lm - a_{1,2}m^2 = 0.$$

If we solve this equation in the variable l we have that the discriminant is

$$\left((a_{2,2} - a_{1,1})\,m\right)^2 + 4a_{1,2}^2 m^2$$

which is always positive. Hence, it always has two solutions that is the direction of two axes, respectively

$$l_1 = \frac{-(a_{2,2}-a_{1,1})m + \sqrt{((a_{2,2}-a_{1,1})m)^2 + 4a_{1,2}^2 m^2}}{2a_{1,2}}$$
$$l_2 = \frac{-(a_{2,2}-a_{1,1})m - \sqrt{((a_{2,2}-a_{1,1})m)^2 + 4a_{1,2}^2 m^2}}{2a_{1,2}}.$$

Since ellipses and hyperbolas have two perpendicular directions of the axes then have two axes perpendicular to each other. □

When the directions of the axes are determined the equation of the axis is obtained by imposing that the axis passes through the center of the conic, which is the center of the family of diameters. Let the direction of an axis be (l,m) and **C** be the center of the conic having coordinates (x_c, y_c) the equation of the corresponding axis is

$$\frac{x - x_c}{l} = \frac{y - y_c}{m}.$$

Corollary 6.1. *In a circumference every diameter is axis.*

Proof. If the conic is a circumference it has equation $x^2 + y^2 = R$, thus $a_{1,1} = a_{2,2}$ and $a_{1,2} = 0$. It follows that the equation

$$a_{1,2}l^2 + (a_{2,2} - a_{1,1})\, lm - a_{1,2}m^2 = 0$$

is always verified. Hence every diameter is axis. □

Example 6.27. Let us consider the conic having equation

$$9x^2 + 4xy + 6y^2 - 10 = 0.$$

The associated matrix

$$\begin{pmatrix} 9 & 2 & 0 \\ 2 & 6 & 0 \\ 0 & 0 & -10 \end{pmatrix}$$

is non-singular. Hence, the conic is non-degenerate.

The submatrix

$$\begin{pmatrix} 9 & 2 \\ 2 & 6 \end{pmatrix}$$

has positive determinant. Hence, the conic is an ellipse.

The equation of the family of diameters is

$$(9x + 2y)\,l + (2x + 6y)\,m = (9l + 2m)\,x + (2l + 6m)\,y = 0.$$

From the equation of the family of diameters we can obtain the coordinates of the center by solving the following system of linear equations

$$\begin{cases} 9x + 2y = 0 \\ 2x + 6y = 0 \end{cases}.$$

The coordinates of the center are $(0,0)$. In order to find the direction of the axes (l,m) it is imposed that

$$-(2l + 6m)\, l + (9l + 2m)\, m = 0 \Rightarrow -2l^2 + 3lm + 2m^2 = 0.$$

Posing $\mu = \frac{m}{l}$ and dividing the equation by l^2 we obtain

$$2\mu^2 + 3\mu - 2 = 0$$

whose solutions are

$$\mu = \tfrac{1}{2}$$
$$\mu = -2.$$

Hence the directions of the axes are $\left(1, \tfrac{1}{2}\right)$ and $(1, -2)$. The corresponding equations of the axes, i.e. equations of lines having these directions and passing through the center of the conic are

$$x - 2y = 0$$
$$2x + y = 0.$$

Theorem 6.9. *Let $\mathbf{A}$ be the matrix associated with a parabola and $(a_{1,1}x + a_{1,2} y + a_{1,3})\, l + (a_{1,2}x + a_{2,2}y + a_{2,3})\, m$ be the equation of the family of its parallel diameters. The axis of a parabola is only one and is parallel to its diameters. The coefficients of the equation $ax + by + c = 0$ of the axis are equal to the result of the following matrix product*

$$\left(a\ b\ c \right) = \left(a_{1,1}\ a_{1,2}\ 0 \right) \mathbf{A}^c.$$

Corollary 6.2. *The coefficients of the equation of the axis of a parabola can also be found as*

$$\left(a\ b\ c \right) = \left(a_{2,1}\ a_{2,2}\ 0 \right) \mathbf{A}^c.$$

Example 6.28. The conic having equation

$$x^2 + 4xy + 4y^2 - 6x + 1 = 0$$

has associated matrix

$$\begin{pmatrix} 1 & 2 & -3 \\ 2 & 4 & 0 \\ -3 & 0 & 1 \end{pmatrix}$$

which is non-singular. The conic is non-degenerate.

The submatrix
$$\begin{pmatrix} 1 & 2 \\ 2 & 4 \end{pmatrix}$$

is singular. The conic is a parabola.

The family of diameters have equation

$$(x+2y-3)\,l + (2x+4y)\,m = 0.$$

To find the axis of the conic let us calculate

$$(1\,2\,0)\begin{pmatrix} 1 & 2 & -3 \\ 2 & 4 & 0 \\ -3 & 0 & 1 \end{pmatrix} = (5\ 10\ -3).$$

The equation of the axis is $5x + 10y - 3 = 0$.

Although an in depth understanding of the axis of the parabola is outside the scopes of this book, it may be useful to observe that the procedure is, at the first glance, very different with respect to that described for ellipse and hyperbola. However, the procedures are conceptually identical after the assumption that the center of the parabola exists and is outside the plane where the parabola lays. This center is a so called *point at infinity* which is a notion of Projective Geometry, see [17].

Definition 6.16. The *vertex of a conic* is the intersection of the conic with its axes.

Example 6.29. Let us find the vertex of the parabola above. We simply need to find the intersections of the conic with its axis, i.e. we need to solve the following system of equations:
$$\begin{cases} x^2 + 4xy + 4y^2 - 6x + 1 = 0 \\ 5x + 10y - 3 = 0. \end{cases}$$

By substitution we find that $x = \frac{17}{75}$ and $y = \frac{14}{75}$.

Definition 6.17. An *asymptote* of a conic is a line passing through its center and having as its direction the asymptotic direction.

An ellipse has no asymptotes because it has no asymptotic directions. An hyperbola has two asymptotes, one per each asymptotic direction. A parabola, although has one asymptotic direction, has no (univocally defined) center, and thus no asymptote.

Example 6.30. Let us consider again the hyperbola having equation

$$-2x^2 + 2y^2 - x + 3xy + 5y + 1 = 0$$

We know that the asymptotic directions are $(1, -2)$ and $\left(1, \frac{1}{2}\right)$.

The center of the conic can be easily found by soling the following system of linear equations:

$$\begin{cases} -2x + \frac{3}{2}y - \frac{1}{2} = 0 \\ \frac{3}{2}x + 2y + \frac{5}{2} = 0 \end{cases}$$

whose solution is $x_c = -\frac{19}{25}$ and $y_c = -\frac{17}{25}$.

The equations of the asymptotes are

$$\frac{x + \frac{19}{25}}{1} = \frac{y + \frac{17}{25}}{-2}$$

and

$$\frac{x + \frac{19}{25}}{1} = \frac{y + \frac{17}{25}}{\frac{1}{2}}$$

Example 6.31. Let us consider the equation of the following conic

$$x^2 - 2y^2 + 4xy - 8x + 6 = 0.$$

The associated matrix is

$$\begin{pmatrix} 1 & 2 & -4 \\ 2 & -2 & 0 \\ -4 & 0 & 6 \end{pmatrix}$$

is non-singular. Hence, the conic is non-degenerate. The determinant of the matrix

$$\begin{pmatrix} 1 & 2 \\ 2 & -2 \end{pmatrix}$$

is -6 that is < 0. Hence the conic is an hyperbola.

Let us search for the asymptotic directions of the conic by solving the equation

$$-2\mu^2 + 4\mu + 1 = 0.$$

The solutions are $1 - \sqrt{\frac{3}{2}}$ and $1 + \sqrt{\frac{3}{2}}$, respectively. The asymptotic directions are then $\left(1, 1 - \sqrt{\frac{3}{2}}\right)$ and $\left(1, 1 + \sqrt{\frac{3}{2}}\right)$.

Let us find the coordinates of the center

$$\begin{cases} x + 2y - 4 = 0 \\ 2x - 2y = 0 \end{cases}$$

which yields to $x_c = \frac{4}{3}$ and $y_c = \frac{4}{3}$. Let us now write the equations of the asymptotes:

$$\frac{x - \frac{4}{3}}{1} = \frac{y - \frac{4}{3}}{1 - \sqrt{\frac{3}{2}}}$$

$$\frac{x - \frac{4}{3}}{1} = \frac{y - \frac{4}{3}}{1 + \sqrt{\frac{3}{2}}}.$$

6.5.6 Canonic Form of a Conic

Let us consider a generic conic having equation

$$a_{1,1}x^2 + 2a_{1,2}xy + 2a_{1,3}x + a_{2,2}y^2 + 2a_{2,3}y + a_{3,3} = 0,$$

and its associated matrices $\mathbf{A^c}$ and $\mathbf{I_{3,3}}$. The expression of a conic in its canonic form is a technique of obtaining the conic in its simplified formulation by changing the reference system, i.e. by rotating and translating the reference system so that the equations in Sect. 6.4 can be used in a straightforward way. In some cases, the transformation into a canonic form can be extremely convenient and lead to a simplified mathematical description of the conic. On the other hand, the transformation itself can be computationally onerous as it requires the solution of a nonlinear system of equation. Two different procedures are illustrated in the following paragraphs, the first is for conics having a center, i.e. ellipses and hyperbolas, the second is for parabolas.

Canonic Form of Ellipse and Hyperbola

If the conic is an ellipse or an hyperbola and thus has a center, the canonic form is obtained by choosing a reference system whose center coincides with the center of the conic and whose axes coincide with the axes of the conic. The equation of the canonic form of an ellipse or an hyperbola is

$$LX^2 + MY^2 + N = 0$$

where

$$\begin{cases} LMN = \det(\mathbf{A^c}) \\ LM = \det(\mathbf{I_{3,3}}) \\ L + M = \mathrm{tr}(\mathbf{I_{3,3}}). \end{cases}$$

Example 6.32. The conic having equation

$$x^2 - 2xy + 3y^2 - 2x + 1 = 0$$

is non-degenerate since $\det(\mathbf{A^c}) = -1$ and is ellipse since $\det(\mathbf{I_{3,3}}) = 2$. The trace of $\mathbf{I_{3,3}}$ is 4. To write it in its canonic form, we may write

$$\begin{cases} LMN = -1 \\ LM = 2 \\ L+M = 4. \end{cases}$$

The solutions of this system are

$$\begin{aligned} L = 2-\sqrt{2}, M = 2+\sqrt{2}, N = -\tfrac{1}{2} \\ L = 2+\sqrt{2}, M = 2-\sqrt{2}, N = -\tfrac{1}{2}. \end{aligned} \qquad (6.17)$$

This means that we have two canonic form for the ellipse, that is

$$\left(2-\sqrt{2}\right)X^2 + \left(2+\sqrt{2}\right)Y^2 - \frac{1}{2} = 0$$

and

$$\left(2+\sqrt{2}\right)X^2 + \left(2-\sqrt{2}\right)Y^2 - \frac{1}{2} = 0.$$

The ellipse of the two equations is obviously the same but it corresponds, in one case, to an ellipse whose longer part is aligned the abscissa's axis and, in the other case, to an ellipse whose longer axis is aligned to the ordinate's axis.

Canonic Form of the Parabola

If the conic is a parabola it does not have a center in the same plane where the conic lies. In order to have the conic written in its canonic form, the reference system is chosen to have its origin coinciding with the vertex of the parabola, the abscissa's axis aligned with the axis of the parabola, and the ordinate's axis perpendicular to the abscissa's axis and tangent to the vertex. The equation of a parabola in its canonic form is

$$MY^2 + 2BX = 0$$

where

$$\begin{cases} -MB^2 = \det(\mathbf{A^c}) \\ M = \text{tr}(\mathbf{I_{3,3}}). \end{cases}$$

Example 6.33. The parabola having equation

$$x^2 - 2xy + y^2 - 2x + 6y - 1 = 0$$

has $\det(\mathbf{A}^c) = -4$ and $\mathrm{tr}(\mathbf{I}_{3,3}) = 2$. To find M and B we write the system

$$\begin{cases} -MB^2 = -4 \\ M = 2. \end{cases}$$

The equations of this parabola in its canonic form are

$$2Y^2 + 2\sqrt{2}X = 0$$
$$2Y^2 - 2\sqrt{2}X = 0.$$

The two equations refer to a parabola which lies on the positive (right) semiplane and on the negative (left) semiplane, respectively.

Exercises

6.1. Identify the direction of the line having equation $4x - 3y + 2 = 0$

6.2. Identify, if they exist, the intersection points of the following two lines:

$$3x - 2y + 4 = 0$$
$$4x + y + 1 = 0.$$

6.3. Identify, if they exist, the intersection points of the following two lines:

$$3x - 2y + 4 = 0$$
$$9x - 6y + 1 = 0.$$

6.4. Check whether or not the conic having equation

$$4x^2 - 2y^2 + 2xy - 4y + 8 = 0$$

is degenerate or not. Classify then the conic.

6.5. Check whether or not the conic having equation

$$4x^2 + 2y^2 + 2xy - 4y - 6 = 0$$

is degenerate or not. Classify then the conic.

6.6. Check whether or not the conic having equation

$$x^2 + y^2 + 2xy - 8x - 6 = 0$$

is degenerate or not. Classify then the conic.

6.7. Check whether or not the conic having equation

$$x^2 + 2xy - 7x - 8y + 12 = 0$$

is degenerate or not. Classify then the conic.

6.8. Check whether or not the conic having equation

$$x^2 - 16y^2 + 6xy + 5x - 40y + 24 = 0$$

is degenerate or not. Classify then the conic.

Part II
Elements of Linear Algebra

Chapter 7
An Overview on Algebraic Structures

7.1 Basic Concepts

This chapter recaps and formalizes concepts used in the previous sections of this book. Furthermore, this chapter reorganizes and describes in depth the topics mentioned at the end of Chap. 1, i.e. a formal characterization of the abstract algebraic structures and their hierarchy. This chapter is thus a revisited summary of concepts previously introduced and used and provides the mathematical basis for the following chapters.

Definition 7.1. Let A be a nonempty set. An *internal binary operation* or *internal composition law* is a function (mapping) $f : A \times A \to A$.

Example 7.1. The sum $+$ is an internal composition law over the set of natural numbers $\mathbb{N}$, i.e. as we know, the sum of two natural numbers is always a natural number.

Definition 7.2. Let A and B be two nonempty sets. An *external binary operation* or *external composition law* is a function (mapping) $f : A \times B \to A$ where the set $B \neq A$.

Example 7.2. The product of a scalar by a vector is an external composition law over the sets $\mathbb{R}$ and $\mathbb{V}_3$. Obviously if $\lambda \in \mathbb{R}$ and $\vec{v} \in \mathbb{V}_3$, $\vec{w} + \lambda \vec{v} \in \mathbb{V}_3$.

Let us focus at the beginning on internal composition laws.

Definition 7.3. The couple composed of a set A and an internal composition law $*$ defined over the set is said *algebraic structure*.

As mentioned in Chap. 1, algebra is the study of the connections among objects. Consequently, an algebraic structure is a collection of objects (a set) which can be related to each other by means of a composition law.

© Springer Nature Switzerland AG 2019
F. Neri, *Linear Algebra for Computational Sciences and Engineering*,
https://doi.org/10.1007/978-3-030-21321-3_7

7.2 Semigroups and Monoids

If we indicate with $*$ a generic operator and a and b two set elements, the internal composition law of a and b is indicated with $a * b$.

Definition 7.4. Let a, b, and c be three elements of A. An internal composition law $*$ is said to be *associative* when

$$(a * b) * c = a * (b * c).$$

An associative composition law is usually indicated with $\cdot$.

Definition 7.5. The couple of a set A with an associative internal composition law $\cdot$, is said *semigroup* and is indicated with $(A, \cdot)$.

It can easily be observed that the composition of three elements a, b, c of a semigroup can be univocally represented as $a \cdot b \cdot c$ without the use of parentheses. Given the string $a \cdot b \cdot c$ we can choose to calculate $a \cdot b$ first, and then $(a \cdot b) \cdot c$ or $b \cdot c$ first, and then $a \cdot (b \cdot c)$. We would attain the same result.

Definition 7.6. Let a and b are two elements of A. An internal composition law $*$ is said to be *commutative* when $a \cdot b = b \cdot a$.

Definition 7.7. A semigroup $(A, \cdot)$ having a commutative internal composition law is said *commutative semigroup*.

Example 7.3. The algebraic structure composed of the set of real numbers $\mathbb{R}$ and the multiplication, $(\mathbb{R}, \cdot)$ is a commutative semigroup since both associative and commutative properties are valid for real numbers.

Example 7.4. The algebraic structure composed of the set of square matrices $\mathbb{R}_{n,n}$ and the multiplication of matrices $(\mathbb{R}_{2,2}, \cdot)$ is a semigroup because the multiplication is associative (it is not commutative, though).

Example 7.5. The algebraic structure composed of the set of real numbers $\mathbb{R}$ and the division, $(\mathbb{R}, /)$ is not a semigroup since the division of numbers is not an associative operator. For example,

$$(6/2)/3 \neq 6(2/3)$$

that is

$$\frac{\left(\frac{6}{2}\right)}{3} \neq \frac{6}{\left(\frac{2}{3}\right)}.$$

Definition 7.8. Let $B \subset A$. The subset B is said to be closed with respect to the composition law $\cdot$ when $\forall b, b' \in B : b \cdot b' \in B$.

Example 7.6. As we know from the Fundamental Theorem of Algebra the set of natural numbers $\mathbb{N}$, which can be seen as a subset of $\mathbb{R}$, is closed with respect to the sum $+$.

Definition 7.9. The neutral element of a semigroup $(A, \cdot)$ is that special element $e \in A$ such that $\forall a \in A : a \cdot e = e \cdot a = a$.

Example 7.7. The neutral element of $\mathbb{N}$ with respect to the sum $+$ is 0. The neutral element of $\mathbb{N}$ with respect to the product is 1.

Proposition 7.1. *Let $(A, \cdot)$ be a semigroup and e its neutral element. The neutral element is unique.*

Proof. Let us assume, by contradiction that also e' is a neutral element. It follows that $e \cdot e' = e' \cdot e = e$. Since also e is neutral element $e' \cdot e = e \cdot e' = e'$. Hence $e = e'$. □

Definition 7.10. A semigroup $(A, \cdot)$ having neutral element is said *monoid* and is indicated with $(M, \cdot)$.

Example 7.8. The semigroup $(\mathbb{R}, +)$ is a monoid where the neutral element is 0.

Example 7.9. The semigroup $(\mathbb{R}_{n,n}, \cdot)$ composed of square matrices and their product is a monoid and its neutral element is the identity matrix.

Definition 7.11. Let $(M, \cdot)$ be a monoid and e its neutral element. If $a \in M$, the *inverse element* of a is that element $b \in M$ such that $a \cdot b = e$ and is indicated with a^{-1}.

Example 7.10. The monoid $(\mathbb{Q}, \cdot)$ has neutral element $e = 1$. For all the elements $a \in \mathbb{Q}$ its inverse is $\frac{1}{a}$ since it always occurs that $a\frac{1}{a} = 1$ regardless of a.

Proposition 7.2. *Let $(M, \cdot)$ be a monoid and e its neutral element. Let $a \in M$. The inverse element of a, if exists, is unique.*

Proof. Let $b \in M$ be the inverse of a. Let us assume, by contradiction, that c is also an inverse element of a. Hence it follows that

$$a \cdot b = e = a \cdot c$$

and

$$b = b \cdot e = b \cdot (a \cdot c) = (b \cdot a) \cdot c = e \cdot c = c. \square$$

Example 7.11. Let us consider the monoid $(\mathbb{Q}, \cdot)$ and one element $a \in \mathbb{Q}$ that is $a = 5$. The only inverse element of a is $\frac{1}{a} = \frac{1}{5}$.

Definition 7.12. Let $(M, \cdot)$ be a monoid and $a \in M$. If the inverse element of a exists, a is said to be *invertible*.

It can be easily observed that neutral element of a monoid is always invertible and its inverse is the neutral element itself.

Example 7.12. If we consider the monoid $(\mathbb{R}, \cdot)$, the neutral element is $e = 1$ and its inverse is $\frac{1}{1} = 1 = e$.

Proposition 7.3. *Let $(M, \cdot)$ be a monoid and e its neutral element. Let a be an invertible element of this monoid and a^{-1} its inverse element. It follows that a^{-1} is also invertible and*

$$\left(a^{-1}\right)^{-1} = a.$$

Proof. Since a^{-1} is the inverse element of a, it follows that $a \cdot a^{-1} = a^{-1} \cdot a = e$. Hence, one element i such that $i \cdot a^{-1} = a^{-1} \cdot i = e$ exists and this element is $i = a$. This means that a^{-1} is invertible and its unique inverse is a, i.e. $\left(a^{-1}\right)^{-1} = a$. □

Example 7.13. Let us consider the monoid $(\mathbb{R}, \cdot)$ and its element $a = 5$. The inverse element of a is $a^{-1} = \frac{1}{5}$. We can easily verify that

$$\left(a^{-1}\right)^{-1} = \left(\frac{1}{5}\right)^{-1} = 5 = a.$$

Example 7.14. Let us consider the monoid $(\mathbb{R}_{2,2}, \cdot)$. It can be easily verified that the inverse matrix of the inverse of every non-singular matrix $\mathbf{A}$ is the matrix $\mathbf{A}$ itself. For example,

$$\mathbf{A} = \begin{pmatrix} 1 & 1 \\ 0 & 2 \end{pmatrix}$$

is non-singular and its inverse is

$$\mathbf{A}^{-1} = \begin{pmatrix} 1 & -0.5 \\ 0 & 0.5 \end{pmatrix}.$$

Let us calculate the inverse of the inverse:

$$\left(\mathbf{A}^{-1}\right)^{-1} = \begin{pmatrix} 1 & 1 \\ 0 & 2 \end{pmatrix} = \mathbf{A}.$$

Proposition 7.4. *Let $(M, \cdot)$ be a monoid and e its neutral element. Let a and b be two invertible elements of this monoid. It follows that $a \cdot b$ is invertible and its inverse is*

$$(a \cdot b)^{-1} = b^{-1} \cdot a^{-1}.$$

Proof. Considering that a and b are invertible let us calculate:

$$\left(b^{-1} \cdot a^{-1}\right) \cdot (a \cdot b) = b^{-1} \cdot \left(a^{-1} \cdot a\right) \cdot b = b^{-1} \cdot e \cdot b = b^{-1} \cdot b = e$$
$$(a \cdot b) \cdot \left(b^{-1} \cdot a^{-1}\right) = a \cdot \left(b \cdot b^{-1}\right) \cdot a^{-1} = a \cdot e \cdot a^{-1} = a \cdot a^{-1} = e.$$

Hence, $(a \cdot b)$ is invertible and its inverse is $\left(b^{-1} \cdot a^{-1}\right)$. □

Example 7.15. In the case of the monoid $(\mathbb{Q}, \cdot)$ the validity of Proposition 7.4 is trivially verified due to the commutativity of the product of numbers. For example for $a = 5$ and $b = 2$:

$$(a \cdot b)^{-1} = (5 \cdot 2)^{-1} = \frac{1}{10} = \frac{1}{2} \cdot \frac{1}{5} = b^{-1} \cdot a^{-1}.$$

Example 7.16. Let us consider again the monoid $(\mathbb{R}_{2,2}, \cdot)$ and the non-singular matrices

$$\mathbf{A} = \begin{pmatrix} 1 & 1 \\ 0 & 2 \end{pmatrix}$$

and

$$\mathbf{B} = \begin{pmatrix} 1 & 0 \\ 4 & 1 \end{pmatrix}.$$

Let us calculate

$$\mathbf{AB} = \begin{pmatrix} 5 & 1 \\ 8 & 2 \end{pmatrix}$$

and

$$(\mathbf{AB})^{-1} = \begin{pmatrix} 1 & -0.5 \\ -4 & 2.5 \end{pmatrix}.$$

Let us now calculate

$$\mathbf{B}^{-1} = \begin{pmatrix} 1 & 0 \\ -4 & 1 \end{pmatrix}$$

and

$$\mathbf{A}^{-1} = \begin{pmatrix} 1 & -0.5 \\ 0 & 0.5 \end{pmatrix}.$$

Let us multiply the two inverse matrices to verify Proposition 7.4

$$\mathbf{B}^{-1}\mathbf{A}^{-1} = \begin{pmatrix} 1 & -0.5 \\ -4 & 2.5 \end{pmatrix}.$$

It can be observed that the product $\mathbf{A}^{-1}\mathbf{B}^{-1}$ would lead to something else.

Example 7.17. Let us show how a monoid can be generated by means of a nonstandard operator. Let us define an operator $*$ over $\mathbb{Z}$. For two elements a and $b \in \mathbb{Z}$, this operator is

$$a * b = a + b - ab.$$

Let us prove that $(\mathbb{Z}, *)$ is a monoid. In order to be a monoid, $(\mathbb{Z}, *)$ must verify associativity and must have a neutral element. The associativity can be verified by checking that

$$(a * b) * c = a * (b * c)$$

which is

$$a * (b * c) = a + (b * c) - a (b * c) = a + (b + c - bc) - a (b + c - bc) =$$
$$= a + b + c - bc - ab - ac + abc = (a + b - ab) + c - c (a + b - ab) =$$
$$= (a * b) + c - c (a * b) = (a * b) * c.$$

Hence, the operator $*$ is associative. The neutral element is 0 since

$$a * 0 = a + 0 - a0 = a = 0 * a.$$

Let us now look for the inverse element of a. This means that the inverse a^{-1} is an element of $\mathbb{Z}$ such that

$$a * a^{-1} = 0.$$

This occurs when

$$a * a^{-1} = a + a^{-1} - aa^{-1} = 0 \Rightarrow a^{-1} = \frac{a}{a-1}.$$

In order to have $a^{-1} \in \mathbb{Z}$, a must be either 0 or 2. Hence the invertible elements of the monoid $(\mathbb{Z}, *)$ are 0 and 2.

This example shows how, in general only some elements of a monoid are invertible. A special case would be that all the elements of the monoid are invertible. which introduce a new algebraic structure.

7.3 Groups and Subgroups

Definition 7.13. A group $(G, \cdot)$ is a monoid such that all its elements are invertible.

This means that a group should have associative property, neutral element and inverse element for all its elements.

Example 7.18. The monoid $(\mathbb{R}, +)$ is a group since the sum is associative, 0 is the neutral element, and for every element a there is an element $b = -a$ such that $a + b = 0$. On the contrary the monoid $(\mathbb{R}, \cdot)$ is not a group because the element $a = 0$ has no inverse.

Definition 7.14. A group $(G, \cdot)$ is said to be *abelian* (or commutative) if the operation $\cdot$ is also commutative.

Example 7.19. The group $(\mathbb{R}, +)$ is abelian because the sum is commutative. Also, $(\mathbb{V}_3, +)$ is an abelian group because the sum of vectors is commutative and associative, has neutral element $\vec{o}$ and for every vector $\vec{v}$ another vector $\vec{w} = -\vec{v}$ such that $\vec{v} + \vec{w} = \vec{o}$ exist.

Proposition 7.5. *Let $(G, \cdot)$ be a group. If $\forall g \in G : g = g^{-1}$, then the group $(G, \cdot)$ is abelian.*

Proof. Let g and $h \in G$ and $g = g^{-1}$ and $h = h^{-1}$. It follows that

$$g \cdot h = g^{-1} \cdot h^{-1} = (h \cdot g)^{-1} = h \cdot g.$$

Hence, the operator it is abelian. $\square$

Definition 7.15. Let $(G, \cdot)$ be a group. If $g \in G$ and $z \in \mathbb{Z}$. The *power* g^z is defined as

$$g^z = g \cdot g \cdot \ldots \cdot g$$

where the $\cdot$ operation is performed $z - 1$ times (g on the right hand sire of the equation appears z times).

Proposition 7.6. *The following properties of the power are valid*

- $g^{m+n} = g^m \cdot g^n$
- $g^{m \cdot n} = (g^m)^n$.

Proposition 7.7. *Let $(G, \cdot)$ be a group. If g and $h \in G$, $(gh)^z = g^z \cdot h^z$.*

Definition 7.16. Let $(S, \cdot)$ be a semigroup and a, b, and $c \in S$. This semigroup satisfies the *cancellation law* if

$$a \cdot b = a \cdot c \Rightarrow b = c$$
$$b \cdot a = c \cdot a \Rightarrow b = c.$$

Example 7.20. The semigroup $(\mathbb{R}, \cdot)$ satisfies the cancellation law. If we consider the monoid $(\mathbb{R}_{n,n}, \cdot)$ composed of square matrices and their product, we can see that the statement $\mathbf{AB} = \mathbf{AC} \Rightarrow \mathbf{B} = \mathbf{C}$ is true only under the condition that $\mathbf{A}$ is invertible. Hence, the statement is in general not true. A similar consideration can be done for the semigroup $(\mathbb{V}_3, \otimes)$.

Proposition 7.8. *A group $(G, \cdot)$ always satisfies the cancellation law.*

Proof. Let a, b, and c be $\in (G, \cdot)$. Since in a group all the elements have an inverse element. It follows that if $a \cdot b = a \cdot c$ then

$$a^{-1} \cdot a \cdot b = a^{-1} \cdot a \cdot c \Rightarrow b = c.$$

Analogously, if $b \cdot a = c \cdot a$ then

$$b \cdot a \cdot a^{-1} = c \cdot a \cdot a^{-1} \Rightarrow b = c.$$

Hence, for all the groups the cancellation law is valid. $\square$

Example 7.21. Let us consider the monoid $(\mathbb{R}_{2,2}, \cdot)$. This monoid is not a group since only non-singular matrices are invertible. We can easily show that the cancellation law is not always verified. For example, let us consider the matrices

$$\mathbf{A} = \begin{pmatrix} 0 & 5 \\ 0 & 5 \end{pmatrix}$$

$$\mathbf{B} = \begin{pmatrix} 0 & 4 \\ 0 & 4 \end{pmatrix}$$

$$\mathbf{C} = \begin{pmatrix} 5 & 0 \\ 0 & 5 \end{pmatrix}$$

The matrix products

$$\mathbf{AB} = \begin{pmatrix} 0 & 20 \\ 0 & 20 \end{pmatrix}$$

$$\mathbf{CB} = \begin{pmatrix} 0 & 20 \\ 0 & 20 \end{pmatrix}$$

lead to the same results. However, $\mathbf{A} \neq \mathbf{C}$. Thus, the cancellation law is violated.

Definition 7.17. Let $(G, \cdot)$ be a group, e its neutral element and H be a nonempty set such that $H \subset G$. The pair $(H, \cdot)$ is said to be a *subgroup of* $(G, \cdot)$ if the following conditions are verified:

- H is closed with respect to $\cdot$, i.e. $\forall x, y \in H : x \cdot y \in H$
- e, neutral element of $(G, \cdot)$, $\in H$
- $\forall x \in H : x^{-1} \in H$

In other words, a subgroup is an algebraic structure derived from a group (by taking a subset) that still is a group. A trivial example of subgroup is composed of only the neutral element e. When the subgroup contains other elements, besides the neutral element, the subgroup is said *proper subgroup*.

Example 7.22. Considering that $(\mathbb{R}, +)$ is a group, the couple $([-5,5], +)$ is not a proper subgroup because the set is not closed with respect to the sum.

Example 7.23. If we consider the group $(\mathbb{Z}, +)$, the structure $(\mathbb{N}, +)$ would not be its subgroup because inverse elements, e.g. $-1, -2, -3, \ldots$ are not natural numbers.

Example 7.24. Let us consider the group $(Z_{12}, +_{12})$ where

$$Z_{12} = \{0, 1, 2, 3, 4, 5, 6, 7, 8, 9, 10, 11\}$$

and $+_{12}$ is a cyclic sum, i.e. if the sum exceeds 1 of a quantity δ the result of $+_{12}$ is δ. For example $11 + 1 = 0$, $10 + 7 = 6$, etc.

A subgroup of this group is $(H, +_{12})$ where $H = \{0, 2, 4, 6, 8, 10\}$. It can be easily shown that the neutral element $0 \in H$, that the set is closed with respect to $+_{12}$ and that the inverse elements are also contained in the set, e.g. the inverse element of 10 is 2 and the inverse element of 8 is 4.

7.3.1 Cosets

Definition 7.18. Let $(G, \cdot)$ be a group and $(H, \cdot)$ its subgroup. Let us fix a certain $g \in G$. The set

$$Hg = \{h \cdot g | h \in H\}$$

is said *right coset* of H in G while the set

$$gH = \{g \cdot h | h \in H\}$$

is said *left coset* of H in G.

Obviously, if the group is abelian, due to the commutativity of the operator, right and left cosets coincide.

Example 7.25. In order to better clarify the notation of cosets, let us assume that the set G of the group $(G, \cdot)$ is

$$G = \{g_1, g_2, \ldots, g_m\}$$

and H is a subset of G ($H \subset G$) indicated as

$$H = \{h_1, h_2, \ldots, h_m\}.$$

A coset is built up by selecting one element $g \in G$ and operating it with all the elements in H. For example, we select g_5 and we combine it with all the elements of H. With reference to the right coset notation we have

$$Hg = \{h_1 \cdot g_5, h_2 \cdot g_5, \ldots h_n \cdot g_5, \cdot\}.$$

Example 7.26. Let us consider the group $(\mathbb{Z}, +)$ and its subgroup $(\mathbb{Z}5, +)$ where $\mathbb{Z}5$ is the set of $\mathbb{Z}$ elements that are divisible by 5:

$$\mathbb{Z}5 = \{0, 5, 10, 15, 20, \ldots, -5, -10, -15, -20, \ldots\}.$$

It must be noted that the neutral element 0 is included in the subgroup. A right coset is a set of the type $g + h$ with a fixed $g \in G$ and $\forall h \in H$. An example of right coset is the set of $2 + h$, $\forall h \in H$ that is

$$\{2, 7, 12, 17, 22 \ldots -3, -8, -13, -18 \ldots\}.$$

Since the operation is commutative ($2 + h = h + 2$) the left coset is the same.

Example 7.27. Let us consider again the group $(Z_{12}, +_{12})$ and its subgroup $(H, +_{12})$ where $H = \{0, 2, 4, 6, 8, 10\}$. Let us now consider $1 \in Z_{12}$ and build the coset $H + 1$: $\{1, 3, 5, 7, 9, 11\}$.

It must be remarked that a coset is not in general a subgroup, as in this case. Furthermore, if we calculate the coset $H + 2$ we obtain H. This occurs because $2 \in H$ and H is closed with respect to $+_{12}$. Therefore we have the same subgroup. The same results would have been achieved also with $H + 0$, $H + 4$, $H + 6$, $H + 8$, and $H + 10$. Also, the operation $H + 1$, $H + 3$, $H + 5$, $H + 7$, $H + 9$, $H + 11$ leads to the same set $\{1, 3, 5, 7, 9, 11\}$. Hence, starting from this group and subgroup only two cosets can be generated. This fact is expressed saying that the index of H in Z_{12} is 2.

7.3.2 Equivalence and Congruence Relation

Before introducing the concept of congruence relation, let us recall the definition of equivalence relation $\equiv$, see Definition 1.20, and of equivalence class, see Definition 1.22. An equivalence relation defined over a set A is a relation $\mathscr{R}$ over A (that is a subset of $A \times A = A^2$) that is

- reflexive: $\forall x \in A$ it follows that $x \equiv x$
- symmetric: $\forall x, y \in A$ it follows that if $x \equiv y$ then $y \equiv x$
- transitive: $\forall x, y, z \in A$ it follows that if $x \equiv y$ and $y \equiv z$ then $x \equiv z$

The equivalence class of an element $a \in A$, indicated with $[a]$, is a subset of A containing all the elements belonging to A that are equivalent to a. More formally, and equivalence class $[a] \subset A$ is

$$[a] = \{x \in A | x \equiv a\}.$$

Example 7.28. It is easy to see that parallelism among vectors is an equivalence relation since

- every vector is parallel to itself
- if $\vec{u}$ is parallel to $\vec{v}$ then $\vec{v}$ is parallel to $\vec{v}$
- if $\vec{u}$ is parallel to $\vec{v}$ and $\vec{v}$ is parallel to $\vec{w}$ then $\vec{u}$ is parallel to $\vec{w}$.

The set of all the possible vectors parallel vectors is an equivalence class.

We can now introduce the following concept.

Definition 7.19. Let us consider a set A. A *partition*

$$\mathscr{P} = \{P_1, P_2, \ldots P_n\}$$

of the set A is a collection of subsets of A (set of subsets) having the following properties.

- Every set in $\mathscr{P}$ is not the empty set. More formally $\forall i$ it follows that $P_i \neq \emptyset$.
- For every element $x \in A$, there is a unique set $P_i \in \mathscr{P}$ such that $x \in P_i$ (the subsets composing the partition never overlap) More formally, $\forall x \in A$ it follows that $\exists! i$ such that $x \in P_i$. Equivalently, this statement can be expressed as $\forall i$ and j distinct indices of the partition, it follows that $P_i \cap P_j = \emptyset$.
- The union of all the elements of the partition P_i is the set A/ This concept is also expressed by saying that the sets in $\mathscr{P}$ *cover* the set A. More formally $\cup_{i=1}^{n} P_i = A$.

It must be remarked that even though for simplicity we have indicated a partition as a set composed of n sets $(P_1, P_2, \ldots P_n)$, if the set is composed of infinite elements, also the partition can be composed of infinite subsets.

Example 7.29. Let us consider the set

$$A = \{1, 2, 3, 4, 5\}.$$

A partition $\mathscr{P}$ of the set A is

$$\mathscr{P} = \{P_1, P_2, P_3\}$$

where

$$P_1 = \{1, 4\}$$
$$P_2 = \{2, 3\}$$
$$P_3 = \{5\}.$$

It can be easily seen that

- $P_1 \neq \emptyset$, $P_2 \neq \emptyset$, and $P_3 \neq \emptyset$
- $P_1 \cap P_2 = \emptyset$, $P_2 \cap P_3 = \emptyset$ and $P_1 \cap P_3 = \emptyset$ (since the intersection is commutative there are no more combinations)
- $P_1 \cup P_2 \cup P_3 = A$.

Figure 7.1 represents a partitioned set composed of sixteen elements. The partition is composed of five subsets containing three, one, two, fours, and the remaining six elements.

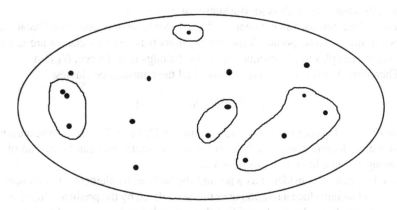

Fig. 7.1 Set of sixteen elements partitioned into five subsets

The following theorem is a general result about equivalence classes. However, it is reported in this section because it is of crucial importance for group theory.

Theorem 7.1. *Let $A = \{a, b, c, d, \ldots\}$ be a set and $\equiv$ be an equivalence relation defined over A. Let $[a], [b], [c], [d] \ldots$ be the all the possible corresponding equivalence classes over the set A. It follows that the equivalence relation $\equiv$ partitions the set A, i.e. $\{[a], [b], [c], [d] \ldots\}$ is a partition of A.*

This means that the two following conditions are simultaneously satisfied.

- *Given two elements $a, b \in A$, either $[a]$ and $[b]$ are disjoint or coincide, i.e. $\forall a, b \in A$ either $[a] = [b]$ or $[a] \cap [b] = \emptyset$.*
- *A is the union of all the equivalence classes, i.e. $[a,] \cup [b] \cup [c] \cup [d] \cup \cdots = A$*

Proof. To prove the first condition, let us consider that the intersection $[a] \cap [b]$ is either empty or is nonempty. If the intersection is the empty set then the classes are disjoint. In the second case, there exists and element x such that $x \in [a] \cap [b]$, hence $x \in [a]$ and $x \in [b]$, i.e. $x \equiv a$ and $x \equiv b$. By symmetry it is true that $a \equiv x$. By transitivity, since $a \equiv x$ and $x \equiv b$, it follows that $a \equiv b$.

From the definition of equivalence class (see Definition 1.22), since $a \equiv b$ it follows that $a \in [b]$. Since it is also true that $a \in [b]$, it follows that $[a] \subset [b]$. Since by symmetry $b \equiv a$, it follows that $b \in [a]$ and $[b] \subset [a]$. Hence $[a] = [b]$. In other words, if two equivalence classes are not disjoint, then the they are coinciding. $\square$

Let $U_A = [a,] \cup [b] \cup [c] \cup [d] \cup \ldots$ be the union of the equivalence classes of the elements of A. To prove the second condition we need to consider that, an element $x \in A$, due to the reflexivity (it is always true that $x \equiv x$) always belongs to at least one class. Without a loss of generality, let us say that $x \in [a]$. Hence, every element of A is also element of U_A. Hence, $A \subset U_A$.

An equivalence class $[a]$ by definition (see Definition 1.22) is

$$[a] = \{x \in A | x \equiv a\}.$$

Hence, the equivalence class $[a]$ is a subset of A ($[a] \subset A$).

Thus, if we consider an element $x \in U_A$, x belongs to only one class (because, as shown in the previous point, all the intersections between two classes are empty), that is for examples $[a]$. Consequently, x also belongs to A. Hence, $U_A \subset A$.

Therefore, $A = U_A$, i.e. A is the union of all the equivalence classes:

$$[a,] \cup [b] \cup [c] \cup [d] \cup \cdots = A. \square$$

Example 7.30. In order to explain the meaning of Theorem 7.1, let us imagine a box filled with coloured balls. The balls are monochromatic and can be of one of the following colours, blue, red, green, or yellow.

The box can be considered as a set and the balls as its elements. Let us name A this set. Let us introduce a relation over this set defined by the predicate "to be of the same colour". It can be easily verified that this relation is an equivalence relation. More specifically, this relation

- is reflexive since every ball is of the colour of itself
- is symmetric since for any two balls $ball_1$ and $ball_2$, if $ball_1$ is of the same colour of $ball_2$ then $ball_2$ is of the same colour of $ball_1$
- is transitive since for any three balls $ball_1$, $ball_2$, $ball_3$, if $ball_1$ and $ball_2$ are of the same colour (let us say blue) and if $ball_2$ and $ball_3$ are of the same colour, then $ball_1$ and $ball_3$ are of the same colour (still blue).

Thus, the relation "to be of the same colour" is an equivalence relation over the elements of the set A.

Let us imagine to select a ball and check its colour. Let us suppose we selected a red ball. Let us now select from the box the balls that are equivalent to it, that is the other red balls. When there are no more red balls in the box, let us imagine that we place all the red balls in a red basket. This basket is an equivalence class of the red balls which we indicate with $[r]$. We can then define one equivalence class (and one basket) per each colour, that is the class of blue balls $[b]$, green balls, $[g]$, and yellow balls $[y]$.

Theorem 7.1 states that the relation "to be of the same colour" partitions the set A. This means that

- there is no ball in more than one basket: a ball is of only one colour, it cannot be at the same time e.g. green and red. In other words, by strictly following the Theorem syntax, if we assume that a ball can be in two baskets, then it is of the colour of the first and of the second basket. This may happen only if we consider the same basket twice. If the balls are of different colour they are in separate baskets and there is no way of having both of them in the same basket ($[g]$ and $[r]$ are disjoint).
- if the balls contained in each basket are placed back to the box (that is the union of the equivalence classes), we obtain the box with its balls, that is the original set A. The proof of this statement can be interpreted in the following way. A ball taken from a box is a ball also belonging to the basket. Conversely, a ball taken from a basket is also a ball belonging to the box.

Definition 7.20. Let $(G, \cdot)$ be a group and $(H, \cdot)$ its subgroup. Let x and y elements $\in G$. The *right congruence relation modulo H* $x \sim y$ (or simply right congruence) is the relation defined as : $\exists h \in H$ such that $y = h \cdot x$. Analogously, the left congruence is $y = x \cdot h$.

Example 7.31. Let us consider the group $(\mathbb{R}, \cdot)$ and as its (improper) subgroup, the group $(\mathbb{R}, \cdot)$ itself. Let us consider two elements $x, y \in \mathbb{R}$ and pose the equation

$$y = h \cdot x.$$

Since we can always find an $h \in \mathbb{R}$ verifying this equation (it is enough to choose $h = \frac{y}{x}$), then $x \sim y$, i.e. x is congruent with y.

Example 7.32. Let us consider the group $(\mathbb{Z}, +)$ and its subgroup $(\mathbb{Z}5, +)$ with

$$\mathbb{Z}5 = \{\ldots, -15, -10, -5, 0, 5, 10, 15, \ldots\}.$$

Let us consider two elements $x, y \in \mathbb{Z}$, i.e. $x = 45$ and $y = 50$. Since $\exists$ an element $h \in \mathbb{Z}5$ such that $y = h + x$, that is $h = 5$, then $x \sim y$. This congruence relation is also known as congruence relation modulo 5. In an analogous way, we may define the congruence relation modulo e.g. 2, 3, 10 etc. More generally, we may define the generic congruence relation modulo h.

Conversely, if $x = 45$ and $y = 47$ it follows that $x \nsim y$ since there is no $h \in \mathbb{Z}5$ such that $47 = h + 45$.

Example 7.33. Let us consider again the group $(Z_{12}, +_{12})$ and its subgroup $(H, +_{12})$ where $H = \{0, 2, 4, 6, 8, 10\}$. Let us consider two elements $x, y \in \mathbb{Z}_{12}$, i.e. $x = 4$ and $y = 2$. We have $x \sim y$ since $\exists h \in H$, that is $h = 10$, such that $2 = h +_{12} 4$.

The following proposition gives an important property of the congruence relation which allows us to make a strong statement about groups.

Proposition 7.9. *Let $\sim$ be the congruence relation modulo H of a group $(G, \cdot)$. It follows that the congruence is an equivalence relation.*

Proof. To prove the equivalence we need to prove reflexivity, symmetry and transitivity.

1. Within H there exists the neutral element e such that $x = e \cdot x \Rightarrow x \sim x$. Hence, the congruence relation is reflexive.
2. If $\exists h \in H$ such that $y = h \cdot x$, then for the definition of group the inverse element $h^{-1} \in H$ also exists. Hence, we can write $x = h^{-1} \cdot y \Rightarrow x = \tilde{h} \cdot y \Rightarrow y \sim x$. Hence, the congruence relation is symmetric.
3. if $x \sim y$, then there exists a value h_1 such that $y = h_1 \cdot x$. if $y \sim z$, then there exists a value h_2 such that $z = h_2 \cdot y$. Hence, $z = h_1 \cdot h_2 \cdot x$. We know, since H is a group, that $k = h_1 \cdot h_2 \in H$. Hence $\exists k \in H$ such that $z = k \cdot x$, which means $x \sim z$. Hence the congruence relation is transitive.

It follows that the congruence relation is an equivalence. $\square$

Example 7.34. Let us consider again the group $(\mathbb{Z}, +)$, its subgroup $(\mathbb{Z}5, +)$ and the congruence relation $x \sim y$. We can verify that the congruence is

- reflexive since $\exists h \in \mathbb{Z}5$ such that $x = h + x$ that is $h = 0$
- symmetric since if $\exists h \in \mathbb{Z}5$ such that $y = h + x$ then we may write $x = -h + y$ with $-h \in \mathbb{Z}5$
- transitive since if $\exists h \in \mathbb{Z}5$ such that $y = h + x$ and $\exists k \in \mathbb{Z}5$ such that $z = k + y$ we can substitute and obtain

$$z = k + h + x$$

with $(k + h) \in \mathbb{Z}5$.

Thus, the congruence relation is an equivalence relation.

7.3.3 Lagrange's Theorem

Let us consider a group $(G, \cdot)$ and its subgroup $(H, \cdot)$. If we fix an element $g \in G$ the congruence relation $\sim$ identifies the equivalence class (all the elements of G that are congruent with g)

$$[g] = \{x \in G | x \sim g\} = \{x \in G | \exists h \in H | x = h \cdot g\} = \{h \cdot g \in G | h \in H\}$$

that is the definition of (right) coset of H in G, i.e. $[g] = Hg$.

Example 7.35. Let us consider again the group $(\mathbb{Z}, +)$ and its subgroup $(\mathbb{Z}5, +)$. Let us arbitrarily select an element from $\mathbb{Z}$, for example $g = 52$. The equivalence class $[g]$ associated with the congruence relation would contain e.g. $-3, 2, 7, 12, 17, \ldots 52, 57 \ldots$.

The equivalence class $[g]$ would also be the coset Hg built up by adding 52 to all the elements of $\mathbb{Z}5$.

In other words, a coset is an equivalence class associated with the congruence relation. In this light, let us revisit the previous results in the context of group theory by stating the following lemmas. More specifically, by applying Theorem 7.1, from the fact that congruence is an equivalence relation the following lemmas immediately follow.

Lemma 7.1. *Let* $(G, \cdot)$ *be a group and* $(H, \cdot)$ *its subgroup. Two right (left) cosets of H in G either coincide or are disjoint.*

Proof. Since two cosets are two equivalence classes of G, for Theorem 7.1, they either coincide or they are disjoint. $\square$

Lemma 7.2. *Let* $(G, \cdot)$ *be a group and* $(H, \cdot)$ *its subgroup. The set G is equal to the union of all the right (left) cosets:*

$$G = \cup_{i=1}^{n} Hg_i.$$

Proof. Since the cosets are equivalence classes of G, for Theorem 7.1, their union is G. $\square$

Example 7.36. Let us consider again the group $(Z_{12}, +_{12})$ and its subgroup $(H, +_{12})$ where $H = \{0, 2, 4, 6, 8, 10\}$.

We know that two cosets can be generated

$$H + 0 = \{0, 2, 4, 6, 8, 10\}$$
$$H + 1 = \{1, 3, 5, 7, 9, 11\}$$

We can easily see that these two cosets are disjoint and their union is Z_{12}.

Definition 7.21. Let $(A, *)$ an algebraic structure. The *order of an algebraic structure* is the cardinality of the set A associated with it and is indicated with $|A|$.

Definition 7.22. The algebraic structure $(A, *)$ is *finite* if its order is finite.

For simplicity of the notation, let us indicate here with $|Hg|$ the cardinality of a coset (which is a set and not an algebraic structure).

Lemma 7.3. *Let* $(G, \cdot)$ *be a finite group and* $(H, \cdot)$ *its subgroup. The cardinality of every right (left) coset Hg of H in G is equal to the order of* $(H, \cdot)$:

$$\forall j : |Hg_j| = |H|.$$

Proof. Let us consider the subgroup $(H, \cdot)$. Since $(G, \cdot)$ is finite and $H \subset G$ then also $(H, \cdot)$ is finite. Thus, the set H has finite cardinality. Let us indicate with n the cardinality of H, i.e. $|H| = n$, and

$$H = \{h_1, h_2, \ldots, h_n\}.$$

Let Hg be one of the right cosets of H in G:

$$Hg = \{h_1 \cdot g, h_2 \cdot g, \ldots, h_n \cdot g\}.$$

Let us define a function $\phi : H \rightarrow Hg$:

$$\phi(h) = h \cdot g.$$

Let us consider two values h_1 and $h_2 \in H$, such that $h_1 \neq h_2$. The values taken by the function $\phi(h)$ in h_1 and h_2 are

$$\phi(h_1) = h_1 \cdot g$$
$$\phi(h_2) = h_2 \cdot g,$$

respectively. Since $h_1 \neq h_2$, it follows that

$$h_1 \cdot g \cdot g^{-1} \neq h_2 \cdot g \cdot g^{-1}.$$

For the cancellation law, it follows that

$$h_1 \cdot g \neq h_2 \cdot g,$$

which means that $\phi(h)$ is an injective function.

Let us consider a generic y element of the coset, i.e. $y \in Hg$. Thus,

$$y \in \{h_1 \cdot g, h_2 \cdot g, \ldots, h_n \cdot g\}.$$

This means that $\exists j$ such that $y = h_j \cdot g$ or, equivalently,

$$\exists h \in H \text{ such that } y = h \cdot g,$$

that is

$$\exists h \in H \text{ such that } y = \phi(h).$$

This means that the function is $\phi(h)$ is also surjective.

Since both injection and surjection properties are verified, it follows that ϕ is bijective. Thus, for Proposition 1.4, the cardinality of H is equal to the cardinality of the right coset Hg. This statement can be rephrased as the cardinality of a right coset of H in G is equal to the order of $(H, \cdot)$. $\square$

The proof for a left coset is analogous.

Example 7.37. By using the same numbers of Example 7.36, we can immediately see that the cardinality of H as well as the cardinality of each coset is 6.

Theorem 7.2. Lagrange's Theorem. *Let $(G, \cdot)$ be a finite group and $(H, \cdot)$ its subgroup. Then, the order of H divides the order of G, i.e. the ratio of the cardinality of*

G by the cardinality of H is an integer number:

$$\frac{|G|}{|H|} = k$$

with $k \in \mathbb{N}$ and $k \neq 0$.

Proof. Let $Hg_1, Hg_2, \ldots, Hg_k$ be all the right cosets of H in G. From Lemma 7.2, it follows that

$$G = Hg_1 \cup Hg_2 \cup \ldots \cup Hg_k$$

and, for Lemma 7.1, the sets composing this union are disjoint. Since they are disjoint we can write that the cardinality of G is equal to the sum of the cardinalities of each coset:

$$|G| = |Hg_1| + |Hg_2| + \ldots + |Hg_k|.$$

For the Lemma 7.3, the cardinality of each coset is equal to the cardinality of the corresponding subgroup. Hence,

$$|G| = k|H| \Rightarrow \frac{|G|}{|H|} = k. \square$$

Example 7.38. Let us consider again the group $(Z_{12}, +_{12})$ and its subgroup $(H, +_{12})$ where $H = \{0, 2, 4, 6, 8, 10\}$. We know that we have two cosets of cardinality 6 that is also the cardinality of H. The cardinality of Z_{12} is 12. The Lagrange's Theorem is hence verified since

$$\frac{|G|}{|H|} = \frac{12}{6} = 2$$

is an integer number.

7.4 Rings

As we know from Chap. 1 a ring is a set equipped with two operators. For convenience, let us report again Definition 1.34.

Definition 7.23. Ring. A *ring R* is a set equipped with two operations called *sum* and *product*. The sum is indicated with a + sign while the product operator is simply omitted (the product of x_1 by x_2 is indicated as $x_1 x_2$). The set R is closed with respect to these two operations and contains neutral elements with respect to both sum and product, indicated with 0_R and 1_R respectively. In addition, the following properties, namely *axioms of a ring*, must be valid.

- commutativity (sum): $x_1 + x_2 = x_2 + x_1$
- associativity (sum): $(x_1 + x_2) + x_3 = x_1 + (x_2 + x_3)$
- neutral element (sum): $x + 0_R = x$
- inverse element (sum): $\forall x \in R : \exists (-x) \, | \, x + (-x) = 0_R$

- associativity (product): $(x_1 x_2) x_3 = x_1 (x_2 x_3)$
- distributivity 1: $x_1 (x_2 + x_3) = x_1 x_2 + x_1 x_3$
- distributivity 2: $(x_2 + x_3) x_1 = x_2 x_1 + x_3 x_1$
- neutral element (product): $x 1_R = 1_R x = x$

Example 7.39. The algebraic structures $(\mathbb{Z}, +,)$, $(\mathbb{Q}, +,)$, $(\mathbb{R}, +,)$, and $(\mathbb{C}, +,)$ are rings.

Example 7.40. The algebraic structure $(\mathbb{R}_{n,n}, +,)$ composed of square matrices with sum and product is a ring. The neutral elements are the zero matrix for the sum and the identity matrix for the product.

A few remarks can immediately be made on the basis only of the ring definition. At first, a ring has two neutral elements. Then, commutativity is a requirement for the sum but not for the product. The structure $(\mathbb{R}_{n,n}, +,)$ is a ring although the product is not commutative. Furthermore the existence of the inverse element with respect to the product for all the elements of the set is also not a requirement. Hence, $(\mathbb{R}_{n,n}, +,)$ is a ring although square matrices are not always invertible.

Finally, a ring can be seen as the combination of a group $(R, +)$ and a monoid $(R,)$.

The latter observations constitute the theoretical foundation of the following proposition.

Proposition 7.10. *Let $(R, +,)$ be a ring. The following properties are valid.*

- *there exists only one neutral element 0_R with respect to the sum*
- *for every element $a \in R$ there exists only one element $-a$ (this element is said opposite element) such that $a + (-a) = 0_R$*
- *cancellation law is valid with respect to the sum : $a + b = c + b \Rightarrow a = c$*
- *there exists only one neutral element 1_R with respect to the product*

Proof. Since $(R, +)$ can be interpreted as a group, the first three statements are immediately proved by applying Proposition 7.1, Proposition 7.2, and Proposition 7.8, respectively.

Since $(R,)$ can be interpreted as a monoid, for Proposition 7.1, the neutral element with respect to the second operator is unique. $\square$

Before entering into ring theory in greater details, let us convey that $a - b = a + (-b)$.

Proposition 7.11. *Let $(R, +,)$ be a ring. It follows that $\forall a \in R$*

$$a 0_R = 0_R a = 0_R$$

Proof. If $c = a 0_R$ then for distributivity $a (0_R + 0_R) = a 0_R + a 0_R = c + c$. Hence $c = c + c$. Considering that it is always true that if the same quantity is added and subtracted the result stays unvaried: $c = c + c - c$, we can write

$$c = c + c - c = c - c = 0_R.$$

Hence, the result of $a0_R$ is always 0_R. □

Proposition 7.12. *Let* $(R,+,)$ *be a ring. It follows that* $\forall a,b \in R$

$$a(-b) = -(ab)$$

Proof. Let us directly check the result of the operation $a(-b)+ab$:

$$a(-b) + ab = a(-b+b) = a0_R = 0_R. \square$$

Analogously it can be checked that $(-a)b = -(ab)$.

By using this proposition the following two corollaries can be easily proved.

Corollary 7.1. *Let* $(R,+,)$ *be a ring. It follows that* $\forall a \in R$

$$a(-1_R) = (-1_R)a = -a.$$

Corollary 7.2. *Let* $(R,+,)$ *be a ring. It follows that*

$$(-1_R)(-1_R) = 1_R.$$

Proposition 7.13. *Let* $(R,+,)$ *be a ring. It follows that* $\forall a,b \in R$

$$(-a)(-b) = ab.$$

Proof.

$$(-a)(-b) = a(-1_R)b(-1_R) = a(-1_R)(-1_R)b = a1_Rb = ab. \square$$

Definition 7.24. Let $(R,+,)$ be a ring. If $0_R = 1_R$ the ring is said to be *degenerate*.

Example 7.41. Let us consider again the set Z_{12} where

$$Z_{12} = \{0,1,2,3,4,5,6,7,8,9,10,11\}$$

and the cyclic sum $+_{12}$. We can also introduce another cyclic sum but with a different cycle, e.g. $+_6$. For example, $4 +_6 2 = 0$.

The algebraic structure $(Z_{12}, +_{12}, +_6)$ is a degenerate ring since 0 would be the neutral element with respect to both the operators.

Definition 7.25. Let $(R,+,)$ be a ring and $a,b \in R$. The ring is said to be *commutative* when $ab = ba$.

Example 7.42. Some examples of commutative rings are $(\mathbb{Z},+,)$, $(\mathbb{Q},+,)$, and $(\mathbb{R},+,)$ while $(\mathbb{R}_{n,n},+,)$ is not a commutative ring.

Example 7.43. Let us indicate with $\mathbb{R}^{\mathbb{R}}$ be a set of all the possible functions defined on $\mathbb{R}$ and having codomain $\mathbb{R}$. It can be verified by checking the ring axioms, that the algebraic structure $(\mathbb{R}^{\mathbb{R}},+,)$ is a commutative ring and $0_R = 0$ and $1_R = 1$.

Example 7.44. Let X be a nonempty set and $P(X)$ be its power set. It can be proved that the algebraic structure $(P(X), \Delta, \cap)$ where Δ is the symmetric difference is a commutative ring. This special structure is called *Boolean Ring* and constitutes the theoretical foundation of Boolean Algebra, see Appendix A.

Definition 7.26. Let $(R, +,)$ be a ring and $a \in R$. The nth *power* of a is the product of a calculated n times:

$$a^n = aaaa \ldots aaaa.$$

Proposition 7.14. *Let* $(R, +,)$ *be a ring,* $a \in R$, *and* $n, m \in \mathbb{N}$. *It follows that*

- $a^{n+m} = a^n a^m$
- $a^{nm} = (a^n)^m$.

In general, if $(R, +,)$ is a ring and $a \in R$, it is not true that $(ab)^n = a^n b^n$. The latter equation is valid only if the ring is commutative.

Proposition 7.15. *et* $(R, +,)$ *be a commutative ring and* $a, b \in R$. *It follows that* $(ab)^n = a^n b^n$.

Proof. Let us consider $(ab)^n$. From the definition of power it follows that

$$(ab)^n = ababababababab \ldots$$

where the term ab appears n times.

Since the ring is commutative we can re-write this equation as

$$(ab)^n = abbaabbaabbaab \ldots = ab^2 a^2 b^2 a^2 b^2 a^2 b \ldots.$$

By iteratively applying commutativity and the definition of power we obtain

$$(ab)^n = ab^{n-1} a^{n-1} b = a^n b^n. \square$$

For commutative rings the following theorem is also valid.

Theorem 7.3. Newton's Binomial. *Let* $(R, +,)$ *be a commutative ring and* $a, b \in R$. *It occurs that* $\forall n \in \mathbb{N}$

$$(a+b)^n = \sum_{i=0}^{n} \binom{n}{i} a^{n-i} b^i$$

where $\binom{n}{i}$ *is said binomial coefficient and defined as*

$$\binom{n}{i} = \frac{n!}{i!(n-i)!}$$

with the initial/boundary condition that

$$\binom{n}{0} = \binom{n}{n} = 1.$$

Example 7.45. Newton's binomial is a powerful formula that allows to represent a binomial of any power as a sum of monomials. For example the square of a binomial $(a+b)^2$ can be written as

$$(a+b)^2 = \binom{2}{0} a^2 b^0 + \binom{2}{1} a^1 b^1 + \binom{2}{2} a^0 b^2 = \frac{2}{2} a^2 + \frac{2}{1} ab + \frac{2}{2} b^2$$

that is the formula of the square of a binomial $a^2 + b^2 + 2ab$.

Definition 7.27. Let $(R, +,)$ be a ring and $S \subset R$ with $S \neq \emptyset$. If $(S, +,)$ is a ring then $(S, +,)$ is said *subring*.

Example 7.46. It can be observed that $(\mathbb{Z}, +,)$ is a subring of $(\mathbb{Q}, +,)$ that is a subring of $(\mathbb{R}, +,)$ that is a subring of $(\mathbb{C}, +,)$.

7.4.1 Cancellation Law for Rings

Proposition 7.10 states the validity of the cancellation law with respect to the sum. However, the properties of rings do not contain any statement about cancellation law with respect to the product. The reason is that, in general, the cancellation law of the product is not valid.

Example 7.47. If $(R, +,)$ and $a, b, c \in R$, the cancellation law of the product would be

$$ab = ac \Rightarrow b = c.$$

Let us consider the ring $(\mathbb{R}_{2,2}, +,)$ of the square matrices of size 2 and let us pose

$$a = \begin{pmatrix} 0 & 1 \\ 0 & 1 \end{pmatrix},$$

$$b = \begin{pmatrix} 0 & 1 \\ 0 & 1 \end{pmatrix},$$

and

$$c = \begin{pmatrix} 1 & 0 \\ 0 & 1 \end{pmatrix}.$$

It results that

$$ab = \begin{pmatrix} 0 & 1 \\ 0 & 1 \end{pmatrix} = ac$$

and still $b \neq c$.

The cancellation law can be not valid also in the case of commutative rings.

Example 7.48. Let us consider the ring of real functions $\left(\mathbb{R}^{\mathbb{R}}, +, \right)$ and two functions belonging to $\mathbb{R}^{\mathbb{R}}$:

$$f(x) = \begin{cases} x \text{ if } x \geq 0 \\ 0 \text{ if } x \leq 0 \end{cases}$$

and

$$g(x) = \begin{cases} x \text{ if } x \leq 0 \\ 0 \text{ if } x \geq 0. \end{cases}$$

If we now pose $a = 0$, $b = f(x)$, and $c = g(x)$ and calculate

$$ca = 0 = cb$$

while $a \neq b$.

The latter example suggests that there is a relation between the cancellation law (more specifically its failure) and the fact that the product of two elements different from the neutral element is null, i.e. 0_R.

Definition 7.28. Let $(R, +,)$ be a ring and $a \in R$ such that $a \neq 0_R$. The element a is said *zero divisor* if there exists an element $b \in R$ with $b \neq 0_R$ such that

$$ab = 0_R.$$

In other words, if a ring contains zero divisors then the cancellation law of the product is not valid.

Example 7.49. Let us consider the ring $(\mathbb{R}_{2,2}, +,)$ and the matrices

$$a = \begin{pmatrix} 0 & 1 \\ 0 & 5 \end{pmatrix}$$

and

$$b = \begin{pmatrix} 7 & 10 \\ 0 & 0 \end{pmatrix}.$$

Obviously, neither a nor b are the null matrix. When we multiply a by b we obtain

$$ab = \begin{pmatrix} 0 & 0 \\ 0 & 0 \end{pmatrix}.$$

This means that although $a \neq 0_R$ and $b \neq 0_R$ still it happens that $ab = 0_R$. Hence, we can conclude that $(\mathbb{R}_{2,2}, +,)$ and b is a zero divisor of a.

A ring without zero divisors is a special ring formally introduced in the following definition.

Definition 7.29. A commutative ring which does not contain any zero divisor is said *integral domain*.

Proposition 7.16. *Let* $(R, +,)$ *be an integral domain. Then it occurs that the cancellation law is valid, i.e. for all* $a, b, c \in R$ *with* $c \neq 0_R$

$$ac = bc \Rightarrow a = b.$$

Proof. Let us consider three elements $a, b, c \in R$ with $c \neq 0_R$ such that $ac = bc$. It follows that

$$ac = bc \Rightarrow (ac - bc) = 0_R \Rightarrow (a - b)c = 0_R.$$

Since $(R, +,)$ is an integral domain (hence without zero divisors) and $c \neq 0_R$ it follows that necessarily $(a - b) = 0_R$. This fact means that

$$a = b. \quad \square$$

7.4.2 Fields

Definition 7.30. Let $(R, +,)$ be a ring and $a \in R$. The element a is said to be invertible if there exists and element $b \in R$ such that $ab = 1_R$. The element b is said inverse of the element a.

Proposition 7.17. *Let* $(R, +,)$ *be a ring and* $a \in R$. *The inverse element of* a, *when it exists, is unique.*

Proof. Let us assume, by contradiction, that a has two inverse elements and b, c are both inverse elements of a. It follows that

$$b = b1_R = b(ac) = (ba)c = 1_R c = c. \quad \square$$

Example 7.50. The only invertible elements of the ring $(\mathbb{Z}, +,)$ are -1 and 1.

Example 7.51. The invertible elements of the ring $(\mathbb{R}_{n,n}, +,)$ are those square matrices of order n that are non-singular.

Example 7.52. In the case of the ring $(\mathbb{Q}, +,)$, all the elements of $\mathbb{Q}$ except 0 can be expressed as a fraction. Therefore, all its elements are invertible.

Definition 7.31. A commutative ring $(F, +,)$ such that all the elements of F except the neutral element with respect to the sum 0_F are invertible is said *field*.

In other words, a field can be seen as the combination of two groups (with one exception about 0_F).

Example 7.53. From the previous example we can easily state that $(\mathbb{Q}, +,)$ is a field. Also $(\mathbb{R}, +,)$ and $(\mathbb{C}, +,)$ are fields.

Proposition 7.18. *Every field is an integral domain.*

Proof. Let $(F, +,)$ be a field and $a \in F$ with $a \neq 0_F$. By definition, an integral domain is a commutative ring that does not contain zero divisors. Let us consider a generic element $b \in F$ such that $ab = 0_F$. Since in fields all the elements are invertible a is also invertible and $a^{-1}a = 1_F$. It follows that

$$b = 1_F b = (a^{-1}a) b = a^{-1}(ab) = a^{-1}0_F = 0_F.$$

Hence, since $b = 0_F$, b is not a zero divisor of a. This means that there are no zero divisors. $\square$

The concept of field is the arrival point of this excursus on algebraic structures and the basic instrument to introduce the topics of the following chapters.

7.5 Homomorphisms and Isomorphisms

Subsequent to this basic introduction to group and ring theories, this section shows how a mapping can be defined over algebraic structures. In particular, this section focuses on a specific class of mappings that has interesting properties and practical implications.

Definition 7.32. Let $(G, \cdot)$ and $(G', *)$ be two groups. A mapping $\phi : G \to G'$ such that for all $x, y \in G$ it follows that

$$\phi(x \cdot y) = \phi(x) * \phi(y)$$

is said *group homomorphism* from G to G' (or from $(G, \cdot)$ to $(G', *)$).

Example 7.54. Let us consider the groups $(\mathbb{Z}, +)$ and $(2\mathbb{Z}, +)$ where $2\mathbb{Z}$ is the set of integer even numbers. The mapping $\phi : \mathbb{Z} \to 2\mathbb{Z}$ defined as

$$\phi(x) = 2x.$$

Let us show that this mapping is an homomorphism:

$$\phi(x \cdot y) = \phi(x + y) = 2(x + y) = 2x + 2y = \phi(x) + \phi(y) = \phi(x) * \phi(y).$$

In a similar way, the concepts of the other algebraic structures endowed with an operator can be defined, e.g. semigroup homomorphism and monoid homomorphism. Regarding algebraic structures endowed with two operators, a separate definition must be given.

Definition 7.33. Let $(R, +,)$ and $(R', \oplus, *)$ be two rings. A mapping $f : R \to R'$ such that for all $x, y \in R$ it follows that

- $f(x + y) = f(x) \oplus f(y)$
- $f(xy) = f(x) * f(y)$

- $f(1_R) = 1_{R'}$

is said *ring homomorphism* from R to R' (or from $(R, +,)$ to $(R', \oplus, *)$).

Homomorphism, from ancient Greek "omos" and "morphé", literally means "the same form". An homomorphism is a transformation that preserves the structure between two algebraic structures. Let us better clarify this fact by means of the following example.

Example 7.55. Let us consider the rings $(\mathbb{R}, +,)$ and $(\mathbb{R}_{2,2}, +,)$. The mapping $\phi : \mathbb{R} \to \mathbb{R}_{2,2}$ defined as

$$\phi(x) = \begin{pmatrix} x & 0 \\ 0 & x \end{pmatrix}$$

where $x \in \mathbb{R}$ is an homomorphism. This fact is very easy to verify in this case. Let us check that $\phi(x+y) = \phi(x) + \phi(y)$:

$$\phi(x+y) = \begin{pmatrix} x+y & 0 \\ 0 & x+y \end{pmatrix} = \begin{pmatrix} x & 0 \\ 0 & x \end{pmatrix} + \begin{pmatrix} y & 0 \\ 0 & y \end{pmatrix} = \phi(x) + \phi(y).$$

Let us now check that $\phi(xy) = \phi(x)\phi(y)$:

$$\phi(xy) = \begin{pmatrix} xy & 0 \\ 0 & xy \end{pmatrix} = \begin{pmatrix} x & 0 \\ 0 & x \end{pmatrix} \begin{pmatrix} y & 0 \\ 0 & y \end{pmatrix} = \phi(x)\phi(y).$$

Finally, considering that $1_{\mathbb{R}} = 1$,

$$\phi(1) = \begin{pmatrix} 1 & 0 \\ 0 & 1 \end{pmatrix} = 1_{\mathbb{R}_{2,2}}.$$

In other words, the main feature of an homomorphism is a mapping that transforms a group into a group, a ring into a ring etc. However, there is no requirement for an homomorphism to be a bijection. Intuitively, we may think of an homomorphism as a transformation defined over the elements of an algebraic structure A that has a results elements of another algebraic structure B. In general, we cannot obtain starting from the algebraic structure B the elements of A. Let us see the following example.

Example 7.56. Let us consider the groups $(\mathbb{R}^3, +)$ and $(\mathbb{R}^2, +)$ and the mapping $\phi : \mathbb{R}^3 \to \mathbb{R}^2$

$$\phi(\mathbf{x}) = \phi(x, y, z) = (x, y).$$

It can be easily shown that this mapping is an homomorphism. If we consider two vectors $\mathbf{x_1} = (x_1, y_1, z_1)$ and $\mathbf{x_2} = (x_2, y_2, z_2)$ it follows that

$$\phi(\mathbf{x_1} + \mathbf{x_2}) = (x_1 + x_2, y_1 + y_2) = (x_1, y_1) + (x_2, y_2) = \phi(\mathbf{x_1}) + \phi(\mathbf{x_2}).$$

Hence, this mapping is an homomorphism. On the other hand, if we start from a vector (x, y) it is impossible to find the vector (x, y, z) that generated it. For example if we consider the vector $(1,1)$ we cannot detect the point in $\mathbb{R}^3$ that generated it because it could be $(1,1,1)$, $(1,1,2)$, $(1,1,8)$, $(1,1,3.56723)$ etc. More formally, this mapping is not injective. Hence, this mapping is not bijective.

The case of a bijective homomorphism is special and undergoes a separate definition.

Definition 7.34. A bijective homomorphism is said isomorphism.

An intuitive way to describe isomorphisms is by means of the following quote of the mathematician Douglas Hofstadter:

> The word 'isomorphism' applies when two complex structures can be mapped onto each other, in such a way that to each part of one structure there is a corresponding part in the other structure, where 'corresponding' means that the two parts play similar roles in their respective structures.

If an isomorphism holds between two algebraic structures, then these algebraic structures are said *isomorphic*. Isomorphism is an extremely important concept in mathematics and is a precious help in problem solving. When a problem is very hard to be solved in its domain (its algebraic structure), the problem can be transformed into and isomorphic one and solved within the isomorphic algebraic structure. The solution in the isomorphic domain is then antitransformed to the original problem domain.

An extensive example of isomorphism is presented in Chap. 12 within the context of graph theory. The following examples give an idea of what an isomorphism is and how it can be the theoretical foundation of several computational techniques.

Example 7.57. Let us consider the groups $(\mathbb{N}, +)$ and $(10^{\mathbb{N}},)$ where $10^{\mathbb{N}}$ is the set of the powers of 10 with natural exponent. The mapping $\phi : \mathbb{N} \rightarrow 10^{\mathbb{N}}$ is

$$\phi(x) = 10^x.$$

Let us show that the mapping is an homomorphism:

$$\phi(x+y) = 10^{x+y} = 10^x 10^y = \phi(x)\phi(y).$$

In order to verify that this homomorphism is an isomorphism let us show that this mapping is an injection and a surjection. It is an injection since if $x_1 \neq x_2$ then $10^{x_1} \neq 10^{x_2}$. It is a surjection since every positive number can be expressed as 10^x. Hence, this mapping is an isomorphism.

Example 7.58. Let us indicate with $\mathbb{R}^+$ the set of positive real numbers. Let us consider the groups $(\mathbb{R}^+,)$ and $(\mathbb{R}, +)$ and the mapping $f : \mathbb{R}^+ \rightarrow \mathbb{R}$, $f(x) = \log(x)$ where log is the logarithm, see [18]. This mapping is an homomorphism since

$$f(xy) = \log(xy) = \log(x) + \log(y) = f(x) + f(y).$$

The mapping is an isomorphism because it is injective as if $x_1 \neq x_2$ then $\log(x_1) \neq \log(x_2)$ and it is surjective as every real number can be expressed as a logarithm of a (positive) number:

$$\forall t \in \mathbb{R} \exists x \in \mathbb{R}^+ \ni {}^\prime t = \log(x).$$

Obviously, it can be remarked that there is a relation between the fact that an homomorphism is an isomorphism and that the mapping is invertible. If we think about the words, the prefix *iso-* is stronger than the prefix *homo-*. While 'homo' means 'same', i.e. of the same kind, the word 'iso' means 'identical'. In this light, while an homomorphism transforms an algebraic structure into a structure of the same kind, an isomorphism transforms an algebraic structure into a structurally identical algebraic structure. For this reason, the latter unlike the first is also reversible.

Example 7.59. The Laplace Transform is an isomorphism between differential equations and complex algebraic equations, see [19].

Exercises

7.1. Considering the set $A = \{0, 1, 2, 4, 6\}$ verify whether or not $(A, +)$ is an algebraic structure, semigroup, monoid, or group.

7.2. Considering the set of the square matrices $\mathbf{R}_{n,n}$ and the product of matrices $\cdot$, verify whether or not $(\mathbf{R}_{n,n}, \cdot)$ is an algebraic structure, semigroup, monoid, or group.

7.3. Let Z_8 be $\{0, 1, 2, 3, 4, 5, 6, 7\}$ and $+_8$ be the cyclic sum with cyclic sum defined as:
$$a +_8 b = a + b \qquad \text{if} (a + b) \leq 7; \forall a, b \in Z_8$$
$$a +_8 b = a + b - 8 \ \text{if} (a + b) > 7; \forall a, b \in Z_8.$$

Determine whether or not $(H, +_8)$ with $H = \{0, 2, 4, 6\}$ is a subgroup. Represent the cosets and verify the Lagrange's Theorem in the present case.

7.4. Let $(\mathbb{Q}, *)$ be an algebraic structure where the operator $*$ is defined as

$$a * b = a + 5b.$$

Verify whether or not $(\mathbb{Q}, *)$ is a monoid and identify, if they exist, its inverse elements. If the structure is a monoid verify whether or not this monoid is a group.

7.5. Let us consider the groups $(\mathbb{R}, +)$ and $(\mathbb{R}^+,)$. Verify whether or not the mapping $f : \mathbb{R} \to \mathbb{R}^+, f(x) = e^x$ is an homomorphism and an isomorphism.

Chapter 8
Vector Spaces

8.1 Basic Concepts

This chapter revisits the concept of vector bringing it to an abstract level. Throughout this chapter, for analogy we will refer to vectors using the same notation as for numeric vectors.

Definition 8.1. Vector Space. Let E to be a non-null set $(E \neq \emptyset)$ and $\mathbb{K}$ to be a *scalar set* (in this chapter and in Chap. 10 we will refer with $\mathbb{K}$ to either the set of real numbers $\mathbb{R}$ or the set of complex numbers $\mathbb{C}$). Let us name *vectors* the elements of the set E. Let "+" be an internal composition law, $E \times E \to E$. Let "·" be an external composition law, $\mathbb{K} \times E \to E$. The triple $(E, +, \cdot)$ is said *vector space* of the vector set E over the *scalar field* $(\mathbb{K}, +,)$ if and only if the following ten axioms, namely *vector space axioms* are verified. As in the case of the product of a scalar by a vector, see Chap. 4, the symbol of external composition law · will be omitted.

- E is closed with respect to the internal composition law: $\forall \mathbf{u}, \mathbf{v} \in E : \mathbf{u} + \mathbf{v} \in E$
- E is closed with respect to the external composition law: $\forall \mathbf{u} \in E$ and $\forall \lambda \in \mathbb{K} :$ $\lambda \mathbf{u} \in E$
- commutativity for the internal composition law: $\forall \mathbf{u}, \mathbf{v} \in E : \mathbf{u} + \mathbf{v} = \mathbf{v} + \mathbf{u}$
- associativity for the internal composition law: $\forall \mathbf{u}, \mathbf{v}, \mathbf{w} \in E \times E : \mathbf{u} + (\mathbf{v} + \mathbf{w}) = (\mathbf{u} + \mathbf{v}) + \mathbf{w}$
- neutral element for the internal composition law: $\forall \mathbf{u} \in E : \exists! \mathbf{o} \in E | \mathbf{u} + \mathbf{o} = \mathbf{u}$
- opposite element for the internal composition law: $\forall \mathbf{u} \in E : \exists! -\mathbf{u} \in E | \mathbf{u} + -\mathbf{u} = \mathbf{o}$
- associativity for the external composition law: $\forall \mathbf{u} \in E$ and $\forall \lambda, \mu \in \mathbb{K} : \lambda (\mu \mathbf{u}) = (\lambda \mu) \mathbf{u} = \lambda \mu \mathbf{u}$
- distributivity 1: $\forall \mathbf{u}, \mathbf{v} \in E$ and $\forall \lambda \in \mathbb{K} : \lambda (\mathbf{u} + \mathbf{v}) = \lambda \mathbf{u} + \lambda \mathbf{v}$
- distributivity 2: $\forall \mathbf{u} \in E$ and $\forall \lambda, \mu \in \mathbb{K} : (\lambda + \mu) \mathbf{u} = \lambda \mathbf{u} + \lambda \mathbf{u}$
- neutral elements for the external composition law: $\forall \mathbf{u} \in E : \exists! 1 \in \mathbb{K} | 1 \mathbf{u} = \mathbf{u}$

where $\mathbf{o}$ is the null vector.

© Springer Nature Switzerland AG 2019

F. Neri, *Linear Algebra for Computational Sciences and Engineering*,
https://doi.org/10.1007/978-3-030-21321-3_8

In order to simplify the notation, in this chapter and in Chap. 10, with $\mathbb{K}$ we will refer to the scalar field $(\mathbb{K}, +,)$.

Example 8.1. The following triples are vector spaces.

- The set of geometric vector $\mathbb{V}_3$, the sum between vectors and the product of a scalar by a geometric vector, $(\mathbb{V}_3, +,)$.
- The set of matrices $\mathbb{R}_{m,n}$, the sum between matrices and the product of a scalar by a matrix, $(\mathbb{R}_{m,n}, +,)$.
- The set of numeric vectors $\mathbb{R}^n$ with $n \in \mathbb{N}$, the sum between vectors and the product of a scalar by a numeric vector, $(\mathbb{R}^n, +,)$. In the latter case if $n = 1$, the set of numeric vectors is the set of real numbers, which still is a vector space.

Thus, numeric vectors with sum and product is a vector space but a vector spaces is a general (abstract) concept that deals with the sets and composition laws that respect the above-mentioned ten axioms.

8.2 Vector Subspaces

Definition 8.2. Vector Subspace. Let $(E, +, \cdot)$ be a vector space, $U \subset E$, and $U \neq \emptyset$. The triple $(U, +, \cdot)$ is a vector subspace of $(E, +, \cdot)$ if $(U, +, \cdot)$ is a vector space over the same field $\mathbb{K}$ with respect to both the composition laws.

Proposition 8.1. *Let $(E, +, \cdot)$ be a vector space, $U \subset E$, and $U \neq \emptyset$. The triple $(U, +, \cdot)$ is a vector subspace of $(E, +, \cdot)$ if and only if U is closed with respect to both the composition laws $+$ and $\cdot$, i.e.*

- $\forall \mathbf{u}, \mathbf{v} \in U : \mathbf{u} + \mathbf{v} \in U$
- $\forall \lambda \in \mathbb{K}$ *and* $\forall \mathbf{u} \in U : \lambda \mathbf{u} \in U$.

Proof. Since the elements of U are also elements of E, they are vectors that satisfy the eight axioms regarding internal and external composition laws. If U is also closed with respect to the composition laws then $(U, +, \cdot)$ is a vector space and since $U \subset E, U$ is vector subspace of $(E, +, \cdot)$. $\square$

If $(U, +, \cdot)$ is a vector subspace of $(E, +, \cdot)$, then it is a vector space. Thus, the ten axioms, including the closure with respect of the composition laws, are valid. $\square$

Proposition 8.2. *Let $(E, +, \cdot)$ be a vector space over a field $\mathbb{K}$. Every vector subspace $(U, +, \cdot)$ of $(E, +, \cdot)$ contains the null vector.*

Proof. Considering that $0 \in \mathbb{K}$, it follows that $\forall \mathbf{u} \in U : \exists \lambda | \lambda \mathbf{u} = \mathbf{o}$. Since $(U, +, \cdot)$ is a vector subspace, the set U is closed with respect to the external composition law, $\mathbf{o} \in U$. $\square$

Proposition 8.3. *For every vector space $(E, +, \cdot)$, at least two vector subspaces exist, i.e. $(E, +, \cdot)$ and $(\{\mathbf{o}\}, +, \cdot)$.*

Example 8.2. Let us consider the vector space $\left(\mathbb{R}^3, +, \cdot\right)$ and its subset $U \subset \mathbb{R}^3$:

$$U = \{(x,y,z) \in \mathbb{R}^3 | 3x + 4y - 5z = 0\}$$

and let us prove that $(U, +, \cdot)$ is a vector subspace of $\left(\mathbb{R}^3, +, \cdot\right)$.

We have to prove the closure with respect to the two composition laws.

1. Let us consider two arbitrary vectors belonging to U, $\mathbf{u_1} = (x_1, y_1, z_1)$ and $\mathbf{u_2} = (x_2, y_2, z_2)$. These two vectors are such that

$$3x_1 + 4y_1 - 5z_1 = 0$$
$$3x_2 + 4y_2 - 5z_2 = 0.$$

Let us calculate

$$\mathbf{u_1} + \mathbf{u_2} = (x_1 + x_2, y_1 + y_2, z_1 + z_2).$$

In correspondence to the vector $\mathbf{u_1} + \mathbf{u_2}$,

$$3(x_1 + x_2) + 4(y_1 + y_2) - 5(z_1 + z_2) =$$
$$= 3x_1 + 4y_1 - 5z_1 + 3x_2 + 4y_2 - 5z_2 = 0 + 0 = 0.$$

This means that $\forall \mathbf{u_1}, \mathbf{u_2} \in U : \mathbf{u_1} + \mathbf{u_2} \in U$.

2. Let us consider an arbitrary vector $\mathbf{u} = (x, y, z) \in U$ and an arbitrary scalar $\lambda \in \mathbb{R}$. We know that $3x + 4y - 5z = 0$. Let us calculate

$$\lambda \mathbf{u} = (\lambda x, \lambda y, \lambda z).$$

In correspondence to the vector $\lambda \mathbf{u}$,

$$3\lambda x + 4\lambda y - 5\lambda z =$$
$$= \lambda (3x + 4y - 5z) = \lambda 0 = 0.$$

This means that $\forall \lambda \in \mathbf{K}$ and $\forall \mathbf{u} \in U : \lambda \mathbf{u} \in U$.

Thus, we proved that $(U, +, \cdot)$ is a vector subspace $\left(\mathbb{R}^3, +, \cdot\right)$.

Example 8.3. Let us consider the vector space $\left(\mathbb{R}^3, +, \cdot\right)$ and its subset $U \subset \mathbb{R}^3$:

$$U = \{(x,y,z) \in \mathbb{R}^3 | 8x + 12y - 7z + 1 = 0\}.$$

Since the null vector $\mathbf{o} \notin U$, $(U, +, \cdot)$ is not a vector space.

Although it is not necessary, let us check that the set U is not closed with respect to the internal composition law. In order to do this, let us consider two vectors (x_1, y_1, z_1) and (x_2, y_2, z_2). The sum vector $(x_1 + x_2, y_1 + y_2, z_1 + z_2)$ is

$$8(x_1 + x_2) + 4(y_1 + y_2) - 5(z_1 + z_2) + 1 =$$
$$= 8x_1 + 4y_1 - 5z_1 + 8x_2 + 4y_2 - 5z_2 + 1.$$

If we consider that $8x_2 + 4y_2 - 5z_2 + 1 = 0$ it follows that $8x_1 + 4y_1 - 5z_1 \neq 0$. Hence,

$$8(x_1 + x_2) + 4(y_1 + y_2) - 5(z_1 + z_2) + 1 \neq 0.$$

This means that $(x_1 + x_2, y_1 + y_2, z_1 + z_2) \notin U$. The set U is not closed with respect to the internal composition law.

Analogously, if $(x, y, z) \in U$, then $(\lambda x, \lambda y, \lambda z)$, with *lambda* scalar, is

$$8(\lambda x) + 4(\lambda y) - 5(\lambda z) + 1$$

which in general is not equal to zero. Hence, $(\lambda x, \lambda y, \lambda z) \notin U$.

Vector spaces are a general concept which applies not only to numeric vectors.

Example 8.4. Let us consider the set $\mathscr{F}$ of the real-valued functions continuous on an interval $[a, b]$. We can show that the triple $(\mathscr{F}, +, \cdot)$ is a vector space over the real field $\mathbb{R}$.

We need to show the closure with respect to the two composition laws. For this aim, let us consider two continuous functions $f(x), g(x) \in \mathscr{F}$.

Considering that the sum of two continuous functions

$$f(x) + g(x)$$

is a continuous function, the set is closed with respect to the internal composition law.

Since $\forall \lambda \in \mathbb{R}$ it follows that

$$\lambda f(x)$$

is also continuous, the set is closed also with respect to the external composition law.

Thus, $(\mathscr{F}, +, \cdot)$ is a vector space.

Theorem 8.1. *Let $(E, +, \cdot)$ be a vector space. If $(U, +, \cdot)$ and $(V, +, \cdot)$ are two vector subspaces of $(E, +, \cdot)$, then $(U \cap V, +, \cdot)$ is a vector subspace of $(E, +, \cdot)$.*

Proof. For the Proposition 8.1 it would be enough to prove the closure of the set $U \cap V$ with respect to the composition laws to prove that $(U \cap V, +, \cdot)$ is a vector subspace of $(E, +, \cdot)$.

1. Let $\mathbf{u}, \mathbf{v}$ be two arbitrary vectors $\in U \cap V$. If $\mathbf{u} \in U \cap V$ then $\mathbf{u} \in U$ and $\mathbf{u} \in V$. Analogously, if $\mathbf{v} \in U \cap V$ then $\mathbf{v} \in U$ and $\mathbf{v} \in V$. Since both $\mathbf{u}, \mathbf{v} \in U$ and $(U, +, \cdot)$ is a vector space, then $\mathbf{u} + \mathbf{v} \in U$. Since $\mathbf{u}, \mathbf{v} \in V$ and $(V, +, \cdot)$ is a vector space, then $\mathbf{u} + \mathbf{v} \in V$. It follows that $\mathbf{u} + \mathbf{v}$ belongs to both U and V, i.e. it belongs to their intersection $U \cap V$. This means that $U \cap V$ is closed with respect to the $+$ operation.

2. Let **u** be an arbitrary vector $\in U \cap V$ and λ an arbitrary scalar $\in \mathbb{K}$. If $\mathbf{u} \in U \cap V$ then $\mathbf{u} \in U$ and $\mathbf{u} \in V$. Since $(U, +, \cdot)$ is a vector space, then $\lambda\mathbf{u} \in U$. Since $(V, +, \cdot)$ is a vector space, then $\lambda\mathbf{u} \in V$. Thus, $\lambda\mathbf{u}$ belongs to both U and V, i.e. it belongs to their intersection $U \cap V$. This means that $U \cap V$ is closed with respect to the $\cdot$ operation.

Thus, since $U \cap V$ is closed with respect to both the composition laws, $(U \cap V, +, \cdot)$ is a vector subspaces of $(E, +, \cdot)$. □

Corollary 8.1. *Let $(E, +, \cdot)$ be a vector space. If $(U, +, \cdot)$ and $(V, +, \cdot)$ are two vector subspaces of $(E, +, \cdot)$, then $U \cap V$ is always a non-empty set as it contains at least the null vector.*

It must be observed than if $(U, +, \cdot)$ and $(V, +, \cdot)$ are two vector subspaces of $(E, +, \cdot)$ their union is in general not a subspace of $(E, +, \cdot)$. On the contrary, it can be easily proved that $(U \cup V, +, \cdot)$ is a vector subspace of $(E, +, \cdot)$ if $U \subset V$ or $V \subset U$.

Example 8.5. Let us consider the vector space $(\mathbb{R}^2, +, \cdot)$ and its subsets $U \subset \mathbb{R}^2$ and $V \subset \mathbb{R}^2$:

$$U = \{(x, y) \in \mathbb{R}^2 | -5x + y = 0\}$$
$$V = \{(x, y) \in \mathbb{R}^2 | 3x + 2y = 0\}.$$

It can be easily shown that both $(U, +, \cdot)$ and $(V, +, \cdot)$ are vector subspaces of $(\mathbb{R}^2, +, \cdot)$. The intersection $U \cap V$ is composed of those (x, y) values belonging to both the sets, i.e. satisfying both the conditions above. This means that $U \cap V$ is composed of those (x, y) values satisfying the following system of linear equations:

$$\begin{cases} -5x + y = 0 \\ 3x + 2y = 0. \end{cases}$$

This is an homogeneous system of linear equations which is determined (the incomplete matrix associated with the system is non-singular). Hence the only solution is $(0, 0)$, that is the null vector **o**. A geometric interpretation of the system above is the intersection of two lines passing through the origin of a reference system and intersecting in it, that is the null vector. As stated by Theorem 8.1, $(U \cap V, +, \cdot)$ is a vector subspace of $(\mathbb{R}^2, +, \cdot)$. In this case, the vector subspace is the special one $(\{\mathbf{o}\}, +, \cdot)$.

Example 8.6. Let us consider the vector space $(\mathbb{R}^2, +, \cdot)$ and its subsets $U \subset \mathbb{R}^2$ and $V \subset \mathbb{R}^2$:

$$U = \{(x, y) \in \mathbb{R}^2 | x - 5y = 0\}$$
$$V = \{(x, y) \in \mathbb{R}^2 | 3x + 2y - 2 = 0\}.$$

It can be easily shown that while $(U, +, \cdot)$ is a vector subspace of $(\mathbb{R}^2, +, \cdot)$, $(V, +, \cdot)$ is not a vector space. Regardless of this fact the intersection $U \cap V$ can be still calculated and the set composed of those (x, y) values satisfying the following system of linear equations:

$$\begin{cases} x - 5y = 0 \\ 3x + 2y = 2 \end{cases}$$

that is $\left(\frac{10}{17}, \frac{2}{17}\right)$. Since $U \cap V$ does not contain the null vector, obviously $(U \cap V, +, \cdot)$ is a not vector space.

Example 8.7. Let us consider the vector space $(\mathbb{R}^2, +, \cdot)$ and its subsets $U \subset \mathbb{R}^2$ and $V \subset \mathbb{R}^2$:

$$U = \{(x, y) \in \mathbb{R}^2 | -5x + y = 0\}$$
$$V = \{(x, y) \in \mathbb{R}^2 | -15x + 3y = 0\}.$$

It can be easily shown that both $(U, +, \cdot)$ and $(V, +, \cdot)$ are vector subspaces of $(\mathbb{R}^2, +, \cdot)$. The intersection $U \cap V$ is composed of those (x, y) satisfying the following system of linear equations:

$$\begin{cases} -5x + y = 0 \\ -15x + 3y = 0. \end{cases}$$

Since the system is homogeneous the system is compatible and at least the null vector is its solution. By applying Rouchè-Capelli Theorem, since the rank is lower than the number of variables the system is undetermined, i.e. it has ∞ solutions. The system can be interpreted as two overlapping lines. This means that all the points in U are also points of V and the two sets coincide $U = V$. Hence, $U \cap V = U = V$ and $(U \cap V, +, \cdot)$ is a vector subspace of $(\mathbb{R}^2, +, \cdot)$.

Example 8.8. Let us consider the vector space $(\mathbb{R}^3, +, \cdot)$ and its subsets $U \subset \mathbb{R}^2$ and $V \subset \mathbb{R}^3$:

$$U = \{(x, y, z) \in \mathbb{R}^3 | 3x + y + z = 0\}$$
$$V = \{(x, y, z) \in \mathbb{R}^3 | x - z = 0\}.$$

It can be easily verified that both $(U, +, \cdot)$ and $(V, +, \cdot)$ are vector subspaces of $(\mathbb{R}^3, +, \cdot)$.

The intersection set $U \cap V$ is given by

$$\begin{cases} 3x + y + z = 0 \\ x - z = 0. \end{cases}$$

This is an homogeneous system of two linear equations in three variables. Besides the null vector, by applying Rouchè-Capelli Theorem, since the rank of the associated matrix is 2 and there are three variables, the system has ∞ solutions. The system above can be interpreted as the intersection of two planes.

The second equations can be written as $x = z$. By substituting the second equation into the first one, we obtain

$$4x + y = 0 \Rightarrow y = -4x.$$

Hence, if we fix a value of $x = a$, a vector $(a, -4a, a)$ satisfies the system of linear equations. This means that the intersection set $U \cap V$ contains ∞ elements and is given by:

$$U \cap V = \{(a, -4a, a) \mid a \in \mathbb{R}\}.$$

The resulting triple $(U \cap V, +, \cdot)$, for Theorem 8.1, is a vector space.

Definition 8.3. Let $(E, +, \cdot)$ be a vector space. Let $(U, +, \cdot)$ and $(V, +, \cdot)$ be two vector subspaces of $(E, +, \cdot)$. The *sum subset* is a set $S = U + V$ defined as

$$S = U + V = \{\mathbf{w} \in E \mid \exists \mathbf{u} \in U, \mathbf{v} \in V \mid \mathbf{w} = \mathbf{u} + \mathbf{v}\}.$$

Theorem 8.2. *Let $(E, +, \cdot)$ be a vector space. If $(U, +, \cdot)$ and $(V, +, \cdot)$ are two vector subspaces of $(E, +, \cdot)$, then $(S = U + V, +, \cdot)$ is a vector subspace of $(E, +, \cdot)$.*

Proof. For the Proposition 8.1 it would be enough to prove the closure of the set S with respect to the composition laws to prove that $(S, +, \cdot)$ is a vector subspace of $(E, +, \cdot)$.

1. Let $\mathbf{w_1}, \mathbf{w_2}$ be two arbitrary vectors belonging to S. From the definition of the sum subset we can write that

$$\exists \mathbf{u_1} \in U \text{ and } \exists \mathbf{v_1} \in V \mid \mathbf{w_1} = \mathbf{u_1} + \mathbf{v_1}$$
$$\exists \mathbf{u_2} \in U \text{ and } \exists \mathbf{v_2} \in V \mid \mathbf{w_2} = \mathbf{u_2} + \mathbf{v_2}.$$

The sum of $\mathbf{w_1}$ and $\mathbf{w_2}$ is equal to

$$\mathbf{w_1} + \mathbf{w_2} = \mathbf{u_1} + \mathbf{v_1} + \mathbf{u_2} + \mathbf{v_2} = (\mathbf{u_1} + \mathbf{u_2}) + (\mathbf{v_1} + \mathbf{v_2}).$$

Since U and V are vector spaces, $\mathbf{u_1} + \mathbf{u_2} \in U$ and $\mathbf{v_1} + \mathbf{v_2} \in V$.
Thus, according to the definition of sum subset $\mathbf{w_1} + \mathbf{w_2} = (\mathbf{u_1} + \mathbf{u_2}) + (\mathbf{v_1} + \mathbf{v_2}) \in S$.

2. Let $\mathbf{w}$ be an arbitrary vector $\in S$ and λ an arbitrary scalar $\in \mathbb{K}$. From the definition of sum set

$$S = U + V = \{\mathbf{w} \in E \mid \exists \mathbf{u} \in U, \mathbf{v} \in V \mid \mathbf{w} = \mathbf{u} + \mathbf{v}\}.$$

If we compute the product of λ by $\mathbf{w}$, we obtain

$$\lambda \mathbf{w} = \lambda (\mathbf{u} + \mathbf{v}) = \lambda \mathbf{u} + \lambda \mathbf{v}$$

where, since U and V are vector spaces, $\lambda \mathbf{u} \in U$ and $\lambda \mathbf{v} \in V$. Thus $\lambda \mathbf{w} = \lambda \mathbf{u} + \lambda \mathbf{v} \in S$. $\square$

The following figure illustrates the concept of sum set. The white circles represent the vectors in the vector subspaces $(U,+,\cdot)$ and $(V,+,\cdot)$ while the black circles are vectors belonging to the sum set S. It can be noticed that a vector $\mathbf{w} \in S$ can be the sum of two (or more) different pairs of vectors, i.e. $\mathbf{w} = \mathbf{u}_1 + \mathbf{v}_1$ and $\mathbf{w} = \mathbf{u}_2 + \mathbf{v}_2$ in figure. Furthermore, at least the null vector $\mathbf{o}$ belongs to both U and V and thus at least $\mathbf{o} \in U \cap V$. Other vectors can also belong to the intersection.

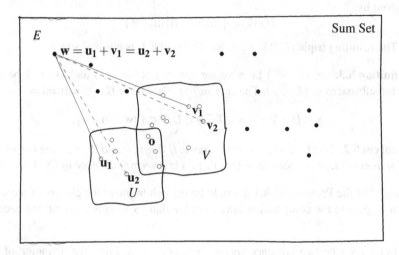

Example 8.9. Let us consider the vector space $(\mathbb{R}^2,+,\cdot)$ and its subsets $U \subset \mathbb{R}^2$ and $V \subset \mathbb{R}^2$:

$$U = \{(x,y) \in \mathbb{R}^2 | y = 0\}$$
$$V = \{(x,y) \in \mathbb{R}^2 | x = 0\}.$$

It can be easily proved that $(U,+,\cdot)$ and $(V,+,\cdot)$ are vector subspaces. The two subsets can be rewritten as

$$U = \{(a,0) | a \in \mathbb{R}\}$$
$$V = \{(0,b) | b \in \mathbb{R}\}.$$

The only intersection vector is the null vector, i.e. $U \cap V = \{\mathbf{o}\}$. If we now calculate the sum subset $S = U + V$ we obtain

$$S = U + V = \{(a,b) | a,b \in \mathbb{R}\}$$

that coincides with the entire $\mathbb{R}^2$. For Theorem 8.2, $(S,+,\cdot)$ is a vector subspace of $(\mathbb{R}^2,+,\cdot)$. In this case the vector subspace is $(\mathbb{R}^2,+,\cdot)$. The latter is obviously a vector space.

Example 8.10. Let us consider the vector space $\left(\mathbb{R}^2, +, \cdot\right)$ and its subsets $U \subset \mathbb{R}^2$ and $V \subset \mathbb{R}^2$:

$$U = \{(x,y) \in \mathbb{R}^2 \,|\, -2x + y = 0\}$$
$$V = \{(x,y) \in \mathbb{R}^2 \,|\, 3x + y = 0\}.$$

The triples $(U, +, \cdot)$ and $(V, +, \cdot)$ are vector subspaces. The two subsets can be rewritten as

$$U = \{(a, 2a) \,|\, a \in \mathbb{R}\}$$
$$V = \{(b, -3b) \,|\, b \in \mathbb{R}\}.$$

Also in this case $U \cap V = \{\mathbf{o}\}$. Let us calculate the sum subset $S = U + V$:

$$S = U + V = \{(a + b, 2a - 3b) \,|\, a, b \in \mathbb{R}\}.$$

For Theorem 8.2, $(S, +, \cdot)$ is a vector subspace of $\left(\mathbb{R}^2, +, \cdot\right)$. Again, by varying $(a,b) \in \mathbb{R}^2$ the entire $\mathbb{R}^2$ is generated. Hence again the sum vector subspace is $\left(\mathbb{R}^2, +, \cdot\right)$.

Example 8.11. Let us consider the vector space $\left(\mathbb{R}^3, +, \cdot\right)$ and its subsets $U \subset \mathbb{R}^3$ and $V \subset \mathbb{R}^3$:

$$U = \{(x,y,0) \in \mathbb{R}^3 \,|\, x, y \in \mathbb{R}\}$$
$$V = \{(x,0,z) \in \mathbb{R}^3 \,|\, x, z \in \mathbb{R}\}.$$

The triples $(U, +, \cdot)$ and $(V, +, \cdot)$ are vector subspaces. The intersection of the two subspaces is $U \cap V = \{(x,0,0) \in \mathbb{R}^3 \,|\, x \in \mathbb{R}\}$, i.e. it is not only the null vector. The sum is again the entire $\mathbb{R}^3$.

Definition 8.4. Let $(E, +, \cdot)$ be a vector space. Let $(U, +, \cdot)$ and $(V, +, \cdot)$ be two vector subspaces of $(E, +, \cdot)$. If $U \cap V = \{\mathbf{o}\}$ the subset sum $S = U + V$ is indicated as $S = U \oplus V$ and named *subset direct sum*.

The triple $(S = U \oplus V, +, \cdot)$ is a vector subspace of $(E, +, \cdot)$, as it is a special case of $(S, +, \cdot)$, and the subspaces $(U, +, \cdot)$ and $(V, +, \cdot)$ are said *supplementary*.

Theorem 8.3. *Let $(E, +, \cdot)$ be a vector space. Let $(U, +, \cdot)$ and $(V, +, \cdot)$ be two vector subspaces of $(E, +, \cdot)$. The sum vector subspace $(U + V, +, \cdot)$ is a direct sum vector subspace $(U \oplus V, +, \cdot)$ $(U \cap V = \{\mathbf{o}\})$ if and only if*

$$S = U + V = \{\mathbf{w} \in E \,|\, \exists! \mathbf{u} \in U \text{ and } \exists! \mathbf{v} \in V \,|\, \mathbf{w} = \mathbf{u} + \mathbf{v}\}.$$

Proof. If S is a subset direct sum then $U \cap V = \{o\}$. By contradiction, let us assume that $\exists u_1, u_2 \in U$ and $\exists v_1, v_2 \in V$ with $u_1 \neq u_2$ and $v_1 \neq v_2$ such that

$$\begin{cases} w = u_1 + v_1 \\ w = u_2 + v_2. \end{cases}$$

Under this hypothesis,

$$o = u_1 + v_1 - u_2 - v_2 = (u_1 - u_2) + (v_1 - v_2).$$

By rearranging this equation we have

$$(u_1 - u_2) = -(v_1 - v_2)$$

where

- $(u_1 - u_2) \in U$ and $(v_1 - v_2) \in V$, respectively, since $(U, +, \cdot)$ and $(V, +, \cdot)$ are two vector spaces (and thus their sets are closed with respect to the internal composition law)
- $-(v_1 - v_2) \in V$ (for the axiom of the opposite element)
- $(u_1 - u_2) \in V$ and $(v_1 - v_2) \in U$ because the vectors are identical

Thus, $(u_1 - u_2) \in U \cap V$ and $(v_1 - v_2) \in U \cap V$.

Since $U \cap V = \{o\}$ we cannot have non-null vectors in the intersection set. Thus,

$$\begin{cases} (u_1 - u_2) = o \\ (v_1 - v_2) = o \end{cases} \Rightarrow \begin{cases} u_1 = u_2 \\ v_1 = v_2 \end{cases} \quad \square$$

If $S = \{w \in E | \exists! u \in U \text{ and } \exists! v \in V | w = u + v\}$ let us assume by contradiction that $U \cap V \neq \{o\}$. Thus

$$\exists t \in U \cap V$$

with $t \neq o$.

Since $(U, +, \cdot)$ and $(V, +, \cdot)$ are two vector subspaces of $(E, +, \cdot)$, for the Theorem 8.1 $(U \cap V, +, \cdot)$ is a vector subspace of $(E, +, \cdot)$.

Since $(U \cap V, +, \cdot)$ is a vector space, if $t \in U \cap V$ also $-t \in U \cap V$.

We may think that

$$t \in U$$
$$-t \in V.$$

Since also $(S, +, \cdot)$ is also a vectors space $o \in S$. Thus, we can express the null vector $o \in S$ as the sum of an element of U and an element of V (as in the definition of sum set)

$$o = t + (-t).$$

On the other hand, o can be also expressed as

$$o = o + o$$

where the first $\mathbf{o}$ is considered an element of U and the second $\mathbf{o}$ is considered an element of V.

For hypothesis $\forall \mathbf{w} \in S : \exists! \mathbf{u} \in U$ *and* $\exists! \mathbf{v} \in V | \mathbf{w} = \mathbf{u} + \mathbf{v}$. We expressed $\mathbf{o}$ that is an element of S as sum of two different pairs of vectors. This is a contradiction. Thus $\mathbf{t}$ must be equal to $\mathbf{o}$ and, in other words, $U \cap V = \mathbf{o}$. $\square$

A graphical representation of the direct sum set is given in the following, every time the intersection of U and V is composed only by the null vector then only one element of U and only one element of V can contribute to the sum.

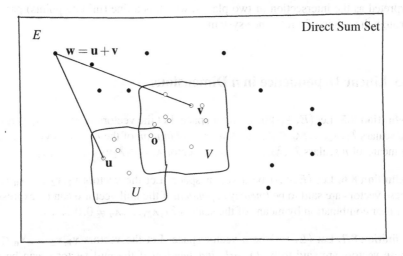

Example 8.12. As shown above the sum associated with the sets

$$U = \{(a,0) \,|\, a \in \mathbb{R}\}$$
$$V = \{(0,b) \,|\, b \in \mathbb{R}\}.$$

is a direct sum since $U \cap V = \{\mathbf{o}\}$. The sum set is the entire $\mathbb{R}^2$ generates as

$$S = U + V = \{(a,b) \,|\, a,b \in \mathbb{R}\}.$$

Obviously, there is only one way to obtain (a,b) from $(a,0)$ and $(0,b)$.
On the contrary, for the sets

$$U = \{(x,y,0) \in \mathbb{R}^3 \,|\, x,y \in \mathbb{R}\}$$
$$V = \{(x,0,z) \in \mathbb{R}^3 \,|\, x,z \in \mathbb{R}\},$$

whose intersection is not only the null vector, each vector (a,b,c) can be obtained in infinite ways. For example the if we consider the vector $(1,2,3)$, we know that it can be expressed as

$$(1,2,3) = (x_1, y, 0) + (x_2, 0, z)$$

which leads to the following system of linear equations:

$$\begin{cases} x_1 + x_2 = 1 \\ y = 2 \\ z = 3. \end{cases}$$

This system has ∞ solutions depending on x_1 and x_2. This example can be interpreted as the intersection of two planes, which is a line (infinite points) passing through the origin of a reference system.

8.3 Linear Dependence in n Dimensions

Definition 8.5. Let $(E, +, \cdot)$ be a vector space. Let the vectors $\mathbf{v_1}, \mathbf{v_2}, \ldots, \mathbf{v_n} \in E$ and the scalars $\lambda_1, \lambda_2, \ldots, \lambda_n \in \mathbb{K}$. The *linear combination* of the n vectors $\mathbf{v_1}, \mathbf{v_2}, \ldots, \mathbf{v_n}$ by means of n scalars $\lambda_1, \lambda_2, \ldots, \lambda_n$ is the vector $\lambda_1 \mathbf{v_1} + \lambda_2 \mathbf{v_2} + \ldots + \lambda_n \mathbf{v_n}$.

Definition 8.6. Let $(E, +, \cdot)$ be a vector space. Let the vectors $\mathbf{v_1}, \mathbf{v_2}, \ldots, \mathbf{v_n} \in E$. These vectors are said to be *linearly dependent* if the null vector $\mathbf{o}$ can be expressed as linear combination by means of the scalars $\lambda_1, \lambda_2, \ldots, \lambda_n \neq 0, 0, \ldots, 0$.

Definition 8.7. Let $(E, +, \cdot)$ be a vector space. Let the vectors $\mathbf{v_1}, \mathbf{v_2}, \ldots, \mathbf{v_n} \in E$. These vectors are said to be *linearly independent* if the null vector $\mathbf{o}$ can be expressed as linear combination only by means of the scalars $0, 0, \ldots, 0$.

Proposition 8.4. *Let $(E, +, \cdot)$ be a vector space. Let the vectors $\mathbf{v_1}, \mathbf{v_2}, \ldots, \mathbf{v_n} \in E$. The vectors $\mathbf{v_1}, \mathbf{v_2}, \ldots, \mathbf{v_n}$ are linearly dependent if and only if at least one of them can be expressed as linear combination of the others.*

Proof. If the vectors are linearly dependent then

$$\lambda_1, \lambda_2, \ldots, \lambda_n \neq 0, 0, \ldots, 0 \ni '$$
$$\mathbf{o} = \lambda_1 \mathbf{v_1} + \lambda_2 \mathbf{v_2} + \ldots + \lambda_n \mathbf{v_n}.$$

Let us assume that $\lambda_n \neq 0$ and rearrange the expression

$$-\lambda_n \mathbf{v_n} = \lambda_1 \mathbf{v_1} + \lambda_2 \mathbf{v_2} + \ldots + \lambda_{n-1} \mathbf{v_{n-1}}.$$

and

$$\mathbf{v_n} = -\frac{\lambda_1}{\lambda_n} \mathbf{v_1} - \frac{\lambda_2}{\lambda_n} \mathbf{v_2} - \ldots - \frac{\lambda_{n-1}}{\lambda_n} \mathbf{v_{n-1}}. \square$$

If one vector can be expressed as a linear combination of the others (let us assume $\mathbf{v_n}$) then

$$\mathbf{v_n} = k_1\mathbf{v_1} + k_2\mathbf{v_2} + \ldots + k_{n-1}\mathbf{v_{n-1}}.$$

We can rearrange the expression as

$$\mathbf{o} = k_1\mathbf{v_1} + k_2\mathbf{v_2} + \ldots + k_{n-1}\mathbf{v_{n-1}} - \mathbf{v_n}$$

that is the null vector expressed by the linear combination of scalars that are not all zeros, i.e.

$$k_1, k_2, \ldots, k_{n-1}, -1 \neq 0, 0, \ldots 0. \quad \square$$

Example 8.13. Let us consider the following vectors $\in \mathbb{R}^3$

$$\mathbf{v_1} = (4, 2, 0)$$
$$\mathbf{v_2} = (1, 1, 1)$$
$$\mathbf{v_3} = (6, 4, 2).$$

These vectors are linearly dependent since

$$(0, 0, 0) = (4, 2, 0) + 2(1, 1, 1) - (6, 4, 2)$$

that is $\mathbf{v_3}$ as a linear combination of $\mathbf{v_1}$ and $\mathbf{v_2}$

$$(6, 4, 2) = (4, 2, 0) + 2(1, 1, 1).$$

Proposition 8.5. *Let $(E, +, \cdot)$ be a vector space. Let the vectors $\mathbf{v_1}, \mathbf{v_2}, \ldots, \mathbf{v_n} \in E$.*

>*If $\mathbf{v_1} \neq \mathbf{o}$*
>*If $\mathbf{v_2}$ is not linear combination of $\mathbf{v_1}$*
>*If $\mathbf{v_3}$ is not linear combination of $\mathbf{v_1}$ and $\mathbf{v_2}$*
>*...*
>*If $\mathbf{v_n}$ is not linear combination of $\mathbf{v_1}, \mathbf{v_2}, \ldots, \mathbf{v_{n-1}}$*

It follows that $\mathbf{v_1}, \mathbf{v_2}, \ldots, \mathbf{v_n}$ are linearly independent.

Proof. Let us assume, by contradiction that the vectors are linearly dependent, i.e. *exists* a n-tuple $\lambda_1, \lambda_2, \ldots, \lambda_n \neq 0, 0 \ldots, 0$ such that $\mathbf{o} = \lambda_1\mathbf{v_1} + \lambda_2\mathbf{v_2} \ldots + \lambda_n\mathbf{v_n}$.

Under this hypothesis, we can guess that $\lambda_n \neq 0$ and write

$$\mathbf{v_n} = -\frac{\lambda_1}{\lambda_n}\mathbf{v_1} - \frac{\lambda_2}{\lambda_n}\mathbf{v_2} + \ldots - \frac{\lambda_{n-1}}{\lambda_n}\mathbf{v_{n-1}}.$$

Since one vector has been expressed as linear combination of the others we reached a contradiction. $\quad \square$

Example 8.14. Linear dependence and independence properties seen for $\mathbb{V}_3$ in Chap. 4 are generally valid and thus also in the context of vector spaces.

Let us consider the following vectors $\in \mathbb{R}^3$:

$$\mathbf{v_1} = (1,0,1)$$
$$\mathbf{v_2} = (1,1,1)$$
$$\mathbf{v_3} = (2,1,2).$$

These vectors are linearly dependent since

$$\mathbf{o} = \lambda_1 \mathbf{v_1} + \lambda_2 \mathbf{v_2} + \lambda_3 \mathbf{v_3}$$

with $\lambda_1, \lambda_2, \lambda_3 = 1, 1, -1$.

In this case we can write

$$\mathbf{v_3} = \lambda_1 \mathbf{v_1} + \lambda_2 \mathbf{v_2}$$

with $\lambda_1, \lambda_2 = 1, 1$.

Let us consider now the following vectors

$$\mathbf{v_1} = (1,0,1)$$
$$\mathbf{v_2} = (1,1,1)$$
$$\mathbf{v_3} = (0,0,2).$$

There is no way to express any of these vectors as a linear combination of the other two. The vectors are linearly independent.

Theorem 8.4. *Let $(E, +, \cdot)$ be a vector space. Let the vectors $\mathbf{v_1}, \mathbf{v_2}, \ldots, \mathbf{v_n} \in E$. If the n vectors are linearly dependent while $n-1$ are linearly independent, there is a unique way to express one vector as linear combination of the others:*

$$\forall \mathbf{v_k} \in E, \exists! \lambda_1, \lambda_2, \ldots, \lambda_{k-1}, \lambda_{k+1}, \ldots, \lambda_n \neq 0, 0, \ldots, 0 \ni \text{`}$$
$$\mathbf{v_k} = \lambda_1 \mathbf{v_1} + \lambda_2 \mathbf{v_2} + \ldots + \lambda_{k-1} \mathbf{v_{k-1}} + \lambda_{k+1} \mathbf{v_{k+1}} + \ldots + \lambda_n \mathbf{v_n}$$

Proof. Let us assume by contradiction that the linear combination is not unique:

- $\exists \lambda_1, \lambda_2, \ldots, \lambda_{k-1}, \lambda_{k+1}, \ldots, \lambda_n \neq 0, 0, \ldots, 0$ such that

$$\mathbf{v_k} = \lambda_1 \mathbf{v_1} + \lambda_2 \mathbf{v_2} \ldots + \lambda_{k-1} \mathbf{v_{k-1}} + \lambda_{k+1} \mathbf{v_{k+1}} + \ldots + \lambda_n \mathbf{v_n}$$

- $\exists \mu_1, \mu_2, \ldots, \mu_{k-1}, \mu_{k+1}, \ldots, \mu_n \neq 0, 0, \ldots, 0$ such that

$$\mathbf{v_k} = \mu_1 \mathbf{v_1} + \mu_2 \mathbf{v_2} \ldots + \mu_{k-1} \mathbf{v_{k-1}} + \mu_{k+1} \mathbf{v_{k+1}} + \ldots + \mu_n \mathbf{v_n}$$

where $\lambda_1, \lambda_2, \ldots, \lambda_{k-1}, \lambda_{k+1}, \ldots, \lambda_n \neq \mu_1, \mu_2, \ldots, \mu_{k-1}, \mu_{k+1}, \ldots, \mu_n \neq 0, 0, \ldots, 0 | \mathbf{v_k}$.

Under this hypothesis, we can write that

$$\mathbf{o} = (\lambda_1 - \mu_1) \mathbf{v_1} + (\lambda_2 - \mu_2) \mathbf{v_2} \ldots + (\lambda_{k-1} - \mu_{k-1}) \mathbf{v_{k-1}} + (\lambda_{k+1} - \mu_{k-1}) \mathbf{v_{k+1}} +$$
$$+ \ldots + (\lambda_n - \mu_n) \mathbf{v_n}$$

Since the $n-1$ vectors are linearly independent

$$\begin{cases} \lambda_1 - \mu_1 = 0 \\ \lambda_2 - \mu_2 = 0 \\ \dots \\ \lambda_{k-1} - \mu_{k-1} = 0 \\ \lambda_{k+1} - \mu_{k+1} = 0 \\ \dots \\ \lambda_n - \mu_n = 0 \end{cases} \Rightarrow \begin{cases} \lambda_1 = \mu_1 \\ \lambda_2 = \mu_2 \\ \dots \\ \lambda_{k-1} = \mu_{k-1} \\ \lambda_{k+1} = \mu_{k+1} \\ \dots \\ \lambda_n = \mu_n. \end{cases}$$

Thus, the linear combination is unique. $\square$

Example 8.15. Let us consider again the following linearly dependent vectors $\in \mathbb{R}^3$

$$\begin{aligned} \mathbf{v_1} &= (1,0,1) \\ \mathbf{v_2} &= (1,1,1) \\ \mathbf{v_3} &= (2,1,2). \end{aligned}$$

Any pair of them is linearly independent. Let us express $\mathbf{v_3}$ as a linear combination of the other two vectors. We write

$$(2,1,2) = \lambda_1 (1,0,1) + \lambda_2 (1,1,1)$$

which results into the system

$$\begin{cases} \lambda_1 + \lambda_2 = 2 \\ \lambda_2 = 1 \\ \lambda_1 + \lambda_2 = 2 \end{cases}$$

which is determined and has only one solution, that is $\lambda_1, \lambda_2 = 1, 1$.

Example 8.16. Let us consider the following vectors

$$\begin{aligned} \mathbf{v_1} &= (1,0,1) \\ \mathbf{v_2} &= (1,1,1) \\ \mathbf{v_3} &= (2,0,2). \end{aligned}$$

Since $\mathbf{v_3} = 2\mathbf{v_1}$, the vectors are linearly dependent. If we try to express $\mathbf{v_2}$ as a linear combination of $\mathbf{v_1}$ and $\mathbf{v_3}$ we obtain

$$(1,1,1) = \lambda_1 (1,0,1) + \lambda_2 (2,0,2)$$

which results into the system

$$\begin{cases} \lambda_1 + 2\lambda_2 = 1 \\ 0 = 1 \\ \lambda_1 + 2\lambda_2 = 1 \end{cases}$$

which is obviously impossible. This fact occurred since the remaining $n-1$ vectors, i.e. $\mathbf{v_1}$ and $\mathbf{v_3}$ were not linearly independent.

We can interpret geometrically this fact by considering $\mathbf{v_1}$ and $\mathbf{v_3}$ as parallel vectors while $\mathbf{v_2}$ has a different direction. We have attempted to express a vector as the sum of two parallel vectors which is obviously impossible unless the three vectors are all parallel.

Proposition 8.6. *Let* $(E,+,\cdot)$ *be a vector space and* $\mathbf{v_1}, \mathbf{v_2}, \ldots, \mathbf{v_n}$ *be its n vectors. If one of these vectors is equal to the null vector* $\mathbf{o}$, *these vectors are linearly dependent.*

Proof. Let us assume that $\mathbf{v_n} = \mathbf{o}$ and let us pose

$$\mathbf{o} = \lambda_1 \mathbf{v_1} + \lambda_2 \mathbf{v_2} + \ldots + \lambda_{n-1} \mathbf{v_{n-1}} + \lambda_n \mathbf{o}.$$

Even if $\lambda_1, \lambda_2, \ldots, \lambda_{n-1} = 0, 0, \ldots, 0$ the equality is verified for any scalar $\lambda_n \in \mathbb{K}$. Thus, the vectors are linearly dependent. $\square$

Example 8.17. Let us consider the following vectors

$$\mathbf{v_1} = (0,0,3)$$
$$\mathbf{v_2} = (3,1,1)$$
$$\mathbf{v_3} = (0,0,0)$$

and pose
$$(0,0,0) = \lambda_1 (0,0,3) + \lambda_2 (3,1,1) + \lambda_3 (0,0,0).$$

For example $\lambda_1, \lambda_2, \lambda_3 = 0, 0, 5$ satisfies this equation.

8.4 Linear Span

Definition 8.8. Let $(E,+,\cdot)$ be a vector space. The set containing the totality of all the possibly linear combinations of the vectors $\mathbf{v_1}, \mathbf{v_2}, \ldots, \mathbf{v_n} \in E$ by means of n scalars is named *linear span* (or simply *span*) and is indicated with $L(\mathbf{v_1}, \mathbf{v_2}, \ldots, \mathbf{v_n}) \subset E$ or synthetically with L:

$$L(\mathbf{v_1}, \mathbf{v_2}, \ldots, \mathbf{v_n}) = \{\lambda_1 \mathbf{v_1} + \lambda_2 \mathbf{v_2} + \ldots + \lambda_n \mathbf{v_n} | \lambda_1, \lambda_2, \ldots, \lambda_n \in \mathbb{K}\}.$$

In the case
$$L(\mathbf{v_1}, \mathbf{v_2}, \ldots, \mathbf{v_n}) = E,$$

the vectors are said to span the set E or, equivalently, are said to span the vector space $(E,+,\cdot)$.

Example 8.18. The vectors $\mathbf{v_1} = (1,0)$; $\mathbf{v_2} = (0,2)$; $\mathbf{v_n} = (1,1)$ span the entire $\mathbb{R}^2$ since any point $(x,y) \in \mathbb{R}^2$ can be generated from

$$\lambda_1 \mathbf{v_1} + \lambda_2 \mathbf{v_2} + \lambda_3 \mathbf{v_3}$$

with

$$\lambda_1, \lambda_2, \lambda_3 \in \mathbb{R}.$$

Theorem 8.5. *The span* $L(\mathbf{v_1}, \mathbf{v_2}, \ldots, \mathbf{v_n})$ *with the composition laws is a vector subspace of* $(E, +, \cdot)$.

Proof. In order to prove that $(L, +, \cdot)$ is a vector subspace, for Proposition 8.1, it is enough to prove the closure of L with respect to the composition laws.

1. Let $\mathbf{u}$ and $\mathbf{w}$ be two arbitrary distinct vectors $\in L$. Thus,

$$\mathbf{u} = \lambda_1 \mathbf{v_1} + \lambda_2 \mathbf{v_2} + \ldots + \lambda_n \mathbf{v_n}$$
$$\mathbf{w} = \mu_1 \mathbf{v_1} + \mu_2 \mathbf{v_2} + \ldots + \mu_n \mathbf{v_n}.$$

Let us compute $\mathbf{u} + \mathbf{w}$,

$$\mathbf{u} + \mathbf{w} = \lambda_1 \mathbf{v_1} + \lambda_2 \mathbf{v_2} + \ldots + \lambda_n \mathbf{v_n} + \mu_1 \mathbf{v_1} + \mu_2 \mathbf{v_2} + \ldots + \mu_n \mathbf{v_n} =$$
$$= (\lambda_1 + \mu_1) \mathbf{v_1} + (\lambda_2 + \mu_2) \mathbf{v_2} + \ldots + (\lambda_n + \mu_n) \mathbf{v_n}.$$

Hence $\mathbf{u} + \mathbf{w} \in L$.
2. Let $\mathbf{u}$ be an arbitrary vector $\in L$ and μ and arbitrary scalar $\in \mathbb{K}$. Thus,

$$\mathbf{u} = \lambda_1 \mathbf{v_1} + \lambda_2 \mathbf{v_2} + \ldots + \lambda_n \mathbf{v_n}.$$

Let us compute $\mu \mathbf{u}$

$$\mu \mathbf{u} = \mu (\lambda_1 \mathbf{v_1} + \lambda_2 \mathbf{v_2} + \ldots + \lambda_n \mathbf{v_n}) =$$
$$= \mu \lambda_1 \mathbf{v_1} + \mu \lambda_2 \mathbf{v_2} + \ldots + \mu \lambda_n \mathbf{v_n}.$$

Hence, $\mu \mathbf{u} \in L$. $\square$

Example 8.19. Let us consider the following vectors $\in \mathbb{R}^3$:

$$\mathbf{v_1} = (1,0,1)$$
$$\mathbf{v_2} = (1,1,1)$$
$$\mathbf{v_3} = (0,1,1)$$
$$\mathbf{v_4} = (2,0,0).$$

These vectors span $\mathbb{R}^3$ and

$$(L(\mathbf{v_1}, \mathbf{v_2}, \mathbf{v_3}, \mathbf{v_4}), +, \cdot)$$

is a vector space.

Theorem 8.6. *Let* $(L(\mathbf{v_1}, \mathbf{v_2}, \ldots, \mathbf{v_n}), +, \cdot)$ *be a vector subspace of* $(E, +, \cdot)$. *Let* $s \in \mathbb{N}$ *with* $s < n$. *If* s *vectors* $\mathbf{v_1}, \mathbf{v_2}, \ldots, \mathbf{v_s}$ *are linearly independent while each of the remaining* $n - s$ *vectors is linear combination of the linearly independent* s *vectors, then* $L(\mathbf{v_1}, \mathbf{v_2}, \ldots, \mathbf{v_n}) = L(\mathbf{v_1}, \mathbf{v_2}, \ldots, \mathbf{v_s})$ *(the two spans coincide).*

Proof. Without a generality loss, let us assume that the first s vectors in the vector subspace $L(\mathbf{v_1}, \mathbf{v_2}, \ldots, \mathbf{v_n})$ are linearly independent. For hypothesis, we can express the remaining vectors as linear combinations of the first s vectors.

$$\begin{cases} \mathbf{v_{s+1}} = h_{s+1,1}\mathbf{v_1} + h_{s+1,2}\mathbf{v_2} + \ldots + h_{s+1,s}\mathbf{v_s} \\ \mathbf{v_{s+2}} = h_{s+2,1}\mathbf{v_1} + h_{s+2,2}\mathbf{v_2} + \ldots + h_{s+2,s}\mathbf{v_s} \\ \ldots \\ \mathbf{v_n} = h_{n,1}\mathbf{v_1} + h_{n,2}\mathbf{v_2} + \ldots + h_{n,s}\mathbf{v_s} \end{cases}$$

For the definition 8.8 $\forall \mathbf{v} \in L(\mathbf{v_1}, \mathbf{v_2}, \ldots, \mathbf{v_n})$:

$$\mathbf{v} = \lambda_1 \mathbf{v_1} + \lambda_2 \mathbf{v_2} + \ldots + \lambda_s \mathbf{v_s} + \lambda_{s+1}\mathbf{v_{s+1}} + \ldots + \lambda_n \mathbf{v_n} =$$
$$= \lambda_1 \mathbf{v_1} + \lambda_2 \mathbf{v_2} + \ldots + \lambda_s \mathbf{v_s} +$$
$$+ \lambda_{s+1} (h_{s+1,1}\mathbf{v_1} + h_{s+1,2}\mathbf{v_2} + \ldots + h_{s+1,s}\mathbf{v_s}) + \ldots$$
$$+ \lambda_n (h_{n,1}\mathbf{v_1} + h_{n,2}\mathbf{v_2} + \ldots + h_{n,s}\mathbf{v_s}) =$$
$$= (\lambda_1 + \lambda_{s+1}h_{s+1,1} + \ldots + \lambda_n h_{n,1}) \mathbf{v_1} +$$
$$+ (\lambda_2 + \lambda_{s+1}h_{s+1,2} + \ldots + \lambda_n h_{s+1,2}) \mathbf{v_2} +$$
$$+ \ldots +$$
$$+ (\lambda_s + \lambda_{s+1}h_{s+1,s} + \ldots + \lambda_n h_{n,s}) \mathbf{v_s}.$$

This means that

$$\forall \mathbf{v} \in L(\mathbf{v_1}, \mathbf{v_2}, \ldots, \mathbf{v_n}) : \mathbf{v} \in L(\mathbf{v_1}, \mathbf{v_2}, \ldots, \mathbf{v_s})$$

i.e.

$$L(\mathbf{v_1}, \mathbf{v_2}, \ldots, \mathbf{v_n}) \subset L(\mathbf{v_1}, \mathbf{v_2}, \ldots, \mathbf{v_s}).$$

Let us consider the definition 8.8 again. $\forall \mathbf{w} \in L(\mathbf{v_1}, \mathbf{v_2}, \ldots, \mathbf{v_s})$:

$$\mathbf{w} = l_1\mathbf{v_1} + l_2\mathbf{v_2} + \ldots + l_s\mathbf{v_s} = l_1\mathbf{v_1} + l_2\mathbf{v_2} + \ldots + l_s\mathbf{v_s} + 0\mathbf{v_{s+1}} + 0\mathbf{v_{s+2}} + \ldots + 0\mathbf{v_n}.$$

This means that

$$\forall \mathbf{w} \in L(\mathbf{v_1}, \mathbf{v_2}, \ldots, \mathbf{v_s}) : \mathbf{w} \in L(\mathbf{v_1}, \mathbf{v_2}, \ldots, \mathbf{v_n}),$$

i.e.

$$L(\mathbf{v_1}, \mathbf{v_2}, \ldots, \mathbf{v_s}) \subset L(\mathbf{v_1}, \mathbf{v_2}, \ldots, \mathbf{v_n}).$$

Since $L(\mathbf{v_1}, \mathbf{v_2}, \ldots, \mathbf{v_n}) \subset L(\mathbf{v_1}, \mathbf{v_2}, \ldots, \mathbf{v_s})$ and $L(\mathbf{v_1}, \mathbf{v_2}, \ldots, \mathbf{v_s}) \subset L(\mathbf{v_1}, \mathbf{v_2}, \ldots, \mathbf{v_n})$, then

$$L(\mathbf{v_1}, \mathbf{v_2}, \ldots, \mathbf{v_s}) = L(\mathbf{v_1}, \mathbf{v_2}, \ldots, \mathbf{v_n}). \square$$

Example 8.20. The following vectors

$$\mathbf{v_1} = (0,0,1)$$
$$\mathbf{v_2} = (0,1,0)$$
$$\mathbf{v_3} = (1,0,0)$$

are linearly independent and span the entire set $\mathbf{R}^3$, i.e. we can generate any vector $\mathbf{w} \in \mathbb{R}^3$ by linear combination

$$\mathbf{w} = \lambda_1 \mathbf{v_1} + \lambda_2 \mathbf{v_2} + \lambda_3 \mathbf{v_3}$$

with $\lambda_1, \lambda_2, \lambda_3 \in \mathbb{R}$.

If we add the vector

$$\mathbf{v_4} = (1,1,1),$$

it results that $\mathbf{v_4}$ is linear combination of $\mathbf{v_1}$, $\mathbf{v_2}$, and $\mathbf{v_3}$ (it is their sum).

It follows that any vector $\mathbf{w} \in \mathbb{R}^3$ can be obtained by linear combination of

$$\mathbf{w} = \lambda_1 \mathbf{v_1} + \lambda_2 \mathbf{v_2} + \lambda_3 \mathbf{v_3} + \lambda_4 \mathbf{v_4}$$

with $\lambda_1, \lambda_2, \lambda_3, \lambda_4 \in \mathbb{R}$. This can be easily verified by posing $\lambda_4 = 0$.

This fact can be written as

$$L(\mathbf{v_1}, \lambda_2 \mathbf{v_2}, \mathbf{v_3}, \mathbf{v_4}) = L(\mathbf{v_1}, \lambda_2 \mathbf{v_2}, \mathbf{v_3}) = \mathbb{R}^3.$$

Theorem 8.7. *Let $1 - 2 - \ldots - n$ be the fundamental permutation of the first n numbers $\in \mathbb{N}$. Let $\sigma(1) - \sigma(n) - \ldots - \sigma(n)$ be another permutation of the same first n numbers $\in \mathbb{N}$. The sets generated by the linearly independent vectors are equal:*

$$L(\mathbf{v_1}, \mathbf{v_2}, \ldots, \mathbf{v_n}) = L\left(\mathbf{v_{\sigma(1)}}, \mathbf{v_{\sigma(2)}}, \ldots, \mathbf{v_{\sigma(n)}}\right).$$

Proof. From the Definition 8.8, we know that $\forall \mathbf{v} \in L(\mathbf{v_1}, \mathbf{v_2}, \ldots, \mathbf{v_n}) : \mathbf{v} = \lambda_1 \mathbf{v_1} + \lambda_2 \mathbf{v_2} + \ldots + \lambda_n \mathbf{v_n}$. Due to the commutativity we can rearrange the terms of the sum in a way such that $\mathbf{v} = \lambda_{\sigma(1)} \mathbf{v_{\sigma(1)}} + \lambda_{\sigma(2)} \mathbf{v_{\sigma(2)}} + \ldots + \lambda_{\sigma(n)} \mathbf{v_{\sigma(n)}}$. This means that

$$\forall \mathbf{v} \in L(\mathbf{v_1}, \mathbf{v_2}, \ldots, \mathbf{v_n}) : \mathbf{v} \in L\left(\mathbf{v_{\sigma(1)}}, \mathbf{v_{\sigma(2)}}, \ldots, \mathbf{v_{\sigma(n)}}\right).$$

Hence, $L(\mathbf{v_1}, \mathbf{v_2}, \ldots, \mathbf{v_n}) \subset L\left(\mathbf{v_{\sigma(1)}}, \mathbf{v_{\sigma(2)}}, \ldots, \mathbf{v_{\sigma(n)}}\right)$.

From the Definition 8.8, we know that $\forall \mathbf{w} \in L\left(\mathbf{v_{\sigma(1)}}, \mathbf{v_{\sigma(2)}}, \ldots, \mathbf{v_{\sigma(n)}}\right) : \mathbf{w} = \mu_{\sigma(1)} \mathbf{v_{\sigma(1)}} + \mu_{\sigma(2)} \mathbf{v_{\sigma(2)}} + \mu_{\sigma(n)} \mathbf{v_{\sigma(n)}}$. Due to the commutativity we can rearrange the terms of the sum in a way such that $\mathbf{v} = \mu_1 \mathbf{v_1} + \mu_2 \mathbf{v_2} + \ldots + \mu_n \mathbf{v_n}$. This means that

$$\forall \mathbf{w} \in L\left(\mathbf{v_{\sigma(1)}}, \mathbf{v_{\sigma(2)}}, \ldots, \mathbf{v_{\sigma(n)}}\right) : \mathbf{w} \in L(\mathbf{v_1}, \mathbf{v_2}, \ldots, \mathbf{v_n}).$$

Hence, $L\left(\mathbf{v_{\sigma(1)}}, \mathbf{v_{\sigma(2)}}, \ldots, \mathbf{v_{\sigma(n)}}\right) \subset L(\mathbf{v_1}, \mathbf{v_2}, \ldots, \mathbf{v_n})$.

Thus, $L(\mathbf{v_1}, \mathbf{v_2}, \ldots, \mathbf{v_n}) = L\left(\mathbf{v_{\sigma(1)}}, \mathbf{v_{\sigma(2)}}, \ldots, \mathbf{v_{\sigma(n)}}\right)$. $\square$

Example 8.21. This theorem simply states that both set of vectors

$$\mathbf{v_1} = (0,0,1)$$
$$\mathbf{v_2} = (0,1,0)$$
$$\mathbf{v_3} = (1,0,0)$$

and

$$\mathbf{v_1} = (1,0,0)$$
$$\mathbf{v_2} = (0,1,0)$$
$$\mathbf{v_3} = (0,0,1)$$

span the same set, that is $\mathbb{R}^3$.

Proposition 8.7. *Let $L(\mathbf{v_1}, \mathbf{v_2}, \ldots, \mathbf{v_n})$ be a span. If $\mathbf{w} \in L(\mathbf{v_1}, \mathbf{v_2}, \ldots, \mathbf{v_n})$ and $\mathbf{w}$ is such that*

$$\mathbf{w} = \lambda_1 \mathbf{v_1} + \lambda_2 \mathbf{v_2} + \ldots + \lambda_i \mathbf{v_i} + \ldots + \lambda_n \mathbf{v_n}$$

with $\lambda_i \neq 0$, then

$$L(\mathbf{v_1}, \mathbf{v_2}, \ldots, \mathbf{v_i}, \ldots, \mathbf{v_n}) = L(\mathbf{v_1}, \mathbf{v_2}, \ldots, \mathbf{w}, \ldots, \mathbf{v_n}).$$

Proof. Since $\mathbf{w} = \lambda_1 \mathbf{v_1} + \lambda_2 \mathbf{v_2} + \ldots + \lambda_i \mathbf{v_i} + \ldots + \lambda_n \mathbf{v_n}$ and $\lambda_i \neq 0$, then

$$\mathbf{v_i} = \frac{1}{\lambda_i}\mathbf{w} - \frac{\lambda_1}{\lambda_i}\mathbf{v_1} - \frac{\lambda_2}{\lambda_i}\mathbf{v_2} - \ldots - \frac{\lambda_{i-1}}{\lambda_i}\mathbf{v_{i-1}} - \frac{\lambda_{i+1}}{\lambda_i}\mathbf{v_{i+1}} \ldots - \frac{\lambda_n}{\lambda_i}\mathbf{v_n}.$$

$\forall \mathbf{v} \in L(\mathbf{v_1}, \mathbf{v_2}, \ldots, \mathbf{v_n})$: there exist scalars $\mu_1, \mu_2, \ldots, \mu_n$ such that we can write $\mathbf{v}$ as $\mathbf{v} = \mu_1 \mathbf{v_1} + \mu_2 \mathbf{v_2} + \ldots + \mu_i \mathbf{v_i} + \ldots + \mu_n \mathbf{v_n}$. We can substitute $\mathbf{v_i}$ with the expression above and thus express any vector $\mathbf{v}$ as linear combination of $\mathbf{v_1}, \mathbf{v_2}, \ldots, \mathbf{w} \ldots, \mathbf{v_n}$:

$$\mathbf{v} = k_1 \mathbf{v_1} + k_2 \mathbf{v_2} + \ldots + k_{i-1}\mathbf{v_{i-1}} + k_{i+1}\mathbf{v_{i+1}} + \cdots + k_n \mathbf{v_n} + k_w \mathbf{w}$$

with $k_j = \mu_j - \frac{\lambda_j}{\lambda_i}$ for $j = 1, 2, \ldots, n$ and $j \neq i$. Moreover, $k_w = \frac{1}{\lambda_i}$.
Hence, $\mathbf{v} \in L(\mathbf{v_1}, \mathbf{v_2}, \ldots, \mathbf{w}, \ldots, \mathbf{v_n})$ and consequently

$$L(\mathbf{v_1}, \mathbf{v_2}, \ldots, \mathbf{v_n}) \subset L(\mathbf{v_1}, \mathbf{v_2}, \ldots, \mathbf{w}, \ldots, \mathbf{v_n}).$$

Analogously, $\forall \mathbf{v} \in L(\mathbf{v_1}, \mathbf{v_2}, \ldots, \mathbf{w}, \ldots, \mathbf{v_n})$: there exist scalars

$$\mu_1, \mu_2, \ldots, \mu_{i-1}, \mu_w, \mu_{i+1}, \ldots, \mu_n$$

such that we can write $\mathbf{v}$ as

$$\mathbf{v} = \mu_1 \mathbf{v_1} + \mu_2 \mathbf{v_2} + \ldots + \mu_w \mathbf{w} + \ldots + \mu_n \mathbf{v_n}.$$

We can substitute $\mathbf{w}$ with the expression of $\mathbf{w}$ above and thus express any vector $\mathbf{v}$ as linear combination of $\mathbf{v_1}, \mathbf{v_2}, \ldots, \mathbf{v_i}, \ldots, \mathbf{v_n}$:

$$\mathbf{v} = k_1 \mathbf{v_1} + k_2 \mathbf{v_2} + \cdots + k_i \mathbf{v_i} + \ldots + k_n \mathbf{v_n}$$

with $k_j = \mu_j + \mu_w \lambda_j$ for $j = 1, 2, \ldots, n$ and $j \neq i$. Moreover, $k_i = \mu_w \lambda_i$.

Hence, $\mathbf{v} \in L(\mathbf{v_1}, \mathbf{v_2}, \ldots, \mathbf{v_i}, \ldots, \mathbf{v_n})$ and consequently

$$L(\mathbf{v_1}, \mathbf{v_2}, \ldots, \mathbf{w}, \ldots, \mathbf{v_n}) \subset L(\mathbf{v_1}, \mathbf{v_2}, \ldots, \mathbf{v_n}).$$

Thus, $L(\mathbf{v_1}, \mathbf{v_2}, \ldots, \mathbf{v_i}, \ldots, \mathbf{v_n}) = L(\mathbf{v_1}, \mathbf{v_2}, \ldots, \mathbf{w}, \ldots, \mathbf{v_n})$.   □

Example 8.22. If we consider again

$$\begin{aligned} \mathbf{v_1} &= (0, 0, 1) \\ \mathbf{v_2} &= (0, 1, 0) \\ \mathbf{v_3} &= (1, 0, 0) \end{aligned}$$

and

$$\mathbf{w} = (1, 1, 1),$$

we have

$$L(\mathbf{v_1}, \mathbf{v_2}, \mathbf{v_3}) = \mathbb{R}^3$$

and

$$\mathbf{w} = \lambda_1 \mathbf{v_1} + \lambda_2 \mathbf{v_2} + \lambda_3 \mathbf{v_3}$$

with $\lambda_1, \lambda_2, \lambda_3 = 1, 1, 1$.

We can check that

$$L(\mathbf{v_1}, \mathbf{v_2}, \mathbf{w}) = \mathbb{R}^3.$$

In order to achieve this aim let us consider a vector $\mathbf{u} = (3, 4, 5)$. We can generate $\mathbf{u}$ by means of $\mathbf{v_1}, \mathbf{v_2}, \mathbf{v_3}$ with $\lambda_1, \lambda_2, \lambda_3 = 5, 4, 3$.

We can obtain $\mathbf{u}$ also by means of $\mathbf{v_1}, \mathbf{v_2}, \mathbf{w}$. It is enough to write

$$(3, 4, 5) = \mu_1 (0, 0, 1) + \mu_2 (0, 1, 0) + \mu_3 (1, 1, 1)$$

which leads to the following system of linear equations

$$\begin{cases} \mu_3 = 3 \\ \mu_2 + \mu_3 = 4 \\ \mu_1 + \mu_3 = 5 \end{cases}$$

whose solution is $\mu_1, \mu_2, \mu_3 = 2, 1, 3$

Example 8.23. Let us consider the following vectors

$$\begin{aligned} \mathbf{v_1} &= (0, 1) \\ \mathbf{v_2} &= (1, 0). \end{aligned}$$

The span $L(\mathbf{v_1}, \mathbf{v_2}) = \mathbb{R}^2$. If we consider the vectors

$$\mathbf{w} = (5, 0)$$

we can write

$$\mathbf{w} = \lambda_1 \mathbf{v_1} + \lambda_2 \mathbf{v_2}$$

with $\lambda_1 = 0$ and $\lambda_2 = 5$.

Proposition 8.7 tells us that

$$L(\mathbf{v_1}, \mathbf{v_2}) = L(\mathbf{v_1}, \mathbf{w})$$

since

$$L((0,1),(1,0)) = L((0,1),(5,0)) = \mathbb{R}^2.$$

On the contrary,

$$L(\mathbf{v_1}, \mathbf{v_2}) \neq L(\mathbf{w}, \mathbf{v_2})$$

since $\lambda_1 = 0$. We may verify this fact considering that

$$L((5,0),(1,0)) \neq \mathbb{R}^2.$$

The latter span is the line of the plane having second coordinate 0 (horizontal line having equation $y = 0$).

Lemma 8.1. First Linear Dependence Lemma. *Let* $\mathbf{v_1}, \mathbf{v_2}, \ldots, \mathbf{v_n}$ *be* n *linearly dependent vectors and* $\mathbf{v_1} \neq \mathbf{o}$, *then* $\exists$ *index* $j \in \{2, 3, \ldots, n\}$ *such that*

$$\mathbf{v_j} \in L\left(\mathbf{v_1}, \mathbf{v_2}, \ldots, \mathbf{v_{j-1}}\right)$$

Proof. Since $\mathbf{v_1}, \mathbf{v_2}, \ldots, \mathbf{v_n}$ are n linearly dependent vectors,

$$\exists \lambda_1, \lambda_2, \ldots, \lambda_n \neq 0, 0, \ldots, 0 \ni$$
$$\mathbf{o} = \lambda_1 \mathbf{v_1} + \lambda_2 \mathbf{v_2} + \ldots + \lambda_n \mathbf{v_n}.$$

Since $\mathbf{v_1} \neq \mathbf{o}$ and $\lambda_1, \lambda_2, \ldots, \lambda_n \neq 0, 0, \ldots, 0$ at least one scalar λ_k among $\lambda_2, \ldots, \lambda_n$ is non-null. This would be the only way we could have that the sum is equal to $\mathbf{o}$.

Let us apply commutativity and arrange the vectors in a way that the j non-null coefficient are sorted from the smallest to the largest. Thus, the largest non-null scalar is λ_j. All the scalars whose index is greater than j are null. Hence, we can write

$$\mathbf{v_j} = -\frac{\lambda_1}{\lambda_j}\mathbf{v_1} - \frac{\lambda_2}{\lambda_j}\mathbf{v_2} - \ldots - \frac{\lambda_{j-1}}{\lambda_j}\mathbf{v_{j-1}},$$

i.e. $\mathbf{v_j} \in L\left(\mathbf{v_1}, \mathbf{v_2}, \ldots, \mathbf{v_{j-1}}\right)$. $\square$

Example 8.24. Let us consider the following linearly dependent vectors

$$\mathbf{v_1} = (0,1)$$
$$\mathbf{v_2} = (1,0)$$
$$\mathbf{v_3} = (1,1).$$

We can observe that

$$\mathbf{v_3} \in L\left(\mathbf{v_1}, \mathbf{v_2}\right).$$

Example 8.25. In order to appreciate the hypothesis $\mathbf{v}_1 \neq \mathbf{o}$ of Lemma 8.1 let us consider the following linearly dependent vectors

$$\mathbf{v}_1 = (0,0)$$
$$\mathbf{v}_2 = (1,0)$$
$$\mathbf{v}_3 = (0,1).$$

In this case

$$\mathbf{v}_3 \notin L(\mathbf{v}_1, \mathbf{v}_2).$$

Lemma 8.2. Second Linear Dependence Lemma. *Let $\mathbf{v}_1, \mathbf{v}_2, \ldots, \mathbf{v}_n$ be n linearly dependent vectors and $\mathbf{v}_1 \neq \mathbf{o}$, then one vector can be removed from the span: $\exists$ index $j \in \{2, 3, \ldots, n\}$ such that*

$$L(\mathbf{v}_1, \mathbf{v}_2, \ldots, \mathbf{v}_j, \ldots, \mathbf{v}_n) = L(\mathbf{v}_1, \mathbf{v}_2, \ldots, \mathbf{v}_{j-1}, \mathbf{v}_{j+1} \ldots, \mathbf{v}_n)$$

Proof. Let us consider a generic vector $\mathbf{v}$. $\forall \mathbf{v} \in L(\mathbf{v}_1, \mathbf{v}_2, \ldots, \mathbf{v}_n)$ it follows that there exist scalars $\mu_1, \mu_2, \ldots, \mu_n$ such that

$$\mathbf{v} = \mu_1 \mathbf{v}_1 + \mu_2 \mathbf{v}_2 + \ldots + \mu_j \mathbf{v}_j + \ldots + \mu_n \mathbf{v}_n.$$

If we substitute the expression of $\mathbf{v}_j$ from the First Linear Dependence Lemma into the span, we obtain that, after the removal of $\mathbf{v}_j$, the vectors $\mathbf{v}_1, \mathbf{v}_2, \ldots, \mathbf{v}_{j-1}$, $\mathbf{v}_{j+1} \ldots, \mathbf{v}_n$ still span the vector space, i.e.

$$\mathbf{v} = k_1 \mathbf{v}_1 + k_2 \mathbf{v}_2 + \ldots + k_{j-1} \mathbf{v}_{j-1} + \mu_{j+1} \mathbf{v}_{j+1} + \ldots + \mu_n \mathbf{v}_n$$

with $k_i = \mu_i - \mu_j \frac{\lambda_i}{\lambda_j}$ for $i = 1, \ldots, j-1$.

This means that $\mathbf{v} \in L(\mathbf{v}_1, \mathbf{v}_2, \ldots, \mathbf{v}_{j-1}, \mathbf{v}_{j+1} \ldots, \mathbf{v}_n)$ and thus

$$L(\mathbf{v}_1, \mathbf{v}_2, \ldots, \mathbf{v}_j, \ldots, \mathbf{v}_n) \subset L(\mathbf{v}_1, \mathbf{v}_2, \ldots, \mathbf{v}_{j-1}, \mathbf{v}_{j+1} \ldots, \mathbf{v}_n).$$

Analogously, $\forall \mathbf{v} \in L(\mathbf{v}_1, \mathbf{v}_2, \ldots, \mathbf{v}_{j-1}, \mathbf{v}_{j+1} \ldots, \mathbf{v}_n)$ it follows that

$$\mathbf{v} = \mu_1 \mathbf{v}_1 + \mu_2 \mathbf{v}_2 + \ldots + \mu_{i-1} \mathbf{v}_{i-1} + \mu_{i+1} \mathbf{v}_{i+1} \ldots + \mu_n \mathbf{v}_n$$

which can be re-written as

$$\mathbf{v} = \mu_1 \mathbf{v}_1 + \mu_2 \mathbf{v}_2 + \ldots + \mu_{i-1} \mathbf{v}_{i-1} + 0 \mathbf{v}_j + \mu_{i+1} \mathbf{v}_{i+1} \ldots + \mu_n \mathbf{v}_n.$$

Hence $\mathbf{v} \in L(\mathbf{v}_1, \mathbf{v}_2, \ldots, \mathbf{v}_j, \ldots, \mathbf{v}_n)$. It follows that

$$L(\mathbf{v}_1, \mathbf{v}_2, \ldots, \mathbf{v}_{j-1}, \mathbf{v}_{j+1} \ldots, \mathbf{v}_n) \subset L(\mathbf{v}_1, \mathbf{v}_2, \ldots, \mathbf{v}_j, \ldots, \mathbf{v}_n).$$

Hence, the two spans coincide:

$$L(\mathbf{v}_1, \mathbf{v}_2, \ldots, \mathbf{v}_j, \ldots, \mathbf{v}_n) = L(\mathbf{v}_1, \mathbf{v}_2, \ldots, \mathbf{v}_{j-1}, \mathbf{v}_{j+1} \ldots, \mathbf{v}_n). \square$$

Example 8.26. Let us consider again the vectors

$$\mathbf{v_1} = (0,0,1)$$
$$\mathbf{v_2} = (0,1,0)$$
$$\mathbf{v_3} = (1,0,0)$$
$$\mathbf{v_4} = (1,1,1).$$

These vectors are linearly dependent.

We know that $\mathbf{v_4}$ can be expressed as a linear combination of the other vectors. This fact means that

$$\mathbf{v_4} \in L(\mathbf{v_1}, \mathbf{v_2}, \mathbf{v_3})$$

that is the first linear dependence lemma.

Furthermore, as seen above

$$L(\mathbf{v_1}, \mathbf{v_2}, \mathbf{v_3}) = L(\mathbf{v_1}, \mathbf{v_2}, \mathbf{v_3}, \mathbf{v_4}) = \mathbb{R}^3,$$

that is the second linear dependence lemma.

8.5 Basis and Dimension of a Vector Space

Definition 8.9. Let $(E, +, \cdot)$ be a vector space. The vector space $(E, +, \cdot)$ is said *finite-dimensional* if $\exists$ a finite number of vectors $\mathbf{v_1}, \mathbf{v_2}, \ldots, \mathbf{v_n}$, such that the vector space $(L, +, \cdot) = (E, +, \cdot)$ where the span L is $L(\mathbf{v_1}, \mathbf{v_2}, \ldots, \mathbf{v_n})$. In this case we say that the vectors $\mathbf{v_1}, \mathbf{v_2}, \ldots, \mathbf{v_n}$ span the vector space.

Example 8.27. A finite-dimensional vector space is $(L((1,0,0), (0,1,0), (0,0,1))$, $+, \cdot)$. This vector span can generate any vector $\in \mathbb{R}^3$. Although, infinite-dimensional sets and vector space do not fall within this scopes of this book, it is important to consider that for example a set $\mathbb{R}^\infty$ can be defined as the Cartesian product of $\mathbb{R}$ performed and infinite amount of times.

Definition 8.10. Let $(E, +, \cdot)$ be a finite-dimensional vector space. A *basis $B = \{\mathbf{v_1}, \mathbf{v_2}, \ldots, \mathbf{v_n}\}$* of $(E, +, \cdot)$ is a set of vectors $\in E$ that verify the following properties.

- $\mathbf{v_1}, \mathbf{v_2}, \ldots, \mathbf{v_n}$ are linearly independent
- $\mathbf{v_1}, \mathbf{v_2}, \ldots, \mathbf{v_n}$ span E, i.e. $E = L(\mathbf{v_1}, \mathbf{v_2}, \ldots, \mathbf{v_n})$.

Example 8.28. Let us consider the vector space $\mathbb{R}^3$. A basis B of $\mathbb{R}^3$

$$(0,0,1)$$
$$(0,1,0)$$
$$(1,0,0)$$

as they are linearly independent and all the numbers in $\mathbb{R}^3$ can be derived by their linear combination.

A set of vectors spanning $\mathbb{R}^3$, i.e. the span L, is given by

$$(0,0,1)$$
$$(0,1,0)$$
$$(1,0,0)$$
$$(1,2,3)$$

as they still allow to generate all the numbers in $\mathbb{R}^3$ but are not linearly independent. Hence, a basis always spans a vector space while a set of vectors spanning a space is not necessarily a basis.

Lemma 8.3. Steinitz's Lemma. *Let $(E,+,\cdot)$ be a finite-dimensional vector space and $\in L(\mathbf{v_1},\mathbf{v_2},\ldots,\mathbf{v_n}) = E$ its span. Let $\mathbf{w_1},\mathbf{w_2},\ldots,\mathbf{w_s}$ be s linearly independent vectors $\in E$.*

It follows that $s \leq n$, i.e. the number of a set of linearly independent vectors cannot be higher than the number of vectors spanning the vector space.

Proof. Let us assume by contradiction that $s > n$.

Since $\mathbf{w_1},\mathbf{w_2},\ldots,\mathbf{w_s}$ are linearly independent, for Proposition 8.6 they are all different from the null vector $\mathbf{o}$. Hence, we know that $\mathbf{w_1} \neq \mathbf{o}$.

Since $L(\mathbf{v_1},\mathbf{v_2},\ldots,\mathbf{v_n})$ spans E and $\mathbf{w_1} \in E$, there exists a tuple $\lambda_1,\lambda_2,\ldots,\lambda_n$ such that

$$\mathbf{w_1} = \lambda_1 \mathbf{v_1} + \lambda_2 \mathbf{v_2} + \ldots + \lambda_n \mathbf{v_n}.$$

Since $\mathbf{w_1} \neq \mathbf{o}$, it follows that $\lambda_1,\lambda_2,\ldots,\lambda_n \neq 0,0,\ldots 0$.

Without a loss of generality let us assume that $\lambda_1 \neq 0$. We can now write

$$\mathbf{v_1} = \frac{1}{\lambda_1}\left(\mathbf{w_1} - \lambda_2 \mathbf{v_2} - \ldots - \lambda_n \mathbf{v_n}\right).$$

Thus, any vector $u \in E$ that would be represented as

$$\mathbf{u} = a_1 \mathbf{v_1} + a_2 \mathbf{v_2} + \ldots + a_n \mathbf{v_n} =$$
$$= \left(\mathbf{w_1} - \lambda_2 \mathbf{v_2} - \ldots - \lambda_n \mathbf{v_n}\right) + a_2 \mathbf{v_2} + \ldots + a_n \mathbf{v_n} =$$
$$= k_1 \mathbf{w_1} + k_2 \mathbf{v_2} + \ldots + k_n \mathbf{v_n}.$$

This means that any vector $\mathbf{u} \in E$ can be represented of linear combination of $\mathbf{w_1},\mathbf{v_2},\ldots,\mathbf{v_n}$. This means that

$$L(\mathbf{w_1},\mathbf{v_2},\ldots,\mathbf{v_n}) = E.$$

We can now express $\mathbf{w_2} \in E$ as

$$\mathbf{w_2} = \mu_1 \mathbf{w_1} + \mu_2 \mathbf{v_2} + \ldots + \mu_n \mathbf{v_n}$$

where $\mu_1,\mu_2,\ldots,\mu_n \neq 0,0,\ldots,0$ (it would happen that $\mathbf{w_2} = \mathbf{o}$ and hence for Proposition 8.6 the vectors linearly dependent). Furthermore, $\mathbf{w_2}$ cannot be expressed as

$\mathbf{w_2} = \mu_1 \mathbf{w_1}$ (they are linearly independent). Hence, there exists $j \in \{2, 3, \ldots, n\}$ such that $\mu_j \neq 0$. We can then assume that $\mu_2 \neq 0$ and state that any vector $\mathbf{u} \in E$ can be expressed as

$$\mathbf{u} = l_1 \mathbf{w_1} + l_2 \mathbf{w_2} + l_3 \mathbf{v_3} \ldots + l_n \mathbf{v_n}.$$

This means that

$$L(\mathbf{w_1}, \mathbf{w_2}, \ldots, \mathbf{v_n}) = E.$$

At the generic k^{th} step we have

$$\mathbf{w_k} = \gamma_1 \mathbf{w_1} + \gamma_2 \mathbf{w_2} + \ldots + \gamma_{k-1} \mathbf{w_{k-1}} + \gamma_k \mathbf{v_k} \cdots + \gamma_n \mathbf{v_n}.$$

Since $\mathbf{w_k}$ cannot be expressed as the linear combination of $\mathbf{w_1}, \mathbf{w_2}, \ldots, \mathbf{w_{k-1}}$ then $\exists j \in \{k, k+1, \ldots n\} \ni \text{'} \gamma_j \neq 0$. Assuming that $\gamma_j = \gamma_k$ we have

$$L(\mathbf{w_1}, \mathbf{w_2}, \ldots, \mathbf{w_k}, \ldots, \mathbf{v_n}) = E.$$

Reiterating until the nth step, we have that any vector $\mathbf{u} \in E$ can be expressed as

$$\mathbf{u} = h_1 \mathbf{w_1} + h_2 \mathbf{w_2} + h_3 \mathbf{w_3} \ldots + h_n \mathbf{w_n}.$$

This means that

$$L(\mathbf{w_1}, \mathbf{w_2}, \ldots, \mathbf{w_n}) = E.$$

Since by contradiction $s > n$ there are no more $\mathbf{v}$ vectors while still there are $s - n$ $\mathbf{w}$ vectors.

In particular, $\mathbf{w_{n+1}} \in E$ and hence can be written as

$$\mathbf{w_{n+1}} = \delta_1 \mathbf{w_1} + \delta_2 \mathbf{w_2} + \delta_3 \mathbf{w_3} \ldots + \delta_n \mathbf{w_n}.$$

Since $\mathbf{w_{n+1}}$ has been expressed as linear combination of the others then for Theorem 8.4 the vectors are linearly dependent. This is against the hypothesis and a contradiction has been reached. $\square$

Example 8.29. As shown above, a set of vectors spanning $\mathbb{R}^3$, i.e. the span L, is given by

$$\mathbf{v_1} = (0, 0, 1)$$
$$\mathbf{v_2} = (0, 1, 0)$$
$$\mathbf{v_3} = (1, 0, 0)$$
$$\mathbf{v_4} = (1, 2, 3).$$

Let us consider a set of linearly independent vectors $\in \mathbb{R}^3$:

$$\mathbf{w_1} = (1, 4, 0)$$
$$\mathbf{w_2} = (0, 5, 0)$$
$$\mathbf{w_3} = (0, 2, 1).$$

We can see that these linearly independent vectors are less than the vectors spanning $\mathbb{R}^3$. This is essentially the sense of Steinitz's lemma.

Taking $\mathbb{R}^3$ as an example, we know from Theorem 4.6 that four vectors are always linearly dependent. Thus, in $\mathbb{R}^3$ at most three linearly independent vectors may exist. Conversely, at least three vectors are needed to span $\mathbb{R}^3$. If we take into consideration $\mathbf{w_1}$ and $\mathbf{w_3}$ would not be enough to generate any vector $\in \mathbb{R}^3$. For example, the vector $\mathbf{t} = (50, 0, 20)$ could not be generated as a linear combination of $\mathbf{w_1}$ and $\mathbf{w_3}$. In order to check this, let us write

$$(50, 0, 20) = \lambda_1 (1, 4, 0) + \lambda_3 (0, 2, 1)$$

which leads to

$$\begin{cases} \lambda_1 = 50 \\ 4\lambda_1 + 2\lambda_3 = 0 \\ \lambda_3 = 20. \end{cases}$$

This system is impossible, i.e. $\mathbf{t}$ cannot be generated by $\mathbf{w_1}$ and $\mathbf{w_3}$. We can conclude that $\mathbf{w_1}$ and $\mathbf{w_3}$ do not span $\mathbb{R}^3$.

Theorem 8.8. *Let $(E, +, \cdot)$ be a finite-dimensional vector space and $\in L(\mathbf{v_1}, \mathbf{v_2}, \ldots, \mathbf{v_n}) = E$ its span and $B = \{\mathbf{w_1}, \mathbf{w_2}, \ldots, \mathbf{w_s}\}$ be its basis. It follows that $s \leq n$.*

Proof. The vectors composing the basis are linearly independent. For the Steinitz's Lemma, it follows immediately that $s \leq n$. □

Example 8.30. From the previous example, $B = \{\mathbf{w_1}, \mathbf{w_2}, \mathbf{w_3}\}$ is a basis since its vectors are linearly independent and span $\mathbb{R}^3$. There is totally three vectors in this basis, which is less than the number of vectors $\mathbf{v_1}, \mathbf{v_2}, \mathbf{v_3}, \mathbf{v_4}$ spanning $\mathbb{R}^3$.

Definition 8.11. The number of vectors composing a basis is said *order of a basis*.

Theorem 8.9. *Let $(E, +, \cdot)$ be a finite-dimensional vector space. All the bases of a vector spaces have the same order.*

Proof. Let $B_1 = \{\mathbf{v_1}, \mathbf{v_2}, \ldots, \mathbf{v_n}\}$ and $B_2 = \{\mathbf{w_1}, \mathbf{w_2}, \ldots, \mathbf{w_s}\}$ be two arbitrary bases of $(E, +, \cdot)$.

If B_1 is a basis of $(E, +, \cdot)$, then $L(\mathbf{v_1}, \mathbf{v_2}, \ldots, \mathbf{v_n}) = E$. If $\mathbf{w_1}, \mathbf{w_2}, \ldots, \mathbf{w_s}$ are linearly independent vectors $\in E$, then $\mathbf{w_1}, \mathbf{w_2}, \ldots, \mathbf{w_s} \in L(\mathbf{v_1}, \mathbf{v_2}, \ldots, \mathbf{v_n})$. For the Lemma 8.3, $s \leq n$.

If B_2 is a basis of $(E, +, \cdot)$, then $L(\mathbf{w_1}, \mathbf{w_2}, \ldots, \mathbf{w_s}) = E$. If $\mathbf{v_1}, \mathbf{v_2}, \ldots, \mathbf{v_n}$ are linearly independent vectors $\in E$, then $\mathbf{v_1}, \mathbf{v_2}, \ldots, \mathbf{v_n} \in L(\mathbf{w_1}, \mathbf{w_2}, \ldots, \mathbf{w_s})$. For the Lemma 8.3, $n \leq s$.

Hence, the bases have the same order $s = n$. □

Example 8.31. If we consider two bases of $\mathbb{R}^3$, they will have the same order. We know that in $\mathbb{R}^3$ at most three vectors can be linearly independent, thus a basis of $\mathbb{R}^3$ can be composed of at most three vectors. We have seen that at least three vectors are needed to span $\mathbb{R}^3$. Thus each basis must have three vectors.

Definition 8.12. Let $(E, +, \cdot)$ be a finite-dimensional vector space. The order of a basis of $(E, +, \cdot)$ is said *dimension* of $(E, +, \cdot)$ and is indicated with $\dim(E, +, \cdot)$ or simply with $\dim(E)$.

Theorem 8.10. *Let $(E, +, \cdot)$ be a finite-dimensional vector space.*
The dimension $\dim(E, +, \cdot) = n$ of a vector space (or simply $\dim(E)$) is

- *the maximum number of linearly independent vectors of E;*
- *the minimum number of vectors spanning E*

Proof. If $\dim(E, +, \cdot) = n$, then $\exists$ a basis

$$B = \{\mathbf{v_1}, \mathbf{v_2}, \ldots, \mathbf{v_n}\}.$$

The basis, by definition, contains n linearly independent vectors spanning E and the order of the basis is the number of vectors n.

Let us assume, by contradiction, that n is not the maximum number of linearly independent vectors.

Let us assume that there exist $n + 1$ linearly independent vectors in E:

$$\mathbf{w_1}, \mathbf{w_2}, \ldots, \mathbf{w_n}, \mathbf{w_{n+1}}.$$

Since B is a basis, its elements span the vector space $(E, +, \cdot)$, i.e.

$$L(\mathbf{v_1}, \mathbf{v_2}, \ldots, \mathbf{v_n}) = E.$$

For the Steinitz Lemma, Lemma 8.3, the number of linearly independent vectors $n + 1$ cannot be higher than the number of vectors n that span the vector subspace. Thus,

$$n + 1 \leq n,$$

that is a clear contradiction. This means that the maximum number of linearly independent vectors is n. $\quad\square$

In order to prove that n is also the minimum number of vectors spanning a space, let us assume, by contradiction, that $n - 1$ vectors span the vector space $(E, +, \cdot)$, i.e.

$$L(\mathbf{u_1}, \mathbf{u_2}, \ldots, \mathbf{u_{n-1}}) = E.$$

Since, for the hypotheses $\dim(E, +, \cdot) = n$, the order of a basis B is n, i.e. $\exists n$ linearly independent vectors:

$$\mathbf{v_1}, \mathbf{v_2}, \ldots, \mathbf{v_n}.$$

For the Lemma 8.3, the number of linearly independent vectors n cannot be higher than the number of vectors $n - 1$ that generate the vector subspace,

$$n \leq n - 1,$$

that is a clear contradiction. This means that n is the minimum number of vectors spanning a space. $\quad\square$

Example 8.32. Again, in the case of $\mathbb{R}^3$, we know that each basis is composed of three vectors. For definition of dimension dim $\left(\mathbb{R}^3\right) = 3$. We already know that in $\mathbb{R}^3$ at most three linearly independent vectors there exist and at least three vectors are needed to span the vector space. Thus, the dimension of the vector space is the maximum number of linearly independent vectors and the minimum number of spanning vectors.

Theorem 8.11. *Let $(E, +, \cdot)$ be a finite-dimensional vector space and let n be its dimension (i.e. dim$(E, +, \cdot) = n$). Let $\mathbf{v_1}, \mathbf{v_2}, \ldots, \mathbf{v_n}$ be n vectors $\in E$. The vectors $\mathbf{v_1}, \mathbf{v_2}, \ldots, \mathbf{v_n}$ span the vector space (i.e. $L(\mathbf{v_1}, \mathbf{v_2}, \ldots, \mathbf{v_n}) = E$) if and only if $\mathbf{v_1}, \mathbf{v_2}, \ldots, \mathbf{v_n}$ are linearly independent.*

Proof. If $\mathbf{v_1}, \mathbf{v_2}, \ldots, \mathbf{v_n}$ span a vector space, then $L(\mathbf{v_1}, \mathbf{v_2}, \ldots, \mathbf{v_n}) = E$. Let us assume, by contradiction, that $\mathbf{v_1}, \mathbf{v_2}, \ldots, \mathbf{v_n}$ are linearly dependent. Amongst these n linearly dependent vectors, $r < n$ of them must be linearly independent vectors. These vectors are indicated with $\mathbf{v}_{\sigma(1)}, \mathbf{v}_{\sigma(2)}, \ldots, \mathbf{v}_{\sigma(r)}$. For the Second Linear Dependence Lemma we can remove one vector from $L(\mathbf{v_1}, \mathbf{v_2}, \ldots, \mathbf{v_n})$ and still have an equal linear span. We can reiterate the reasoning until only linearly independent vectors compose the linear span:

$$L\left(\mathbf{v}_{\sigma(1)}, \mathbf{v}_{\sigma(2)}, \ldots, \mathbf{v}_{\sigma(r)}\right) = L(\mathbf{v_1}, \mathbf{v_2}, \ldots, \mathbf{v_n}).$$

This means that also that the r vectors $\mathbf{v}_{\sigma(1)}, \mathbf{v}_{\sigma(2)}, \ldots, \mathbf{v}_{\sigma(r)}$ span E:

$$L\left(\mathbf{v}_{\sigma(1)}, \mathbf{v}_{\sigma(2)}, \ldots, \mathbf{v}_{\sigma(r)}\right) = E.$$

On the other hand, since dim $(E, +, \cdot) = n$, for Theorem 8.10, the minimum number of vectors spanning the vector subspace is n. Thus, it is impossible that $r < n$. $\square$

By hypothesis, we now consider $\mathbf{v_1}, \mathbf{v_2}, \ldots, \mathbf{v_n}$ linearly independent. Let us assume, by contradiction, that the vectors $\mathbf{v_1}, \mathbf{v_2}, \ldots, \mathbf{v_n}$ do not span E, i.e. $L(\mathbf{v_1}, \mathbf{v_2}, \ldots, \mathbf{v_n}) \neq E$. This fact can be stated as the vectors $\mathbf{v_1}, \mathbf{v_2}, \ldots, \mathbf{v_n}$ are not enough to span E and more vectors are needed (at least one more vector is needed).

Thus, we add a vector $\mathbf{u} \in E$ to obtain span of the vector space:

$$L(\mathbf{v_1}, \mathbf{v_2}, \ldots, \mathbf{v_n}, \mathbf{u}) = E.$$

We have then $n + 1$ vectors spanning E. These vectors must be linearly independent because if $\mathbf{u}$ is a linear combination of the others, for the Second Linear Dependence Lemma it can be removed from the span:

$$L(\mathbf{v_1}, \mathbf{v_2}, \ldots, \mathbf{v_n}, \mathbf{u}) = L(\mathbf{v_1}, \mathbf{v_2}, \ldots, \mathbf{v_n}).$$

Since dim $(E, +, \cdot) = n$, , for Theorem 8.10, the maximum number of linearly independent vectors in E is n. Thus, it is impossible that $n + 1$ linearly independent vectors exist. $\square$

Example 8.33. Let us consider the following linearly independent vectors $\in \mathbb{R}^3$

$$\mathbf{v_1} = (1,0,1)$$
$$\mathbf{v_2} = (0,2,0)$$
$$\mathbf{v_3} = (1,0,2).$$

Any vector $\in \mathbb{R}^3$ can be generated as linear combination of $\mathbf{v_1}, \mathbf{v_2}$ and $\mathbf{v_3}$. For example the vector $\mathbf{t} = (21,8,2)$ can be expressed as

$$(21,8,2) = \lambda_1 (1,0,1) + \lambda_2 (0,2,0) + \lambda_3 (1,0,2)$$

which leads to the following system

$$\begin{cases} \lambda_1 + \lambda_3 = 21 \\ \lambda_2 = 8 \\ \lambda_1 + 2\lambda_3 = 2 \end{cases}$$

whose solution is $\lambda_1, \lambda_2, \lambda_3 = 40, 8, -19$. We can always represent a vector of $\mathbb{R}^3$ since $\mathbf{v_1}, \mathbf{v_2}$ and $\mathbf{v_3}$ are linearly independent.

Let us now consider the following linearly dependent vectors

$$\mathbf{w_1} = (1,0,1)$$
$$\mathbf{w_2} = (0,2,0)$$
$$\mathbf{w_3} = (1,2,1)$$

and let us attempt to express $\mathbf{t} = (21,8,2)$

$$(21,8,2) = \lambda_1 (1,0,1) + \lambda_2 (0,2,0) + \lambda_3 (1,2,1)$$

which leads to the following system

$$\begin{cases} \lambda_1 + \lambda_3 = 21 \\ \lambda_2 + \lambda_3 = 8 \\ \lambda_1 + \lambda_3 = 2 \end{cases}$$

which is impossible. Three linearly dependent vectors cannot span $\mathbb{R}^3$.

Example 8.34. Let us consider the vector space $(E, +, \cdot)$ where

$$E = \{(x,y,z) \in \mathbb{R}^3 \,|\, x - 3y - 7z = 0\}.$$

In order to determine span and basis of this vector space, let us assume that $y = \alpha$, $z = \beta$, and let us solve the equation with respect to $x : x = 3\alpha + 7\beta$. Hence, there are ∞^2 solutions of the kind $(3\alpha + 7\beta, \alpha, \beta)$. This expression can be written as

$$(3\alpha + 7\beta, \alpha, \beta) = (3\alpha, \alpha, 0) + (7\beta, 0, \beta) =$$
$$= \alpha (3,1,0) + \beta (7,0,1).$$

Hence, $E = L((3,1,0),(7,0,1))$.

By applying the Proposition 8.5, $(3,1,0) \neq \mathbf{o}$ and thus linearly independent. In addition, $(7,0,1)$ is not linear combination of $(3,1,0)$. Thus, the two vectors are linearly independent and compose a basis $B = \{(3,1,0),(7,0,1)\}$. Thus, $\dim(E,+,\cdot) = 2$.

Example 8.35. Let us consider now the vector space $(E,+,\cdot)$ where

$$E = \left\{ (x,y,z) \in \mathbb{R}^3 \mid \begin{cases} x - 3y + 2z = 0 \\ x + y - z = 0 \end{cases} \right\}.$$

Since the rank of the system is 2, it has ∞^1 solutions. A solution is:

$$\left(\det \begin{pmatrix} -3 & 2 \\ 1 & -1 \end{pmatrix}, -\det \begin{pmatrix} 1 & 2 \\ 1 & -1 \end{pmatrix}, \det \begin{pmatrix} 1 & -3 \\ 1 & 1 \end{pmatrix} \right) = (1,3,4).$$

Hence, $L((1,3,4)) = E$ and $B = \{(1,3,4)\}$. It follows that $\dim(E,+,\cdot) = 1$.

Example 8.36. Let us consider now the vector space $(E,+,\cdot)$ where

$$E = \left\{ (x,y,z) \in \mathbb{R}^3 \mid \begin{cases} x - y = 0 \\ y + z = 0 \\ 3x + z = 0 \end{cases} \right\}.$$

Since the incomplete matrix is non-singular, the only solution of the linear system is $(0,0,0)$. Hence, $L((0,0,0)) = E$. In this special case, the vector space is composed only of the null vector. Since there are no linearly independent vectors, $\dim(E,+,\cdot) = 0$.

More generally, if a vector subspace of $\mathbb{R}^3$ is identified by three linear equations in three variables,

- if the rank of the system is 3, then $\dim(E,+,\cdot) = 0$. The geometrical interpretation of this vector subspace is the origin of a system of coordinates in the three dimensional space (only one point).
- if the rank of the system is 2, the system has ∞^1 solutions and $\dim(E,+,\cdot) = 1$. The geometrical interpretation of this vector subspace is a line passing through the origin.
- if the rank of the system is 1, the system has ∞^2 solutions and $\dim(E,+,\cdot) = 2$. The geometrical interpretation of this vector subspace is a plane passing through the origin.

As previously seen in Proposition 8.2, if the origin, i.e. the null vector, was not included in the vector subspace, the vector space axioms would not be verified.

Example 8.37. Let us consider the vector space $(E,+,\cdot)$ where

$$E = \left\{ (x,y,z) \in \mathbb{R}^3 \mid \begin{cases} x + 2y + 3z = 0 \\ x + y + 3z = 0 \\ x - y = 0 \end{cases} \right\}.$$

The matrix associated with this system of linear equations is

$$\begin{pmatrix} 1 & 2 & 3 \\ 1 & 1 & 3 \\ 1 & -1 & 0 \end{pmatrix}$$

whose determinant is null. The rank of this matrix 2. Hence the system has ∞^1 solutions. In order to find a general solution, from the last equation we can write $x = y = \alpha$. By substituting in the first equation we have

$$\alpha + 2\alpha + 3z = 0 \Rightarrow z = -\alpha.$$

The infinite solutions of the system are proportional to $(\alpha, \alpha, -\alpha) = \alpha(1, 1, -1)$.

This means that $L((1, 1, -1)) = E$ and a basis of this vector space is $B = \{(1, 1, -1)\}$. The latter statement is true because this vector is not the null vector (and hence linearly independent). The dimension of this vector space is $\dim(E, +, \cdot) = 1$.

Example 8.38. Let us consider the vector space $(E, +, \cdot)$ where

$$E = \left\{ (x, y, z) \in \mathbb{R}^3 \mid \begin{cases} x + 2y + 3z = 0 \\ 2x + 4y + 6z = 0 \\ 3x + 6y + 9z = 0 \end{cases} \right\}.$$

The matrix associated with this system of linear equations

$$\begin{pmatrix} 1 & 2 & 3 \\ 2 & 4 & 6 \\ 3 & 6 & 9 \end{pmatrix}$$

is singular as well as all its order 2 submatrices. Hence the rank of the matrix is 1 and the system has ∞^2 solutions. If we pose $y = \alpha$ and $z = \beta$ we have $x = -2\alpha - 3\beta$. Hence, the solutions are proportional to

$$(-2\alpha - 3\beta, \alpha, \beta)$$

which can written as

$$(-2\alpha - 3\beta, \alpha, \beta) = (-2\alpha, \alpha, 0) + (-3\beta, 0, \beta) = \alpha(-2, 1, 0) + \beta(-3, 0, 1).$$

These couples of vectors solve the system of linear equations. In order to show that these two vectors compose a basis we need to verify that they are linearly independent. By definition, this means that

$$\lambda_1 \mathbf{v_1} + \lambda_2 \mathbf{v_2} = \mathbf{o}$$

only if $\lambda_1, \lambda_2 = 0, 0$.

In our case this means that

$$\lambda_1(-2,1,0) + \lambda_2(-3,0,1) = \mathbf{o}$$

only if $\lambda_2, \lambda_2 = 0, 0$. This is equivalent to say that the system of linear equations

$$\begin{cases} -2\lambda_1 - 3\lambda_2 = 0 \\ \lambda_1 = 0 \\ \lambda_2 = 0 \end{cases}$$

is determined. The only values that satisfy the equations are $\lambda_2, \lambda_2 = 0, 0$. This means that the vectors are linearly independent and hence compose a basis $B = \{(-2,1,0), (-3,0,1)\}$. The dimension is then $\dim(E,+,\cdot) = 2$.

Example 8.39. Let us consider the following vectors $\in \mathbb{R}^3$:

$$\mathbf{u} = (2,-1,1) \, \mathbf{v} = (3,1,2)$$

The associated matrix $\mathbf{A}$ has rank 2. Hence, the vectors are linearly independent. These two vectors could compose a basis $B = \{\mathbf{u}, \mathbf{v}\}$ and thus could span a vector space having dimension 2. These vectors cannot span $\mathbb{R}^3$.

Example 8.40. Let us consider the following vectors $\in \mathbb{R}^3$:

$$\mathbf{u} = (2,-1,3)$$
$$\mathbf{v} = (1,0,-1)$$
$$\mathbf{w} = (2,1,-2)$$

The associated matrix is

$$\mathbf{A} = \begin{pmatrix} 2 & -1 & 3 \\ 1 & 0 & -1 \\ 2 & 1 & -2 \end{pmatrix}$$

whose rank is 3. Hence, these vectors are linearly independent and compose a basis B.

Let us now consider the vector $\mathbf{t} = (0,2,0)$. Let us express $\mathbf{t}$ in the basis B. This means that we need to find the coefficients λ, μ, ν such that $\mathbf{t} = \lambda \mathbf{u} + \mu \mathbf{v} + \nu \mathbf{w}$:

$$\mathbf{t} = \lambda \mathbf{u} + \mu \mathbf{v} + \nu \mathbf{w} =$$
$$= \lambda(2,-1,3) + \mu(1,0,-1) + \nu(2,1,-2) =$$
$$= (2\lambda + \mu + 2\nu, -\lambda + \nu, 3\lambda - \mu - 2\nu).$$

Thus, we can write the following system of linear equations

$$\begin{cases} 2\lambda + \mu + 2\nu = 0 \\ -\lambda + \nu = 2 \\ 3\lambda - \mu - 2\nu = 0. \end{cases}$$

The matrix associated with this system is non-singular. Hence, the system has only one solution λ, μ, ν that allows to express $\mathbf{t}$ in the basis B.

Theorem 8.12. Basis Reduction Theorem. *Let* $(E, +, \cdot)$ *be a finite-dimensional vector space and* $L(\mathbf{v_1}, \mathbf{v_2}, \ldots, \mathbf{v_m}) = E$ *one of its spans. If some vectors are removed a basis of* $(E, +, \cdot)$ *is obtained.*

Proof. Let us consider the span $L(\mathbf{v_1}, \mathbf{v_2}, \ldots, \mathbf{v_m})$ and apply the following iterative procedure.

- **Step 1**: If $\mathbf{v_1} = \mathbf{o}$, it is removed, otherwise left in the span
- **Step k**: if $\mathbf{v_k} \in L(\mathbf{v_1}, \mathbf{v_2}, \ldots, \mathbf{v_{k-1}})$, it is removed otherwise left in the span

The procedure is continued until there are vectors available. For the First Linear Dependence Lemma the n remaining vectors span the vector space, i.e. $L(\mathbf{v_1}, \mathbf{v_2}, \ldots, \mathbf{v_n}) = E$. For the Proposition 8.5, the vectors are linearly independent. Hence, they compose a basis. $\square$

Example 8.41. Let us consider the following vectors $\in \mathbb{R}^4$:

$$\mathbf{u} = (2, -1, 1, 3)$$
$$\mathbf{v} = (0, 2, 1, -1)$$
$$\mathbf{w} = (1, 2, 0, 1)$$
$$\mathbf{a} = (3, 4, 2, 3)$$
$$\mathbf{b} = (2, 4, 0, 2)$$

Let us consider the span $L(\mathbf{u}, \mathbf{v}, \mathbf{w}, \mathbf{a}, \mathbf{b})$. The first vector, $\mathbf{u} \neq \mathbf{o}$ and therefore it is left in the span. The second vector, $\mathbf{v}$ is not a linear combination of $\mathbf{u}$ or, in other words $\mathbf{v} \notin L(\mathbf{u})$. Hence $\mathbf{v}$ is left in the span. We can verify that $\mathbf{w}$ is not a linear combination of $\mathbf{u}$ and $\mathbf{v}$. Also, $\mathbf{a}$ is not a linear combination of $\mathbf{u}$, $\mathbf{v}$, and $\mathbf{w}$.

On the contrary, we can observe that $\mathbf{b}$ is a linear combination of the other four vectors:

$$\mathbf{b} = \lambda_1 \mathbf{u} + \lambda_2 \mathbf{v} + \lambda_3 \mathbf{w} + \lambda_4 \mathbf{a}$$

where $\lambda_1 = \lambda_2 = \lambda_4 = 0$ and $\lambda_3 = 2$. Hence, $\mathbf{b}$ is removed from the span. The updated span is

$$L(\mathbf{u}, \mathbf{v}, \mathbf{w}, \mathbf{a}) = E.$$

The vectors are linearly independent and hence compose a basis $B = \{\mathbf{u}, \mathbf{v}, \mathbf{w}, \mathbf{a}\}$ of $\mathbb{R}^4$.

Theorem 8.13. Basis Extension Theorem. *Let* $(E, +, \cdot)$ *be a finite-dimensional vector space and* $\mathbf{w_1}, \mathbf{w_2}, \ldots, \mathbf{w_s}$ *be s linearly independent vectors of the vector space. If* $\mathbf{w_1}, \mathbf{w_2}, \ldots, \mathbf{w_s}$ *are not already a basis, they can be extended to a basis (by adding other linearly independent vectors).*

Proof. Let us consider a list of linearly independent vectors $\mathbf{w_1}, \mathbf{w_2}, \ldots, \mathbf{w_s}$. Since $(E, +, \cdot)$ is finite-dimensional $\exists$ a list of vectors $\mathbf{v_1}, \mathbf{v_2}, \ldots, \mathbf{v_n}$ that spans E, i.e. $\exists \mathbf{v_1}, \mathbf{v_2}, \ldots, \mathbf{v_n}$ such that $L(\mathbf{v_1}, \mathbf{v_2}, \ldots, \mathbf{v_n}) = E$. Let us apply the following iterative procedure.

- **Step 1**: If $\mathbf{v_1} \in L(\mathbf{w_1}, \mathbf{w_2}, \ldots, \mathbf{w_s})$ then the span is left unchanged otherwise the span is updated to $L(\mathbf{w_1}, \mathbf{w_2}, \ldots, \mathbf{w_s}, \mathbf{v_1})$
- **Step k**: If $\mathbf{v_k} \in L(\mathbf{w_1}, \mathbf{w_2}, \ldots, \mathbf{w_s}, \mathbf{v_1}, \mathbf{v_2}, \ldots \mathbf{v_{k-1}})$ (after having properly renamed the indices) then the span is left unchanged otherwise the span is updated to $L(\mathbf{w_1}, \mathbf{w_2}, \ldots, \mathbf{w_s}, \mathbf{v_1}, \mathbf{v_2}, \ldots \mathbf{v_k})$

Considering how the span has been constructed, the new span is composed of linearly independent vectors. Since $L(\mathbf{v_1}, \mathbf{v_2}, \ldots, \mathbf{v_n}) = E$ (the vectors were already spanning E), from the construction procedure, the new list of vectors also spans E, $L(\mathbf{w_1}, \mathbf{w_2}, \ldots, \mathbf{w_s}, \mathbf{v_1}, \mathbf{v_2}, \ldots \mathbf{v_k}) = E$ (the indices have been properly arranged). Hence, we found a new basis. $\square$

Example 8.42. Let us consider the following vectors belonging to $\mathbb{R}^4$:

$$\mathbf{v_1} = (1, 0, 0, 0)$$
$$\mathbf{v_2} = (0, 1, 0, 0)$$
$$\mathbf{v_3} = (0, 0, 1, 0)$$
$$\mathbf{v_4} = (0, 0, 0, 1).$$

It can be easily shown that these vectors are linearly independent and compose a basis of $\mathbb{R}^4$. Let us now consider the following vectors belonging to $\mathbb{R}^4$:

$$\mathbf{w_1} = (5, 2, 0, 0)$$
$$\mathbf{w_2} = (0, 6, 0, 0).$$

These two vectors are linearly independent. Let us now apply the Basis Extension Theorem to find another basis. Let us check whether or not $\mathbf{v_1} \in L(\mathbf{w_1}, \mathbf{w_2})$. The vector $\mathbf{v_1} \in L(\mathbf{w_1}, \mathbf{w_2})$ since $\exists \lambda_1, \lambda_2 \in \mathbb{R}$ such that $\mathbf{v_1} = \lambda_1 \mathbf{w_1} + \lambda_2 \mathbf{w_2}$, that is $\lambda_1, \lambda_2 = \frac{1}{5}, -\frac{1}{15}$.

This result was found by simply imposing

$$(1, 0, 0, 0) = \lambda_1 (5, 2, 0, 0) + \lambda_2 (0, 6, 0, 0)$$

which leads to the following system of linear equations

$$\begin{cases} 5\lambda_1 + 0\lambda_2 = 1 \\ 2\lambda_1 + 6\lambda_2 = 0 \\ 0\lambda_1 + 0\lambda_2 = 0 \\ 0\lambda_1 + 0\lambda_2 = 0. \end{cases}$$

The last two equations are always verified. Hence, this is a system of two equations in two variables which is determined and its solutions are $\lambda_1, \lambda_2 = \frac{1}{5}, -\frac{1}{15}$.

Thus, we do not add $\mathbf{v_1}$ to the span.

We check now whether or not $\mathbf{v_2} \in L(\mathbf{w_1}, \mathbf{w_2})$. The vector belongs to the span since $\mathbf{v_2} = \lambda_1 \mathbf{w_1} + \lambda_2 \mathbf{w_2}$ with $\lambda_1, \lambda_2 = 0, \frac{1}{6}$. Thus we do not add it to the span.

We check now whether or not $\mathbf{v_3} \in L(\mathbf{w_1}, \mathbf{w_2})$. In order to achieve this aim we need to check whether or not $\exists \lambda_1, \lambda_2 \in \mathbb{R}$ such that $\mathbf{v_3} = \lambda_1 \mathbf{w_1} + \lambda_2 \mathbf{w_2}$:

$$(0,0,1,0) = \lambda_1 (5,2,0,0) + \lambda_2 (0,6,0,0)$$

which leads to the following system of linear equations

$$\begin{cases} 5\lambda_1 + 0\lambda_2 = 0 \\ 2\lambda_1 + 6\lambda_2 = 0 \\ 0\lambda_1 + 0\lambda_2 = 1 \\ 0\lambda_1 + 0\lambda_2 = 0. \end{cases}$$

The last equation is always verified. The third equation is never verified. Hence, the system is impossible, those λ_1, λ_2 values do not exist. This means that $\mathbf{w_1}, \mathbf{w_2}, \mathbf{v_3}$ are linearly independent. The vector $\mathbf{v_3}$ can be added to the span which becomes

$$\mathbf{w_1} = (5,2,0,0)$$
$$\mathbf{w_2} = (0,6,0,0)$$
$$\mathbf{v_3} = (0,0,1,0).$$

By applying the same reasoning we can easily find that also $\mathbf{w_1}, \mathbf{w_2}, \mathbf{v_3}, \mathbf{v_4}$ are linearly independent and can be added to the span. Since we added all the vectors of the span (unless already contained in it through linear combinations of the others), $L(\mathbf{w_1}, \mathbf{w_2}, \mathbf{v_3}, \mathbf{v_4}) = E$. Hence, we have found a new basis, that is $B = \{\mathbf{w_1}, \mathbf{w_2}, \mathbf{v_3}, \mathbf{v_4}\}$.

Theorem 8.14. Grassmann's Formula. *Let* $(E, +, \cdot)$ *be a finite-dimensional vector space. Let* $(U, +, \cdot)$ *and* $(V, +, \cdot)$ *be vector subspaces of* $(E, +, \cdot)$. *Then,*

$$\dim(U + V) + \dim(U \cap V) = \dim(U) + \dim(V).$$

Proof. Let us suppose that $\dim(U) = r$ and $\dim(V) = s$, i.e. $\exists$ two bases

$$B_U = \{\mathbf{u_1}, \mathbf{u_2}, \ldots, \mathbf{u_r}\}$$
$$B_V = \{\mathbf{v_1}, \mathbf{v_2}, \ldots, \mathbf{v_s}\}$$

of $(U, +, \cdot)$ and $(V, +, \cdot)$, respectively. For the Theorem 8.1, $(U \cap V, +, \cdot)$ is a vector subspace of $(E, +, \cdot)$. Let us suppose that one of its bases is $B_{U \cap V} = \{\mathbf{t_1}, \mathbf{t_2}, \ldots, \mathbf{t_l}\}$.

Since all the vectors contained in $B_{U \cap V}$ are also vectors in U, for the Theorem of Basis extension, we can obtain B_U from $B_{U \cap V}$, by adding one by one the vectors from B_U:

$$B_U = \{\mathbf{t_1}, \mathbf{t_2}, \ldots, \mathbf{t_l}, \mathbf{u_{l+1}}, \mathbf{u_{l+2}}, \ldots, \mathbf{u_r}\},$$

where $\mathbf{u}_{l+1}, \mathbf{u}_{l+2}, \ldots, \mathbf{u_r}$ are some vectors from B_U after the indices have been rearranged.

Since all the vectors contained in $B_{U \cap V}$ are also vectors in V, for the Theorem of Basis extension, we can obtain B_V from $B_{U \cap V}$, by adding one by one the vectors from B_V:

$$B_V = \{\mathbf{t_1}, \mathbf{t_2}, \ldots, \mathbf{t_l}, \mathbf{v_{l+1}}, \mathbf{v_{l+2}}, \ldots, \mathbf{v_s}\},$$

where $\mathbf{v_{l+1}}, \mathbf{v_{l+2}}, \ldots, \mathbf{v_r}$ are some vectors from B_V after the indices have been rearranged.

For the Definition 8.3,

$$S = U + V = \{\mathbf{w} \in E | \exists \mathbf{u} \in U, \mathbf{v} \in V | \mathbf{w} = \mathbf{u} + \mathbf{v}\}.$$

Thus, we can write the generic $\mathbf{w} = \mathbf{u} + \mathbf{v}$ by means of infinite scalars.

$$\mathbf{w} = \mathbf{u} + \mathbf{v} = \lambda_1 \mathbf{t_1} + \lambda_2 \mathbf{t_2} + \ldots + \lambda_l \mathbf{t_l} + a_{l+1}\mathbf{u_{l+1}} + a_{l+2}\mathbf{u_{l+2}} + \ldots + a_r\mathbf{u_r}$$
$$+ \mu_1 \mathbf{t_1} + \mu_2 \mathbf{t_2} + \ldots + \mu_l \mathbf{t_l} + b_{l+1}\mathbf{v_{l+1}} + b_{l+2}\mathbf{v_{l+2}} + \ldots + b_s\mathbf{v_s} =$$
$$= (\lambda_1 + \mu_1)\mathbf{t_1} + (\lambda_2 + \mu_2)\mathbf{t_2} + \ldots (\lambda_l + \mu_l)\mathbf{t_l} +$$
$$+ a_{l+1}\mathbf{u_{l+1}} + a_{l+2}\mathbf{u_{l+2}} + \ldots + a_r\mathbf{u_r} + b_{l+1}\mathbf{v_{l+1}} + b_{l+2}\mathbf{v_{l+2}} + \ldots + b_s\mathbf{v_s}$$

Hence, by means of a linear combination we can represent all the vectors $\mathbf{w} \in U + V$. In other words, the $r + s - l$ vectors $\mathbf{t_1}, \mathbf{t_2}, \ldots, \mathbf{t_l}, \mathbf{u_{l+1}}, \mathbf{u_{l+2}}, \ldots, \mathbf{u_r}, \mathbf{v_{l+1}}, \mathbf{v_{l+2}}, \ldots, \mathbf{v_s}$ span $(U + V, +, \cdot)$:

$$L(\mathbf{t_1}, \mathbf{t_2}, \ldots, \mathbf{t_l}, \mathbf{u_{l+1}}, \mathbf{u_{l+2}}, \ldots, \mathbf{u_r}, \mathbf{v_{l+1}}, \mathbf{v_{l+2}}, \ldots, \mathbf{v_s}) = U + V.$$

Let us check now the linear independence of these $r + s - l$ vectors. Let us impose that

$$\alpha_1 \mathbf{t_1} + \alpha_2 \mathbf{t_2} + \ldots + \alpha_l \mathbf{t_l} +$$
$$+ \beta_{l+1}\mathbf{u_{l+1}} + \beta_{l+2}\mathbf{u_{l+2}} + \ldots + \beta_r\mathbf{u_r} +$$
$$+ \gamma_{l+1}\mathbf{v_{l+1}} + \gamma_{l+2}\mathbf{v_{l+2}} + \ldots + \gamma_s\mathbf{v_s} = \mathbf{o}.$$

Hence, we could write where

$$\alpha_1 \mathbf{t_1} + \alpha_2 \mathbf{t_2} + \ldots + \alpha_l \mathbf{t_l} +$$
$$+ \beta_{l+1}\mathbf{u_{l+1}} + \beta_{l+2}\mathbf{u_{l+2}} + \ldots + \beta_r\mathbf{u_r} = \mathbf{d} =$$
$$= -(\gamma_{l+1}\mathbf{v_{l+1}} + \gamma_{l+2}\mathbf{v_{l+2}} + \ldots + \gamma_s\mathbf{v_s})$$

Since $\mathbf{d}$ can be expressed as linear combination of $\mathbf{v_{l+1}}, \mathbf{v_{l+2}}, \ldots, \mathbf{v_s}, \mathbf{d} \in V$. Since $\mathbf{d}$ can be expressed as linear combination of the elements of a basis of U, $\mathbf{d} \in U$. Hence, $\mathbf{d} \in U \cap V$. This means that $\mathbf{d}$ can be expressed as linear combination of $\mathbf{t_1}, \mathbf{t_2}, \ldots, \mathbf{t_l}$:

$$\mathbf{d} = \alpha_1' \mathbf{t_1} + \alpha_2' \mathbf{t_2} + \ldots + \alpha_l' \mathbf{t_l}$$

Since $t_1, t_2, \ldots, t_l$ are linearly independent, there is only one way, for the Theorem 8.4, to represent d as linear combination of them. Thus, $\beta_{l+1} = \beta_{l+2} = \ldots = \beta_r = 0$ and $\alpha_1 = \alpha'_1$, $\alpha_2 = \alpha'_2$, $\ldots \alpha_l = \alpha'_l$.

Hence, we can write the expression above as

$$\alpha_1 t_1 + \alpha_2 t_2 + \ldots + \alpha_l t_l +$$
$$+ \gamma_{l+1} v_{l+1} + \gamma_{l+2} v_{l+2} + \ldots + \gamma_s v_s = o.$$

Since $t_1, t_2, \ldots, t_l, v_{l+1}, v_{l+2}, \ldots, v_s$ compose a basis they are linearly independent. Thus, $\alpha_1 = \alpha_2 = \ldots = \alpha_l = \gamma_{l+1} = \gamma_{l+2} = \ldots = \gamma_s = 0$.

Since the null vector o can be expressed as linear combination of the vectors $t_1, t_2, \ldots, t_l, u_{l+1}, u_{l+2}, \ldots, u_r, v_{l+1}, v_{l+2}, \ldots, v_s$ only by means of null coefficients, the above $r + s - l$ vectors are linearly independent. Thus these vectors compose a basis B_{U+V}:

$$B_{U+V} = \{t_1, t_2, \ldots, t_l, u_{l+1}, u_{l+2}, \ldots, u_r, v_{l+1}, v_{l+2}, \ldots, v_s\}.$$

It follows that

$$\dim(U + V) = r + s - l$$

where $r = \dim(U)$, $s = \dim(V)$, and $l = \dim(U \cap V)$, i.e.

$$\dim(U + V) = \dim(U) + \dim(V) - \dim(U \cap V). \square$$

Example 8.43. Let us consider the vector space $(\mathbb{R}^3, +, \cdot)$ and two vector subspaces: $(U, +, \cdot)$ and $(V, +, \cdot)$, respectively.

Let us verify/interpret the Grassmann's formula in the following cases:

- if $U = \{(0, 0, 0)\}$ and $V = \mathbb{R}^3$ then $\dim(U) = 0 \dim(V) = 3$. In this case $U \cap V = (0, 0, 0) = U$ and $U + V = \mathbb{R}^3 = V$. It follows that $\dim(U + V) + \dim(U \cap V) = 3 + 0 = \dim(U) + \dim(V) = 0 + 3$.
- If the dimension of both U and V is 1, i.e. only one linearly independent vector and thus one line, we can distinguish two subcases

  - the two vectors in U and V, respectively, represent two lines passing through the origin. The intersection $U \cap V = (0, 0, 0)$ is the origin, while the sum $U + V$ is the plane that contains the two vectors. It follows that $\dim(U + V) + \dim(U \cap V) = 2 + 0 = \dim(U) + \dim(V) = 1 + 1$
  - the two vectors in U and V, respectively, represent two coinciding lines. Both intersection and sum coincide with the vector, i.e. $U \cap V = U + V = U = V$. It follows that $\dim(U + V) + \dim(U \cap V) = 1 + 1 = \dim(U) + \dim(V) = 1 + 1$

- if the dimension of U is 1 while that of V is 2, i.e. one line passing through the origin and one plane passing through the origin, we can distinguish two subcases:

  - the line does not lay in the plane. It follows that $U \cap V = (0, 0, 0)$ and $U + V = \mathbb{R}^3$. Hence, $\dim(U + V) + \dim(U \cap V) = 3 + 0 = \dim(U) + \dim(V) = 1 + 2$

  – the line lays in the plane. It follows that $U \cap V = U$ and $U + V = V$. Hence, $\dim(U + V) + \dim(U \cap V) = 2 + 1 = \dim(U) + \dim(V) = 1 + 2$

- If the dimension of both U and V is 2, i.e. two linearly independent vectors and thus two planes passing through the origin, we can distinguish two subcases

  – the planes do not coincide. It follows that $U \cap V$ is a line while $U + V = \mathbb{R}^3$. Hence, $\dim(U + V) + \dim(U \cap V) = 3 + 1 = \dim(U) + \dim(V) = 2 + 2$
  – the planes coincide. It follows that $U \cap V = U + V$ and $U + V = U = V$, i.e. intersection and sum are the same coinciding plane. Hence, $\dim(U + V) + \dim(U \cap V) = 2 + 2 = \dim(U) + \dim(V) = 2 + 2$

8.6 Row and Column Spaces

Let us consider a matrix $\mathbf{A} \in \mathbb{K}_{m,n}$:

$$\mathbf{A} = \begin{pmatrix} a_{1,1} & a_{1,2} & \ldots & a_{1,n} \\ a_{2,1} & a_{2,2} & \ldots & a_{2,n} \\ \ldots & \ldots & \ldots & \ldots \\ a_{m,1} & a_{m,2} & \ldots & a_{m,n} \end{pmatrix}.$$

This matrix contains m row vectors $\mathbf{r_1}, \mathbf{r_2}, \ldots, \mathbf{r_m} \in \mathbb{K}^n$

$$\mathbf{r_1} = (a_{1,1}, a_{1,2}, \ldots, a_{1,n})$$
$$\mathbf{r_2} = (a_{2,1}, a_{2,2}, \ldots, a_{2,n})$$
$$\ldots$$
$$\mathbf{r_n} = (a_{m,1}, a_{m,2}, \ldots, a_{m,n})$$

and n column vectors $\mathbf{c^1}, \mathbf{c^2}, \ldots, \mathbf{c^n} \in \mathbb{K}^m$

$$\mathbf{c^1} = (a_{1,1}, a_{2,1}, \ldots, a_{m,1})$$
$$\mathbf{c^2} = (a_{1,2}, a_{2,2}, \ldots, a_{m,2})$$
$$\ldots$$
$$\mathbf{c^n} = (a_{1,n}, a_{2,n}, \ldots, a_{m,n})$$

Definition 8.13. The *row space* and *column space* of a matrix $A \in \mathbb{K}_{m,n}$ are the vector spaces $(\mathbb{K}^n, +, \cdot)$ and $(\mathbb{K}^m, +, \cdot)$ generated by row vectors and column vectors, respectively, of the matrix.

Theorem 8.15. *Let a matrix* $\mathbf{A} \in \mathbb{K}_{m,n}$. *The maximum number* p *of linearly independent row vectors is at most equal to the maximum number* q *of linearly column vectors.*

Proof. Without a loss of generality let us consider a matrix $\mathbf{A} \in \mathbb{K}_{4,3}$

$$\mathbf{A} = \begin{pmatrix} a_{1,1} & a_{1,2} & a_{1,3} \\ a_{2,1} & a_{2,2} & a_{2,3} \\ a_{3,1} & a_{3,2} & a_{3,3} \\ a_{4,1} & a_{4,2} & a_{4,3} \end{pmatrix}$$

and let us suppose that the maximum number of linearly independent column vectors is $q = 2$ and the third is linear combination of the other two. Let us suppose that $\mathbf{c}^1$ and $\mathbf{c}^2$ are linearly independent while $\mathbf{c}^3$ is linear combination of them:

$$\mathbf{c}^3 = \lambda \mathbf{c}^1 + \mu \mathbf{c}^2 \Rightarrow$$

$$\Rightarrow \begin{pmatrix} a_{1,3} \\ a_{2,3} \\ a_{3,3} \\ a_{4,3} \end{pmatrix} = \lambda \begin{pmatrix} a_{1,1} \\ a_{2,1} \\ a_{3,1} \\ a_{4,1} \end{pmatrix} + \mu \begin{pmatrix} a_{1,2} \\ a_{2,2} \\ a_{3,2} \\ a_{4,2} \end{pmatrix}.$$

This equation can be written as

$$\begin{cases} a_{1,3} = \lambda a_{1,1} + \mu a_{1,2} \\ a_{2,3} = \lambda a_{2,1} + \mu a_{2,2} \\ a_{3,3} = \lambda a_{3,1} + \mu a_{3,2} \\ a_{3,3} = \lambda a_{4,1} + \mu a_{4,2}. \end{cases}$$

We can write the row vectors as

$$\mathbf{r}_1 = (a_{1,1}, a_{1,2}, \lambda a_{1,1} + \mu a_{1,2}) = (a_{1,1}, 0, \lambda a_{1,1}) + (0, a_{1,2}, \mu a_{1,2})$$
$$\mathbf{r}_2 = (a_{2,1}, a_{2,2}, \lambda a_{2,1} + \mu a_{2,2}) = (a_{2,1}, 0, \lambda a_{2,1}) + (0, a_{2,2}, \mu a_{2,2})$$
$$\mathbf{r}_3 = (a_{3,1}, a_{3,2}, \lambda a_{3,1} + \mu a_{3,2}) = (a_{3,1}, 0, \lambda a_{3,1}) + (0, a_{3,2}, \mu a_{3,2})$$
$$\mathbf{r}_4 = (a_{4,1}, a_{4,2}, \lambda a_{4,1} + \mu a_{4,2}) = (a_{4,1}, 0, \lambda a_{4,1}) + (0, a_{4,2}, \mu a_{4,2}).$$

These row vectors can be written as

$$\mathbf{r}_1 = a_{1,1}(1, 0, \lambda) + a_{1,2}(0, 1, \mu)$$
$$\mathbf{r}_2 = a_{2,1}(1, 0, \lambda) + a_{2,2}(0, 1, \mu)$$
$$\mathbf{r}_3 = a_{3,1}(1, 0, \lambda) + a_{3,2}(0, 1, \mu)$$
$$\mathbf{r}_4 = a_{4,1}(1, 0, \lambda) + a_{4,2}(0, 1, \mu).$$

Hence, the row vectors are linear combination of $(1,0,\lambda)$ and $(0,1,\mu)$. This is due to the fact that one column vector was supposed to be linear combination of the other two column vectors. We have $p = q = 2$. □

Theorem 8.16. *Let a matrix* $\mathbf{A} \in \mathbb{K}_{m,n}$. *The maximum number q of linearly independent column vectors is at the most equal to the maximum number p of linearly independent row vectors.*

Corollary 8.2. *The dimension of a row (column) space is equal to the rank of the associated matrix.*

Example 8.44. Let us consider the following matrix:

$$\mathbf{A} = \begin{pmatrix} 3 & 2 & 5 \\ 1 & 1 & 2 \\ 0 & 1 & 1 \\ 1 & 0 & 1 \end{pmatrix}$$

where the third column is the sum (and thus linear combination) of the first two columns.

The rows of the matrix can be written as

$$\mathbf{A} = \begin{pmatrix} 3 & 2 & 3+2 \\ 1 & 1 & 1+1 \\ 0 & 1 & 0+1 \\ 1 & 0 & 1+0 \end{pmatrix}$$

and then

$$\mathbf{A} = \begin{pmatrix} 3 & 0 & 3 \\ 1 & 0 & 1 \\ 0 & 0 & 0 \\ 1 & 0 & 1 \end{pmatrix} + \begin{pmatrix} 0 & 2 & 2 \\ 0 & 1 & 1 \\ 0 & 1 & 0 \\ 0 & 0 & 0 \end{pmatrix}.$$

This can written as

$$\mathbf{A} = \begin{pmatrix} 3(1,0,1) \\ 1(1,0,1) \\ 0(1,0,1) \\ 1(1,0,1) \end{pmatrix} + \begin{pmatrix} 2(0,1,1) \\ 1(0,1,1) \\ 1(0,1,1) \\ 0(0,1,1) \end{pmatrix}.$$

This means that all the rows can be expressed as linear combination of two vectors, i.e. $(1,0,1)$ and $(0,1,1)$, respectively. Hence, the matrix has two linearly independent columns and two linearly independent rows. It can be appreciated that 2 is also the rank of the matrix $\mathbf{A}$.

Exercises

8.1. Determine whether or not $(U,+,\cdot)$, $(V,+,\cdot)$, and $(U \cap V,+,\cdot)$ are vector spaces where

$$U = \{(x,y,z) \in \mathbb{R}^3 | 5x + 5y + 5z = 0\}$$
$$V = \{(x,y,z) \in \mathbb{R}^3 | 5x + 5y + 5z + 5 = 0\}.$$

8.2. Let $(U,+,\cdot)$ and $(V,+,\cdot)$ be two vector subspaces of $(E,+,\cdot)$. Prove that if $U \subset V$ or $V \subset U$, then $(U \cup V,+,\cdot)$ is a vector subspace of $(E,+,\cdot)$.

8.3. Let us consider the vector spaces $(U,+,\cdot)$ and $(V,+,\cdot)$ where

$$U = \{(x,y,z) \in \mathbb{R}^3 | x - y + 4z = 0\}$$

$$V = \{(x,y,z) \in \mathbb{R}^3 | y - z = 0\}$$

where $+$ and $\cdot$ are sum of vectors and product of a vector by a scalar, respectively.

1. Find the intersection set $U \cap V$ of the vector space $(U \cap V,+,\cdot)$
2. Find the sum set $S = U + V$ of the vector space $(S,+,\cdot)$
3. Determine whether or not the vector space $(S,+,\cdot)$ is a direct sum $(S,\oplus,\cdot)$
4. Verify the Grassman's Formula for the vector spaces $(U,+,\cdot)$ and $(V,+,\cdot)$.

8.4. Let us consider the vector space $(E,+,\cdot)$ where

$$E = \left\{ (x,y,z) \in \mathbb{R}^3 | \begin{cases} x + 2y + 3z = 0 \\ 2x + 4y + 6z = 0 \\ 3x + 6y + 9z = 0 \end{cases} \right\}.$$

Identify a basis of the vector space. Determine the dimension of the vector space.

8.5. Find a basis and dimension of the vector space $(U,+,\cdot)$ where

$$U = \left\{ (x,y,z) \in \mathbb{R}^3 | \begin{cases} x + y + 2z = 0 \\ 2x + y + 3z = 0 \\ y + z = 0 \end{cases} \right\}.$$

and $+$ and $\cdot$ are sum of vectors and product of a vector by a scalar, respectively.

8.6. Let us consider the vector space $(E,+,\cdot)$ where

$$E = \left\{ (x,y,z) \in \mathbb{R}^3 | \begin{cases} 5x + 2y + z = 0 \\ 15x + 6y + 3z = 0 \\ 10x + 4y + 2z = 0 \end{cases} \right\}.$$

Identify a basis of the vector space and verify that the vectors spanning the vector space are linearly independent. Determine the dimension of the vector space.

8.7. Let us consider the following vectors $\in \mathbb{R}^4$:

$$\mathbf{u} = (2,2,1,3)$$
$$\mathbf{v} = (0,2,1,-1)$$
$$\mathbf{w} = (2,4,2,2)$$
$$\mathbf{a} = (3,1,2,1)$$
$$\mathbf{b} = (0,5,0,2).$$

Find a basis by eliminating a linearly dependent vector.

Chapter 9
An Introduction to Inner Product Spaces: Euclidean Spaces

9.1 Basic Concepts: Inner Products

Definition 9.1. Inner Product. Let $(E, +, \cdot)$ be a finite-dimensional vector space over the field $\mathbb{K}$. The mapping

$$\phi : E \times E \to \mathbb{K}$$

is said *inner product* and is indicated with

$$\langle \mathbf{x}, \mathbf{y} \rangle$$

when, $\forall$ vectors $\mathbf{x}$, $\mathbf{y}$ and $\mathbf{z}$, it verifies the following properties

- conjugate symmetry: $\langle \mathbf{x}, \mathbf{y} \rangle = \overline{\langle \mathbf{y}, \mathbf{x} \rangle}$
- distributivity: $\langle \mathbf{x} + \mathbf{z}, \mathbf{y} \rangle = \langle \mathbf{x}, \mathbf{y} \rangle + \langle \mathbf{z}, \mathbf{y} \rangle$
- homogeneity: $\forall \lambda \in \mathbb{K} \ \lambda \langle \mathbf{x}, \mathbf{y} \rangle = \langle \lambda \mathbf{x}, \mathbf{y} \rangle$
- positivity: $\langle \mathbf{x}, \mathbf{x} \rangle \geq 0$
- definiteness: $\langle \mathbf{x}, \mathbf{x} \rangle = 0$ if and only if $\mathbf{x} = \mathbf{o}$

The inner product is a general concept which includes but is not limited to numeric vectors.

Example 9.1. Let us consider a vector space $(\mathscr{F}, +, \cdot)$ whose vectors are continuous real valued functions on the interval $[-1, 1]$. If $f(x)$ and $g(x)$ are two continuous functions on the interval $[-1, 1]$, an example of inner product is the integral of their product

$$\langle f, g \rangle = \int_{-1}^{1} f(x) g(x) \, dx.$$

Definition 9.2. Hermitian Product. Let $(E, +, \cdot)$ be a finite-dimensional vector space over the complex field $\mathbb{C}$ with $E \subseteq \mathbb{C}^n$. The *Hermitian product* is the inner product

$$\phi : E \times E \to \mathbb{C}$$

© Springer Nature Switzerland AG 2019

F. Neri, *Linear Algebra for Computational Sciences and Engineering*,

https://doi.org/10.1007/978-3-030-21321-3_9

defined as

$$\langle \mathbf{x}, \mathbf{y} \rangle = \mathbf{x}^{\dot{\mathrm{T}}} \mathbf{y} = \sum_{i=1}^{n} \dot{x}_i y_i.$$

where $\mathbf{x}$ and $\mathbf{y} \in E$.

Example 9.2. Let us consider the following two vectors

$$\mathbf{x} = \begin{pmatrix} 2 \\ 1+j \\ 5-7j \end{pmatrix}$$

and

$$\mathbf{y} = \begin{pmatrix} -3 \\ 1 \\ 0 \end{pmatrix}.$$

The Hermitian product is

$$\langle \mathbf{x}, \mathbf{y} \rangle = \mathbf{x}^{\dot{\mathrm{T}}} \mathbf{y} = \left(2 \ (1-j) \ (5+7j) \right) \begin{pmatrix} -3 \\ 1 \\ 0 \end{pmatrix} = -6 + 1 - j + 0 = -5 - j.$$

9.2 Euclidean Spaces

The Hermitian product is an example of inner product. A restriction to real numbers of the Hermitian product is the scalar product that is a special case of inner product.

Definition 9.3. Scalar Product. Let $(E, +, \cdot)$ be a finite-dimensional vector space over the field $\mathbb{R}$. The *Scalar Product* is the inner product

$$\phi : E \times E \to \mathbb{R}$$

defined as

$$\langle \mathbf{x}, \mathbf{y} \rangle = \mathbf{x}^{\mathrm{T}} \mathbf{y} = \mathbf{x}\mathbf{y} = \sum_{i=1}^{n} x_i y_i.$$

The properties of the inner product in the special case of the scalar product can be written as:

- commutativity: $\forall \mathbf{x}, \mathbf{y} \in E : \mathbf{x}\mathbf{y} = \mathbf{y}\mathbf{x}$
- distributivity: $\forall \mathbf{x}, \mathbf{y}, \mathbf{z} \in E : \mathbf{x}(\mathbf{y} + \mathbf{z}) = \mathbf{x}\mathbf{y} + \mathbf{x}\mathbf{z}$
- homogeneity: $\forall \mathbf{x}, \mathbf{y} \in E$ and $\forall \lambda \in \mathbb{R} \ (\lambda \mathbf{x})\mathbf{y} = \mathbf{x}(\lambda \mathbf{y}) = \lambda \mathbf{x}\mathbf{y}$
- identity property: $\forall \mathbf{x} \in E : \mathbf{x}\mathbf{x} \geq 0$ and $\mathbf{x}\mathbf{x} = 0$ if and only if $\mathbf{x} = \mathbf{o}$

The scalar product can be seen as a special case of the Hermitian product where all the complex numbers have null imaginary part.

In order to distinguish the symbol of the scalar product from the symbol used for the product of a scalar by a vector · (external composition law), let us indicate with • the scalar product operator.

Definition 9.4. The triple $(E_n, +, •)$ is said *Euclidean space*.

The symbol • will be used when a Euclidean space is mentioned. However, when scalar product operation of two vectors is described, e.g. **xy**, no symbol will be used.

It must be remarked that since the scalar product is not an external composition law, a Euclidean space is not a vector space. In vector spaces the result of the external composition law is a vector, i.e. an element of the set E. In the case of Euclidean spaces the result of a scalar product is a scalar and thus not a vector (not an element of E).

Example 9.3. The triple $(\mathbb{V}_3, +,)$, i.e. the set of geometric vectors with the operations of (1) sum between vectors; (2) scalar product of geometric vectors, is a Euclidean space.

Analogously, $(\mathbb{R}^2, +, •)$ and $(\mathbb{R}^3, +, •)$ are also Euclidean spaces.

Definition 9.5. Let $\mathbf{x}, \mathbf{y} \in E_n$ where E_n is a Euclidean space. These two vectors are *orthogonal* if $\mathbf{xy} = 0$.

Proposition 9.1. *Let $(E_n, +, •)$ be a Euclidean space. Every vector belonging to a Euclidean space is orthogonal to the null vector:* $\forall \mathbf{x} \in E_n : \mathbf{xo} = 0$.

Proof. Considering that $\mathbf{o} = \mathbf{o} - \mathbf{o}$, it follows that

$$\mathbf{xo} = \mathbf{x}(\mathbf{o} - \mathbf{o}) = \mathbf{xo} - \mathbf{xo} = \mathbf{o}. \qquad \square$$

Proposition 9.2. *Let $(E_n, +, •)$ be a Euclidean vector space and $\mathbf{x_1}, \mathbf{x_2}, \ldots, \mathbf{x_n}$ be n non-null vectors belonging to it. If the vectors $\mathbf{x_1}, \mathbf{x_2}, \ldots, \mathbf{x_n}$ are all orthogonal to each other, i.e. $\forall$ indexes $i, j : \mathbf{x_i}$ is orthogonal to $\mathbf{x_j}$, then the vectors $\mathbf{x_1}, \mathbf{x_2}, \ldots, \mathbf{x_n}$ are linearly independent.*

Proof. Let us consider the linear combination of the vectors $\mathbf{x_1}, \mathbf{x_2}, \ldots, \mathbf{x_n}$ by means of the scalars $\lambda_1, \lambda_2, \ldots, \lambda_n \in \mathbb{R}$ and let us impose that

$$\lambda_1 \mathbf{x_1} + \lambda_2 \mathbf{x_2} + \ldots + \lambda_n \mathbf{x_n} = \mathbf{o}.$$

Let us now multiply (scalar product) this linear combination by $\mathbf{x_1}$ and we find that

$$\mathbf{x_1}(\lambda_1 \mathbf{x_1} + \lambda_2 \mathbf{x_2} + \ldots + \lambda_n \mathbf{x_n}) = 0.$$

Due to the orthogonality, this expression is equal to

$$\lambda_1 \mathbf{x_1} \mathbf{x_1} = 0.$$

For hypothesis $\mathbf{x_1} \neq \mathbf{o}$, then the expression can be equal to 0 only if $\lambda_1 = 0$. The linear combination can be multiplied by $\mathbf{x_2}$ to find that the expression is equal to 0 only if $\lambda_2 = 0$, ..., by $\mathbf{x_n}$ to find that the expression is equal to 0 only if $\lambda_n = 0$. If we impose that the linear combination is equal to the null vector, we find that all the scalars are null. Hence, the vectors are linearly independent. $\quad\square$

Example 9.4. The following vectors of the Euclidean space $\left(\mathbb{R}^2, +, \bullet\right)$ are orthogonal:

$$\mathbf{v_1} = (1, 5)$$
$$\mathbf{v_2} = (-5, 1)$$

Let us check the linear dependence:

$$(0, 0) = \lambda_1 (1, 5) + \lambda_2 (-5, 1)$$

which leads to

$$\begin{cases} \lambda_1 - 5\lambda_2 = 0 \\ 5\lambda_1 + \lambda_2 = 0. \end{cases}$$

The system is determined and the vectors are linearly independent.

Example 9.5. The following vectors of the Euclidean space $\left(\mathbb{R}^3, +, \bullet\right)$

$$\mathbf{v_1} = (1, 0, 0)$$
$$\mathbf{v_2} = (0, 1, 0)$$
$$\mathbf{v_3} = (0, 0, 1)$$

are obviously all orthogonal (each pair of vectors is orthogonal) and linearly independent. If we added another arbitrary vector $\mathbf{v_4}$, we would have linearly dependent vectors and that not all the pairs would be orthogonal.

Definition 9.6. Let $(E_n, +, \bullet)$ be a Euclidean space and $U \subset E_n$ with $U \neq \emptyset$. If $(U, +, \bullet)$ is also a Euclidean space then it is said *Euclidean subspace* of $(E_n, +, \bullet)$.

Definition 9.7. Let $(E_n, +, \bullet)$ be a Euclidean space and $(U, +, \bullet)$ be its sub-space. Let us indication with $\mathbf{u}$ be the generic element (vector) of U. The set of all the vectors orthogonal to all the vectors of U

$$U^o = \{\mathbf{x} \in E_n | \mathbf{xu} = 0, \forall \mathbf{u} \in U\}$$

is said *orthogonal set* to the set U.

Proposition 9.3. *Let* $(E_n, +, \bullet)$ *be a Euclidean space and* $(U, +, \cdot)$ *be its sub-space. The orthogonal set* U^o *with sum and product of a scalar by a vector is a vector space* $(U^o, +, \cdot)$ *namely* orthogonal space *to* $(U, +, \bullet)$.

Proof. In order to prove that $(U^o, +, \cdot)$ is a vector space we have to prove that the set U^o is closed with respect to sum and product of a scalar by a vector.

Let us consider two vectors $\mathbf{x_1}, \mathbf{x_2} \in U^o$. Let us calculate $(\mathbf{x_1} + \mathbf{x_2})\mathbf{u} = 0 + 0 = 0$ $\forall \mathbf{u} \in U$. This means that $(\mathbf{x_1} + \mathbf{x_2}) \in U^o$.

Let $\lambda \in \mathbb{R}$ and $\mathbf{x} \in U^o$. Let us calculate $\lambda \mathbf{x} \mathbf{u} = 0 \ \forall \mathbf{u} \in U$. This means that $\lambda \mathbf{u} \in U^o$.

Hence, $(U^o, +, \cdot)$ is a vector space. $\quad \square$

Example 9.6. Let us consider the vectors

$$\mathbf{u_1} = (1,0,0)$$
$$\mathbf{u_2} = (0,1,0)$$

and the set spanned by them

$$U = L(\mathbf{u_1}, \mathbf{u_2}).$$

The set U can be interpreted as a plane in the space. We may now think about a set U^o composed of all those vectors that are orthogonal to U, i.e. orthogonal to the plane spanned by $\mathbf{u_1}$ and $\mathbf{u_2}$. The set U^o would be

$$U^o = \alpha (0,0,1)$$

with $\alpha \in \mathbf{R}$.

We can easily verify that $(U^o, +, \cdot)$ is a vector space.

Definition 9.8. Let a vector $\mathbf{x} \in E_n$. The *module* of the vector $\mathbf{x}$, indicated with $\| \mathbf{x} \|$, is equal to $\sqrt{\mathbf{x}\mathbf{x}}$.

Example 9.7. The module of the following vector of $(\mathbb{R}^2, +, \bullet)$

$$(1,2)$$

is

$$\sqrt{1+4} = \sqrt{5}.$$

9.3 Euclidean Spaces in Two Dimensions

This section refers to the special Euclidean space $(\mathbb{R}^2, +, \bullet)$.

Lemma 9.1. *Let us consider two orthogonal vectors* $\mathbf{e}$ $\mathbf{y}$ *belonging to* $\mathbb{R}^2$ *and such that*

$$\mathbf{e} \neq \mathbf{o}$$
$$\mathbf{y} \neq \mathbf{o}.$$

It follows that for every vector $\mathbf{x} \in \mathbb{R}^2$, $\exists$ *two scalars* $\alpha, \beta \in \mathbb{R}$ *such that* $\mathbf{x}$ *can be expressed as*

$$\mathbf{x} = \alpha \mathbf{e} + \beta \mathbf{y}.$$

Proof. Let us write the equation $\mathbf{x} = \alpha \mathbf{e} + \beta \mathbf{y}$ as

$$\mathbf{x} - \alpha \mathbf{e} = \beta \mathbf{y}.$$

Let us multiply the equation by $\mathbf{e}$:

$$\mathbf{xe} - \alpha \mathbf{ee} = \beta \mathbf{ye} = 0$$

since $\mathbf{y}$ is orthogonal to $\mathbf{e}$.

It follows that

$$\mathbf{xe} = \alpha \parallel \mathbf{e} \parallel^2$$

and thus

$$\alpha = \frac{\mathbf{xe}}{\parallel \mathbf{e} \parallel^2}.$$

Since $\mathbf{e} \neq \mathbf{o}$ then $\mathbf{ee} = \parallel \mathbf{e} \parallel^2 \neq 0$. Thus, α exists.

Let us write again the equation

$$\mathbf{x} = \alpha \mathbf{e} + \beta \mathbf{y}$$

and write it as

$$\mathbf{x} - \beta \mathbf{y} = \alpha \mathbf{e}.$$

Let us multiply the equation by $\mathbf{y}$ and obtain

$$\mathbf{xy} - \beta \mathbf{yy} = \alpha \mathbf{ey} = 0$$

since $\mathbf{e}$ and $\mathbf{y}$ are orthogonal.

It follows that

$$\mathbf{xy} = \beta \parallel \mathbf{y} \parallel^2$$

and

$$\beta = \frac{\mathbf{xy}}{\parallel \mathbf{y} \parallel^2}.$$

Since $\mathbf{y} \neq \mathbf{o}$, it follows that $\mathbf{yy} = \parallel \mathbf{y} \parallel^2 \neq 0$ and that the scalar β also exists.

Thus there exist two scalars α and β such that $\mathbf{x} = \alpha \mathbf{e} + \beta \mathbf{y}$ always exist. $\square$

In essence Lemma 9.1 says that a vector can always be decomposed along two orthogonal directions.

Definition 9.9. Let us consider a Euclidean space $\left(\mathbb{R}^2, +, \bullet\right)$ and a vector $\mathbf{x} \in \mathbb{R}^2$. The equation

$$\mathbf{x} = \alpha \mathbf{e} + \beta \mathbf{y}$$

is named *decomposition* of the vector $\mathbf{x}$ along the orthogonal directions of $\mathbf{e}$ and $\mathbf{y}$.

Example 9.8. Let us consider the vector of $\left(\mathbb{R}^2, +, \bullet\right)$: $\mathbf{x} = (5, 2)$.

If we want to decompose this vector along the orthogonal directions of $\mathbf{e} = (1, 1)$ and $\mathbf{y} = (-1, 1)$ we need to find

$$\alpha = \frac{\mathbf{xe}}{\parallel \mathbf{e} \parallel^2} = \frac{7}{2}$$

and

$$\beta = \frac{\mathbf{x}\mathbf{y}}{\|\mathbf{y}\|^2} = -\frac{3}{2}.$$

It follows that

$$(5,2) = \frac{7}{2}(1,1) - \frac{3}{2}(-1,1)$$

Definition 9.10. Let us consider a Euclidean space $(\mathbb{R}^2, +, \bullet)$ and a vector $\mathbf{x} \in \mathbb{R}^2$. Let us decompose $\mathbf{x}$ along the directions of two orthogonal vectors $\mathbf{e}$ and $\mathbf{y}$

$$\mathbf{x} = \alpha\mathbf{e} + \beta\mathbf{y}. \tag{9.1}$$

The vector $\alpha\mathbf{e}$ is named *orthogonal projection* of the vector $\mathbf{x}$ on the direction of $\mathbf{e}$.

Remark 9.1. The n-dimensional version of the orthogonal decomposition would be

$$\mathbf{x} = \sum_{i=1}^{n} \alpha_i\mathbf{e_i}$$

with all the pairs of vectors orthogonal, i.e. $\mathbf{e_i^T}\mathbf{e_j} = 0 \; \forall i, j$.

Proposition 9.4. *The maximum length of the orthogonal projection of a vector* $\mathbf{x}$ *is its module* $\|\mathbf{x}\|$:

$$\|\alpha\mathbf{e}\| \leq \|\mathbf{x}\|.$$

Proof. Considering that $\mathbf{x} = \alpha\mathbf{e} + \beta\mathbf{y}$ and $\mathbf{e}\mathbf{y} = 0$, we can write

$$\|\mathbf{x}\|^2 = \mathbf{x}\mathbf{x} = (\alpha\mathbf{e} + \beta\mathbf{y})(\alpha\mathbf{e} + \beta\mathbf{y}) =$$
$$= \|\alpha\mathbf{e}\|^2 + \|\beta\mathbf{y}\|^2 \geq \|\alpha\mathbf{e}\|^2.$$

This inequality leads to

$$\|\alpha\mathbf{e}\| \leq \|\mathbf{x}\|. \; \square$$

Example 9.9. Let us consider again the vector $\mathbf{x} = (5,2)$ and its orthogonal projection $\alpha\mathbf{e}\frac{7}{2}(1,1)$. The two modules are

$$\sqrt{5^2 + 2^2} = \sqrt{29} \geq \sqrt{\left(\frac{7}{2}\right)^2 + \left(\frac{7}{2}\right)^2} = \sqrt{\frac{49}{2}} = \sqrt{24.5}.$$

Theorem 9.1. Cauchy-Schwarz Inequality. *Let* $(E_n, +, \bullet)$ *be a Euclidean vector space. For every* $\mathbf{x}$ *and* $\mathbf{y} \in E_n$, *the following inequality holds:*

$$\|\mathbf{x}\mathbf{y}\| \leq \|\mathbf{x}\|\|\mathbf{y}\|.$$

Proof. If either $\mathbf{x}$ or $\mathbf{y}$ is equal to the null vector, then the Cauchy-Schwarz inequality becomes $0 = 0$, that is always true.

In the general case, let us suppose that neither $\mathbf{x}$ nor $\mathbf{y}$ are equal to the null vector. Let us now indicate with $\alpha\mathbf{y}$ the orthogonal projection of the vector $\mathbf{x}$ onto the direction of **y**. For the Lemma 9.1 we can always express $\mathbf{x} = \alpha\mathbf{y} + \mathbf{z}$ where $\mathbf{yz} = 0$. On the basis of this statement we can write

$$\mathbf{xy} = (\alpha\mathbf{y} + \mathbf{z})\,\mathbf{y} = \alpha\mathbf{yy} = \alpha \parallel \mathbf{y} \parallel^2 = \alpha \parallel \mathbf{y} \parallel \parallel \mathbf{y} \parallel .$$

By computing the module of this equation we obtain

$$\parallel \mathbf{xy} \parallel = \parallel \alpha \parallel \parallel \mathbf{y} \parallel \parallel \mathbf{y} \parallel$$

where the module of a scalar is the absolute value of the scalar itself, i.e.

$$\parallel \alpha \parallel = \mid \alpha \mid$$
$$\parallel \mathbf{xy} \parallel = \mid \mathbf{xy} \mid .$$

Furthermore, since α is a scalar it follows that

$$\parallel \alpha \parallel \parallel \mathbf{y} \parallel \parallel \mathbf{y} \parallel = \parallel \alpha\mathbf{y} \parallel \parallel \mathbf{y} \parallel .$$

For the Proposition 9.4 $\parallel \alpha\mathbf{y} \parallel \leq \parallel \mathbf{x} \parallel$. Hence,

$$\parallel \mathbf{xy} \parallel \leq \parallel \mathbf{x} \parallel \parallel \mathbf{y} \parallel . \quad \square$$

The Cauchy-Schwartz inequality is generally true but the proof provided above refers to the bidimensional case since it makes use of the orthogonal decomposition.

Example 9.10. Let us consider the following vectors: $\mathbf{x} = (3,2,1)$ and $\mathbf{y} = (1,1,0)$. The modules of these two vectors are $\parallel \mathbf{x} \parallel = \sqrt{14}$ and $\parallel \mathbf{y} \parallel = \sqrt{2}$, respectively.

The scalar product is $\mathbf{xy} = 5$. Considering that $\parallel 5 \parallel = 5$. It follows that $5 \leq \sqrt{2}\sqrt{14} \approx 5.29$.

Theorem 9.2. Minkowski's Inequality. *Let $(E_n, +, \bullet)$ be a Euclidean space. For every $\mathbf{x}$ and $\mathbf{y} \in E_n$, the following inequality holds:*

$$\parallel \mathbf{x} + \mathbf{y} \parallel \leq \parallel \mathbf{x} \parallel + \parallel \mathbf{y} \parallel .$$

Proof. By the definition of module

$$\parallel \mathbf{x} + \mathbf{y} \parallel^2 = (\mathbf{x} + \mathbf{y})\,(\mathbf{x} + \mathbf{y}) = \mathbf{xx} + \mathbf{xy} + \mathbf{yx} + \mathbf{yy} =$$
$$\parallel \mathbf{x} \parallel^2 + \parallel \mathbf{y} \parallel^2 + 2\mathbf{xy}$$

Still considering that the module of a scalar is its absolute value and thus

$$\parallel \mathbf{xy} \parallel = \mid \mathbf{xy} \mid = \begin{cases} \mathbf{xy} \text{ if } \mathbf{xy} \geq 0 \\ -\mathbf{xy} \text{ if } \mathbf{xy} < 0. \end{cases}$$

In other words,

$$\mathbf{xy} \leq \mid \mathbf{xy} \mid = \parallel \mathbf{xy} \parallel .$$

In this light, let us now compute the module of the equation:

$$\| \mathbf{x} + \mathbf{y} \|^2 = \| \mathbf{x} \|^2 + \| \mathbf{y} \|^2 + 2\mathbf{xy} \leq \| \mathbf{x} \|^2 + \| \mathbf{y} \|^2 + 2 \| \mathbf{xy} \| .$$

For the Cauchy-Schwarz inequality $2 \| \mathbf{xy} \| \leq \| 2 \| \mathbf{x} \| \| \mathbf{y} \|$. Hence,

$$\| \mathbf{x} + \mathbf{y} \|^2 \leq \| \mathbf{x} \|^2 + \| \mathbf{y} \|^2 + 2 \| \mathbf{x} \| \| \mathbf{y} \| = (\| \mathbf{x} \| + \| \mathbf{y} \|)^2 \Rightarrow$$
$$\Rightarrow \| \mathbf{x} + \mathbf{y} \| \leq \| \mathbf{x} \| + \| \mathbf{y} \| . \square$$

Example 9.11. Let us consider again the vectors: $\mathbf{x} = (3,2,1)$ and $\mathbf{y} = (1,1,0)$. As shown above, the modules of these two vectors are $\| \mathbf{x} \| = \sqrt{14}$ and $\| \mathbf{y} \| = \sqrt{2}$, respectively.

The sum $\mathbf{x} + \mathbf{y} = (4,3,1)$ has module $\sqrt{26}$. It follows that $\sqrt{26} \approx 5.10 \leq \sqrt{2} + \sqrt{14} \approx 5.15$.

Theorem 9.3. Pythagorean Formula. *Let $(E_n, +, \bullet)$ be a Euclidean space. For every $\mathbf{x}$ and $\mathbf{y} \in E_n$, and $\mathbf{x}$, $\mathbf{y}$ orthogonal, the following equality holds:*

$$\| \mathbf{x} + \mathbf{y} \|^2 = \| \mathbf{x} \|^2 + \| \mathbf{y} \|^2$$

Proof. By the definition of module

$$\| \mathbf{x} + \mathbf{y} \|^2 = (\mathbf{x} + \mathbf{y})(\mathbf{x} + \mathbf{y}) = \mathbf{xx} + \mathbf{xy} + \mathbf{yx} + \mathbf{yy} =$$
$$\| \mathbf{x} \|^2 + \| \mathbf{y} \|^2$$

since for the perpendicularity $\mathbf{xy} = 0$.

Example 9.12. Let us consider the vectors $\mathbf{x} = (2,-1,1)$ and $\mathbf{y} = (1,2,0)$. It can be easily verified that the scalar product $\mathbf{xy} = 0$, i.e. the vectors are orthogonal. If we calculate the modules of these vectors we obtain, $\| \mathbf{x} \|^2 = 6$ and $\| \mathbf{y} \|^2 = 5$, respectively.

The sum is $\mathbf{x} + \mathbf{y} = (3,1,1)$ which has square module $\| \mathbf{x} + \mathbf{y} \| = 11$, which is obviously equal to $6 + 5$.

9.4 Gram-Schmidt Orthonormalization

Definition 9.11. Let $\mathbf{a_1}, \mathbf{a_2}, \ldots, \mathbf{a_n}$ be n vectors belonging to a Euclidean space. The Gram determinant or Gramian is the determinant of the following matrix, namely Gram matrix,

$$\begin{pmatrix} \mathbf{a_1 a_1} & \mathbf{a_1 a_2} & \ldots & \mathbf{a_1 a_n} \\ \mathbf{a_2 a_1} & \mathbf{a_2 a_2} & \ldots & \mathbf{a_2 a_n} \\ \ldots & \ldots & \ldots & \ldots \\ \mathbf{a_n a_1} & \mathbf{a_n a_2} & \ldots & \mathbf{a_n a_n} \end{pmatrix} .$$

The Gram determinant is indicated with $G(\mathbf{a_1}, \mathbf{a_2}, \ldots, \mathbf{a_n})$.

Theorem 9.4. *If the vectors* $\mathbf{a_1}, \mathbf{a_2}, \ldots, \mathbf{a_n}$ *are linearly dependent, then the Gram determinant* $G(\mathbf{a_1}, \mathbf{a_2}, \ldots, \mathbf{a_n})$ *is equal to* 0.

Proof. If the vectors are linearly dependent for the Proposition 8.4 one vector can be expressed as linear combination of the others. Thus, let us write

$$\mathbf{a_n} = \lambda_1 \mathbf{a_1} + \lambda_2 \mathbf{a_2} + \ldots + \lambda_{n-1} \mathbf{a_{n-1}}$$

where $\lambda_1, \lambda_2, \ldots, \lambda_{n-1} \in \mathbb{R}$.

If $\mathbf{a_n}$ is multiplied by a generic $\mathbf{a_i}$ for $i = 1, 2, \ldots, n-1$ we obtain

$$\mathbf{a_n} \mathbf{a_i} = \lambda_1 \mathbf{a_1} \mathbf{a_i} + \lambda_2 \mathbf{a_2} \mathbf{a_i} + \ldots + \lambda_{n-1} \mathbf{a_{n-1}} \mathbf{a_i}.$$

Thus, if we substitute the elements of the last line of the Gram matrix with the linear combination above, we have expressed the nth row as a linear combination of all the other rows by means of the scalars $\lambda_1, \lambda_2, \ldots, \lambda_{n-1}$. Hence, the Gramian is 0. $\square$

Example 9.13. Let us consider the following vectors:

$$\mathbf{v_1} = (0, 0, 1)$$
$$\mathbf{v_2} = (0, 1, 0)$$
$$\mathbf{v_3} = (0, 2, 2).$$

It can be easily checked that these vectors are linearly dependent since

$$\mathbf{v_3} = \lambda_1 \mathbf{v_1} + \lambda_2 \mathbf{v_2}$$

with $\lambda_1, \lambda_2 = 2, 2$.

Let us calculate the Gram matrix

$$\begin{pmatrix} \mathbf{v_1}\mathbf{v_1} & \mathbf{v_1}\mathbf{v_2} & \mathbf{v_1}\mathbf{v_3} \\ \mathbf{v_2}\mathbf{v_1} & \mathbf{v_2}\mathbf{v_2} & \mathbf{v_2}\mathbf{v_3} \\ \mathbf{v_3}\mathbf{v_1} & \mathbf{v_3}\mathbf{v_2} & \mathbf{v_3}\mathbf{v_3} \end{pmatrix} = \begin{pmatrix} 1 & 0 & 2 \\ 0 & 1 & 2 \\ 2 & 2 & 8 \end{pmatrix}.$$

The Gramian determinant is

$$G(\mathbf{v_1}, \mathbf{v_2}, \mathbf{v_3}) = 8 + 0 + 0 - 4 - 4 - 0 = 0.$$

Thus, as stated in the theorem above, the Gramian determinant associated with linearly dependent vectors is null.

Theorem 9.5. *If the vectors* $\mathbf{a_1}, \mathbf{a_2}, \ldots, \mathbf{a_n}$ *are linearly independent the Gram determinant* $G(\mathbf{a_1}, \mathbf{a_2}, \ldots, \mathbf{a_n}) > 0$.

The proof of this theorem is given in Appendix B.

Example 9.14. Let us consider the following linearly independent vectors:

$$\mathbf{v_1} = (0,0,1)$$
$$\mathbf{v_2} = (0,1,0)$$
$$\mathbf{v_3} = (1,0,0).$$

The Gram matrix is

$$\begin{pmatrix} \mathbf{v_1v_1} \ \mathbf{v_1v_2} \ \mathbf{v_1v_3} \\ \mathbf{v_2v_1} \ \mathbf{v_2v_2} \ \mathbf{v_2v_3} \\ \mathbf{v_3v_1} \ \mathbf{v_3v_2} \ \mathbf{v_3v_3} \end{pmatrix} = \begin{pmatrix} 1\ 0\ 0 \\ 0\ 1\ 0 \\ 0\ 0\ 1 \end{pmatrix},$$

whose determinant is $1 > 0$.

Remark 9.2. In addition, it must be observed that the latter two theorems give a generalization of the Cauchy-Schwarz inequality. It can be seen that the Gramian associated with two vectors $\mathbf{x}$ and $\mathbf{y}$ is

$$\det \begin{pmatrix} \mathbf{xx} \ \mathbf{xy} \\ \mathbf{xy} \ \mathbf{yy} \end{pmatrix} = \parallel \mathbf{x} \parallel^2 \parallel \mathbf{y} \parallel^2 - \parallel \mathbf{xy} \parallel^2 \geq 0 \Rightarrow \parallel \mathbf{xy} \parallel \leq \parallel \mathbf{x} \parallel \parallel \mathbf{y} \parallel.$$

Definition 9.12. Let a vector $\mathbf{x} \in E_n$. The *versor* of the vector $\mathbf{x}$ is

$$\hat{\mathbf{x}} = \left(\frac{\mathbf{x}}{||\mathbf{x}||} \right).$$

Definition 9.13. Let $(E_n, +, \bullet)$ be a Euclidean space. An *orthonormal basis* of a Euclidean space is a basis composed of versors where each arbitrary pair of vectors is orthogonal.

Obviously the Gramian associated with an orthonormal basis is 1 since the Gram matrix would be the identity matrix.

Theorem 9.6. *Every Euclidean space has an orthonormal basis.*

Gram-Schmidt Orthonormalization Algorithm

In a Euclidean space having dimension equal to n, every basis $B = \{\mathbf{x_1}, \mathbf{x_2} \ldots, \mathbf{x_n}\}$ can be transformed into an orthonormal basis $B_e = \{\mathbf{e_1}, \mathbf{e_2} \ldots, \mathbf{e_n}\}$ by applying the so-called *Gram-Schmidt* method. This method consists of the following steps:

- The first versor can be obtained as

$$\mathbf{e_1} = \frac{\mathbf{x_1}}{\parallel \mathbf{x_1} \parallel}$$

- The second versor will have the direction of

$$\mathbf{y_2} = \mathbf{x_2} + \lambda_1 \mathbf{e_1}.$$

We impose the orthogonality:

$$0 = \mathbf{e_1 y_2} = \mathbf{e_1 x_2} + \lambda_1 \mathbf{e_1 e_1} = \mathbf{e_1 x_2} + \lambda_1 \parallel \mathbf{e_1} \parallel^2.$$

Considering that the module a versor is equal to 1 ($\parallel \mathbf{e_1} \parallel = 1$) If follows that

$$\lambda_1 = -\mathbf{e_1 x_2}$$

and

$$\mathbf{y_2} = \mathbf{x_2} - \mathbf{e_1 x_2 e_1}.$$

Hence,

$$\mathbf{e_2} = \frac{\mathbf{y_2}}{\parallel \mathbf{y_2} \parallel}.$$

- The third versor will have the direction of

$$\mathbf{y_3} = \mathbf{x_3} + \lambda_1 \mathbf{e_1} + \lambda_2 \mathbf{e_2}$$

We impose the orthogonality:

$$0 = \mathbf{y_3 e_2} = \mathbf{x_3 e_2} + \lambda_1 \mathbf{e_1 e_2} + \lambda_2 \mathbf{e_2 e_2} = \mathbf{x_3 e_2} + \lambda_2. \qquad (9.2)$$

It follows that

$$\lambda_2 = -\mathbf{e_2 x_3}$$

and

$$\mathbf{y_3} = \mathbf{x_3} + \lambda_1 \mathbf{e_1} - \mathbf{e_2 x_3 e_2}.$$

Hence,

$$\mathbf{e_3} = \frac{\mathbf{y_3}}{\parallel \mathbf{y_3} \parallel}.$$

- The nth versor has direction given by

$$\mathbf{y_n} = \mathbf{x_n} + \lambda_1 \mathbf{e_1} + \lambda_2 \mathbf{e_2} + \ldots + \lambda_{n-1} \mathbf{e_{n-1}} - \mathbf{e_{n-1} x_n e_{n-1}}.$$

By imposing the orthogonality we obtain

$$\lambda_{n-1} = -\mathbf{e_{n-1} x_n}.$$

Hence,

$$\mathbf{e_n} = \frac{\mathbf{y_n}}{\parallel \mathbf{y_n} \parallel}.$$

This method is a simple and yet powerful instrument since orthogonal bases are in general much more convenient to handle and make mathematical models simpler (all the angles among reference axes is the same and the scalar product of any pair is null) but not less accurate.

The pseudocode describing the Gram-Schmidt orthonormalization given in Algorithm 9.

Algorithm 9 Gram-Schmidt Orthonormalization

Input $\mathbf{x_1}, \mathbf{x_2}, \ldots, \mathbf{x_n}$
$\mathbf{e_1} = \frac{\mathbf{x_1}}{\|\mathbf{x_1}\|}$
for $j = 2 : n$ **do**
   $\lambda_{j-1} = -\mathbf{e_{j-1}}\mathbf{x_j}$
   $\Sigma_k = 0$
   for $k = 1 : j-1$ **do**
      $\Sigma_k = \Sigma_k + \lambda_k \mathbf{e_k}$
   end for
   $\mathbf{y_j} = \mathbf{x_j} + \Sigma_k$
end for

Example 9.15. Let us consider the following two vectors of $\mathbb{R}^2$: $\mathbf{x_1} = (3,1)$ and $\mathbf{x_2} = (2,2)$. It can be easily shown that these two vectors span the entire $\mathbb{R}^2$ and that they are linearly independent. The latter can be shown by checking that

$$\lambda_1(3,1) + \lambda_2(2,2) = \mathbf{o}$$

only if $\lambda_1, \lambda_2 = 0, 0$. This statement is equivalent to say that the homogeneous system of linear equations

$$\begin{cases} 3\lambda_1 + 2\lambda_2 = 0 \\ \lambda_1 + 2\lambda_2 = 0 \end{cases}$$

is determined.

Thus, $B = \{\mathbf{x_1}, \mathbf{x_2}\}$ is a basis of the vector space $(\mathbb{R}^2, +, \cdot)$.

Furthermore, we know that in $\mathbb{R}^2$ the scalar product can be defined (as a scalar product of vectors). Hence, $(\mathbb{R}^2, +, \bullet)$ is a Euclidean space.

Let us now apply the Gram-Schmidt method to find an orthonormal basis $B_U = \{\mathbf{e_1}, \mathbf{e_2}\}$. The first vector is

$$\mathbf{e_1} = \frac{\mathbf{x_1}}{\|\mathbf{x_1}\|} = \frac{(3,1)}{\sqrt{10}} = \left(\frac{3}{\sqrt{10}}, \frac{1}{\sqrt{10}}\right).$$

For the calculation of the second vector the direction must be detected at first. The direction is that given by

$$\mathbf{y_2} = \mathbf{x_2} + \lambda_1 \mathbf{e_1} = (2,2) + \lambda_1\left(\frac{3}{\sqrt{10}}, \frac{1}{\sqrt{10}}\right).$$

Let us find the orthogonal direction by imposing that $\mathbf{y_2 e_1} = 0$. Hence,

$$\mathbf{y_2 e_1} = 0 = \mathbf{x_2 e_1} + \lambda_1 \mathbf{e_1 e_1} = (2,2)\left(\frac{3}{\sqrt{10}}, \frac{1}{\sqrt{10}}\right) + \lambda_1,$$

which leads to

$$\lambda_1 = -\frac{6}{\sqrt{10}} - \frac{2}{\sqrt{10}} = -\frac{8}{\sqrt{10}}.$$

The vector $\mathbf{y_2}$ is

$$\mathbf{y_2} = \mathbf{x_2} + \lambda_1 \mathbf{e_1} = (2,2) + \left(-\frac{8}{\sqrt{10}}\right)\left(\frac{3}{\sqrt{10}}, \frac{1}{\sqrt{10}}\right) = (2,2) + \left(-\frac{24}{10}, -\frac{8}{10}\right)$$

$$= (-0.4, 1.2).$$

Finally, we can calculate the versor

$$\mathbf{e_2} = \frac{\mathbf{y_2}}{\parallel \mathbf{y_2} \parallel} = \frac{(-0.4, 1.2)}{1.26} = \left(-\frac{0.4}{1.26}, \frac{1.2}{1.26}\right).$$

The vectors $\mathbf{e_1}, \mathbf{e_2}$ are an orthonormal basis of $(\mathbb{R}^2, +, \cdot)$.

Exercises

9.1. Let us consider the following two vectors

$$\mathbf{x} = (2,5,7)$$
$$\mathbf{y} = (-4,0,12).$$

Verify Cauchy-Schwarz and Minkowski's inequalities.

9.2. Let us consider the following two linearly independent vectors of $\mathbb{R}^2$: $\mathbf{x_1} = (2,1)$ and $\mathbf{x_2} = (0,2)$. Apply the Gram-Schmidt orthogonalization algorithm to find an orthonormal basis of $\mathbb{R}^2$.

Chapter 10
Linear Mappings

10.1 Introductory Concepts

Although the majority of the topics in this book (all the topics taken into account excluding only complex polynomials) are related to linear algebra, the subject "linear algebra" has never been introduced in the previous chapters. More specifically, while the origin of the term algebra has been mentioned in Chap. 1, the use of the adjective linear has never been discussed. Before entering into the formal definitions of linearity, let us illustrate the subject at the intuitive level. Linear algebra can be seen as a subject that studies vectors. If we consider that vector spaces are still vectors endowed with composition laws, that matrices are collections of row (or column) vectors, that systems of linear equations are vector equations, and that a number can be interpreted as a single element vector, we see that the concept of vector is the elementary entity of linear algebra. As seen in Chap. 4, a vector is generated by a segment of a line. Hence, the subject linear algebra studies "portions" of lines and their interactions, which justifies the adjective "linear".

Before entering into the details of this subject let us define again the concept of mapping shown in Chap. 1 by using the notions of vector and vector space.

Definition 10.1. Let $(E, +, \cdot)$ and $(F, +, \cdot)$ be two vector spaces defined over the scalar field $\mathbb{K}$. Let $f : E \to F$ be a relation. Let U be a set such that $U \subset E$. The relation f is said *mapping* when

$$\forall \mathbf{u} \in U : \exists! \mathbf{w} \in F \ni ` f(\mathbf{u}) = \mathbf{w}.$$

The set U is said *domain* and is indicated it with $\text{dom}(f)$.

A vector $\mathbf{w}$ such that

$$\mathbf{w} = f(\mathbf{u})$$

is said to be the *mapped* (or *transformed*) of $\mathbf{u}$ through f.

© Springer Nature Switzerland AG 2019
F. Neri, *Linear Algebra for Computational Sciences and Engineering*,
https://doi.org/10.1007/978-3-030-21321-3_10

Definition 10.2. Let f be a mapping $E \to F$, where E and F are sets associated with the vector spaces $(E, +, \cdot)$ and $(F, +, \cdot)$. The *image* of f, indicated with $\text{Im}(f)$, is a set defined as

$$\text{Im}(f) = \{\mathbf{w} \in F | \exists \mathbf{u} \in E \ni \text{`} f(\mathbf{u}) = \mathbf{w}\}.$$

Example 10.1. Let $(\mathbb{R}, +, \cdot)$ be a vector space. An example of mapping $\mathbb{R} \to \mathbb{R}$ is $f(x) = 2x + 2$. The domain of the mapping $\text{dom}(f) = \mathbb{R}$ while the image $\text{Im}(f) = \mathbb{R}$. The reverse image of $\mathbb{R}$ through f is $\mathbb{R}$.

Example 10.2. Let $(\mathbb{R}, +, \cdot)$ be a vector space. An example of mapping $\mathbb{R} \to \mathbb{R}$ is $f(x) = e^x$. The domain of the mapping $\text{dom}(f) = \mathbb{R}$ while the image $\text{Im}(f) =]0, \infty[$. The reverse image of $]0, \infty[$ through f is $\mathbb{R}$.

Example 10.3. Let $(\mathbb{R}, +, \cdot)$ be a vector space. An example of mapping $\mathbb{R} \to \mathbb{R}$ is $f(x) = x^2 + 2x + 2$. The domain of the mapping $\text{dom}(f) = \mathbb{R}$ while the image $\text{Im}(f) = [1, \infty[$. The reverse image of $[1, \infty[$ through f is $\mathbb{R}$.

Example 10.4. Let $(\mathbb{R}^2, +, \cdot)$ and $(\mathbb{R}, +, \cdot)$ be two vector spaces. An example of mapping $\mathbb{R}^2 \to \mathbb{R}$ is $f(x, y) = x + 2y + 2$. The domain of the mapping $\text{dom}(f) = \mathbb{R}^2$ while the image $\text{Im}(f) = \mathbb{R}$. The reverse image of $\mathbb{R}$ through f is $\mathbb{R}^2$.

Example 10.5. From the vector space $(\mathbb{R}^2, +, \cdot)$ an example of mapping $\mathbb{R}^2 \to \mathbb{R}^2$ is $f(x, y) = (x + 2y + 2, 8y - 3)$.

Example 10.6. From the vector spaces $(\mathbb{R}^2, +, \cdot)$ and $(\mathbb{R}^3, +, \cdot)$ an example of mapping $\mathbb{R}^3 \to \mathbb{R}^2$ is $f(x, y, z) = (x + 2y + -z + 2, 6y - 4z + 2)$.

Example 10.7. From the vector spaces $(\mathbb{R}^2, +, \cdot)$ and $(\mathbb{R}^3, +, \cdot)$ an example of mapping $\mathbb{R}^2 \to \mathbb{R}^3$ is $f(x, y) = (6x - 2y + 9, -4x + 6y + 8, x - y)$.

Definition 10.3. The mapping f is said *surjective* if the image of f coincides with F:

$$\text{Im}(f) = F.$$

Example 10.8. The mapping $\mathbb{R} \to \mathbb{R}$, $f(x) = 2x + 2$ is surjective because $\text{Im}(f) = \mathbb{R}$, i.e. its image is equal to the entire set of the vector space.

Example 10.9. The mapping $\mathbb{R} \to \mathbb{R}$, $f(x) = e^x$ is not surjective because $\text{Im}(f) =]0, \infty[$, i.e. its image is not equal to the entire set of the vector space.

Definition 10.4. The mapping f is said *injective* if

$$\forall \mathbf{u}, \mathbf{v} \in E \text{ with } \mathbf{u} \neq \mathbf{v} \Rightarrow f(\mathbf{u}) \neq f(\mathbf{v}).$$

An alternative and equivalent definition of injective mapping is: f is injective if

$$\forall \mathbf{u}, \mathbf{v} \in E \text{ with } f(\mathbf{u}) = f(\mathbf{v}) \Rightarrow \mathbf{u} = \mathbf{v}.$$

Example 10.10. The mapping $\mathbb{R} \to \mathbb{R}$, $f(x) = e^x$ is injective because $\forall x_1, x_2$ such that $x_1 \neq x_2$, it follows that $e^{x_1} \neq e^{x_2}$.

Example 10.11. The mapping $\mathbb{R} \to \mathbb{R}$, $f(x) = x^2$ is not injective because $\exists x_1, x_2$ with $x_1 \neq x_2$ such that $x_1^2 = x_2^2$. For example if $x_1 = 3$ and $x_2 = -3$, thus $x_1 \neq x_2$, it occurs that $x_1^2 = x_2^2 = 9$.

Example 10.12. The mapping $\mathbb{R} \to \mathbb{R}$, $f(x) = x^3$ is injective because $\forall x_1, x_2$ such that $x_1 \neq x_2$, it follows that $x_1^3 \neq x_2^3$.

Definition 10.5. The mapping f is said *bijective* if f is injective and surjective.

Example 10.13. The mapping $\mathbb{R} \to \mathbb{R}$, $f(x) = e^x$ is injective but not surjective. Hence, this mapping is not bijective.

Example 10.14. The mapping $\mathbb{R} \to \mathbb{R}$, $f(x) = 2x + 2$ is injective and surjective. Hence, this mapping is bijective.

Example 10.15. As shown above, the mapping $\mathbb{R} \to \mathbb{R}$, $f(x) = x^3$ is injective. The mapping is also surjective as its image is $\mathbb{R}$. Hence, this mapping is bijective.

Definition 10.6. Let f be a mapping $E \to F$, where E and F are sets associated with the vector spaces $(E, +, \cdot)$ and $(F, +, \cdot)$. The mapping f is said *linear mapping* if the following properties are valid:

- additivity: $\forall \mathbf{u}, \mathbf{v} \in E : f(\mathbf{u} + \mathbf{v}) = f(\mathbf{u}) + f(\mathbf{v})$
- homogeneity: $\forall \lambda \in \mathbb{K}$ and $\forall \mathbf{v} \in E : f(\lambda \mathbf{v}) = \lambda f(\mathbf{v})$

The two properties of linearity can be combined and written in the following compact way:

$$\forall \lambda_1, \lambda_2 \in \mathbb{K}; \forall \mathbf{u}, \mathbf{v} \in E : f(\lambda_1 \mathbf{u} + \lambda_2 \mathbf{v}) = \lambda_1 f(\mathbf{u}) + \lambda_2 f(\mathbf{v})$$

or extended in the following way

$$\begin{aligned}
&\forall \lambda_1, \lambda_2, \ldots, \lambda_n \\
&\forall \mathbf{v_1}, \mathbf{v_2}, \ldots, \mathbf{v_n} \\
&f(\lambda_1 \mathbf{v_1} + \lambda_2 \mathbf{v_2} + \ldots + \lambda_n \mathbf{v_n}) = \\
&= \lambda_1 f(\mathbf{v_1}) + \lambda_2 f(\mathbf{v_2}) + \ldots + \lambda_n f(\mathbf{v_n}).
\end{aligned}$$

Example 10.16. Let us consider again the following mapping $f : \mathbb{R} \to \mathbb{R}$

$$\forall x : f(x) = e^x$$

and let us check its linearity.

Let us consider two vectors (numbers in this case) x_1 and x_2. Let us calculate $f(x_1 + x_2) = e^{x_1 + x_2}$. From basic calculus we know that

$$e^{x_1 + x_2} \neq e^{x_1} + e^{x_2}.$$

The additivity is not verified. Hence the mapping is not linear. We know from calculus that an exponential function is not linear.

Example 10.17. Let us consider the following mapping $f : \mathbb{R} \to \mathbb{R}$

$$\forall x : f(x) = 2x$$

and let us check its linearity.

Let us consider two vectors (numbers in this case) x_1 and x_2. We have that

$$f(x_1 + x_2) = 2(x_1 + x_2)$$
$$f(x_1) + f(x_2) = 2x_1 + 2x_2.$$

It follows that $f(x_1 + x_2) = f(x_1) + f(x_2)$. Hence, this mapping is additive. Let us check the homogeneity by considering a generic scalar λ. We have that

$$f(\lambda x) = 2\lambda x$$
$$\lambda f(x) = \lambda 2x.$$

It follows that $f(\lambda x) = \lambda f(x)$. Hence, since also homogeneity is verified this mapping is linear.

Definition 10.7. Let f be a mapping $E \to F$, where E and F are sets associated with the vector spaces $(E, +, \cdot)$ and $(F, +, \cdot)$. The mapping f is said *affine mapping* if the mapping

$$g(\mathbf{v}) = f(\mathbf{v}) - f(\mathbf{o})$$

is linear.

Example 10.18. Let us consider the following mapping $f : \mathbb{R} \to \mathbb{R}$

$$\forall x : f(x) = x + 2$$

and let us check its linearity.

Let us consider two vectors (numbers in this case) x_1 and x_2. We have that

$$f(x_1 + x_2) = x_1 + x_2 + 2$$
$$f(x_1) + f(x_2) = x_1 + 2 + x_2 + 2 = x_1 + x_2 + 4.$$

It follows that $f(x_1 + x_2) \neq f(x_1) + f(x_2)$. Hence, this mapping is not linear. Still, $f(0) = 2$ and

$$g(x) = f(x) - f(0) = x$$

which is a linear mapping. This means that $f(x)$ is an affine mapping.

Example 10.19. Let us consider the following mapping $f : \mathbb{R}^3 \to \mathbb{R}^2$

$$\forall x, y, z : f(x, y, z) = (2x - 5z, 4y - 5z)$$

and let us check its linearity.

If it is linear then the two properties of additivity and homogeneity are valid. If two vectors are $\mathbf{v} = (x, y, z)$ and $\mathbf{v}' = (x', y', z')$, then

$$f(\mathbf{v} + \mathbf{v}') = f(x + x', y + y', z + z') = (2(x + x') - 5(z + z'), 4(y + y') - 5(z + z')) =$$
$$((2x - 5z, 4y - 5z) + (2x' - 5z', 4y' - 5z')) = f(\mathbf{v}) + f(\mathbf{v}')$$

and

$$f(\lambda \mathbf{v}) = f(\lambda x, \lambda y, \lambda z) = (\lambda 2x - \lambda 5z, \lambda 4y - \lambda 5z) =$$
$$= \lambda(2x - 5z, 4y - 5z) = \lambda f(\mathbf{v}).$$

The mapping is linear.

Example 10.20. Let us consider the following mapping $f : \mathbb{V}_3 \to \mathbb{R}$

$$\forall \vec{v} : f(\vec{v}) = (\vec{u}\,\vec{v})$$

Let us check the additivity of this mapping:

$$f(\vec{v} + \vec{v}') = \vec{u}(\vec{v} + \vec{v}') =$$
$$= \vec{u}\,\vec{v} + \vec{u}\,\vec{v}' = f(\vec{v}) + f(\vec{v}).$$

Let us check now the homogeneity

$$f(\lambda \vec{v}) = \vec{u}(\lambda \vec{v}) = \lambda(\vec{u}\,\vec{v}) = \lambda f(\vec{v}).$$

Hence this mapping is linear.

Proposition 10.1. *Let f be a linear mapping $E \to F$. Let us indicate with $\mathbf{o_E}$ and $\mathbf{o_F}$ the null vectors of the vector spaces $(E, +, \cdot)$ and $(F, +, \cdot)$, respectively. It follows that*

$$f(\mathbf{o_E}) = \mathbf{o_F}.$$

Proof. Simply let us write

$$f(\mathbf{o_E}) = f(0\mathbf{o_E}) = 0f(\mathbf{o_E}) = \mathbf{o_F}. \square$$

Example 10.21. Let us consider now the following mapping $f : \mathbb{R}^3 \to \mathbb{R}^4$

$$\forall x, y, z : f(x, y, z) = (23x - 51z, 3x + 4y - 5z, 32x + 5y - 6z + 1, 5x + 5y + 5z)$$

and let us check its linearity.
    If we calculate
$$f(0, 0, 0) = (0, 0, 1, 0) \neq \mathbf{o_{\mathbb{R}^4}}.$$

This is against Proposition 10.1. Hence the mapping is not linear. Nonetheless,

$$g(x,y,z) = f(x,y,z) - f(0,0,0) =$$
$$= (23x - 51z, 3x + 4y - 5z, 32x + 5y - 6z + 1, 5x + 5y + 5z) - (0,0,1,0) =$$
$$= (23x - 51z, 3x + 4y - 5z, 32x + 5y - 6z, 5x + 5y + 5z)$$

is linear. Hence, the mapping is affine.

Proposition 10.2. *Let f be a linear mapping $E \to F$. It follows that*

$$f(-\mathbf{v}) = -f(\mathbf{v}).$$

Proof. We can write

$$f(-\mathbf{v}) = f(-1\mathbf{v}) = -1f(\mathbf{v}) = -f(\mathbf{v}). \quad \square$$

Example 10.22. Let us consider the following linear mapping $f : \mathbb{R} \to \mathbb{R}$, $f(x) = 2x$. With reference to the notation of Proposition 10.1, we can easily see that $\mathbf{o_E} = 0$ and $\mathbf{o_F} = 0$. Then, if we calculate

$$f(\mathbf{o_E}) = f(0) = (2)(0) = 0 = \mathbf{o_F}.$$

Furthermore, considering the vector $\mathbf{v} = x$, which in this case is a number we have

$$f(-\mathbf{v}) = f(-x) = (2)(-x) = -2x = -f(\mathbf{v}).$$

Example 10.23. Let us consider the following linear mapping $f : \mathbb{R}^2 \to \mathbb{R}$, $f(x,y) = 2x + y$. Considering that $\mathbf{o_E} = (0,0)$, $\mathbf{o_F} = 0$, and $\mathbf{v} = (x,y)$, let us check the two propositions:

$$f(\mathbf{o_E}) = f(0,0) = (2)(0) + 0 = 0 = \mathbf{o_F}$$

and

$$f(-\mathbf{v}) = f(-x,-y) = (2)(-x) - y = -2x - y = -(2x+y) = -f(\mathbf{v}).$$

10.2 Linear Mappings and Vector Spaces

Definition 10.8. Let f be a mapping $U \subset E \to F$. The *image of U through f*, indicated with $f(U)$, is a set defined as

$$f(U) = \{\mathbf{w} \in F | \exists \mathbf{u} \in U \ni \text{`} f(\mathbf{u}) = \mathbf{w}\}.$$

Theorem 10.1. *Let $f : E \to F$ be a linear mapping and $(U,+,\cdot)$ be a vector subspace of $(E,+,\cdot)$. Let $f(U)$ be the set of all the transformed vectors of U. It follows that the triple $(f(U),+,\cdot)$ is a vector subspace of $(F,+,\cdot)$.*

Proof. In order to prove that $(f(U),+,\cdot)$ is a vector subspace of $(F,+,\cdot)$ we have to show that the set $f(U)$ is closed with respect to the two composition laws. By definition, the fact that a vector $\mathbf{w} \in (U)$ means that $\exists \mathbf{v} \in U$ such that $f(\mathbf{v}) = \mathbf{w}$.

Thus, if we consider two vectors $\mathbf{w}, \mathbf{w}' \in f(U)$ then

$$\mathbf{w} + \mathbf{w}' = f(\mathbf{v}) + f(\mathbf{v}') = f(\mathbf{v} + \mathbf{v}').$$

Since for hypothesis $(U,+,\cdot)$ is a vector space, then $\mathbf{v} + \mathbf{v}' \in U$. Hence, $f(\mathbf{v} + \mathbf{v}') \in f(U)$. The set $f(U)$ is closed with respect to the internal composition law.

Let us now consider a generic scalar $\lambda \in \mathbb{K}$ and calculate

$$\lambda \mathbf{w} + \lambda f(\mathbf{v}) = f(\lambda \mathbf{v}).$$

Since for hypothesis $(U,+,\cdot)$ is a vector space, then $\lambda \mathbf{v} \in U$. Hence, $f(\lambda \mathbf{v}) \in f(U)$. The set $f(U)$ is closed with respect to the external composition law.

Since the set $f(U)$ is closed with respect to both the composition laws the triple $(f(U),+,\cdot)$ is a vector subspace of $(F,+,\cdot)$. $\square$

Example 10.24. Let us consider the vector space $(\mathbb{R},+,\cdot)$ and the linear mapping $f : \mathbb{R} \to \mathbb{R}$ defined as $f(x) = 5x$.

The theorem above states that $(f(\mathbb{R}),+,\cdot)$ is also a vector space. In this case, the application of the theorem is straightforward since $f(\mathbb{R}) = \mathbb{R}$ and $(\mathbb{R},+,\cdot)$ is clearly a vector space.

Example 10.25. Let us consider the vector space $(\mathbb{R}^2,+,\cdot)$ and the linear mapping $f : \mathbb{R}^2 \to \mathbb{R}$ defined as $f(x) = 6x + 4y$.

In this case $f(\mathbb{R}^2) = \mathbb{R}$ and $(\mathbb{R},+,\cdot)$ is a vector space.

Example 10.26. Let us consider the vector space $(U,+,\cdot)$ where

$$U = \{(x,y,z) \in \mathbb{R}^3 \mid x + 2y + z = 0\}$$

and the linear mapping $f : \mathbb{R}^3 \to \mathbb{R}^2$ defined as

$$f(x,y,z) = (3x + 2y, 4y + 5z).$$

The set U, as we know from Chap. 8, can be interpreted as a plane of the space passing through the origin of the reference system. The linear mapping f simply projects the points of this plane into another plane, that is the two-dimensional space $\mathbb{R}^2$. Thus, $f(U)$ is a plane passing through the origin $(0,0,0)$ which clearly is a vector space.

Example 10.27. Let us consider the mapping $f : \mathbb{R}^2 \to \mathbb{R}^2$

$$f(x,y) = (x + y, x - y)$$

and the set

$$U = \{(x,y) \in \mathbb{R}^2 \mid 2x - y = 0\}.$$

It can be easily seem that U is a line passing through the origin and then $(U, +, \cdot)$ is a vector space.

This set can be represented by means of the vector

$$U = \alpha (1, 2)$$

with $\alpha \in \mathbb{R}$.

Let us now calculate $f(U)$ by replacing (x, y) with $(\alpha, 2\alpha)$:

$$f(U) = (\alpha + 2\alpha, \alpha - 2\alpha) = (3\alpha, -\alpha) = \alpha(3, -1)$$

that is a line passing through the origin. Hence also

$$(f(U), +, \cdot)$$

is a vector space.

Corollary 10.1. *Let $f : E \to F$ be a linear mapping. If $f(E) = \text{Im}(f)$ then $\text{Im}(f)$ is a vector subspace of $(F, +, \cdot)$.*

Definition 10.9. Let f be a mapping $E \to W \subset F$. The *inverse image of W through* f, indicated with $f^{-1}(W)$, is a set defined as

$$f^{-1}(W) = \{\mathbf{u} \in E | f(\mathbf{u}) \in W\}.$$

Theorem 10.2. *Let $f : E \to F$ be a linear mapping. If $(W, +, \cdot)$ is a vector subspace of $(F, +, \cdot)$, then $(f^{-1}(W), +, \cdot)$ is a vector subspace of $(E, +, \cdot)$.*

Proof. In order to prove that $(f^{-1}(W), +, \cdot)$ is a vector subspace of $(E, +, \cdot)$ we have to prove the closure of $f^{-1}(W)$ with respect to the two composition laws. If a vector $\mathbf{v} \in f^{-1}(W)$ then $f(\mathbf{v}) \in W$.

We can write for the linearity of f

$$f(\mathbf{v} + \mathbf{v}') = f(\mathbf{v}) + f(\mathbf{v}').$$

Since $(W, +, \cdot)$ is a vector space, $f(\mathbf{v}) + f(\mathbf{v}') \in W$. Hence, $f(\mathbf{v} + \mathbf{v}') \in W$ and $\mathbf{v} + \mathbf{v}' \in f^{-1}(W)$. Thus, the set $f^{-1}(W)$ is closed with respect to the first composition law.

Let us consider a generic scalar $\lambda \in \mathbb{K}$ and calculate

$$f(\lambda \mathbf{v}) = \lambda f(\mathbf{v}).$$

Since $(W, +, \cdot)$ is a vector space, $\lambda f(\mathbf{v}) \in W$. Since $f(\lambda \mathbf{v}) \in W$ the $\lambda \mathbf{v} \in f^{-1}(W)$. Thus, the set $f^{-1}(W)$ is closed with respect to the second composition law.

Hence, $(f^{-1}(W), +, \cdot)$ is a vector subspace of $(E, +, \cdot)$. $\square$

10.3 Endomorphisms and Kernel

Definition 10.10. Let f be a linear mapping $E \to F$. If $E = F$, i.e. $f : E \to E$, the linear mapping is said *endomorphism*.

Example 10.28. The linear mapping $f : \mathbb{R} \to \mathbb{R}$, $f(x) = 2x$ is an endomorphism since both the sets are $\mathbb{R}$.

Example 10.29. The linear mapping $f : \mathbb{R}^2 \to \mathbb{R}$, $f(x,y) = 2x - 4y$ is not an endomorphism since $\mathbb{R}^2 \neq \mathbb{R}$.

Example 10.30. The linear mapping $f : \mathbb{R}^2 \to \mathbb{R}^2$, $f(x,y) = (2x + 3y, 9x - 2y)$ is an endomorphism.

Definition 10.11. A *null mapping* $O : E \to F$ is a mapping defined in the following way:
$$\forall \mathbf{v} \in E : O(\mathbf{v}) = \mathbf{o_F}.$$

It can easily be proved that this mapping is linear.

Example 10.31. The linear mapping $f : \mathbb{R} \to \mathbb{R}$, $f(x) = 0$ is a null mapping.

Example 10.32. The linear mapping $f : \mathbb{R}^2 \to \mathbb{R}$, $f(x,y) = 0$ is a null mapping.

Example 10.33. The linear mapping $f : \mathbb{R}^2 \to \mathbb{R}^2$, $f(x,y) = (0,0)$ is a null mapping.

Definition 10.12. An *identity mapping* $I : E \to F$ is a mapping defined in the following way:
$$\forall \mathbf{v} \in E : I(\mathbf{v}) = \mathbf{v}.$$

It can easily be proved that this mapping is linear and is an endomorphism.

Example 10.34. The linear mapping $f : \mathbb{R} \to \mathbb{R}$, $f(x) = x$ is an identity mapping.

Example 10.35. The linear mapping $f : \mathbb{R}^2 \to \mathbb{R}^2$, $f(x,y) = (x,y)$ is an identity mapping.

Example 10.36. If we consider a linear mapping $f : \mathbb{R}^2 \to \mathbb{R}$, we cannot define an identity mapping since. Of course, for a vector $(x,y) \in \mathbb{R}^2$ we cannot have $f(x,y) = (x,y)$ as it would not be defined within $\mathbb{R}$. This explains why identity mappings make sense only for endomorphisms.

Proposition 10.3. *Let $f : E \to E$ be an endomorphism. If $\mathbf{v_1}, \mathbf{v_2}, \ldots, \mathbf{v_n} \in E$ are linearly dependent then $f(\mathbf{v_1}), f(\mathbf{v_2}), \ldots, f(\mathbf{v_n}) \in F$ are also linearly dependent.*

Proof. If $\mathbf{v_1}, \mathbf{v_2}, \ldots, \mathbf{v_n} \in E$ are linearly dependent then $\exists$ scalars $\lambda_1, \lambda_2, \ldots, \lambda_n \neq 0, 0, \ldots, 0$ such that
$$\mathbf{o_E} = \lambda_1 \mathbf{v_1} + \lambda_2 \mathbf{v_2} + \ldots + \lambda_n \mathbf{v_n}.$$

Let us apply the linear mapping to this equation:

$$f(\mathbf{o_E}) = f(\lambda_1 \mathbf{v_1} + \lambda_2 \mathbf{v_2} + \ldots + \lambda_n \mathbf{v_n}).$$

For the linearity of the mapping we can write

$$f(\lambda_1 \mathbf{v_1} + \lambda_2 \mathbf{v_2} + \ldots + \lambda_n \mathbf{v_n}) =$$
$$= f(\lambda_1 \mathbf{v_1}) + f(\lambda_2 \mathbf{v_2}) + \ldots + f(\lambda_n \mathbf{v_n}) =$$
$$= \lambda_1 f(\mathbf{v_1}) + \lambda_2 f(\mathbf{v_2}) + \ldots + \lambda_n f(\mathbf{v_n}) = f(\mathbf{o_E}).$$

Since for Proposition 10.1 $f(\mathbf{o_E}) = \mathbf{o_F}$ we have

$$\mathbf{o_F} = \lambda_1 f(\mathbf{v_1}) + \lambda_2 f(\mathbf{v_2}) + \ldots + \lambda_n f(\mathbf{v_n})$$

with $\lambda_1, \lambda_2, \ldots, \lambda_n \neq 0, 0, \ldots, 0$, i.e. $f(\mathbf{v_1}), f(\mathbf{v_2}), \ldots, f(\mathbf{v_n})$ are linearly dependent.
□

Example 10.37. To understand the meaning of the proposition above let us consider the following vectors of $\mathbb{R}^2$:

$$\mathbf{v_1} = (0, 1)$$
$$\mathbf{v_2} = (0, 4)$$

and the linear mapping $f : \mathbb{R}^2 \to \mathbb{R}^2$ defined as

$$f(x, y) = (x + y, x + 2y).$$

The two vectors are clearly linearly dependent as $\mathbf{v_2} = 4\mathbf{v_1}$. Let us check the linear dependence of the mapped vectors:

$$f(\mathbf{v_1}) = (1, 2)$$
$$f(\mathbf{v_2}) = (4, 8).$$

These vectors are also linearly dependent since $f(\mathbf{v_2}) = 4f(\mathbf{v_1})$.

Remark 10.1. It must be observed that there is no dual proposition for linearly independent vectors, i.e. if $\mathbf{v_1}, \mathbf{v_2}, \ldots, \mathbf{v_n} \in E$ are linearly independent then we cannot draw conclusions on the linear dependence of $f(\mathbf{v_1}), f(\mathbf{v_2}), \ldots, f(\mathbf{v_n}) \in F$.

Example 10.38. Let us now consider the following linearly independent vectors of $\mathbb{R}^2$:

$$\mathbf{v_1} = (0, 1)$$
$$\mathbf{v_2} = (1, 0)$$

and the linear mapping: $f : \mathbb{R}^2 \to \mathbb{R}^2$ defined as

$$f(x, y) = (x + y, 2x + 2y).$$

We obtain
$$f(\mathbf{v_1}) = (1,2)$$
$$f(\mathbf{v_2}) = (1,2).$$

We have a case where the transformed of linearly independent vectors are linearly dependent.

Definition 10.13. Let $f : E \to F$ be a linear mapping. The *kernel* of f is the set

$$\ker(f) = \{\mathbf{v} \in E | f(\mathbf{v}) = \mathbf{o_F}\}.$$

The following examples better clarify the concept of kernel.

Example 10.39. Let us find the kernel of the linear mapping $f : \mathbb{R} \to \mathbb{R}$ defined as $f(x) = 5x$ In this case the kernel ker $= \{0\}$. This can be easily calculated imposing $f(x) = 0$, i.e. $5x = 0$.

Example 10.40. Let us consider the linear mapping $f : \mathbb{R}^2 \to \mathbb{R}$ defined as $f(x,y) = 5x - y$. To find the kernel means to find the (x,y) values such that $f(x,y) = 0$, i.e. those (x,y) values that satisfy the equation

$$5x - y = 0.$$

This is an equation in two variables. For the Rouché Capelli Theorem this equation has ∞^1 solutions. These solutions are all proportional to $(\alpha, 5\alpha)$, $\forall \alpha \in \mathbb{R}$. This means that the kernel of this linear mapping is the composed of the point a line belonging to $\mathbb{R}^2$ and passing through the origin.

More formally, the kernel is

$$\ker(f) = \{(\alpha, 5\alpha), \alpha \in \mathbb{R}\}.$$

Example 10.41. Let us consider the linear mapping $f : \mathbb{R}^3 \to \mathbb{R}^3$ defined as

$$f(x,y,z) = (x+y+z, x-y-z, x+y+2z).$$

The kernel of this mapping is the set of (x,y,z) such that the mapping is equal to $\mathbf{o_F}$. This means that the kernel of this mapping is the set of (x,y,z) such that

$$\begin{cases} x+y+z = 0 \\ x-y-z = 0 \\ x+y+2z = 0. \end{cases}$$

It can be easily verified that this homogeneous system of linear equations is determined. Thus,

$$\ker(f) = \{(0,0,0)\} = \{\mathbf{o_E}\}.$$

Example 10.42. Let us consider now the linear mapping $f : \mathbb{R}^3 \to \mathbb{R}^3$ defined as

$$f(x,y,z) = (x+y+z, x-y-z, 2x+2y+2z).$$

To find the kernel means to solve the following system of linear equations:

$$\begin{cases} x+y+z=0 \\ x-y-z=0 \\ 2x+2y+2z=0. \end{cases}$$

We can easily verify that

$$\det \begin{pmatrix} 1 & 1 & 1 \\ 1 & -1 & -1 \\ 2 & 2 & 2 \end{pmatrix} = 0$$

and the rank of the system is $\rho = 2$. Thus, this system is undetermined and has ∞^1 solutions. If we pose $x = \alpha$ we find out that the infinite solutions of the system are $\alpha(0,-1,1), \forall \alpha \in \mathbb{R}$. Thus, the kernel of the mapping is

$$\ker(f) = \{\alpha(0,1,-1), \alpha \in \mathbb{R}\}.$$

Theorem 10.3. Let $f : E \to F$ be a linear mapping. The triple $(\ker(f),+,\cdot)$ is a vector subspace of $(E,+,\cdot)$.

Proof. Let us consider two vectors $\mathbf{v}, \mathbf{v}' \in \ker(f)$. If a vector $\mathbf{v} \in \ker(f)$ then $f(\mathbf{v}) = \mathbf{o_F}$. Thus,

$$f(\mathbf{v}+\mathbf{v}') = f(\mathbf{v}) + f(\mathbf{v}') = \mathbf{o_F} + \mathbf{o_F} = \mathbf{o_F}$$

and $f(\mathbf{v}+\mathbf{v}') \in \ker(f)$. Thus, $\ker(f)$ is closed with respect to the first composition law. Let us consider a generic scalar $\lambda \in \mathbb{K}$ and calculate

$$f(\lambda \mathbf{v}) = \lambda f(\mathbf{v}) = \lambda \mathbf{o_F} = \mathbf{o_F}.$$

Hence, $f(\lambda \mathbf{v}) \in \ker(f)$ and $\ker(f)$ is closed with respect to the second composition law.

This means that $(\ker(f),+,\cdot)$ is a vector subspace of $(E,+,\cdot)$. $\square$

As shown in the examples above, the calculation of $\ker f$ can always be considered as the solution of a homogeneous system of linear equations. Thus, $\ker f$ always contains the null vector. This confirms the statement of the theorem above, i.e. $(\ker(f),+,\cdot)$ is a vector space. The next example further clarifies this fact.

Example 10.43. Let us consider again the linear mapping $f : \mathbb{R}^3 \to \mathbb{R}^3$ defined as

$$f(x,y,z) = (x+y+z, x-y-z, 2x+2y+2z).$$

We already know that

$$\ker(f) = \{\alpha(0,1,-1), \alpha \in \mathbb{R}\}.$$

This means at first that $\ker(f) \subset \mathbb{R}^3$ and, more specifically, can be seen as a line of the space passing through the origin of the reference system. From its formulation, it can be observed that $(\ker(f), +, \cdot)$ is a vector space having dimension one.

Theorem 10.4. *Let $f : E \to F$ be a linear mapping and $\mathbf{u}, \mathbf{v} \in E$. It follows that*

$$f(\mathbf{u}) = f(\mathbf{v})$$

if and only if

$$\mathbf{u} - \mathbf{v} \in \ker(f).$$

Proof. If $f(\mathbf{u}) = f(\mathbf{v})$ then

$$f(\mathbf{u}) - f(\mathbf{v}) = \mathbf{o_F} \Rightarrow f(\mathbf{u}) + f(-\mathbf{v}) = \mathbf{o_F} \Rightarrow$$
$$\Rightarrow f(\mathbf{u} - \mathbf{v}) = \mathbf{o_F}.$$

For the definition of kernel $\mathbf{u} - \mathbf{v} \in \ker(f)$. $\square$
    If $\mathbf{u} - \mathbf{v} \in \ker(f)$ then

$$f(\mathbf{u} - \mathbf{v}) = \mathbf{o_F} \Rightarrow f(\mathbf{u}) - f(\mathbf{v}) = \mathbf{o_F} \Rightarrow$$
$$\Rightarrow f(\mathbf{u}) = f(\mathbf{v}). \square$$

Example 10.44. Again for the linear mapping $f : \mathbb{R}^3 \to \mathbb{R}^3$ defined as

$$f(x,y,z) = (x+y+z, x-y-z, 2x+2y+2z)$$

we consider the following two vectors $\mathbf{u}, \mathbf{v}$, e.g.

$$\mathbf{u} = (6,4,-7)$$
$$\mathbf{v} = (6,5,-8).$$

Let us calculate their difference

$$\mathbf{u} - \mathbf{v} = (6,4,-7) - (6,5,-8) = (0,-1,1)$$

which belongs to $\ker(f)$.
    Let us calculate the mapping of these two vectors:

$$f(\mathbf{v}) = (6+4-7, 6-4+7, 12+8-14) = (3,9,6)$$
$$f(\mathbf{v'}) = (6+5-8, 6-5+8, 12+10-16) = (3,9,6).$$

As stated from the theorem the mapped values are the same.

Theorem 10.5. *Let $f : E \to F$ be a linear mapping. The mapping f is injective if and only if*

$$\ker(f) = \{\mathbf{o_E}\}.$$

Proof. Let us assume that f is injective and, by contradiction, let us assume that

$$\exists \mathbf{v} \in \ker(f)$$

with $\mathbf{v} \neq \mathbf{o_E}$.

For definition of kernel

$$\forall \mathbf{v} \in \ker(f) : f(\mathbf{v}) = \mathbf{o_F}.$$

On the other hand, for Proposition 10.1, $f(\mathbf{o_E}) = \mathbf{o_F}$. Thus,

$$f(\mathbf{v}) = f(\mathbf{o_E}).$$

Since f is injective, for definition of injective mapping this means that $\mathbf{v} = \mathbf{o_E}$. We have reached a contradiction.

Hence, every vector $\mathbf{v}$ in the kernel is $\mathbf{o_E}$, i.e.

$$\ker(f) = \{\mathbf{o_E}\} . \square$$

Let us assume that $\ker(f) = \{\mathbf{o_E}\}$ and let us consider two vectors $\mathbf{u}, \mathbf{v} \in E$ such that $f(\mathbf{u}) = f(\mathbf{v})$. It follows that

$$f(\mathbf{u}) = f(\mathbf{v}) \Rightarrow f(\mathbf{u}) - f(\mathbf{v}) = \mathbf{o_F}.$$

It follows from the linearity of f that

$$f(\mathbf{u} - \mathbf{v}) = \mathbf{o_F}.$$

For the definition of kernel

$$\mathbf{u} - \mathbf{v} \in \ker(f).$$

However, since for hypothesis

$$\ker(f) = \{\mathbf{o_E}\}$$

then

$$\mathbf{u} - \mathbf{v} = \mathbf{o_E}.$$

Hence, $\mathbf{u} = \mathbf{v}$.

Since, $\forall \mathbf{u}, \mathbf{v} \in E$ such that $f(\mathbf{u}) = f(\mathbf{v})$ it follows that $\mathbf{u} = \mathbf{v}$ then f is injective. $\square$

Example 10.45. Let us consider once again, the mapping $f : \mathbb{R}^3 \to \mathbb{R}^3$ defined as

$$f(x, y, z) = (x + y + z, x - y - z, 2x + 2y + 2z).$$

We know that

$$\ker(f) = \{\alpha(0, 1, -1)\} \neq \mathbf{o_E} = (0, 0, 0).$$

From the theorem above we expect that this mapping is not injective. A mapping is injective if it always occurs that the transformed of different vectors are different.

In this case if we consider the following vectors

$$\mathbf{u} = (0, 8, -8)$$
$$\mathbf{v} = (0, 9, -9)$$

it follows that $\mathbf{u} \neq \mathbf{v}$. The transformed vectors are

$$f(\mathbf{u}) = (0, 0, 0)$$
$$f(\mathbf{v}) = (0, 0, 0).$$

Thus, in correspondence to $\mathbf{u} \neq \mathbf{v}$ we have $f(\mathbf{u}) = f(\mathbf{v})$. In other words this mapping is not injective, as expected.

Example 10.46. If we consider the mapping $f : \mathbb{R}^3 \to \mathbb{R}^3$ defined as

$$f(x, y, z) = (x + y + z, x - y - z, x + y + 2z)$$

we know that its kernel is $\mathbf{o_E}$. We can observe that if we take two different vectors and calculate their transformed vectors we will never obtain the same vector. This mapping is injective.

As a further comment we can say that if the homogeneous system of linear equations associated with a mapping is determined (i.e. its only solution is the null vector) then the kernel is only the null vector and the mapping is injective.

Example 10.47. A linear mapping $f : \mathbb{R} \to \mathbb{R}$ defined as

$$f(x) = mx$$

with m finite and $m \neq 0$ is injective. It can be observed that its kernel is

$$\ker(f) = \{0\}.$$

In the special case $m = \infty$, f is not a mapping while in the case $m = 0$, f is a mapping and is also linear. The function is not injective and its kernel is

$$\ker(f) = \{\alpha(1), \alpha \in \mathbb{R}\},$$

i.e. the entire set $\mathbb{R}$.

This result can be achieved looking at the equation $mx = 0$ as a system of one equation in one variable. If $m = 0$ the matrix associated with system is singular and has null rank. Hence the system has ∞^1 solutions

Intuitively, we can observe that for $m \neq 0$ the mapping transforms a line into a line while for $m = 0$ the mapping transforms a line into a constant (a point). Hence we can see that there is a relation between the kernel and the deep meaning of the mapping. This topic will be discussed more thoroughly below.

Theorem 10.6. *Let $f : E \to F$ be a linear mapping. Let $\mathbf{v_1}, \mathbf{v_2}, \ldots, \mathbf{v_n}$ be n linearly independent vectors $\in E$. If f is injective then $f(\mathbf{v_1}), f(\mathbf{v_2}), \ldots, f(\mathbf{v_n})$ are also linearly independent vectors $\in F$.*

Proof. Let us assume, by contradiction that $\exists \lambda_1, \lambda_2, \ldots, \lambda_n \neq 0, 0, \ldots, 0$ such that

$$\mathbf{o_F} = \lambda_1 f(\mathbf{v_1}) + \lambda_2 f(\mathbf{v_2}) + \cdots + \lambda_n f(\mathbf{v_n}).$$

For the Proposition 10.1 and linearity of f we can write this expression as

$$f(\mathbf{o_E}) = f(\lambda_1 \mathbf{v_1} + \lambda_2 \mathbf{v_2} + \cdots + \lambda_n \mathbf{v_n}).$$

Since for hypothesis f is injective, it follows that

$$\mathbf{o_E} = \lambda_1 \mathbf{v_1} + \lambda_2 \mathbf{v_2} + \cdots + \lambda_n \mathbf{v_n}$$

with $\lambda_1, \lambda_2, \ldots, \lambda_n \neq 0, 0, \ldots, 0$.

This is impossible because $\mathbf{v_1}, \mathbf{v_2}, \ldots, \mathbf{v_n}$ are linearly independent. Hence we reached a contradiction and $f(\mathbf{v_1}), f(\mathbf{v_2}), \ldots, f(\mathbf{v_n})$ must be linearly independent.
□

Example 10.48. Let us consider the injective mapping $f : \mathbb{R}^3 \to \mathbb{R}^3$ defined as

$$f(x,y,z) = (x+y+z, x-y-z, x+y+2z)$$

and the following linearly independent vectors of $\mathbb{R}^3$:

$$\mathbf{u} = (1,0,0)$$
$$\mathbf{v} = (0,1,0)$$
$$\mathbf{w} = (0,0,1).$$

The transformed of these vectors are

$$f(\mathbf{u}) = (1,1,1)$$
$$f(\mathbf{v}) = (1,-1,1)$$
$$f(\mathbf{w}) = (1,-1,2).$$

Let us check their linear dependence by finding, if they exist, the values of λ, μ, ν such that

$$\mathbf{o} = \lambda f(\mathbf{u}) + \mu f(\mathbf{v}) + \nu f(\mathbf{w}).$$

This is equivalent to solving the following homogeneous system of linear equations:

$$\begin{cases} \lambda + \mu + \nu = 0 \\ \lambda - \mu - \nu = 0 \\ \lambda + \mu + 2\nu = 0. \end{cases}$$

The system is determined; thus, its only solution is $(0,0,0)$. It follows that the vectors are linearly independent.

We can put into relationship Proposition 10.3 and Theorem 10.6: while linearly dependent is always preserved by a linear mapping, linearly independent is preserved only by injective mapping, i.e. by mappings whose kernel is null.

10.4 Rank and Nullity of Linear Mappings

Definition 10.14. Let $f : E \to F$ be a linear mapping and $\text{Im}\,(f)$ its image. The dimension of the image $\dim\,(\text{Im}\,(f))$ is said *rank* of a mapping.

Definition 10.15. Let $f : E \to F$ be a linear mapping and $\ker\,(f)$ its kernel. The dimension of the image $\dim\,(\ker\,(f))$ is said *nullity* of a mapping.

Example 10.49. If we consider again the linear mapping $f : \mathbb{R}^3 \to \mathbb{R}^3$ defined as

$$f\,(x,y,z) = (x+y+z, x-y-z, 2x+2y+2z)$$

and its kernel

$$\ker\,(f) = \{\alpha\,(0,1,-1)\,, \alpha \in \mathbb{R}\}$$

we can immediately see that $(\ker\,(f)\,, +, \cdot)$ is a vector space having dimension one. Hence, the nullity of the mapping is one.

In order to approach the calculation of the image, let us intuitively consider that this mapping transforms vectors of $\mathbb{R}^3$ into vectors that are surely linearly dependent as the third component is always twice the value of the first component. This means that this mapping transforms points of the space into points of a plane, i.e. $(\text{Im}\,(f)\,, +, \cdot)$ is a vector space having dimension two. The rank of this mapping is two.

Theorem 10.7. Rank-Nullity Theorem. *Let $f : E \to F$ be a linear mapping where $(E, +, \cdot)$ and $(F, +, \cdot)$ are vector spaces defined on the same scalar field $\mathbb{K}$. Let $(E, +, \cdot)$ be a finite-dimensional vector space whose dimension is $\dim\,(E) = n$.*

Under these hypotheses the sum of rank and nullity of a mapping is equal to the dimension of the vector space $(E, +, \cdot)$:

$$\dim\,(\ker\,(f)) + \dim\,(\text{Im}\,(f)) = \dim\,(E)\,.$$

Proof. This proof is structured into three parts:

- Well-posedness of the equality
- Special (degenerate) cases
- General case

Well-posedness of the Equality

At first, let us prove that the equality considers only finite numbers. In order to prove this fact, since

$$\dim\,(E) = n$$

is a finite number we have to prove that also $\dim\,(\ker\,(f))$ and $\dim\,(\text{Im}\,(f))$ are finite numbers.

Since, by definition of kernel, the $\ker\,(f)$ is a subset of E, then

$$\dim\,(\ker\,(f)) \leq \dim\,(E) = n\,.$$

Hence, $\dim\,(\ker\,(f))$ is a finite number.

Since $(E, +, \cdot)$ is finite-dimensional,

$$\exists \text{ a basis } B = \{\mathbf{e_1}, \mathbf{e_2}, \ldots, \mathbf{e_n}\}$$

such that every vector $\mathbf{v} \in E$ can be expressed as

$$\mathbf{v} = \lambda_1 \mathbf{e_1} + \lambda_2 \mathbf{e_2} + \ldots + \lambda_n \mathbf{e_n}.$$

Let us apply the linear transformation f to both the terms in the equation

$$f(\mathbf{v}) = f(\lambda_1 \mathbf{e_1} + \lambda_2 \mathbf{e_2} + \ldots + \lambda_n \mathbf{e_n}) =$$
$$= \lambda_1 f(\mathbf{e_1}) + \lambda_2 f(\mathbf{e_2}) + \ldots + \lambda_n f(\mathbf{e_n}).$$

Thus,

$$\operatorname{Im}(f) = L(f(\mathbf{e_1}), f(\mathbf{e_2}), \ldots, f(\mathbf{e_n})).$$

It follows that

$$\dim(\operatorname{Im}(f)) \leq n.$$

Hence, the equality contains only finite numbers.

Special Cases

Let us consider now two special cases:

1. $\dim(\ker(f)) = 0$
2. $\dim(\ker(f)) = n$

If $\dim(\ker(f)) = 0$, i.e. $\ker(f) = \{\mathbf{o_E}\}$, then f injective. Hence, if a basis of $(E, +, \cdot)$ is

$$B = \{\mathbf{e_1}, \mathbf{e_2}, \ldots, \mathbf{e_n}\}$$

also the vectors

$$f(\mathbf{e_1}), f(\mathbf{e_2}), \ldots, f(\mathbf{e_n}) \in \operatorname{Im}(f)$$

are linearly independent for Theorem 10.6. Since these vectors also span $(\operatorname{Im}(f), +, \cdot)$, they compose a basis.

It follows that

$$\dim(\operatorname{Im}(f)) = n$$

and $\dim(\ker(f)) + \dim(\operatorname{Im}(f)) = \dim(E)$.

If $\dim(\ker(f)) = n$, i.e. $\ker(f) = E$. Hence,

$$\forall \mathbf{v} \in E : f(\mathbf{v}) = \mathbf{o_F}$$

and

$$\operatorname{Im}(f) = \{\mathbf{o_F}\}.$$

Thus,

$$\dim(\operatorname{Im}(f)) = 0$$

and $\dim(\ker(f)) + \dim(\operatorname{Im}(f)) = \dim(E)$.

General Case

In the remaining cases, $\dim(\ker(f))$ is $\neq 0$ and $\neq n$. We can write

$$\dim(\ker(f)) = r \Rightarrow \exists B_{\ker} = \{\mathbf{u_1}, \mathbf{u_2}, \ldots \mathbf{u_r}\}$$
$$\dim(\mathrm{Im}(f)) = s \Rightarrow \exists B_{\mathrm{Im}} = \{\mathbf{w_1}, \mathbf{w_2}, \ldots \mathbf{w_s}\}$$

with $0 < r < n$ and $0 < s < n$.

By definition of image

$$\mathbf{w_1} \in \mathrm{Im}(f) \Rightarrow \exists \mathbf{v_1} \in E | f(\mathbf{v_1}) = \mathbf{w_1}$$
$$\mathbf{w_2} \in \mathrm{Im}(f) \Rightarrow \exists \mathbf{v_2} \in E | f(\mathbf{v_2}) = \mathbf{w_2}$$
$$\cdots$$
$$\mathbf{w_s} \in \mathrm{Im}(f) \Rightarrow \exists \mathbf{v_s} \in E | f(\mathbf{v_s}) = \mathbf{w_s}.$$

Moreover, $\forall \mathbf{x} \in E$, the corresponding linear mapping $f(\mathbf{x})$ can be expressed as linear combination of the elements of B_{Im} by means of the scalars $h_1, h_2, \ldots, h_s$

$$f(\mathbf{x}) = h_1 \mathbf{w_1} + h_2 \mathbf{w_2} + \cdots + h_s \mathbf{w_s} =$$
$$= h_1 f(\mathbf{v_1}) + h_2 f(\mathbf{v_2}) + \cdots + h_s f(\mathbf{v_s}) =$$
$$= f(h_1 \mathbf{v_1} + h_2 \mathbf{v_2} + \cdots + h_s \mathbf{v_s}).$$

We know that f is not injective because $r \neq 0$. On the other hand, for the Theorem 10.4

$$\mathbf{u} = \mathbf{x} - h_1 \mathbf{v_1} - h_2 \mathbf{v_2} - \cdots - h_s \mathbf{v_s} \in \ker(f).$$

If we express $\mathbf{u}$ as a linear combination of the elements of $B_{\ker}$ by means of the scalars $l_1, l_2, \ldots, l_r$, we can rearrange the equality as

$$\mathbf{x} = h_1 \mathbf{v_1} + h_2 \mathbf{v_2} + \cdots + h_s \mathbf{v_s} + l_1 \mathbf{u_1} + l_2 \mathbf{u_2} + \cdots + l_r \mathbf{u_r}.$$

Since $\mathbf{x}$ has been arbitrarily chosen, we can conclude that the vectors

$$\mathbf{v_1}, \mathbf{v_2}, \ldots, \mathbf{v_s}, \mathbf{u_1}, \mathbf{u_2}, \ldots, \mathbf{u_r}$$

span E:

$$E = L(\mathbf{v_1}, \mathbf{v_2}, \ldots, \mathbf{v_s}, \mathbf{u_1}, \mathbf{u_2}, \ldots, \mathbf{u_r}).$$

Let us check the linear independence of these vectors. Let us consider the scalars $a_1, a_2, \ldots a_s, b_1, b_2, \ldots, b_r$ and let us express the null vector as linear combination of the other vectors

$$\mathbf{o_E} = a_1 \mathbf{v_1} + a_2 \mathbf{v_2} + \cdots + a_s \mathbf{v_s} + b_1 \mathbf{u_1} + b_2 \mathbf{u_2} + \cdots + b_r \mathbf{u_r}.$$

Let us calculate the linear mapping of this equality and apply the linear properties

$$f(\mathbf{o_E}) = \mathbf{o_F} = f(a_1 \mathbf{v_1} + a_2 \mathbf{v_2} + \cdots + a_s \mathbf{v_s} + b_1 \mathbf{u_1} + b_2 \mathbf{u_2} + \cdots + b_r \mathbf{u_r}) =$$
$$= a_1 f(\mathbf{v_1}) + a_2 f(\mathbf{v_2}) + \cdots + a_s f(\mathbf{v_s}) + b_1 f(\mathbf{u_1}) + b_2 f(\mathbf{u_2}) + \cdots + b_r f(\mathbf{u_r}) =$$
$$= a_1 \mathbf{w_1} + a_2 \mathbf{w_2} + \cdots + a_s \mathbf{w_s} + b_1 f(\mathbf{u_1}) + b_2 f(\mathbf{u_2}) + \cdots + b_r f(\mathbf{u_r}).$$

We know that since $\mathbf{u}_1, \mathbf{u}_2, \ldots, \mathbf{u}_r \in \ker(f)$ then

$$f(\mathbf{u}_1) = \mathbf{o}_F$$
$$f(\mathbf{u}_2) = \mathbf{o}_F$$
$$\ldots$$
$$f(\mathbf{u}_r) = \mathbf{o}_F.$$

It follows that

$$f(\mathbf{o}_E) = \mathbf{o}_F = a_1\mathbf{w}_1 + a_2\mathbf{w}_2 + \cdots + a_s\mathbf{w}_s.$$

Since $\mathbf{w}_1, \mathbf{w}_2, \ldots, \mathbf{w}_s$ compose a basis, they are linearly independent. It follows that $a_1, a_2, \ldots, a_s = 0, 0, \ldots, 0$ and that

$$\mathbf{o}_E = a_1\mathbf{v}_1 + a_2\mathbf{v}_2 + \cdots + a_s\mathbf{v}_s + b_1\mathbf{u}_1 + b_2\mathbf{u}_2 + \cdots + b_r\mathbf{u}_r =$$
$$= b_1\mathbf{u}_1 + b_2\mathbf{u}_2 + \cdots + b_r\mathbf{u}_r.$$

Since $\mathbf{u}_1, \mathbf{u}_2, \ldots, \mathbf{u}_r$ compose a basis, they are linearly independent. Hence, also $b_1, b_2, \ldots, b_r = 0, 0, \ldots, 0$.

It follows that $\mathbf{v}_1, \mathbf{v}_2, \ldots, \mathbf{v}_s, \mathbf{u}_1, \mathbf{u}_2, \ldots, \mathbf{u}_r$ are linearly independent. Since these vectors also span E, they compose a basis. We know, for the hypothesis, that $\dim(E) = n$ and we know that this basis is composed of $r + s$ vectors, that is $\dim(\ker(f)) + \dim(\mathrm{Im}(f))$. Hence,

$$\dim(\ker(f)) + \dim(\mathrm{Im}(f)) = r + s = n = \dim(E). \quad \square$$

Example 10.50. The rank-nullity theorem expresses a relation among $\dim(\ker(f))$, $\dim(\mathrm{Im}(f))$, and $\dim(E)$. Usually, $\dim(\mathrm{Im}(f))$ is the hardest to calculate and this theorem allows an easy way to find it.

Let us consider the two linear mappings $\mathbb{R}^3 \to \mathbb{R}^3$ studied above, which will be renamed f_1 and f_2 to avoid confusion in the notation, i.e.

$$f_1(x,y,z) = (x+y+z, x-y-z, x+y+2z)$$

and

$$f_2(x,y,z) = (x+y+z, x-y-z, 2x+2y+2z).$$

We know that $\ker(f_1) = \{(0,0,0)\}$, i.e. $\dim(\ker(f)) = 0$ and $\dim(\mathbb{R}^3) = 3$. For the rank-nullity theorem $\dim(\mathrm{Im}(f)) = 3$. As a geometrical interpretation, this mapping transforms points (vectors) of the (three-dimensional) space into points of the space.

Regarding f_2, we know that $\ker(f_2) = \alpha(0,-1,1)$, i.e. $\dim(\ker(f)) = 1$ and $\dim(\mathbb{R}^3) = 3$. For the rank-nullity theorem $\dim(\mathrm{Im}(f)) = 2$. As stated above, the mapping f_2 transforms points of the space into points of a plane in the space.

Example 10.51. Let us check the rank-nullity theorem for the following mappings:

$$f_1 : \mathbb{R} \to \mathbb{R}$$
$$f_1(x) = 5x$$

and

$$f_2 : \mathbb{R}^2 \to \mathbb{R}$$
$$f_2(x,y) = x+y.$$

Regarding f_1, the detection of the kernel is very straightforward:

$$5x = 0 \Rightarrow x = 0 \Rightarrow \ker(f_1) = \{0\}.$$

It follows that the rank of f_1 is zero. Since $\dim(\mathbb{R}, +, \cdot) = 1$, we have that the nullity, i.e. $\dim(\mathrm{Im}(f_1), +, \cdot)$, is one. This mapping transforms the points of a line (x axis) into another line (having equation $5x$).

Regarding f_2, the kernel is calculated as

$$x+y = 0 \Rightarrow (x,y) = \alpha(1,-1), \alpha \in \mathbb{R}$$

from which it follows that

$$\ker(f_2) = \alpha(1,-1).$$

We can observe that $\dim(\ker(f_2)) = 1$. Since $\dim(\mathbb{R}^2) = 2$, it follows that $\dim(\mathrm{Im}(f_2)) = 1$. This means that the mapping f_2 transforms the points of the plane ($\mathbb{R}^2$) into the points of a line in the plane.

Example 10.52. Let us consider the linear mapping $\mathbb{R}^3 \to \mathbb{R}^3$ defined as

$$f(x,y,z) = (x+2y+z, 3x+6y+3z, 5x+10y+5z).$$

The kernel of this linear mapping is the set of points (x,y,z) such that

$$\begin{cases} x+2y+z = 0 \\ 3x+6y+3z = 0 \\ 5x+10y+5z = 0. \end{cases}$$

It can be checked that the rank of this homogeneous system of linear equations is $\rho = 1$. Thus ∞^2 solutions exists. If we pose $x = \alpha$ and $z = \gamma$ with $\alpha, \gamma \in \mathbb{R}$ we have that the solution of the system of linear equations is

$$(x,y,z) = \left(\alpha, -\frac{\alpha+\gamma}{2}, \gamma \right),$$

that is also the kernel of the mapping:

$$\ker(f) = \left(\alpha, -\frac{\alpha+\gamma}{2}, \gamma \right).$$

It follows that $\dim(\ker(f), +, \cdot) = 2$. Since $\dim(\mathbb{R}^3, +, \cdot) = 3$, it follows from the rank-nullity theorem that $\dim(\operatorname{Im}(f)) = 1$. We can conclude that the mapping f transforms the points of the space $(\mathbb{R}^3)$ into the points of a line of the space.

If we consider endomorphisms $f : \mathbb{R}^3 \to \mathbb{R}^3$ we can have four possible cases:

- the corresponding system of linear equations is determined (has rank $\rho = 3$): the dimension of the kernel is $\dim(\ker(f)) = 0$ and the mapping transforms points of the space into points of the space
- the corresponding system of linear equations has rank $\rho = 2$: the dimension of the kernel is $\dim(\ker(f)) = 1$ and the mapping transforms points of the space into points of a plane in the space
- the corresponding system of linear equations has rank $\rho = 1$: the dimension of the kernel is $\dim(\ker(f)) = 2$ and the mapping transforms points of the space into points of a line in the space
- the corresponding system of linear equations has rank $\rho = 0$ (the mapping is the null mapping): the dimension of the kernel is $\dim(\ker(f)) = 3$ and the mapping transforms points of the space into a constant, that is $(0,0,0)$

Corollary 10.2. *Let $f : E \to E$ be an endomorphism where $(E, +, \cdot)$ is a finite-dimensional vector space.*

- *If f is injective then it is also surjective.*
- *If f is surjective then it is also injective.*

Proof. Let $\dim(E) = n$. If f is injective $\ker(f) = \{\mathbf{o_E}\}$. Thus,

$$\dim(\ker(f)) = 0$$

and for the Rank-Nullity Theorem

$$n = \dim(E) = \dim(\ker(f)) + \dim(\operatorname{Im}(f)) = \dim(\operatorname{Im}(f))$$

Since $f : E \to E$ is an endomorphism, $\operatorname{Im}(f) \subseteq E$, the fact that $\dim(E) = \dim(\operatorname{Im}(f))$ implicates that $\operatorname{Im}(f) = E$, i.e. f is surjective. $\square$

If f is surjective

$$\dim(\operatorname{Im}(f)) = n = \dim(E)$$

For the Rank-Nullity Theorem

$$n = \dim(E) = \dim(\ker(f)) + \dim(\operatorname{Im}(f)).$$

It follows that $\dim(\ker(f)) = 0$ that is equivalent to say that f is injective. $\square$

Example 10.53. In order to better understand the meaning of this corollary let us remind the meaning of injective and surjective mappings.

In general, a mapping $f : A \to B$ is injective when it occurs that $\forall \mathbf{v_1}, \mathbf{v_2} \in A$, if $\mathbf{v_1} \neq \mathbf{v_2}$ then $f(\mathbf{v_1}) \neq f(\mathbf{v_2})$.

If the mapping is linear, it is injective if and only if its kernel is the null vector, see Theorem 10.5. For example, the mapping $f : \mathbb{R}^3 \to \mathbb{R}^3$ defined as

$$f(x,y,z) = (x - y + 2z, x + y + z, -5x + 2y + z)$$

is injective (the matrix associated with the system is non-singular and then the system is determined).

We know that $\dim(\mathrm{Im}(f), +, \cdot) = 3$ and that this mapping transforms points of the space ($\mathbb{R}^3$) into points of the space ($\mathbb{R}^3$). In our case $B = \mathbb{R}^3$ and the image of the mapping is also $\mathrm{Im}(f) = \mathbb{R}^3$, i.e. $B = \mathrm{Im}(f)$. This statement is the definition of surjective mapping.

Thus, an injective linear mapping is always also surjective. On the other hand, the equivalence of injection and surjection would not be valid for non-linear mappings.

Example 10.54. Let us give an example for a non-injective mapping $f : \mathbb{R}^3 \to \mathbb{R}^3$

$$f(x,y,z) = (x - y + 2z, x + y + z, 2x - 2y + 4z).$$

We can easily see that this mapping is not injective since $\dim(\ker(f), +, \cdot) = 1$. It follows that $\dim(\mathrm{Im}(f), +, \cdot) = 2 \neq \dim(\mathbb{R}^3, +\cdot) = 3$. This mapping is then not surjective.

Example 10.55. Let us consider the mapping $f : \mathbb{R}^2 \to \mathbb{R}$

$$f(x,y) = (x + y).$$

Of course, this mapping is not an endomorphism and not injective since its kernel is $\ker(f) = \alpha(1,2)$ with $\alpha \in \mathbb{R}$. Hence, $\dim(\ker(f)) = 1$ and for the rank-nullity theorem $\dim(E) = 2 = \dim(\ker(f)) + \dim(\mathrm{Im}(f)) = 1 + 1$. This means that $\dim(\mathrm{Im}(f)) = 1$. The mapping is surjective. In other words, the Corollary above states that injective endomorphisms are also surjective and vice-versa. On the other hand, a mapping which is not an endomorphism could be surjective and not injective or injective and not surjective.

Example 10.56. Let us consider the mapping $f : \mathbb{R}^2 \to \mathbb{R}^3$

$$f(x,y) = (x + y, x - y, 3x + 2y).$$

Let us check the injection of this mapping by determining its kernel:

$$\begin{cases} x + y = 0 \\ x - y = 0 \\ 3x + 2y = 0. \end{cases}$$

The rank of the system is $\rho = 2$. Hence, only $(0,0)$ is solution of the system. The kernel is then $\ker(f) = \{(0,0)\}$ and its dimension is $\dim(\ker(f)) = 0$. This means that the mapping is injective. From the rank-nullity theorem we know that

$\dim(E) = 2 = \dim(\text{Im}(f)) \neq \dim(F) = 3$ where $F = \mathbb{R}^3$. Thus the mapping is not surjective.

The latter two examples naturally lead to the following corollaries.

Corollary 10.3. *Let $f : E \to F$ be a linear mapping with $(E, +, \cdot)$ and $(F, +, \cdot)$ finite-dimensional vector spaces. Let us consider that $\dim(E) > \dim(F)$. It follows that the mapping is not injective.*

Proof. Since $\text{Im}(f) \subset F$, it follows that $\dim(\text{Im}(f)) \leq \dim(F)$. Hence from the rank-nullity theorem

$$\dim(E) = \dim(\ker(f)) + \dim(\text{Im}(f)) \Rightarrow$$
$$\dim(\ker(f)) = \dim(E) - \dim(\text{Im}(f)) \geq \dim(E) - \dim(F) > \dim(F) - \dim(F) = 0.$$

In other words, $\dim(\ker(f)) > 0$. Thus, $\ker(f)$ cannot be only the null vector. This means that the mapping cannot be injective. $\square$

Corollary 10.4. *Let $f : E \to F$ be a linear mapping with $(E, +, \cdot)$ and $(F, +, \cdot)$ finite-dimensional vector spaces. Let us consider that $\dim(E) < \dim(F)$. It follows that the mapping is not surjective.*

Proof. For the rank-nullity theorem we know that

$$\dim(E) = \dim(\ker(f)) + \dim(\text{Im}(f)) \Rightarrow$$
$$\Rightarrow \dim(\text{Im}(f)) = \dim(E) - \dim(\ker(f)) \leq \dim(E) < \dim(F).$$

Thus, $\dim(\text{Im}(f)) < \dim(F)$. The mapping cannot be surjective. $\square$

10.4.1 Matrix Representation of a Linear Mapping

Proposition 10.4. *Every linear mapping is a multiplication of a matrix by a vector.*

Proof. Let $f : E \to F$ be a linear mapping where $(E, +, \cdot)$ and $(F, +, \cdot)$ are finite-dimensional vector spaces defined on the same field $\mathbb{K}$ and whose dimension is n and m, respectively. Let us consider a vector $\mathbf{x} \in E$

$$\mathbf{x} = (x_1, x_2, \ldots x_n)$$

and a vector $\mathbf{y} \in F$

$$\mathbf{y} = (y_1, y_2, \ldots y_m).$$

Let us consider now the expression $\mathbf{y} = f(\mathbf{x})$ which can be written as

$$(y_1, y_2, \ldots y_m) = f(x_1, x_2, \ldots x_n).$$

Since f is a linear mapping it can be written as

$$(y_1, y_2, \ldots y_m) = \begin{pmatrix} a_{1,1}x_1 + a_{1,2}x_2 \ldots a_{1,n}x_n, \\ a_{2,1}x_1 + a_{2,2}x_2 \ldots a_{2,n}x_n, \\ \ldots \\ a_{m,1}x_1 + a_{m,2}x_2 \ldots a_{m,n}x_n \end{pmatrix}.$$

This means that

$$y_1 = a_{1,1}x_1 + a_{1,2}x_2 \ldots a_{1,n}x_n,$$
$$y_2 = a_{2,1}x_1 + a_{2,2}x_2 \ldots a_{2,n}x_n,$$
$$\ldots$$
$$y_m = a_{m,1}x_1 + a_{m,2}x_2 \ldots a_{m,n}x_n.$$

Furthermore, since all these equations need to be simultaneously verified, these equations compose a system of linear equations

$$\begin{cases} y_1 = a_{1,1}x_1 + a_{1,2}x_2 \ldots a_{1,n}x_n, \\ y_2 = a_{2,1}x_1 + a_{2,2}x_2 \ldots a_{2,n}x_n, \\ \ldots \\ y_m = a_{m,1}x_1 + a_{m,2}x_2 \ldots a_{m,n}x_n. \end{cases}$$

This is a matrix equation

$$\mathbf{y} = \mathbf{A}\mathbf{x}$$

where

$$\mathbf{A} = \begin{pmatrix} a_{1,1} & a_{1,2} & \cdots & a_{1,n} \\ a_{2,1} & a_{2,2} & \cdots & a_{2,n} \\ \ldots \\ a_{m,1} & a_{m,2} & \cdots & a_{m,n} \end{pmatrix}. \quad \square$$

Corollary 10.5. *The matrix* $\mathbf{A}$ *characterizing a linear mapping* $\mathbf{y} = f(\mathbf{x}) = \mathbf{A}\mathbf{x}$ *is unique.*

Example 10.57. Let us consider the linear mapping $f : \mathbb{R}^3 \to \mathbb{R}^3$, $f(x+y-z, x-z, 3x+2y+z)$. Let us consider a vector $(1,2,1)$. The mapped vector $f(1,2,1) = (2,0,8)$.

Let us calculate the same mapped value as a product of a matrix by a vector:

$$\begin{pmatrix} 1 & 1 & -1 \\ 1 & 0 & -1 \\ 3 & 2 & 1 \end{pmatrix} \begin{pmatrix} 1 \\ 2 \\ 1 \end{pmatrix} = \begin{pmatrix} 2 \\ 0 \\ 8 \end{pmatrix}.$$

They vectors are clearly the same.

Proposition 10.5. *Let* $f : E \to F$ *be a linear mapping where* $(E, +, \cdot)$ *and* $(F, +, \cdot)$ *are finite-dimensional vector spaces defined on the same field* $\mathbb{K}$ *and whose dimen-*

sion is n and m, respectively. The mapping $\mathbf{y} = f(\mathbf{x})$ *is expressed as a matrix equation* $\mathbf{y} = \mathbf{Ax}$. *It follows that the image of the mapping* $\mathrm{Im}(f)$ *is spanned by the column vectors of the matrix* $\mathbf{A}$:

$$\mathrm{Im}(f) = L\left(\mathbf{a}^1, \mathbf{a}^2, \ldots \mathbf{a}^n\right)$$

where

$$\mathbf{A} = \left(\mathbf{a}^1, \mathbf{a}^2, \ldots \mathbf{a}^n\right).$$

Proof. Without a loss of generality let us assume that f is an endomorphism. Let us consider a vector $\mathbf{x} \in E$

$$\mathbf{x} = (x_1, x_2, \ldots x_n)$$

and a vector $\mathbf{y} \in F = E$

$$\mathbf{y} = (y_1, y_2, \ldots y_n).$$

Let us consider now the expression $\mathbf{y} = f(\mathbf{x})$ which can be written as

$$(y_1, y_2, \ldots y_n) = f(x_1, x_2, \ldots x_n).$$

The linear mapping f can be written as

$$(y_1, y_2, \ldots y_m) = \begin{pmatrix} a_{1,1}x_1 + a_{1,2}x_2 \ldots a_{1,n}x_n, \\ a_{2,1}x_1 + a_{2,2}x_2 \ldots a_{2,n}x_n, \\ \cdots \\ a_{n,1}x_1 + a_{n,2}x_2 \ldots a_{n,n}x_n \end{pmatrix}$$

that is

$$y_1 = a_{1,1}x_1 + a_{1,2}x_2 \ldots a_{1,n}x_n,$$
$$y_2 = a_{2,1}x_1 + a_{2,2}x_2 \ldots a_{2,n}x_n,$$
$$\cdots$$
$$y_m = a_{n,1}x_1 + a_{n,2}x_2 \ldots a_{n,n}x_n.$$

These equations can be written in the vectorial form

$$\begin{pmatrix} a_{1,1} \\ a_{2,1} \\ \cdots \\ a_{n,1} \end{pmatrix} x_1 + \begin{pmatrix} a_{1,2} \\ a_{2,2} \\ \cdots \\ a_{n,2} \end{pmatrix} x_2 + \ldots + \begin{pmatrix} a_{1,n} \\ a_{2,n} \\ \cdots \\ a_{n,n} \end{pmatrix} x_n = \begin{pmatrix} y_1 \\ y_1 \\ \cdots \\ y_n \end{pmatrix}.$$

This means that

$$\mathbf{a}^1 x_1 + \mathbf{a}^2 x_2 + \cdots + \mathbf{a}^n x_n = \mathbf{y}.$$

Every time a vector

$$\mathbf{x} = \begin{pmatrix} x_1 \\ x_2 \\ \cdots \\ x_n \end{pmatrix}$$

is chosen, a corresponding vector

$$\mathbf{y} = \begin{pmatrix} y_1 \\ y_2 \\ \dots \\ y_n \end{pmatrix} \in \mathrm{Im}\,(f)$$

is identified.

In other words, the vector space $(\mathrm{Im}\,(f), +, \cdot)$ is spanned by the column of the matrix $\mathbf{A}, \mathbf{a}^1, \mathbf{a}^2, \dots, \mathbf{a}^n$:

$$\mathrm{Im}\,(f) = L\left(\mathbf{a}^1, \mathbf{a}^2, \dots, \mathbf{a}^n\right). \square$$

Example 10.58. Let us consider the linear mapping $f : \mathbb{R}^3 \to \mathbb{R}^2$ defined by the bases $B_{\mathbb{R}^3} = \{\mathbf{e}_1, \mathbf{e}_2, \mathbf{e}_3\}$ and $B_{\mathbb{R}^2} = \{\mathbf{e}_1', \mathbf{e}_2'\}$ as well as the matrix

$$\mathbf{A} = \begin{pmatrix} 1 & -2 & 1 \\ 3 & 1 & -1 \end{pmatrix}.$$

This representation of the mapping is equivalent to $f : \mathbb{R}^3 \to \mathbb{R}^2$ defined as

$$f(x_1, x_2, x_3) = (x_1 - 2x_2 + x_3, 3x_1 + x_2 - x_3).$$

Let us find the $\ker(f)$. By definition of kernel

$$\ker(f) = \left\{ \mathbf{x} \in \mathbb{R}^3 | f(\mathbf{x}) = \mathbf{o}_{\mathbb{R}^2} \right\} =$$
$$= \left\{ (x_1, x_2, x_3) \in \mathbb{R}^3 | f(x_1, x_2, x_3) = (0, 0) \right\}.$$

This means

$$\begin{cases} x_1 - 2x_2 + x_3 = 0 \\ 3x_1 + x_2 - x_3 = 0 \end{cases}$$

This system has rank 2. Thus, it has ∞^1 solutions proportional to $(1, 4, 7)$. Hence,

$$\ker(f) = L((1, 4, 7))$$

and it has dimension equal to 1. Since the dimension of the kernel is not 0, then the mapping is not injective.

Let us find the $\mathrm{Im}\,(f)$. As shown above the image is spanned by the columns of the associated matrix:

$$\mathrm{Im}\,(f) = L(f(\mathbf{e}_1), f(\mathbf{e}_2), f(\mathbf{e}_3)) =$$
$$= L((1, 3), (-2, 1), (1, -1)).$$

It can be easily seen that out of these three vectors, two are linearly independent. Hence, the dimension of $\text{Im}(f)$ is equal to 2. This is in agreement with the rank-nullity theorem as $1+2=3=\dim\left(\mathbb{R}^3\right)$. Since the dimension of $\text{Im}(f)$ is 2 as well as the dimension of $\mathbb{R}^2$, the mapping is surjective.

Let us compute the mapping in the point $(-2,1,0)$:

$$f(-2,1,0) = ((1)(-2)+(-2)1+(1)0,(3)(-2)+(1)1+(-1)0) =$$
$$(-4,-5)$$

The same result can be achieved by calculating the linear combination of the columns of the matrix associated with the mapping:

$$-2\begin{pmatrix}1\\3\end{pmatrix}+1\begin{pmatrix}-2\\1\end{pmatrix}+0\begin{pmatrix}1\\-1\end{pmatrix}=\begin{pmatrix}-4\\-5\end{pmatrix}$$

Example 10.59. Let us consider now the linear mapping $f:\mathbb{R}^2\to\mathbb{R}^3$ defined by the bases $B_{\mathbb{R}^3}=\{\mathbf{e_1},\mathbf{e_2}\}$ and $B_{\mathbb{R}^2}=\{\mathbf{e'_1},\mathbf{e'_2},\mathbf{e'_3}\}$ as well as the matrix

$$\mathbf{A}=\begin{pmatrix}2 & -4\\3 & -1\\-2 & 4\end{pmatrix}.$$

The mapping can equivalently be expressed as

$$\begin{cases}y_1=2x_1-4x_2\\y_2=3x_1-x_2\\y_3=-2x_1+4x_2\end{cases}$$

or

$$\forall(x_1,x_2):f(x_1,x_2)=(y_1,y_2,y_3)=(2x_1-4x_2,3x_1-x_2,2x_1+4x_2)$$

that leads to

$$f(\mathbf{e_1})=2\mathbf{e'_1}+3\mathbf{e'_2}-2\mathbf{e'_3}$$
$$f(\mathbf{e_2})=-4\mathbf{e'_1}-\mathbf{e'_2}+4\mathbf{e'_3}.$$

Let us find the kernel of this mapping, i.e. by applying the definition, let us solve the following homogeneous system of linear equations.

$$\begin{cases}2x_1-4x_2=0\\3x_1-x_2=0\\-2x_1+4x_2=0\end{cases}.$$

It can be easily seen that only $\mathbf{o}_{\mathbb{R}^2}$ solves the system. Hence, $\ker(f) = \{\mathbf{o}_{\mathbb{R}^2}\}$ and its dimension is 0. This means that the mapping is injective.

Example 10.60. Let us find the image of the mapping:

$$\text{Im}(f) = (f(\mathbf{e}_1), f(\mathbf{e}_2)) = ((2, 3, -2), (-4, -1, 4)).$$

These two vectors are linearly independent. Thus, these two vectors compose a basis. Hence, the dimension of the image is 2 unlike the dimension of $\mathbb{R}^3$ (that is 3). The mapping is not surjective.

Example 10.61. Let us consider the linear mapping $f : \mathbb{R}^3 \to \mathbb{R}^3$ defined as

$$f(x_1, x_2, x_3) = (x_1 - x_2 + 2x_3, x_2 + 2x_3, x_1 + 4x_3).$$

The matrix associated with this mapping is

$$\mathbf{A} = \begin{pmatrix} 1 & -1 & 2 \\ 0 & 1 & 2 \\ 1 & 0 & 4 \end{pmatrix}.$$

Let us find the kernel of f. This means that we have to solve the following system of linear equation

$$\begin{cases} x_1 - x_2 + 2x_3 = 0 \\ x_2 + 2x_3 = 0 \\ x_1 + 4x_3 = 0 \end{cases}.$$

The system is undetermined and has ∞^1 solutions proportional to $(4, 2, -1)$. In other words,

$$\ker(f) = L((4, 2, -1))$$

and

$$B_{\ker(f)} = B\{(4, 2, -1)\}.$$

It follows that the dimension of the kernel is 1. Hence, f is not injective.

If we consider the image

$$\text{Im}(f) = L((1, 0, 1), (-1, 1, 0))$$

has dimension 2 as the three columns of the matrix $\mathbf{A}$ are linearly dependent (the third can be obtained as the sum of the first multiplied by 4 and the second column multiplied by 2). The mapping is not surjective.

10.4.2 A Linear Mapping as a Matrix: A Summarizing Scheme

For a given linear mapping $f : E \to F$, with $(E, +, \cdot)$ and $(F, +, \cdot)$ vector spaces, if we pose

$$\dim(E) = n$$
$$\dim(F) = m$$

the mapping is identified by a matrix $\mathbf{A}$ whose size is $m \times n$.

This matrix can be represented in terms of row vectors as

$$\mathbf{A} = \begin{pmatrix} \mathbf{K_1} \\ \mathbf{K_2} \\ \dots \\ \mathbf{K_m} \end{pmatrix}$$

and in terms of column vectors as

$$\mathbf{A} = \left(\mathbf{I_1}, \mathbf{I_2}, \dots, \mathbf{I_n} \right).$$

If we consider a vector $\mathbf{x}$ of n elements, it follows that

$$f(\mathbf{x}) = \mathbf{A}\mathbf{x}$$

which, equivalently, can be expressed as

$$\text{Im}(f) = L(\mathbf{I_1}, \mathbf{I_2}, \dots, \mathbf{I_n})$$

that is the vector space $(\text{Im}(f), +, \cdot)$ image of the mapping is spanned by the column vectors of the matrix $\mathbf{A}$.

In order to find the set kernel $\ker(f)$ the following system of linear equations must be solved

$$\mathbf{A}\mathbf{x} = \mathbf{o}.$$

This means that the vector space $(\ker(f), +, \cdot)$ kernel of the mapping is spanned by the solutions of $\mathbf{A}\mathbf{x} = \mathbf{o}$.

Let us indicate with ρ the rank of the matrix $\mathbf{A}$. Among the vectors spanning the $\ker(f)$ only $n - \rho$ are linearly independent. This means that

$$\dim(\ker(f)) = n - \rho.$$

From the rank-nullity theorem we can immediately check that

$$\dim(\text{Im}(f)) = \rho$$

that is the number of linearly independent column vectors in the matrix $\mathbf{A}$ and the number of linearly independent vectors in $(\text{Im}(f), +, \cdot)$. Of course, it is not a coincidence that the term *rank* recurs to describe the rank of a matrix and the dimension of the of the image space of a linear mapping. As shown, although the two concepts may appear distinct, they indeed coincide.

10.4.3 Invertible Mappings

Definition 10.16. Let $f : E \to F$ be a linear mapping where $(E, +, \cdot)$ and $(F, +, \cdot)$ are finite-dimensional vector spaces. The mapping is said to be invertible if there exists a mapping $g : F \to E$ such that

$$\forall \mathbf{x} \in E : \mathbf{x} = g\left((f(\mathbf{x}))\right)$$

and

$$\forall \mathbf{y} \in F : \mathbf{y} = f\left((g(\mathbf{y}))\right).$$

Proposition 10.6. *Let $f : E \to E$ be an endomorphism where $(E, +, \cdot)$ is a finite-dimensional vector space and let f be identified to the matrix $\mathbf{A}$. The mapping f is invertible if and only if f is bijective and its inverse g is identified by the matrix $\mathbf{A}^{-1}$.*

Proof. The mapping f can be seen as the matrix equation $\mathbf{y} = \mathbf{A}\mathbf{x}$. We know from Corollary 10.2 that an injective endomorphism is always also surjective and thus bijective.

If f is injective it follows that $\ker(f) = \{\mathbf{o_E}\}$. This occurs when the homogeneous system of linear equations $\mathbf{A}\mathbf{x} = \mathbf{o_F}$ is determined and has the null vector as its only solution. This means that the matrix $\mathbf{A}$ is non-singular and thus invertible. It follows that there exists an inverse matrix $\mathbf{A}^{-1}$ such that

$$\mathbf{x} = \mathbf{A}^{-1}\mathbf{A}\mathbf{x}$$

and

$$\mathbf{y} = \mathbf{A}\mathbf{A}^{-1}\mathbf{y}.$$

Thus, if f is a mapping

$$\mathbf{y} = f(\mathbf{x})$$

equivalent to

$$\mathbf{y} = \mathbf{A}\mathbf{x}$$

then, we have found a mapping g

$$\mathbf{x} = g(\mathbf{y})$$

that is

$$\mathbf{x} = \mathbf{A}^{-1}\mathbf{y}. \quad \square$$

If f is invertible there exists an inverse function g. Let us take the function g as $\mathbf{x} = \mathbf{A}^{-1}\mathbf{y}$. The inverse matrix $\mathbf{A}^{-1}$ exists only if $\mathbf{A}$ is non-singular. Under these conditions, $\mathbf{Ax} = \mathbf{o_F}$ is a determined system of linear equations. Its only solution, kernel of the mapping, is the null-vector. This means that $\ker(f) = \{\mathbf{o_E}\}$ and thus f is injective and thus bijective.   □

Corollary 10.6. *Let* $f : E \to E$ *be an endomorphism. If the inverse mapping exists, then it is unique.*

Proof. The mapping f is identified by the matrix $\mathbf{A}$ and its inverse is g by its inverse $\mathbf{A}^{-1}$. Since the inverse of a matrix is unique the inverse mapping g is unique.   □

Example 10.62. Let us consider again the endomorphism $f : \mathbb{R}^3 \to \mathbb{R}^3$, $f(x,y,z) = (x+y-z, x-z, 3x+2y+z)$. The matrix associated with the mapping is

$$\mathbf{A} = \begin{pmatrix} 1 & 1 & -1 \\ 1 & 0 & -1 \\ 3 & 2 & 1 \end{pmatrix}.$$

The homogeneous system of linear equations

$$\begin{cases} x+y-z = 0 \\ x-z = 0 \\ 3x+2y+z = 0 \end{cases}$$

is determined and thus $\ker\{\mathbf{o_E}\}$, i.e. the mapping is injective. It follows that the matrix $\mathbf{A}$ is invertible and its inverse is

$$\mathbf{A}^{-1} = -\frac{1}{4} \begin{pmatrix} 2 & -3 & -1 \\ -4 & 4 & 0 \\ 2 & 1 & -1 \end{pmatrix}.$$

Equivalently, we may state that the inverse mapping of f is $g : \mathbb{R}^3 \to \mathbb{R}^3$

$$g(x,y,z) = \left(-\frac{1}{4}(2x-3y-z), -\frac{1}{4}(-4x+4y), -\frac{1}{4}(2x+y-z) \right).$$

We know that $f(1,2,1) = (2,0,8)$. Let us calculate the inverse vector by means of the inverse mapping g of f:

$$\mathbf{A}^{-1} = -\frac{1}{4} \begin{pmatrix} 2 & -3 & -1 \\ -4 & 4 & 0 \\ 2 & 1 & -1 \end{pmatrix} \begin{pmatrix} 2 \\ 0 \\ 8 \end{pmatrix} = \begin{pmatrix} 1 \\ 2 \\ 1 \end{pmatrix}.$$

Example 10.63. Let us consider again the endomorphism $f : \mathbb{R}^3 \to \mathbb{R}^3$, $f(x+y-z, x-z, 2x+y-2z)$. The matrix associated with the mapping is

$$\mathbf{A} = \begin{pmatrix} 1 & 1 & -1 \\ 1 & 0 & -1 \\ 2 & 1 & -2 \end{pmatrix}.$$

Since the matrix $\mathbf{A}$ is singular the $\ker(f) \neq \{\mathbf{o_E}\}$, i.e. f is not injective. Equivalently, since $\mathbf{A}$ is singular, then $\mathbf{A}$ is not invertible and g does not exist.

Corollary 10.7. *Let $f : E \to F$ be a linear mapping with $\dim(E) \neq \dim(F)$. The inverse mapping of f does not exist.*

Proof. Since the matrix associated with the mapping is rectangular, it has no inverse. Thus, the mapping is not invertible. $\square$

Example 10.64. Let us consider again the endomorphism $f : \mathbb{R}^3 \to \mathbb{R}^2$, $f(x+y-z, x+5y-z)$. The mapping is not invertible.

Remark 10.2. As a further observation, if $\dim(E) < \dim(F)$ is surely not surjective and if $\dim(E) > \dim(F)$ the mapping is surely not injective. A basic requirement for a mapping to be bijective and thus invertible is that it is an endomorphism.

10.4.4 Similar Matrices

Proposition 10.7. *A change of basis is a bijective linear mapping.*

Proof. Let $(E, +, \cdot)$ be a fine-dimensional vector space and $\mathbf{x} \in E$ be its vector. Let the vector $\mathbf{x}$ have its components

$$\mathbf{x} = (x_1, x_2, \ldots, x_n)$$

in the basis of E

$$B = \{\mathbf{e_1}, \mathbf{e_2}, \ldots, \mathbf{e_n}\}.$$

Thus, we can then express $\mathbf{x}$ as

$$\mathbf{x} = x_1 \mathbf{e_1} + x_2 \mathbf{e_2} + \ldots + x_n \mathbf{e_n}.$$

Let us consider now another basis of the same vector space, i.e.

$$B' = \left\{ \mathbf{p^1}, \mathbf{p^2}, \ldots, \mathbf{p^n} \right\}.$$

where the vectors $\mathbf{p^1}, \mathbf{p^2}, \ldots, \mathbf{p^n}$ in the basis B have the following components

$$\mathbf{p^1} = \begin{pmatrix} p_{1,1} \\ p_{2,1} \\ \ldots \\ p_{n,1} \end{pmatrix}$$

$$\mathbf{p^2} = \begin{pmatrix} p_{1,2} \\ p_{2,2} \\ \ldots \\ p_{n,2} \end{pmatrix}$$

$$\mathbf{p^n} = \begin{pmatrix} p_{1,n} \\ p_{2,n} \\ \cdots \\ p_{n,n} \end{pmatrix}.$$

Let us now represent $\mathbf{x}$ in the basis B'. This means that we should find the components

$$(x_1', x_2', \ldots, x_n')$$

such that

$$\mathbf{x} = x_1' \mathbf{p^1} + x_2' \mathbf{p^2} + \ldots + x_n' \mathbf{p^n}.$$

This means that

$$\begin{cases} x_1 = x_1' p_{1,1} + x_2' p_{1,2} + \ldots + x_n' p_{1,n} \\ x_2 = x_1' p_{2,1} + x_2' p_{2,2} + \ldots + x_n' p_{2,n} \\ \cdots \\ x_n = x_1' p_{n,1} + x_2' p_{n,2} + \ldots + x_n' p_{n,n} \end{cases}$$

that is

$$\mathbf{x} = \mathbf{P}\mathbf{x}'$$

where

$$\mathbf{P} = \begin{pmatrix} p_{1,1} & p_{1,2} & \cdots & p_{1,n} \\ p_{2,1} & p_{2,2} & \cdots & p_{2,n} \\ \cdots & \cdots & \cdots & \cdots \\ p_{n,1} & p_{n,2} & \cdots & p_{n,n} \end{pmatrix} = \begin{pmatrix} \mathbf{p^1} & \mathbf{p^2} & \cdots & \mathbf{p^n} \end{pmatrix}$$

$$\mathbf{x}' = \begin{pmatrix} x_1' \\ x_2' \\ \cdots \\ x_n' \end{pmatrix}.$$

Thus, a basis transformation of a vector $\mathbf{x}$ in the basis B into a vector $\mathbf{x}'$ in the basis B' is achieved by a matrix multiplication

$$\mathbf{x} = \mathbf{P}\mathbf{x}' = g(\mathbf{x}')$$
$$\mathbf{x}' = \mathbf{P}^{-1}\mathbf{x} = f(\mathbf{x})$$

where the columns of the matrix $\mathbf{P}$ are the vectors of the basis B' (the matrix $\mathbf{P}$ is thus invertible). Thus, the basis of change consists of an application of a linear mapping. □

Example 10.65. Let us consider the vector space $(\mathbb{R}^3, +, \cdot)$ and its orthonormal basis $B\{\mathbf{e}_1, \mathbf{e}_2, \mathbf{e}_3\}$ where

$$\begin{aligned} \mathbf{e}_1 &= (1,0,0) \\ \mathbf{e}_2 &= (0,1,0) \\ \mathbf{e}_3 &= (0,0,1). \end{aligned}$$

The following three vectors

$$\begin{aligned} \mathbf{p}^1 &= (1,0,2) \\ \mathbf{p}^2 &= (0,5,1) \\ \mathbf{p}^3 &= (1,0,0) \end{aligned}$$

in the basis B are linearly independent and span $\mathbb{R}^3$. Thus, $\mathbf{p}^1, \mathbf{p}^2, \mathbf{p}^3$ are a basis of $\mathbb{R}^3$.

The transformation matrix is

$$\mathbf{P} = \begin{pmatrix} 1 & 0 & 1 \\ 0 & 5 & 0 \\ 2 & 1 & 0 \end{pmatrix}$$

whose inverse matrix is

$$\mathbf{P}^{-1} = \begin{pmatrix} 0 & -0.1 & 0.5 \\ 0 & 0.2 & 0 \\ 1 & 0.1 & -0.5 \end{pmatrix}.$$

Let $\mathbf{x} = (1,1,1)$ be a vector of $\mathbb{R}^3$ in the basis B.

If we want to express the vector $\mathbf{x} = (1,1,1)$ in the new basis $B' = \{\mathbf{p}^1, \mathbf{p}^2, \mathbf{p}^3\}$, we can simply write

$$\mathbf{x}' = \mathbf{P}^{-1}\mathbf{x} = (0.4, 0.2, 0.6).$$

This fact was known already from Chap. 4 where, in the case of $\mathbb{V}_3$, it was shown how to express a vector in a new basis. We already knew that to express a vector in a new basis a solution of a system of linear equations that is a matrix inversion was required.

By using the same notation of Chap. 4 if we want to express $\mathbf{x}$ in the basis of $\mathbf{p}^1$, $\mathbf{p}^2$, and $\mathbf{p}^3$ we may write

$$(1,1,1) = x_1'(1,0,2) + x_2'(0,5,1) + x_3'(1,0,0)$$

which leads to the following system of linear equations

$$\begin{cases} x_1' + x_3' = 1 \\ 5x_2' = 1 \\ 2x_1' + x_2' = 1 \end{cases}$$

that is, essentially, the inversion of the matrix associated with the system:

$$\mathbf{P} = \begin{pmatrix} 1 & 0 & 1 \\ 0 & 5 & 0 \\ 2 & 1 & 0 \end{pmatrix}.$$

The vector in the new basis is the solution of the system of linear equations. This vector is $\lambda, \mu, \nu = 0.4, 0.2, 0.6 = \mathbf{x}'$.

Thus, we can write

$$\mathbf{P}\mathbf{x}' = \mathbf{x} \Rightarrow \mathbf{x}' = \mathbf{P}^{-1}\mathbf{x}.$$

In conclusion, the change of basis by matrix transformations is a generalization to all vector sets and an extension to n variables of what has been shown for vectors in the space.

Remark 10.3. Let us consider an endomorphism $f : E \to E$ and a finite-dimensional vector space $(E, +\cdot)$ whose dimension is n. Let $\mathbf{x}, \mathbf{y} \in E$ and the linear mapping f be such that

$$\mathbf{y} = f(\mathbf{x}).$$

We know that the endomorphism can be expressed as a product of a matrix $\mathbf{A}$ by the vector $\mathbf{x}$:

$$\mathbf{y} = f(\mathbf{x}) = \mathbf{A}\mathbf{x}.$$

Let us apply the change of basis by means of a matrix $\mathbf{P}$:

$$\mathbf{y} = \mathbf{P}\mathbf{y}'$$
$$\mathbf{x} = \mathbf{P}\mathbf{x}'.$$

The endomorphism can be expressed as

$$\mathbf{P}\mathbf{y}' = \mathbf{A}\mathbf{P}\mathbf{x}'.$$

This means

$$\mathbf{y}' = \mathbf{P}^{-1}\mathbf{A}\mathbf{P}\mathbf{x}' = \mathbf{A}'\mathbf{x}'$$

where $\mathbf{A}' = \mathbf{P}^{-1}\mathbf{A}\mathbf{P}$.

Definition 10.17. Let $\mathbf{A}$ and $\mathbf{A}'$ be two square matrices. These two matrices are said *similar* and are indicated with

$$\mathbf{A} \sim \mathbf{A}'$$

when exists a non-singular matrix $\mathbf{P}$ such that

$$\mathbf{A}' = \mathbf{P}^{-1}\mathbf{A}\mathbf{P}.$$

Considering that the matrices $\mathbf{A}$ and $\mathbf{A}'$ represent two endomorphisms, also the two endomorphis are said to be similar.

Theorem 10.8. *Let $\mathbf{A}$ and $\mathbf{A}'$ be two similar matrices. These two matrices have same determinant and trace.*

Proof. Considering the similarity and remembering that $\det\left(\mathbf{P}^{-1}\right) = \frac{1}{\det(\mathbf{P})}$

$$\det\left(\mathbf{A}'\right) = \det\left(\mathbf{P}^{-1}\mathbf{AP}\right) = \det\left(\mathbf{P}^{-1}\right)\det\left(\mathbf{A}\right)\det\left(\mathbf{P}\right) =$$

$$= \det\left(\mathbf{A}\right)\det\left(\mathbf{P}^{-1}\right)\det\left(\mathbf{P}\right) = \det\left(\mathbf{A}\right)\frac{1}{\det\left(\mathbf{P}\right)}\det\left(\mathbf{P}\right) = \det\left(\mathbf{A}\right). \quad \square$$

Regarding the trace, we can remember that $\mathrm{tr}\left(\mathbf{AB}\right) = \mathrm{tr}\left(\mathbf{BA}\right)$

$$\mathrm{tr}\left(\mathbf{A}'\right) = \mathrm{tr}\left(\mathbf{P}^{-1}\mathbf{AP}\right) = \mathrm{tr}\left(\mathbf{P}^{-1}\left(\mathbf{AP}\right)\right) =$$

$$= \mathrm{tr}\left(\left(\mathbf{AP}\right)\mathbf{P}^{-1}\right) = \mathrm{tr}\left(\mathbf{A}\left(\mathbf{P}\right)\mathbf{P}^{-1}\right) = \mathrm{tr}\left(\mathbf{A}\right). \quad \square$$

Example 10.66. Let us consider the following matrix:

$$\mathbf{A} = \begin{pmatrix} 2 & 2 & 0 \\ 0 & 5 & 0 \\ 0 & 0 & 1 \end{pmatrix}$$

corresponding to the endomorphism $f : \mathbb{R}^3 \rightarrow \mathbb{R}^3$

$$\mathbf{y} = f\left(\mathbf{x}\right) = \left(2x_1 + 2x_2, 5x_2, x_3\right)$$

where $\mathbf{y} = \left(y_1, y_2, y_3\right)$.

Let us calculate the function value for $\mathbf{x} = \left(1, 1, 1\right)$. The corresponding function value is $\mathbf{y} = \left(4, 5, 1\right)$.

Let us consider again the transformation matrix

$$\mathbf{P} = \begin{pmatrix} 1 & 0 & 1 \\ 0 & 5 & 0 \\ 2 & 1 & 0 \end{pmatrix}$$

and

$$\mathbf{P}^{-1} = \begin{pmatrix} 0 & -0.1 & 0.5 \\ 0 & 0.2 & 0 \\ 1 & 0.1 & -0.5 \end{pmatrix}.$$

We can calculate $\mathbf{x}'$ and $\mathbf{y}'$

$$\mathbf{x}' = \mathbf{P}^{-1}\mathbf{x} = \left(0.4, 0.2, 0.6\right)$$
$$\mathbf{y}' = \mathbf{P}^{-1}\mathbf{y} = \left(0, 1, 4\right).$$

Let us calculate now

$$\mathbf{A}' = \mathbf{P}^{-1}\mathbf{AP} = \begin{pmatrix} 1 & -2 & 0 \\ 0 & 5 & 0 \\ 1 & 12 & 2 \end{pmatrix}.$$

This means that the endomorphism $\mathbf{y} = f(\mathbf{x})$ can be represented in the new basis (in a new reference system) as $\mathbf{y}' = f'(\mathbf{x}')$, that is

$$\mathbf{y}' = f'(\mathbf{x}') = \mathbf{A}'\mathbf{x}' = \left(x_1' - 2x_2', 5x_2', x_1' + 12x_2' + 2x_3'\right).$$

We can verify that for $\mathbf{x}' = (0.4, 0.2, 0.6)$ we have

$$\mathbf{y}' = \mathbf{A}'\mathbf{x}' = \begin{pmatrix} 1 & -2 & 0 \\ 0 & 5 & 0 \\ 1 & 12 & 2 \end{pmatrix} \begin{pmatrix} 0.4 \\ 0.2 \\ 0.6 \end{pmatrix} = \begin{pmatrix} 0 \\ 1 \\ 4 \end{pmatrix}.$$

Since the matrices $\mathbf{A}$ and $\mathbf{A}'$ are similar, as we can easily check,

$$\det(\mathbf{A}) = \det(\mathbf{A}') = 10$$

and

$$\operatorname{tr}(\mathbf{A}) = \operatorname{tr}(\mathbf{A}') = 8.$$

Proposition 10.8. *The similarity of matrices is an equivalence relation.*

Proof. Let us consider three matrices $\mathbf{A}$, $\mathbf{A}'$, and $\mathbf{A}''$ and let us check the three properties of the equivalence relations.

- Reflexivity: $\mathbf{A} \sim \mathbf{A}$.
  if $\exists$ a non-singular matrix $\mathbf{P} \ni '$

$$\mathbf{A} = \mathbf{P}^{-1}\mathbf{A}\mathbf{P}.$$

If we take the matrix $\mathbf{P}$ equal to the identity matrix $\mathbf{I}$ ($\mathbf{P} = \mathbf{I}$) this equation is always verified. Thus, the similarity is reflexive.
- Symmetry: $\mathbf{A} \sim \mathbf{A}' \Rightarrow \mathbf{A}' \sim \mathbf{A}$.
  Since $\mathbf{A} \sim \mathbf{A}'$ then it $\exists$ a non-singular matrix $\mathbf{P} \ni '$

$$\mathbf{A}' = \mathbf{P}^{-1}\mathbf{A}\mathbf{P}.$$

We can then rewrite this equation as

$$\mathbf{P}\mathbf{A}'\mathbf{P}^{-1} = \mathbf{A}.$$

Let us pose $\mathbf{Q} = \mathbf{P}^{-1}$ and write

$$\mathbf{A} = \mathbf{Q}^{-1}\mathbf{A}'\mathbf{Q},$$

i.e. $\mathbf{A}' \sim \mathbf{A}$. Since $\mathbf{Q}$ is surely invertible (it is the inverse of $\mathbf{P}$), this relation is symmetric.
- Transitivity: $(\mathbf{A}'' \sim \mathbf{A}')$ and $(\mathbf{A}' \sim \mathbf{A}) \Rightarrow \mathbf{A}'' \sim \mathbf{A}$.
  We may then write that $\exists$ a non-singular matrix $\mathbf{P}_1 \ni '$

$$\mathbf{A}' = \mathbf{P}_1^{-1}\mathbf{A}\mathbf{P}_1.$$

and $\exists$ a non-singular matrix $\mathbf{P_2} \ni$ '

$$\mathbf{A}'' = \mathbf{P_2}^{-1}\mathbf{A}'\mathbf{P_2}.$$

By manipulating the latter two equations we have

$$\mathbf{A}'' = \mathbf{P_2}^{-1}\mathbf{P_1}^{-1}\mathbf{A}\mathbf{P_1}\mathbf{P_2}.$$

On the basis of Proposition 2.17, let us re-write this equation as

$$\mathbf{A}'' = (\mathbf{P_1}\mathbf{P_2})^{-1}\mathbf{A}\mathbf{P_1}\mathbf{P_2}.$$

Let us pose $\mathbf{R} = \mathbf{P_1}\mathbf{P_2}$ and then

$$\mathbf{A}'' = \mathbf{R}^{-1}\mathbf{A}\mathbf{R}.$$

Both the matrices $\mathbf{P_1}$ and $\mathbf{P_2}$ are non-singular, i.e. $\det(\mathbf{P_1}) \neq 0$ and $\det(\mathbf{P_2}) \neq 0$. The matrix $\mathbf{R}$ is surely non-singular since

$$\det(\mathbf{R}) = \det(\mathbf{P_1}\mathbf{P_2}) = \det(\mathbf{P_1})\det(\mathbf{P_2}) \neq 0.$$

Thus, the relation is transitive.
The similarity of matrices (and linear mappings) is an equivalence relation. $\square$

This fact means that if a linear mapping is difficult to solve, it can be transformed into an equivalent one, solved and anti-transformed to find the solution of the original problem.

10.4.5 Geometric Mappings

Let us consider a mapping $f : \mathbb{R}^2 \to \mathbb{R}^2$. This mapping can be interpreted as an operators that transforms a point in the plane into another point in the plane. Under these conditions, the mapping is said *geometric mapping in the plane*.

Let us now consider the following mapping $f : \mathbb{R}^2 \to \mathbb{R}^2$:

$$\begin{pmatrix} y_1 \\ y_2 \end{pmatrix} = \begin{pmatrix} s & 0 \\ 0 & s \end{pmatrix} \begin{pmatrix} x_1 \\ x_2 \end{pmatrix} = \begin{pmatrix} sx_1 \\ sx_2 \end{pmatrix}.$$

It can be easily seen that this mapping is linear. This mapping is called *uniform scaling*. If the diagonal elements of the matrix are not equal this linear mapping is called *non-uniform scaling* and is represented by:

$$\begin{pmatrix} y_1 \\ y_2 \end{pmatrix} = \begin{pmatrix} s_1 & 0 \\ 0 & s_2 \end{pmatrix} \begin{pmatrix} x_1 \\ x_2 \end{pmatrix} = \begin{pmatrix} s_1 x_1 \\ s_2 x_2 \end{pmatrix}.$$

In the following figures, the basic points are indicated with a solid line while the transformed points are indicated with a dashed line.

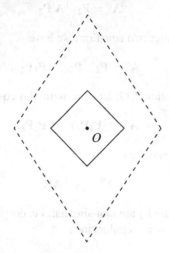

The following linear mapping is called *rotation* and is represented by

$$\begin{pmatrix} y_1 \\ y_2 \end{pmatrix} = \begin{pmatrix} \cos\theta & -\sin\theta \\ \sin\theta & \cos\theta \end{pmatrix} \begin{pmatrix} x_1 \\ x_2 \end{pmatrix} = \begin{pmatrix} x_1\cos\theta - x_2\sin\theta \\ x_1\sin\theta + x_2\cos\theta \end{pmatrix}.$$

The following linear mapping is called *shearing* and is represented by

$$\begin{pmatrix} y_1 \\ y_2 \end{pmatrix} = \begin{pmatrix} 1 & s_1 \\ s_2 & 1 \end{pmatrix} \begin{pmatrix} x_1 \\ x_2 \end{pmatrix} = \begin{pmatrix} x_1 + s_1 x_2 \\ s_2 x_1 + x_2 \end{pmatrix}.$$

If, as in figure, the coefficient $s_2 = 0$ then this mapping is said *horizontal shearing*. If $s_1 = 0$ the mapping is a *vertical shearing*.

The following two linear mappings are said *reflection* with respect to vertical and horizontal axes.

$$\begin{pmatrix} y_1 \\ y_2 \end{pmatrix} = \begin{pmatrix} -1 & 0 \\ 0 & 1 \end{pmatrix} \begin{pmatrix} x_1 \\ x_2 \end{pmatrix} = \begin{pmatrix} -x_1 \\ x_2 \end{pmatrix}$$

and

$$\begin{pmatrix} y_1 \\ y_2 \end{pmatrix} = \begin{pmatrix} 1 & 0 \\ 0 & -1 \end{pmatrix} \begin{pmatrix} x_1 \\ x_2 \end{pmatrix} = \begin{pmatrix} x_1 \\ -x_2 \end{pmatrix}.$$

The reflection with respect to the origin of the reference system is given by

$$\begin{pmatrix} y_1 \\ y_2 \end{pmatrix} = \begin{pmatrix} -1 & 0 \\ 0 & -1 \end{pmatrix} \begin{pmatrix} x_1 \\ x_2 \end{pmatrix} = \begin{pmatrix} -x_1 \\ -x_2 \end{pmatrix}.$$

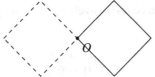

Let us consider now the following mapping:

$$\mathbf{y} = f(\mathbf{x}) = \mathbf{x} + \mathbf{t}$$

where

$$y_1 = x_1 + t_1$$
$$y_2 = x_2 + t_2$$

where $\mathbf{t} = (t_1, t_2)$. This operation, namely *translation* moves the points a constant distance in a specific direction (see figure below). Unlike the previous geometric mapping, a translation is not a linear mapping as the linearity properties are not valid and a matrix representation by means of $\mathbb{R}_{2,2}$ matrices is not possible. More specifically, a translation is an affine mapping.

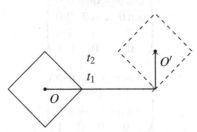

In order to give a matrix representation to affine mappings let us introduce the concept of *homogeneous coordinates*, i.e. we algebraically represent each point $\mathbf{x}$ of the plane by means of three coordinates where the third is identically equal to 1:

$$\mathbf{x} = \begin{pmatrix} x_1 \\ x_2 \\ x_3 \end{pmatrix} = \begin{pmatrix} x_1 \\ x_2 \\ 1 \end{pmatrix}.$$

We can now give a matrix representation to the affine mapping translation in a plane:

$$\begin{pmatrix} y_1 \\ y_2 \\ y_3 \end{pmatrix} = \begin{pmatrix} 1 & 0 & t_1 \\ 0 & 1 & t_2 \\ 0 & 0 & 1 \end{pmatrix} \begin{pmatrix} x_1 \\ x_2 \\ 1 \end{pmatrix} = \begin{pmatrix} x_1 + t_1 \\ x_2 + t_2 \\ 1 \end{pmatrix}.$$

All the linear mappings can be written in homogeneous coordinates simply adding a row and a column to the matrix representing the mapping. For example, the scaling and rotation can be respectively performed by multiplying the following matrices by a point $\mathbf{x}$:

$$\begin{pmatrix} s_1 & 0 & 0 \\ 0 & s_2 & 0 \\ 0 & 0 & 1 \end{pmatrix}$$

and

$$\begin{pmatrix} \cos\theta & -\sin\theta & 0 \\ \sin\theta & \cos\theta & 0 \\ 0 & 0 & 1 \end{pmatrix}$$

If we indicate with $\mathbf{M}$ the 2×2 matrix representing a linear mapping in the plane and with $\mathbf{t}$ the translation vector of the plane, the generic geometric mapping is given by a matrix

$$\begin{pmatrix} \mathbf{M} & \mathbf{t} \\ 0 & 1 \end{pmatrix}.$$

The geometric mappings in the space can be operated in a similar way by adding one dimension. For example, the rotations around the three axes are given by

$$\begin{pmatrix} \cos\theta & -\sin\theta & 0 & 0 \\ \sin\theta & \cos\theta & 0 & 0 \\ 0 & 0 & 1 & 0 \\ 0 & 0 & 0 & 1 \end{pmatrix},$$

$$\begin{pmatrix} \cos\theta & 0 & \sin\theta & 0 \\ 0 & 1 & 0 & 0 \\ -\sin\theta & 0 & \cos\theta & 0 \\ 0 & 0 & 0 & 1 \end{pmatrix},$$

and

$$\begin{pmatrix} 1 & 0 & 0 & 0 \\ 0 & \cos\theta & -\sin\theta & 0 \\ 0 & \sin\theta & \cos\theta & 0 \\ 0 & 0 & 0 & 1 \end{pmatrix}.$$

The translation in the space is given by the following matrix

$$\begin{pmatrix} 1 & 0 & 0 & t_1 \\ 0 & 1 & 0 & t_2 \\ 0 & 0 & 1 & t_3 \\ 0 & 0 & 0 & 1 \end{pmatrix}.$$

10.5 Eigenvalues, Eigenvectors, and Eigenspaces

In order to introduce the new concept of this section, let us consider the following example.

Example 10.67. Let us consider the following linear mapping $f : \mathbb{R}^2 \to \mathbb{R}^2$

$$f(2x - y, 3y)$$

corresponding to the matrix

$$A = \begin{pmatrix} 2 & -1 \\ 0 & 3 \end{pmatrix}.$$

Let us now consider a vector

$$\mathbf{x} = \begin{pmatrix} 1 \\ 1 \end{pmatrix}$$

and let us calculate

$$f \begin{pmatrix} 1 \\ 1 \end{pmatrix} = \mathbf{A}\mathbf{x} = \begin{pmatrix} 1 \\ 3 \end{pmatrix}$$

which can be graphically represented as

For analogy, we may think that a linear mapping (at least $\mathbb{R}^2 \to \mathbb{R}^2$) can be represented by a clock where one pointer is the input and the other is the output. Since both the vectors are applied in the origin, a linear mapping varies the input in length and rotates it around the origin, that is the null vector of the vector space.

Definition 10.18. Let $f : E \rightarrow E$ be an endomorphism where $(E, +, \cdot)$ is a finite-dimensional vector space defined on the scalar field $\mathbb{K}$ whose dimension is n. Every vector $\mathbf{x}$ such that $f(\mathbf{x}) = \lambda \mathbf{x}$ with λ scalar and $\mathbf{x} \in E \setminus \{\mathbf{o_E}\}$ is said *eigenvector* of the endomorphism f related to the *eigenvalue* λ.

The concepts of eigenvectors and eigenvalues are extremely important as well as tricky in mathematics. These concepts appear in various contexts and have practical implementation in various engineering problems. The most difficult aspect of these concepts is that they can be observed from different perspectives and take, on each occasion, a slightly different meaning.

As an initial interpretation by imposing $f(\mathbf{x}) = \lambda \mathbf{x}$ we are requiring that a linear mapping simply scales the input vector. In other words, if we think to vectors in the space, we are requiring that the mapping changes only the module of a vector while it keeps its original direction.

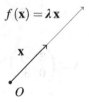

The search of eigenvectors is the search of those vectors belonging to the domain (the identification of a subset of the domain) whose linear mapping application behaves like a multiplication of a scalar by a vector. This scalar is the eigenvalue.

For endomorphisms $\mathbb{R} \rightarrow \mathbb{R}$, the detection eigenvectors and eigenvalues is trivial because the endomorphisms are already in the form $f(\mathbf{x}) = \lambda \mathbf{x}$.

Example 10.68. Let us consider the endomorphism $f : \mathbb{R} \rightarrow \mathbb{R}$ defined as

$$f(x) = 5x.$$

In this case, any vector x (number in this specific case) is a potential eigenvector and $\lambda = 5$ would be the eigenvalue.

Example 10.69. When the endomorphism is between multidimensional vector spaces, the search of eigenvalues and eigenvectors is not trivial.

Let us consider the following endomorphism $f : \mathbb{R}^2 \rightarrow \mathbb{R}^2$

$$f(x, y) = (x + y, 2x).$$

By definition, an eigenvector (x, y) and an eigenvalue λ, respectively, verify the following equation

$$f(x, y) = \lambda (x, y).$$

By combining the last two equations we have

$$\begin{cases} x+y = \lambda x \\ 2x = \lambda y \end{cases} \Rightarrow \begin{cases} (1-\lambda)x+y = 0 \\ 2x-\lambda y = 0. \end{cases}$$

A scalar λ with a corresponding vector (x,y) that satisfy the homogeneous system of linear equations are an eigenvalue and its eigenvector, respectively.

Since the system is homogeneous the only way for it to be determined is if $(x,y) = (0,0)$. If this situation occurs, regardless of the value of λ, the equations of the system are verified. Since by definition an eigenvector $\mathbf{x} \in E \setminus \{\mathbf{o_E}\}$, it follows that $(x,y) = (0,0) = \mathbf{o_E}$ is not an eigenvector.

On the other hand, if we fix the value of λ such that the matrix associated with the system is singular, we have infinite eigenvectors associated with λ.

Theorem 10.9. *Let $f : E \to E$ be an endomorphism. The set $V(\lambda) \subset E$ with $\lambda \in \mathbb{K}$ defined as*

$$V(\lambda) = \{\mathbf{o_E}\} \cup \{\mathbf{x} \in E | f(\mathbf{x}) = \lambda \mathbf{x}\}$$

with the composition laws is a vector subspace of $(E,+,\cdot)$.

Proof. Let us prove the closure of $V(\lambda)$ with respect to the composition laws.

Let us consider two generic vectors $\mathbf{x_1}, \mathbf{x_2} \in V(\lambda)$. For the definition of $V(\lambda)$

$$\mathbf{x_1} \in V(\lambda) \Rightarrow f(\mathbf{x_1}) = \lambda \mathbf{x_1}$$
$$\mathbf{x_2} \in V(\lambda) \Rightarrow f(\mathbf{x_2}) = \lambda \mathbf{x_2}.$$

It follows that

$$f(\mathbf{x_1}+\mathbf{x_2}) = f(\mathbf{x_1}) + f(\mathbf{x_2}) = \lambda \mathbf{x_1} + \lambda \mathbf{x_2} = \lambda(\mathbf{x_1}+\mathbf{x_2}).$$

Hence, since $(\mathbf{x_1}+\mathbf{x_2}) \in V(\lambda)$, the set $V(\lambda)$ is closed with respect to the internal composition law.

Let us consider a scalar $h \in \mathbb{K}$. From the definition of $V(\lambda)$ we know that

$$\mathbf{x} \in V(\lambda) \Rightarrow f(\mathbf{x}) = \lambda \mathbf{x}.$$

It follows that

$$f(h\mathbf{x}) = hf(\mathbf{x}) = h(\lambda \mathbf{x}) = \lambda(h\mathbf{x}).$$

Hence, since $(h\mathbf{x}) \in V(\lambda)$, the set $V(\lambda)$ is closed with respect to the external composition law.

We can conclude that $(V(\lambda),+,\cdot)$ is a vector subspace of $(E,+,\cdot)$. $\square$

Definition 10.19. The vector subspace $(V(\lambda),+,\cdot)$ defined as above is said *eigenspace* of the endomorphism f related to the eigenvalue λ. The dimension of the eigenspace is said *geometric multiplicity* of the eigenvalue λ and is indicated with γ_m.

Example 10.70. Let us consider again the endomorphism $f : \mathbb{R}^2 \to \mathbb{R}^2$

$$f(x,y) = (x+y, 2x).$$

We know that the condition for determining eigenvalues and eigenvectors is given by the system of linear equations

$$\begin{cases} (1-\lambda)x + y = 0 \\ 2x - \lambda y = 0. \end{cases}$$

In order to identify an eigenvalue we need to pose that the matrix associated with the system is singular:

$$\det \begin{pmatrix} (1-\lambda) & 1 \\ 2 & -\lambda \end{pmatrix} = 0.$$

This means that

$$(1-\lambda)(-\lambda) - 2 = 0 \Rightarrow \lambda^2 - \lambda - 2 = 0.$$

The solutions of this polynomial would be the eigenvalues of this endomorphism. The solutions are $\lambda_1 = -1$ and $\lambda_2 = 2$, that are the eigenvalues of the endomorphism.

Let us choose λ_1 for the homogeneous system above:

$$\begin{cases} (1-\lambda_1)x + y = 0 \\ 2x - \lambda_1 y = 0 \end{cases} \Rightarrow \begin{cases} 2x + y = 0 \\ 2x + y = 0. \end{cases}$$

As expected, this system is undetermined and ha ∞^1 solutions of the type $(\alpha, -2\alpha) = \alpha(1, -2)$ with the parameter $\alpha \in \mathbb{R}$. The generic solution $\alpha(1, -2)$ can be interpreted as a set. More specifically this can be interpreted as a line within the plane ($\mathbb{R}^2$).

The theorem above says that $(\alpha(1, -2), +, \cdot)$ is a vector space (and referred to as eigenspace) and a subspace of $(\mathbb{R}^2, +, \cdot)$. The set $\alpha(1, -2)$ is indicated with $V(\lambda_1)$ to emphasize that it has been built after having chosen the eigenvalue λ_1.

10.5.1 Method for Determining Eigenvalues and Eigenvectors

This section conceptualizes in a general fashion the method for determining eigenvalues for any $\mathbb{R}^n \to \mathbb{R}^n$ endomorphisms.

Let $f : E \to E$ be an endomorphism defined over $\mathbb{K}$ and let $(E, +, \cdot)$ be a finite-dimensional vector space having dimension n. A matrix $A \in \mathbb{R}_{n,n}$ is associated with the endomorphism:

$$\mathbf{y} = f(\mathbf{x}) = \mathbf{A}\mathbf{x}$$

where the matrix A is

$$\mathbf{A} = \begin{pmatrix} a_{1,1} & a_{1,2} & \ldots & a_{1,n} \\ a_{2,1} & a_{2,2} & \ldots & a_{2,n} \\ \ldots & \ldots & \ldots & \ldots \\ a_{n,1} & a_{n,2} & \ldots & a_{n,n} \end{pmatrix},$$

the vector $\mathbf{x}$ is

$$\mathbf{x} = \begin{pmatrix} x_1 \\ x_2 \\ \ldots \\ x_n \end{pmatrix}$$

and the vector $\mathbf{y}$ is

$$\mathbf{y} = \begin{pmatrix} y_1 \\ y_2 \\ \ldots \\ y_n \end{pmatrix}.$$

Thus we can write

$$\begin{pmatrix} y_1 \\ y_2 \\ \ldots \\ y_n \end{pmatrix} = \begin{pmatrix} a_{1,1} & a_{1,2} & \ldots & a_{1,n} \\ a_{2,1} & a_{2,2} & \ldots & a_{2,n} \\ \ldots & \ldots & \ldots & \ldots \\ a_{n,1} & a_{n,2} & \ldots & a_{n,n} \end{pmatrix} \begin{pmatrix} x_1 \\ x_2 \\ \ldots \\ x_n \end{pmatrix}.$$

Let us impose that

$$f(x) = \lambda \mathbf{x} = \mathbf{y}$$

that is

$$\begin{pmatrix} a_{1,1} & a_{1,2} & \ldots & a_{1,n} \\ a_{2,1} & a_{2,2} & \ldots & a_{2,n} \\ \ldots & \ldots & \ldots & \ldots \\ a_{n,1} & a_{n,2} & \ldots & a_{n,n} \end{pmatrix} \begin{pmatrix} x_1 \\ x_2 \\ \ldots \\ x_n \end{pmatrix} = \lambda \begin{pmatrix} x_1 \\ x_2 \\ \ldots \\ x_n \end{pmatrix}.$$

This matrix equation corresponds to the following system of linear equations

$$\begin{cases} a_{1,1}x_1 + a_{1,2}x_2 + \ldots + a_{1,n}x_n = \lambda x_1 \\ a_{2,1}x_1 + a_{2,2}x_2 + \ldots + a_{2,n}x_n = \lambda x_2 \\ \ldots \\ a_{n,1}x_1 + a_{n,2}x_2 + \ldots + a_{n,n}x_n = \lambda x_n \end{cases} \Rightarrow$$

$$\Rightarrow \begin{cases} (a_{1,1} - \lambda)x_1 + a_{1,2}x_2 + \ldots + a_{1,n}x_n = 0 \\ a_{2,1}x_1 + (a_{2,2} - \lambda)x_2 + \ldots + a_{2,n}x_n = 0 \\ \ldots \\ a_{n,1}x_1 + a_{n,2}x_2 + \ldots + (a_{n,n} - \lambda)x_n = 0. \end{cases}$$

This is a homogeneous system of linear equations in the components of the eigenvectors related to the eigenvalue λ. Since it is homogeneous, this system always has one solution that is $0,0,\ldots,0$. The null solution is not relevant in this context as eigenvectors cannot be null. If this system has infinite solutions then we can find eigenvectors and eigenvalues. In order to have more than a solution, the determinant of the matrix associated with this system must be null:

$$\det \begin{pmatrix} a_{1,1} - \lambda & a_{1,2} & \ldots & a_{1,n} \\ a_{2,1} & a_{2,2} - \lambda & \ldots & a_{2,n} \\ \ldots & \ldots & \ldots & \ldots \\ a_{n,1} & a_{n,2} & \ldots & a_{n,n} - \lambda \end{pmatrix} = \det\left(\mathbf{A} - \mathbf{I}\lambda\right) = 0.$$

If the calculations are performed a n order polynomial in the variable λ is obtained:

$$p(\lambda) = (-1)^n \lambda^n + (-1)^{n-1} k_{n-1} \lambda^{n-1} + \ldots + (-1) k_1 \lambda + k_0.$$

This polynomial is said *characteristic polynomial* of the endomorphism f. In order to find the eigenvalues we need to find those values of λ such that $p(\lambda) = 0$ and $\lambda \in \mathbb{K}$. This means that although the equation has n roots, some values of λ satisfying the identity to 0 of the characteristic polynomial can be $\notin \mathbb{K}$ and thus not eigenvalues. This can be the case for an endomorphism defined over the $\mathbb{R}$ field. In this case, some roots of the characteristic polynomial can be complex and thus not be eigenvalues.

As a further remark, since the characteristic polynomial is the determinant associated with a homogeneous system of linear equations (that always has at least one solution), this polynomial has at least one root (there is at least one value of λ such that the determinant is null).

Equivalently to what written above, a vector $\mathbf{x}$ is an eigenvector if $\mathbf{A}\mathbf{x} = \lambda\mathbf{x}$. Hence, when the eigenvalue has been determined the corresponding eigenvector is found by solving the homogeneous system of linear equations

$$(\mathbf{A} - \mathbf{I}\lambda)\mathbf{x} = \mathbf{o}$$

where the eigenvalue λ is a constant.

As it can be observed by the examples and on the basis of the vector space theory, if the system $(\mathbf{A} - \mathbf{I}\lambda)\mathbf{x} = \mathbf{o}$ has ∞^k solutions, the geometric multiplicity of the eigenvalue λ is $\gamma_m = k$.

Let us say that If we indicate with ρ the rank of $(\mathbf{A} - \mathbf{I}\lambda)$ then for a fixed eigenvalue λ it follows that

$$\gamma_m = n - \rho.$$

This statement can be reformulated considering that $(\mathbf{A} - \mathbf{I}\lambda)$ represents a matrix associated with an endomorphism $\mathbb{R}^n \to \mathbb{R}^n$. The kernel of this endomorphism is the solution of the system $(\mathbf{A} - \mathbf{I}\lambda)\mathbf{x} = \mathbf{o}$. Thus, the geometric multiplicity γ_m is the dimension of the kernel.

This is the number of linearly independent eigenvectors associated with the eigenvalue λ that is the dimension of the associated eigenspace.

Definition 10.20. Let $f : E \to E$ be an endomorphism. Let $p(\lambda)$ be the order n characteristic polynomial related to the endomorphism. Let λ_0 be a zero of the characteristic polynomial. This characteristic polynomial is said of *algebraic multiplicity* $r \le n$ if it is divisible by $(\lambda - \lambda_0)^r$ and not by $(\lambda - \lambda_0)^{r+1}$.

Example 10.71. Let us consider an endomorphism $f : \mathbb{R}^2 \to \mathbb{R}^2$ over the field $\mathbb{R}$ represented by

$$f(x,y) = (3x + 2y, 2x + y)$$

corresponding to the following matrix

$$\mathbf{A} = \begin{pmatrix} 3 & 2 \\ 2 & 1 \end{pmatrix}.$$

Let us compute the eigenvalues. At first let us calculate the characteristic polynomial

$$\det \begin{pmatrix} 3 - \lambda & 2 \\ 2 & 1 - \lambda \end{pmatrix} = (3 - \lambda)(1 - \lambda) - 4 =$$
$$= 3 - 3\lambda - \lambda + \lambda^2 - 4 = \lambda^2 - 4\lambda - 1.$$

The roots of the polynomial are $\lambda_1 = 1 + \frac{\sqrt{3}}{2}$ and $\lambda_2 = 1 - \frac{\sqrt{3}}{2}$. They are both eigenvalues. In order to find the eigenvectors we need to solve the two following systems of linear equations:

$$\begin{pmatrix} 2 - \frac{\sqrt{3}}{2} & 2 \\ 2 & -\frac{\sqrt{3}}{2} \end{pmatrix} \begin{pmatrix} x_1 \\ x_2 \end{pmatrix} = \begin{pmatrix} 0 \\ 0 \end{pmatrix}$$

and

$$\begin{pmatrix} 2 + \frac{\sqrt{3}}{2} & 2 \\ 2 & \frac{\sqrt{3}}{2} \end{pmatrix} \begin{pmatrix} x_1 \\ x_2 \end{pmatrix} = \begin{pmatrix} 0 \\ 0 \end{pmatrix}.$$

Example 10.72. Let us consider now the endomorphism $f : \mathbb{R}^3 \to \mathbb{R}^3$ defined over the field $\mathbb{R}$ and associated with the following matrix

$$\mathbf{A} = \begin{pmatrix} 0 & 1 & 0 \\ -2 & -3 & \frac{1}{2} \\ 0 & 0 & 0 \end{pmatrix}.$$

In order to find the eigenvalues we have to calculate

$$\det \begin{pmatrix} -\lambda & 1 & 0 \\ -2 & -3 - \lambda & \frac{1}{2} \\ 0 & 0 & -\lambda \end{pmatrix} = -\lambda \left(\lambda^2 + 3\lambda + 2 \right).$$

Hence, the eigenvalues are $\lambda_1 = 0$, $\lambda_2 = -2$, and $\lambda_3 = -1$. By substituting the eigenvalues into the matrix $(\mathbf{A} - \mathbf{I}\lambda)$ we obtain three homogeneous systems of linear equations whose solutions are the eigenvectors.

Theorem 10.10. *Let $f : E \to E$ be an endomorphism and let $(E, +, \cdot)$ be a finite-dimensional vector space having dimension n. Let $\mathbf{x_1}, \mathbf{x_2}, \ldots, \mathbf{x_p}$ be p eigenvectors of the endomorphism f related to the eigenvalues $\lambda_1, \lambda_2, \ldots, \lambda_p$. Let these eigenvalues be all distinct roots of the characteristic polynomial $p(\lambda)$. It follows that the eigenvectors are all linearly independent.*

Proof. Let us assume, by contradiction, that the eigenvectors are linearly dependent. Without a loss of generality, let us assume that the first $r < p$ eigenvectors are the minimum number of linearly dependent vectors. Thus, we can express one of them as lineal combination of the others by means of $l_1, l_2, \ldots, l_{r-1}$ scalars:

$$\mathbf{x_r} = l_1 \mathbf{x_1} + l_2 \mathbf{x_2} + \ldots + l_{r-1} \mathbf{x_{r-1}}.$$

From this equation, we derive two more equations. The first is obtained by multiplying the terms by λ_r

$$\lambda_r \mathbf{x_r} = \lambda_r l_1 \mathbf{x_1} + \lambda_r l_2 \mathbf{x_2} + \ldots + \lambda_r l_{r-1} \mathbf{x_{r-1}}$$

while the second is obtained by calculating the linear mapping of the terms

$$f(\mathbf{x_r}) = \lambda_r \mathbf{x_r} = f(l_1 \mathbf{x_1} + l_2 \mathbf{x_2} + \ldots + l_{r-1} \mathbf{x_{r-1}}) =$$
$$l_1 f(\mathbf{x_1}) + l_2 f(\mathbf{x_2}) + \ldots + l_{r-1} f(\mathbf{x_{r-1}}) =$$
$$\lambda_1 l_1 \mathbf{x_1} + \lambda_2 l_2 \mathbf{x_2} + \ldots + \lambda_{r-1} l_{r-1} \mathbf{x_{r-1}}.$$

Thus, the last two equations are equal. If we subtract the second equation from the first one, we obtain

$$\mathbf{o_E} = l_1 (\lambda_r - \lambda_1) \mathbf{x_1} + l_2 (\lambda_r - \lambda_2) \mathbf{x_2} + \ldots + l_1 (\lambda_r - \lambda_{r-1}) \mathbf{x_{r-1}}.$$

Thus, the null vector is expressed as linear combination of $r - 1$ linearly independent vectors. This may occur only if the scalars are all null:

$$l_1 (\lambda_r - \lambda_1) = 0$$
$$l_2 (\lambda_r - \lambda_2) = 0$$
$$\ldots$$
$$l_{r-1} (\lambda_r - \lambda_{r-1}) = 0.$$

Since for hypothesis the eigenvalues are all distinct

$$(\lambda_r - \lambda_1) \neq 0$$
$$(\lambda_r - \lambda_2) \neq 0$$
$$\cdots$$
$$(\lambda_r - \lambda_{r-1}) \neq 0.$$

Hence, it must be that $l_1, l_2, \ldots, l_{r-1} = 0, 0, \ldots, 0$. It follows that

$$\mathbf{x_r} = l_1 \mathbf{x_1} + l_2 \mathbf{x_2} + \ldots + l_{r-1} \mathbf{x_{r-1}} = \mathbf{o_E}.$$

This is impossible because an eigenvector is non-null by definition. Hence, we reached a contradiction. □

Example 10.73. We know that for the endomorphism $f : \mathbb{R}^2 \to \mathbb{R}^2$

$$f(x,y) = (x+y, 2x)$$

$V(\lambda_1) = \alpha(1, -2)$ and $(1, -2)$ is an eigenvector. Let us determine $V(\lambda_2)$. For $\lambda_2 = 2$ the system is

$$\begin{cases} (1-\lambda_2)x + y = 0 \\ 2x - \lambda_2 y = 0 \end{cases} \Rightarrow \begin{cases} -x + y = 0 \\ 2x - 2y = 0 \end{cases}$$

whose generic solution is $\alpha(1,1)$ with $\alpha \in \mathbb{R}$. Thus, $(1,1)$ is an eigenvector associated with λ_2. Theorem 10.10 states that since $\lambda_1 \neq \lambda_2$ (are distinct roots of the same polynomial) then the corresponding eigenvectors are linearly independent.

We can easily check that $(1, -2)$ and $(1,1)$ are linearly independent.

Let us verify that these vectors are eigenvectors

$$\begin{pmatrix} 1 & 1 \\ 2 & 0 \end{pmatrix} \begin{pmatrix} 1 \\ -2 \end{pmatrix} = \begin{pmatrix} -1 \\ 2 \end{pmatrix} = \lambda_1 \begin{pmatrix} 1 \\ -2 \end{pmatrix}$$

and

$$\begin{pmatrix} 1 & 1 \\ 2 & 0 \end{pmatrix} \begin{pmatrix} 1 \\ 1 \end{pmatrix} = \begin{pmatrix} 2 \\ 2 \end{pmatrix} = \lambda_2 \begin{pmatrix} 1 \\ 1 \end{pmatrix}.$$

These two vectors, separately, span two eigenspaces:

$$V(-1) = L((1, -2))$$
$$V(2) = L((1,1)).$$

The geometric multiplicity of both the eigenvalues is equal to 1.

Let us graphically visualize some of these results. If we consider the eigenvector $\mathbf{x} = (1,1)$ (solid line) and the transformed vector $f(\mathbf{x}) = (2,2)$ (dashed line) we have

In general, we can interpret an eigenvector $\mathbf{x}$ as a vector such that its transformed $f(\mathbf{x})$ is parallel to $\mathbf{x}$. The vector space $(V(\lambda_2), +, \cdot)$, with $\lambda_2 = 2$, is the infinite line of the plane $\mathbb{R}^2$ having the same direction of the eigenvector and its transformed (dotted line):

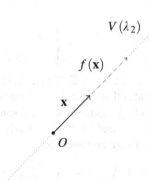

This fact can be expressed by stating that all the vectors on the dotted line, such as $(0.1, 0.1)$, $(3, 3)$, $(30, 30)$, and $(457, 457)$, are all eigenvectors.

If we consider a vector which is not an eigenvector and its transformed they are not parallel. For example for $\mathbf{v} = (1,3)$ its transformed is $f(\mathbf{v}) = (4,2)$. The two vectors are not parallel:

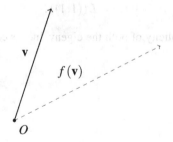

Example 10.74. Let us consider the following endomorphism $f : \mathbb{R}^2 \rightarrow \mathbb{R}^2$

$$f(x,y) = (x+3y, x-y).$$

An eigenvalue of this endomorphism is a value λ such that $f(x,y) = \lambda(x,y)$:

$$\begin{cases} x+3y = \lambda x \\ x-y = \lambda y \end{cases} \Rightarrow \begin{cases} (1-\lambda)x+3y = 0 \\ x+(-1-\lambda)y = 0. \end{cases}$$

The system in undetermined (the associated incomplete matrix is singular) when

$$(1-\lambda)(-1-\lambda)-3 = \lambda^2 - 4 = 0.$$

This equation is verified for $\lambda_1 = 2$ and $\lambda_2 = -2$.
In order to calculate the eigenvector associated with $\lambda_1 = 2$ we write

$$\begin{cases} (1-\lambda_1)x+3y = 0 \\ x+(-1-\lambda_1)y = 0 \end{cases} \Rightarrow \begin{cases} -x+3y = 0 \\ x-3y = 0 \end{cases}$$

whose solution is $\alpha(3,1)$ with $\alpha \in \mathbb{R}$.
For the calculation of the eigenvector associated with $\lambda_2 = -2$ we write

$$\begin{cases} (1-\lambda_2)x+3y = 0 \\ x+(-1-\lambda_2)y = 0 \end{cases} \Rightarrow \begin{cases} 3x+3y = 0 \\ x+y = 0 \end{cases}$$

whose solution is $\alpha(1,-1)$ with $\alpha \in \mathbb{R}$.
Again, we have distinct eigenvalues and linearly independent eigenvectors $(3,1)$ and $(1,-1)$. These two vectors, separately, span two eigenspaces:

$$V(2) = L((3,1))$$
$$V(-2) = L((1,-1)).$$

The geometric multiplicity of λ_1 and λ_2 is 1.

Example 10.75. For the following mapping $f : \mathbb{R}^2 \rightarrow \mathbb{R}^2$ defined as

$$f(x,y) = (3x, 3y)$$

let us impose that $f(x,y) = \lambda(x,y)$.
The following system of linear equation is imposed

$$\begin{cases} 3x = \lambda x \\ 3y = \lambda y \end{cases} \Rightarrow \begin{cases} (3-\lambda)x = 0 \\ (3-\lambda)y = 0. \end{cases}$$

The determinant of the matrix associated with the system, $(3-\lambda)^2$, is null only for $\lambda = 3$, which is a double eigenvalue. If we substitute $\lambda = 3$ we obtain

$$\begin{cases} 0x = 0 \\ 0y = 0 \end{cases}$$

which is always verified. The system is still of two equations in two variables. The rank of the associated matrix is $\rho = 0$. Hence, there exist ∞^2 solutions of the type (α, β) with $\alpha, \beta \in \mathbb{R}$.

The generic solution of the system is $\alpha(1,0) + \beta(0,1)$. The eigenspace is spanned by the two vectors $(1,0)$ and $(0,1)$:

$$V(3) = L((1,0),(0,1))$$

that is the entire plane $\mathbb{R}^2$. This means that every vector, except for the null vector, is an eigenvector.

The geometric multiplicity of the eigenvalue $\lambda = 3$ is 2.

Example 10.76. Let us consider the following linear mapping $f : \mathbb{R}^3 \to \mathbb{R}^3$ defined as

$$f(x,y,z) = (x+y, 2y+z, -4z).$$

Let us search the eigenvalues

$$\begin{cases} x+y = \lambda x \\ 2y+z = \lambda y \\ -4z = \lambda z \end{cases} \Rightarrow \begin{cases} (1-\lambda)x + y = 0 \\ (2-\lambda)y + z = 0 \\ (-4-\lambda)z = 0 \end{cases}$$

whose determinant of the associated matrix is

$$\det \begin{pmatrix} (1-\lambda) & 1 & 0 \\ 0 & (2-\lambda) & 1 \\ 0 & 0 & (-4-\lambda) \end{pmatrix} = (1-\lambda)(2-\lambda)(-4-\lambda).$$

The eigenvalues are all distinct $\lambda_1 = 1$, $\lambda_2 = 2$, and $\lambda_3 = -4$.

Let us substitute $\lambda_1 = 1$ into the system:

$$\begin{cases} y = 0 \\ y+z = 0 \\ -5z = 0. \end{cases}$$

The associate solution is $\alpha(1,0,0)$, with $\alpha \in \mathbb{R} \setminus (0,0,0)$. Nonetheless, the null vector belongs to the eigenspace:

$$V(1) = L((1,0,0)).$$

We can conclude that the geometric multiplicity of the eigenvalue $\lambda_1 = 1$ is 0.

Let us substitute $\lambda_2 = 2$ into the system:

$$\begin{cases} -x+y=0 \\ z=0 \\ -6z=0 \end{cases}$$

whose solution is $\alpha(1,1,0)$ with $\alpha \in \mathbb{R}$. The eigenspace is

$$V(2) = L((1,1,0)).$$

Let us substitute $\lambda_3 = -4$ into the system:

$$\begin{cases} -3x+y=0 \\ -2y+z=0 \\ 0z=0. \end{cases}$$

The last equation is always verified. By posing $x = \alpha$ we have $y = 3\alpha$ and $z = 6\alpha$, i.e. the solution is $\alpha(1,3,6)$ and the eigenspace, is

$$V(-4) = L((1,3,6)).$$

Example 10.77. Let us analyse another case where the eigenvalues are not distinct, i.e. the following linear mapping $f : \mathbb{R}^3 \to \mathbb{R}^3$

$$f(x,y,z) = (x+z, 2y, -x+3z).$$

By applying the definition of eigenvalue we write

$$\begin{cases} x+z = \lambda x \\ 2y = \lambda y \\ -x+3z = \lambda z \end{cases} \Rightarrow \begin{cases} (1-\lambda)x+z=0 \\ (2-\lambda)y=0 \\ -x+(3-\lambda)z=0. \end{cases}$$

This system is undetermined when

$$\det \begin{pmatrix} (1-\lambda) & 0 & 1 \\ 0 & (2-\lambda) & 0 \\ -1 & 0 & (3-\lambda) \end{pmatrix} = (2-\lambda)^3 = 0$$

that is when $\lambda = 2$. This means that only one eigenvector can be calculated, i.e. the three eigenvectors are linearly dependent.

By substituting into the system we have

$$\begin{cases} -x+z=0 \\ 0y=0 \\ -x+z=0. \end{cases}$$

The second equation is always verified while the first and the third say that $x = z$. Equivalently, we can see that this system has rank $\rho = 1$ and thus ∞^2 solutions. If we pose $x = \alpha$ and $y = \beta$ we have that the generic solution $(\alpha, \beta, \alpha) = \alpha(1, 0, 1) + \beta(0, 1, 0)$. The eigenspace is thus spanned by the vectors $(1, 0, 1)$ and $(0, 1, 0)$, that can be interpreted as a plane of the space.

Example 10.78. Finally, let us calculate eigenvalues, eigenvectors, and eigenspace for the following mapping $f : \mathbb{R}^3 \to \mathbb{R}^3$ defined as

$$f(x, y, z) = (5x, 5y, 5z).$$

From the system of linear equations

$$\begin{cases} 5x = \lambda x \\ 5y = \lambda y \\ 5z = \lambda z \end{cases} \Rightarrow \begin{cases} (5 - \lambda)x = 0 \\ (5 - \lambda)y = 0 \\ (5 - \lambda)z = 0 \end{cases}$$

we can see that the determinant associated with the matrix is $(\lambda - 5)^3 = 0$. It follows that the only eigenvalue is triple and $\lambda = 5$.

By substituting the eigenvalue into the system we have

$$\begin{cases} 0x = 0 \\ 0y = 0 \\ 0z = 0 \end{cases}$$

whose rank is $\rho = 0$. We have ∞^3 solutions of the type (α, β, γ) with $\alpha, \beta, \gamma \in \mathbb{R}$ that is

$$\alpha(1, 0, 0) + \beta(0, 1, 0) + \gamma(0, 0, 1).$$

The span of the resulting eigenspace is

$$V(5) = L((1, 0, 0), (0, 1, 0), (0, 0, 1)).$$

This means that every vector of the space, except for the null vector, is an eigenvector. This means that the eigenspace is the entire space $\mathbb{R}^3$.

The geometric multiplicity of the eigenvalue $\lambda = 5$ as well as its algebraic multiplicity is 3.

Proposition 10.9. *Let $f : E \to E$ be an endomorphism. Let $p(\lambda)$ be the order n characteristic polynomial related to the endomorphism. Let γ_m and α_m be the geometric and algebraic multiplicity, respectively, of the characteristic polynomial. It follows that*

$$1 \leq \gamma_m \leq \alpha_m \leq n.$$

Definition 10.21. Let $f : E \to E$ be and endomorphism and $\lambda_1, \lambda_2, \ldots, \lambda_n$ the eigenvalues associated with it. The set of the eigenvalues

$$Sp = \{\lambda_1, \lambda_2, \ldots, \lambda_n\}$$

is said *spectrum* of the endomorphism while

$$sr = \max_i |\lambda_i|$$

is said *spectral radius*.

Theorem 10.11. Gerschgorin's Theorem. *Let* $\mathbf{A} \in \mathbb{C}_{n,n}$ *be a matrix associated with an endomorphism* $f : \mathbb{C}^n \to \mathbb{C}^n$. *Let us consider the following circular sets in the complex plane:*

$$C_i = \left\{ z \in \mathbb{C} \middle| |z - a_{i,j}| \leq \sum_{j=1, i \neq j}^{n} |a_{i,j}| \right\}$$

$$D_i = \left\{ z \in \mathbb{C} \middle| |z - a_{i,j}| \leq \sum_{i=1, i \neq j}^{n} |a_{i,j}| \right\}.$$

For each eigenvalue λ *it follows that*

$$\lambda \in (\cup_i^n = C_i) \cap (\cup_i^n = D_i).$$

Although the proof and details of the Gerschgorin's Theorem are out of the scope of this book, it is worthwhile picturing the meaning of this result. This theorem, for a given endomorphism and thus its associated matrix, allows to make an estimation of the region of the complex plane where the eigenvalues will be located. Of course, since real numbers are a special case of complex numbers, this result is valid also for matrices $\in \mathbb{R}_{n,n}$.

10.6 Matrix Diagonalization

Theorem 10.12. *Let* $\mathbf{A}$ *and* $\mathbf{A}'$ *be two similar matrices. These two matrices have the same characteristic polynomial:*

$$\det\left(\mathbf{A}' - \mathbf{I}\lambda\right) = \det\left(\mathbf{A} - \mathbf{I}\lambda\right).$$

Proof. If the two matrices are similar then $\exists$ a non-singular matrix $\mathbf{P}$ such that

$$\mathbf{A}' = \mathbf{P}^{-1}\mathbf{A}\mathbf{P}.$$

Hence, considering that the identity matrix, being the neutral element of the product between matrices, is similar to itself ($\mathbf{I} = \mathbf{P}^{-1}\mathbf{I}\mathbf{P}$ with $\mathbf{P}$ arbitrary non-singular matrix),

$$\det\left(\mathbf{A}' - \mathbf{I}\lambda\right) = \det\left(\mathbf{P}^{-1}\mathbf{A}\mathbf{P} - \mathbf{I}\lambda\right) =$$

$$= \det\left(\mathbf{P}^{-1}\mathbf{A}\mathbf{P} - \left(\mathbf{P}^{-1}\mathbf{I}\mathbf{P}\right)\lambda\right) = \det\left(\mathbf{P}^{-1}\left(\mathbf{A} - \mathbf{I}\lambda\right)\mathbf{P}\right) =$$

$$= \frac{1}{\det\left(\mathbf{P}\right)}\det\left(\mathbf{A} - \mathbf{I}\lambda\right)\det\left(\mathbf{P}\right) = \det\left(\mathbf{A} - \mathbf{I}\lambda\right). \quad \square$$

This means that two similar matrices have also the same eigenvalues.

Example 10.79. From Example 10.66 we know that the following two matrices are similar:

$$\mathbf{A} = \begin{pmatrix} 2 & 2 & 0 \\ 0 & 5 & 0 \\ 0 & 0 & 1 \end{pmatrix}$$

and

$$\mathbf{A}' = \begin{pmatrix} 1 & -2 & 0 \\ 0 & 5 & 0 \\ 1 & 12 & 2 \end{pmatrix}.$$

By calculating

$$\det\left(\mathbf{A} - \lambda\mathbf{I}\right)$$

and

$$\det\left(\mathbf{A}' - \lambda\mathbf{I}\right)$$

we may verify that the characteristic polynomial in both cases is

$$-\lambda^3 + 8\lambda^2 - 17\lambda + 10.$$

The eigenvalues of both the matrices are $\lambda_1 = 2$, $\lambda_2 = 1$, $\lambda_3 = 5$.

As mentioned in Chap. 2 a *diagonal matrix* is a matrix whose extra-diagonal elements are all zeros. Hence, a diagonal matrix is of the form

$$\begin{pmatrix} \gamma_1, 0, 0, \ldots, 0 \\ 0, \gamma_2, 0, \ldots, 0 \\ 0, 0, \gamma_3, \ldots, 0 \\ \cdots \\ 0, 0, 0, \ldots, \gamma_n \end{pmatrix}.$$

Definition 10.22. Let $f : E \to E$ be an endomorphism. The endomorphism is *diagonalizable* if there exists a basis spanning $(E, +, \cdot)$ with respect to which the matrix defining this mapping is diagonal.

In other words, the endomorphism is *diagonalizable* if it can transformed, by a basis transformation (a coordinate transformation), into a diagonal mapping.

If we remember that a basis spanning $(E, +, \cdot)$ means a transformation matrix **P** whose columns are the vectors composing the basis, we can give an equivalent definition from the perspective of matrices.

Definition 10.23. A square matrix **A** is *diagonalizable* if it is similar to a diagonal matrix: $\exists$ a non-singular matrix **P** such that

$$\mathbf{D} = \mathbf{P}^{-1}\mathbf{A}\mathbf{P}$$

where **D** is a diagonal matrix and **P** is said *transformation matrix*.

By combining the latter definition with Theorem 10.12, for a diagonalizable matrix **A**, there exists a diagonal matrix **D** with the same eigenvalues.

Theorem 10.13. *Let $f : E \to E$ be a finite-dimensional endomorphism being associated with a matrix **A** having order n. Let $\lambda_1, \lambda_2, \ldots, \lambda_n$ be a set of scalars and $\mathbf{x}_1, \mathbf{x}_2, \ldots, \mathbf{x}_n$ be n vectors having n dimensions.*

*Let **P** be a $n \times n$ matrix whose columns are the vectors $\mathbf{x}_1, \mathbf{x}_2, \ldots, \mathbf{x}_n$:*

$$\mathbf{P} = \begin{pmatrix} \mathbf{x}_1 & \mathbf{x}_2 & \ldots & \mathbf{x}_n \end{pmatrix}.$$

*and let **D** be a diagonal matrix whose diagonal elements are the scalars $\lambda_1, \lambda_2, \ldots, \lambda_n$:*

$$\mathbf{D} = \begin{pmatrix} \lambda_1, 0, 0, \ldots, 0 \\ 0, \lambda_2, 0, \ldots, 0 \\ 0, 0, \lambda_3, \ldots, 0 \\ \ldots \\ 0, 0, 0, \ldots, \lambda_n \end{pmatrix}.$$

*It follows that if **A** and **D** are similar, i.e.*

$$\mathbf{D} = \mathbf{P}^{-1}\mathbf{A}\mathbf{P},$$

then $\lambda_1, \lambda_2, \ldots, \lambda_n$ are the eigenvalues of the mapping and $\mathbf{x}_1, \mathbf{x}_2, \ldots, \mathbf{x}_n$ are the corresponding eigenvectors.

Proof. If **A** and **D** are similar then

$$\mathbf{D} = \mathbf{P}^{-1}\mathbf{A}\mathbf{P} \Rightarrow \mathbf{P}\mathbf{D} = \mathbf{A}\mathbf{P}$$

it follows that

$$\mathbf{A}\mathbf{P} = \mathbf{A} \begin{pmatrix} \mathbf{x}_1 & \mathbf{x}_2 & \ldots & \mathbf{x}_n \end{pmatrix} = \begin{pmatrix} \mathbf{A}\mathbf{x}_1 & \mathbf{A}\mathbf{x}_2 & \ldots & \mathbf{A}\mathbf{x}_n \end{pmatrix}$$

and

$$PD = \begin{pmatrix} x_1 & x_2 & \dots & x_n \end{pmatrix} \begin{pmatrix} \lambda_1 & 0 & \dots & 0 \\ 0 & \lambda_2 & \dots & 0 \\ \dots & \dots & \dots & \dots \\ 0 & 0 & \dots & \lambda_n \end{pmatrix} = \begin{pmatrix} \lambda_1 x_1 & \lambda_2 x_2 & \dots & \lambda_n x_n \end{pmatrix}.$$

Since $AP = PD$, it follows that

$$\begin{pmatrix} Ax_1 & Ax_2 & \dots & Ax_n \end{pmatrix} = \begin{pmatrix} \lambda_1 x_1 & \lambda_2 x_2 & \dots & \lambda_n x_n \end{pmatrix}.$$

that is

$$Ax_1 = \lambda_1 x_1$$
$$Ax_2 = \lambda_2 x_2$$
$$\dots$$
$$Ax_n = \lambda_2 x_n.$$

This means that $\lambda_1, \lambda_2, \dots, \lambda_n$ are the eigenvalues of the mapping and $x_1, x_2, \dots, x_n$ are the corresponding eigenvectors. $\square$

The diagonalization can be seen from two equivalent perspectives. According to the first one, the diagonalization is a matrix transformation that aims at generating a diagonal matrix. According to the second one, the diagonalization is a basis transformation into a reference system where the contribution of each component is only along one axis.

As a combination of these two perspectives, if the matrix A is seen like a matrix associated with a system of linear equations, the diagonalization transforms the original system into an equivalent system (i.e. having the same solutions) where all the variables are uncoupled, i.e. every equation is in only one variable. The transformation matrix associated with the basis transformation is a matrix P whose columns are the eigenvectors of the mapping. Then the diagonal matrix D is calculated as $D = P^{-1}AP$.

The result above can be rephrased in the following theorem.

Theorem 10.14. *Let $f : E \to E$ be an endomorphism having n dimensions (it is thus finite-dimensional) associated with the matrix A. If the mapping is diagonalizable then*

$$D = P^{-1}AP$$

where

$$D = \begin{pmatrix} \lambda_1 & 0 & \dots & 0 \\ 0 & \lambda_2 & \dots & 0 \\ \dots & \dots & \dots & \dots \\ 0 & 0 & \dots & \lambda_n \end{pmatrix}$$

with $\lambda_1, \lambda_2, \dots, \lambda_n$ eigenvalues of the mapping (not necessarily distinct) and

$$P = \begin{pmatrix} x_1 & x_2 & \dots & x_n \end{pmatrix}$$

with $x_1, x_2, \dots, x_n$ eigenvectors corresponding to the eigenvalues $\lambda_1, \lambda_2, \dots, \lambda_n$.

The following theorems describe the conditions under which an endomorphism is diagonalizable.

Proposition 10.10. *Let* $f : E \rightarrow E$ *be an endomorphism defined by an order n matrix* **A**. *If the matrix (and thus the endomorphism) is diagonalizable, then its eigenvectors are linearly independent.*

Proof. If the matrix is diagonalizable, then there exists a non-singular matrix **P** such that $\mathbf{D} = \mathbf{P}^{-1}\mathbf{A}\mathbf{P}$, with **D** diagonal.

For Theorem 10.13, the columns of the matrix **P** are the eigenvectors. Since the matrix **P** is invertible, then it is non-singular. Thus, the columns of **P**, i.e. the eigenvectors of the mapping, are linearly independent.   □

If the *n* eigenvectors are all linearly independent, they can be placed as columns of the matrix **P** which would result non-singular. Thus, for Theorem 10.13 it results that $\mathbf{D} = \mathbf{P}^{-1}\mathbf{A}\mathbf{P}$ is diagonal. Thus, **A** is diagonalizable.

Theorem 10.15. Diagonalization Theorem. *Let* $f : E \rightarrow E$ *be an endomorphism defined by a matrix* **A**. *The matrix (and thus the endomorphism) is diagonalizable if and only if has n linearly independent eigenvectors, i.e. one of the following condition occurs:*

- *all the eigenvalues are distinct*
- *the algebraic multiplicity of each eigenvalue coincides with its geometric multiplicity.*

The first condition can be easily proved by considering that the eigenvectors associated with distinct eigenvalues are linearly independent and thus the matrix **P** would be non-singular. It would easily follow that $\mathbf{D} = \mathbf{P}^{-1}\mathbf{A}\mathbf{P}$ is diagonal and **A** diagonalizable.

Example 10.80. Let us consider an endomorphism associated with the following matrix

$$\mathbf{A} = \begin{pmatrix} 1 & 2 & 0 \\ 0 & 3 & 0 \\ 2 & -4 & 2 \end{pmatrix}$$

Let us find the eigenvalues

$$\det \begin{pmatrix} 1-\lambda & 2 & 0 \\ 0 & 3-\lambda & 0 \\ 2 & -4 & 2-\lambda \end{pmatrix} = (1-\lambda)(2-\lambda)(3-\lambda).$$

The roots of the characteristic polynomial are $\lambda_1 = 3$, $\lambda_2 = 2$, and $\lambda_3 = 1$. Since the eigenvalues are all distinct, the matrix is diagonalizable.

In order to find the corresponding eigenvectors we need to solve the following systems of linear equations

$$\begin{pmatrix} -2 & 2 & 0 \\ 0 & 0 & 0 \\ 2 & -4 & -1 \end{pmatrix} \begin{pmatrix} x_1 \\ x_2 \\ x_3 \end{pmatrix} = \begin{pmatrix} 0 \\ 0 \\ 0 \end{pmatrix}$$

whose ∞^1 solutions are proportional to

$$\begin{pmatrix} -1 \\ -1 \\ 2 \end{pmatrix},$$

then,

$$\begin{pmatrix} -1 & 2 & 0 \\ 0 & 1 & 0 \\ 2 & -4 & 0 \end{pmatrix} \begin{pmatrix} x_1 \\ x_2 \\ x_3 \end{pmatrix} = \begin{pmatrix} 0 \\ 0 \\ 0 \end{pmatrix}$$

whose ∞^1 solutions are proportional to

$$\begin{pmatrix} 0 \\ 0 \\ 1 \end{pmatrix},$$

and

$$\begin{pmatrix} 0 & 2 & 0 \\ 0 & 2 & 0 \\ 2 & -4 & 1 \end{pmatrix} \begin{pmatrix} x_1 \\ x_2 \\ x_3 \end{pmatrix} = \begin{pmatrix} 0 \\ 0 \\ 0 \end{pmatrix}$$

whose ∞^1 solutions are proportional to

$$\begin{pmatrix} -1 \\ 0 \\ 2 \end{pmatrix}.$$

Hence, the transformation matrix $\mathbf{P}$ is given by

$$\mathbf{P} = \begin{pmatrix} -1 & 0 & -1 \\ -1 & 0 & 0 \\ 2 & 1 & 2 \end{pmatrix}.$$

The inverse of this matrix is

$$\mathbf{P}^{-1} = \begin{pmatrix} 0 & -1 & 0 \\ 2 & 0 & 1 \\ -1 & 1 & 0 \end{pmatrix}.$$

It can be observed that

$$\mathbf{P}^{-1}\mathbf{A}\mathbf{P} =$$

$$= \begin{pmatrix} 0 & -1 & 0 \\ 2 & 0 & 1 \\ -1 & 1 & 0 \end{pmatrix} \begin{pmatrix} 1 & 2 & 0 \\ 0 & 3 & 0 \\ 2 & -4 & 2 \end{pmatrix} \begin{pmatrix} -1 & 0 & -1 \\ -1 & 0 & 0 \\ 2 & 1 & 2 \end{pmatrix} =$$

$$= \mathbf{D} = \begin{pmatrix} 3 & 0 & 0 \\ 0 & 2 & 0 \\ 0 & 0 & 1 \end{pmatrix}.$$

Example 10.81. Let us consider an endomorphism associated with the following matrix

$$\mathbf{A} = \begin{pmatrix} -8 & 18 & 2 \\ -3 & 7 & 1 \\ 0 & 0 & 1 \end{pmatrix}$$

The characteristic polynomial

$$p(\lambda) = \det(\mathbf{A} - \mathbf{I}\lambda) = (2 + \lambda)(\lambda - 1)^2$$

has roots $\lambda_1 = -2$ with multiplicity 1 and $\lambda_2 = 1$ with multiplicity 2.

In order to find the eigenvector associated with $\lambda_1 = -2$ we need to solve the system of linear equations

$$\begin{pmatrix} -6 & 18 & 2 \\ -3 & 9 & 1 \\ 0 & 0 & 3 \end{pmatrix} \begin{pmatrix} x_1 \\ x_2 \\ x_3 \end{pmatrix} = \begin{pmatrix} 0 \\ 0 \\ 0 \end{pmatrix}.$$

The ∞^1 solutions of this system are proportional to

$$\begin{pmatrix} 3 \\ 1 \\ 0 \end{pmatrix}.$$

Hence, the eigenvalue λ_1 has geometric multiplicity 1.

In order to find the eigenvectors associated with $\lambda_2 = 1$ we need to solve the system of linear equations

$$\begin{pmatrix} -9 & 18 & 2 \\ -3 & 6 & 1 \\ 0 & 0 & 0 \end{pmatrix} \begin{pmatrix} x_1 \\ x_2 \\ x_3 \end{pmatrix} = \begin{pmatrix} 0 \\ 0 \\ 0 \end{pmatrix}.$$

The ∞^1 solutions of this system are proportional to

$$\begin{pmatrix} \frac{1}{6} \\ \frac{1}{36} \\ 1 \end{pmatrix}$$

Hence, the eigenvalue λ_2 has geometric multiplicity 1. Since the algebraic and geometric multiplicities are not the same, the endomorphism is not diagonalizable.

Example 10.82. Let us finally consider an endomorphism associated with the following matrix

$$\mathbf{A} = \begin{pmatrix} 8 & -18 & 0 \\ 3 & -7 & 0 \\ 0 & 0 & -1 \end{pmatrix}.$$

The characteristic polynomial is $p(\lambda) = (1 + \lambda)^2 (\lambda - 2)$. Hence the eigenvalues are $\lambda_1 = 2$ with algebraic multiplicity 1 and $\lambda_2 = -1$ with algebraic multiplicity 2.

In order to find the eigenvectors associated with $\lambda_1 = 2$ we need to solve the system of linear equations

$$\begin{pmatrix} 6 & -18 & 0 \\ 3 & -9 & 0 \\ 0 & 0 & -3 \end{pmatrix} \begin{pmatrix} x_1 \\ x_2 \\ x_3 \end{pmatrix} = \begin{pmatrix} 0 \\ 0 \\ 0 \end{pmatrix}.$$

The ∞^1 solutions of this system are proportional to

$$\begin{pmatrix} 3 \\ 1 \\ 0 \end{pmatrix}.$$

Thus, the geometric multiplicity of λ_1 is 1.

In order to find the eigenvectors associated with $\lambda_2 = -1$ we need to solve the system of linear equations

$$\begin{pmatrix} 9 & -18 & 0 \\ 3 & -6 & 0 \\ 0 & 0 & 0 \end{pmatrix} \begin{pmatrix} x_1 \\ x_2 \\ x_3 \end{pmatrix} = \begin{pmatrix} 0 \\ 0 \\ 0 \end{pmatrix}.$$

Since the first two rows are linearly dependent and the third is null, the rank of the system is 1 and the system has ∞^2 solutions. The generic solution depends on two parameters $\alpha, \beta \in \mathbb{R}$ and is

$$\begin{pmatrix} 2\alpha \\ \alpha \\ \beta \end{pmatrix}.$$

The eigenvectors can be written as $(2, 1, 0)$ and $(0, 0, 1)$. Hence, the geometric multiplicity of λ_2 is 2. Thus, this endomorphism is diagonalizable.

A diagonal matrix of the endomorphism is

$$\mathbf{D} = \begin{pmatrix} 2 & 0 & 0 \\ 0 & -1 & 0 \\ 0 & 0 & -1 \end{pmatrix}$$

and the corresponding transformation matrix is

$$\mathbf{P} = \begin{pmatrix} 3 & 2 & 0 \\ 1 & 1 & 0 \\ 0 & 0 & 1 \end{pmatrix}$$

10.6.1 Diagonalization of a Symmetric Mapping

We know that a symmetric matrix $\mathbf{A}$ is a square matrix such that

$$\forall i, j : a_{i,j} = a_{j,i}$$

or, equivalently such that $\mathbf{A} = \mathbf{A}^T$. We can give the following definition.

Definition 10.24. Let $f : E \to E$ be an endomorphism which can be expressed as

$$\mathbf{y} = f(\mathbf{x}) = \mathbf{A}\mathbf{x}$$

where $\mathbf{A}$ is a symmetric matrix. This endomorphism is said *symmetric mapping.*

Example 10.83. The following mapping

$$\mathbf{y} = f(2x_1 + x_2 - 4x_3, x_1 + x_2 + 2x_3, -4x_1 + 2x_2 - 5x_3)$$

is associated with the matrix

$$\mathbf{A} = \begin{pmatrix} 2 & 1 & -4 \\ 1 & 1 & 2 \\ -4 & 2 & -5 \end{pmatrix}$$

is a symmetric mapping.

Let us recall the scalar product as in Definition 9.3. Given two vectors of real numbers $\mathbf{x}$ and $\mathbf{y}$ their scalar product is

$$\langle \mathbf{x}, \mathbf{y} \rangle = \mathbf{x}^T \mathbf{y}.$$

Theorem 10.16. *Let $\mathbf{x}$ and $\mathbf{y}$ be two n-dimensional real vectors and $\mathbf{A}$ be an $n \times n$ symmetric matrix. It follows that*

$$\langle \mathbf{A}\mathbf{x}, \mathbf{y} \rangle = \langle \mathbf{x}, \mathbf{A}\mathbf{y} \rangle.$$

Proof. Let us apply the definition of scalar product

$$\langle \mathbf{A}\mathbf{x}, \mathbf{y} \rangle = (\mathbf{A}\mathbf{x})^T \mathbf{y}.$$

For Theorem 2.1 this expression can be written as

$$(\mathbf{Ax})^{\mathbf{T}}\mathbf{y} = \left(\mathbf{x}^{\mathbf{T}}\mathbf{A}^{\mathbf{T}}\right)\mathbf{y}.$$

For associativity and symmetry of $\mathbf{A}$ we can write

$$\left(\mathbf{x}^{\mathbf{T}}\mathbf{A}^{\mathbf{T}}\right)\mathbf{y} = \mathbf{x}^{\mathbf{T}}\left(\mathbf{Ay}\right) = \langle\mathbf{x},\mathbf{Ay}\rangle\,.\square$$

Example 10.84. Let us consider the following symmetric matrix $\mathbf{A}$ and the vectors $\mathbf{x}$ and $\mathbf{y}$:

$$\mathbf{A} = \begin{pmatrix} 1\;0\;4 \\ 0\;2\;3 \\ 4\;3\;5 \end{pmatrix},$$

$$\mathbf{x} = \begin{pmatrix} 1 \\ 3 \\ 2 \end{pmatrix}$$

and

$$\mathbf{y} = \begin{pmatrix} 5 \\ -2 \\ 1 \end{pmatrix}.$$

Let us calculate

$$\mathbf{Ax} = \begin{pmatrix} 9 \\ 12 \\ 23 \end{pmatrix}$$

and

$$\langle\mathbf{Ax},\mathbf{y}\rangle = \begin{pmatrix} 9\;12\;23 \end{pmatrix}\begin{pmatrix} 5 \\ -2 \\ 1 \end{pmatrix} = 44.$$

Now let us calculate

$$\mathbf{Ay} = \begin{pmatrix} 9 \\ -1 \\ 19 \end{pmatrix}$$

and

$$\langle\mathbf{x},\mathbf{Ay}\rangle = \begin{pmatrix} 1\;3\;2 \end{pmatrix}\begin{pmatrix} 9 \\ -1 \\ 19 \end{pmatrix} = 44.$$

Let us recall now the definition of Hermitian product as in Definition 9.2 and let us extend this result to complex symmetric matrices. Given two vectors of complex numbers $\mathbf{x}$ and $\mathbf{y}$, the Hermitian product is

$$\langle\mathbf{x},\mathbf{y}\rangle = \mathbf{x}^{\dot{\mathbf{T}}}\mathbf{y}$$

where $\mathbf{x}^{\dot{\mathbf{T}}}$ is the transpose conjugate vector.

Theorem 10.17. *Let* **x** *and* **y** *be two n-dimensional complex vectors and* **A** *be an* $n \times n$ *symmetric matrix. It follows that the Hermitian product*

$$\langle \mathbf{Ax}, \mathbf{y} \rangle = \langle \mathbf{x}, \dot{\mathbf{A}}\mathbf{y} \rangle.$$

Proof. By applying the definition

$$\langle \mathbf{Ax}, \mathbf{y} \rangle = (\dot{\mathbf{Ax}})^{\mathbf{T}} \mathbf{y}$$

Let us consider that

$$\dot{\mathbf{Ax}} = \dot{\mathbf{A}}\dot{\mathbf{x}}$$

and apply Theorem 2.1:

$$(\dot{\mathbf{A}}\dot{\mathbf{x}})^{\mathbf{T}} \mathbf{y} = (\dot{\mathbf{x}}^{\mathbf{T}}\dot{\mathbf{A}}^{\mathbf{T}}) \mathbf{y}.$$

Thus, for associativity and symmetry of the matrix **A**

$$(\dot{\mathbf{x}}^{\mathbf{T}}\dot{\mathbf{A}}^{\mathbf{T}}) \mathbf{y} = \dot{\mathbf{x}}^{\mathbf{T}} (\dot{\mathbf{A}}\mathbf{y}) = \langle \mathbf{x}, \dot{\mathbf{A}}\mathbf{y} \rangle. \quad \square$$

Example 10.85. Let us consider the following symmetric matrix **A** and the vectors **x** and **y**:

$$\mathbf{A} = \begin{pmatrix} 2 & 1+j & 2 \\ 1+j & 5 & 5 \\ 2 & 5 & 4 \end{pmatrix},$$

$$\mathbf{x} = \begin{pmatrix} 4 \\ 5j \\ 1 \end{pmatrix}$$

and

$$\mathbf{y} = \begin{pmatrix} 1 \\ 1 \\ 2-j \end{pmatrix}.$$

Let us calculate

$$\mathbf{Ax} = \begin{pmatrix} 5+5j \\ 29+29j \\ 12+25j \end{pmatrix}$$

Let us apply the Hermitian product

$$\langle \mathbf{Ax}, \mathbf{y} \rangle = (\dot{\mathbf{Ax}})^{\mathbf{T}} \mathbf{y} = 13 - 96j$$

and

$$\langle \mathbf{x}, \dot{\mathbf{A}}\mathbf{y} \rangle = \dot{\mathbf{x}}^{\mathbf{T}} (\dot{\mathbf{A}}\mathbf{y}) = 13 - 96j.$$

Theorem 10.18. *The eigenvalues of a symmetric mapping* $f : E \to E$ *associated with a matrix* $\mathbf{A} \in \mathbb{R}_{n,n}$ *are all real.*

Proof. Let us assume by contradiction that λ is a complex eigenvalue of the symmetric mapping. Let $\dot{\lambda}$ be the conjugate of this eigenvalue. Let $\mathbf{x}$ be the eigenvector associated with the eigenvalue λ ($\mathbf{Ax} = \lambda\mathbf{x}$). Let us consider the Hermitian product $\langle \mathbf{x}, \mathbf{x} \rangle$ and write

$$\dot{\lambda} \langle \mathbf{x}, \mathbf{x} \rangle = \langle \dot{\lambda}\mathbf{x}, \mathbf{x} \rangle = \langle \mathbf{Ax}, \mathbf{x} \rangle = \langle \mathbf{x}, \mathbf{Ax} \rangle = \langle \mathbf{x}, \lambda\mathbf{x} \rangle = \lambda \langle \mathbf{x}, \mathbf{x} \rangle.$$

It follows that

$$\dot{\lambda} = \lambda.$$

A complex number is equal to its conjugate only if the imaginary part is null, i.e. when it is a real number. Thus λ is a real number.   □

It should be remarked that a symmetric mapping has only real eigenvalues only if all the elements of the associated matrix are all real. Thus, a symmetric matrix whose elements are complex numbers can have complex eigenvalues.

Example 10.86. The following symmetric endomorphism $f : \mathbb{R}^2 \to \mathbb{R}^2$

$$f(x,y) = (2x + 6y, 6x + 3y)$$

is associated with the matrix

$$\mathbf{A} = \begin{pmatrix} 2 & 6 \\ 6 & 3 \end{pmatrix}$$

has real eigenvalues $\lambda_1 = -3.52$, $\lambda_2 = 8.52$.

Example 10.87. The following symmetric endomorphism $f : \mathbb{R}^2 \to \mathbb{R}^2$

$$f(x,y) = (4x + 3y, 3x + 2y)$$

is associated with the matrix

$$\mathbf{A} = \begin{pmatrix} 4 & 3 \\ 3 & 2 \end{pmatrix}$$

has real eigenvalues

$$\lambda_1 = 3 - \sqrt{10} = -0.1623$$
$$\lambda_2 = 3 + \sqrt{10} = 6.1623.$$

Theorem 10.19. *Let $f : E \to E$ be a symmetric mapping associated with the symmetric matrix $\mathbf{A} \in \mathbb{R}_{n,n}$. Each pair of eigenvectors $\mathbf{x_i}$ and $\mathbf{x_j}$ corresponding to two distinct eigenvalues λ_i and λ_j are orthogonal.*

Proof. Let $f : E \to E$ be a symmetric mapping associated with the symmetric matrix $\mathbf{A} \in \mathbb{R}_{n,n}$. Let us consider two arbitrary distinct eigenvectors of this mapping, $\mathbf{x_i}$ and $\mathbf{x_j}$ and the corresponding distinct eigenvalues λ_i and λ_j.

Let us calculate the scalar product

$$\lambda_i \langle \mathbf{x_i}, \mathbf{x_j} \rangle = \langle \lambda_i\mathbf{x_i}, \mathbf{x_j} \rangle = \langle \mathbf{Ax_i}, \mathbf{x_j} \rangle = \langle \mathbf{x_i}, \mathbf{Ax_j} \rangle = \langle \mathbf{x_i}, \lambda_j\mathbf{x_j} \rangle = \lambda_j \langle \mathbf{x_i}, \mathbf{x_j} \rangle.$$

Since these two eigenvalues are distinct, i.e. $\lambda_i \neq \lambda_j$, it follows that the equation holds only if

$$\langle \mathbf{x_i}, \mathbf{x_j} \rangle = 0$$

that is, $\mathbf{x_i}$ and $\mathbf{x_j}$ are orthogonal. $\square$

Another way to express this concept is the following.

Remark 10.4. if $\mathbf{A}$ is a symmetric matrix, every pair of eigenvectors belonging to different eigenspaces are orthogonal.

Example 10.88. Let us consider the following symmetric mapping $f : \mathbb{R}^2 \to \mathbb{R}^2$

$$f(x,y) = (5x+y, x+5y)$$

which is associated with the following symmetric matrix

$$\mathbf{A} = \begin{pmatrix} 5 & 1 \\ 1 & 5 \end{pmatrix}.$$

The characteristic polynomial is

$$\det(\mathbf{A} - \lambda\mathbf{I}) = \lambda^2 - 10\lambda + 24$$

whose roots are

$$\lambda_1 = 4$$
$$\lambda_2 = 6.$$

It can be noticed that the eigenvalues are real numbers. Let us calculate the corresponding eigenspaces. For $\lambda_1 = 4$ we have

$$V(4) = \begin{cases} x+y = 0 \\ x+y = 0 \end{cases}$$

whose solution is $\alpha(1, -1)$.

$$V(6) = \begin{cases} -x+y = 0 \\ x-y = 0 \end{cases}$$

whose solution is $\alpha(1, 1)$.

Let us name $\mathbf{x_1}$ and $\mathbf{x_2}$ the eigenvectors associated with the eigenvalues λ_1 and λ_2 respectively:

$$\mathbf{x_1} = \begin{pmatrix} 1 \\ -1 \end{pmatrix}$$

and

$$\mathbf{x_2} = \begin{pmatrix} 1 \\ 1 \end{pmatrix}.$$

It can be easily checked that

$$\langle \mathbf{x_1}, \mathbf{x_2} \rangle = \mathbf{x_1^T} \mathbf{x_2} = 0,$$

i.e. the two vectors belonging to different eigenspaces are orthogonal.

Example 10.89. Let us consider the following symmetric mapping $f : \mathbb{R}^3 \to \mathbb{R}^3$

$$f(x,y,z) = (y+z, x+z, x+y)$$

which is associated with the following symmetric matrix

$$\mathbf{A} = \begin{pmatrix} 0 & 1 & 1 \\ 1 & 0 & 1 \\ 1 & 1 & 0 \end{pmatrix}.$$

The characteristic polynomial

$$\det \begin{pmatrix} -\lambda & 1 & 1 \\ 1 & -\lambda & 1 \\ 1 & 1 & -\lambda \end{pmatrix} = -\lambda^3 + 3\lambda + 2$$

whose roots are

$$\lambda_1 = 2$$
$$\lambda_2 = -1$$

where λ_2 has algebraic multiplicity 2.

Let us determine the corresponding eigenspaces.

$$V(2) = \begin{cases} -2x + y + z = 0 \\ x - 2y + z = 0 \\ x + y - 2z = 0 \end{cases}$$

has general solution $\alpha(1,1,1)$.

$$V(-1) = \begin{cases} x + y + z = 0 \\ x + y + z = 0 \\ x + y + z = 0 \end{cases}$$

has general solution $\alpha(-1,0,1) + \beta(-1,1,0)$.

Thus a transformation matrix would be

$$\mathbf{P} = \begin{pmatrix} \mathbf{x_1} & \mathbf{x_2} & \mathbf{x_3} \end{pmatrix} = \begin{pmatrix} 1 & -1 & -1 \\ 1 & 0 & 1 \\ 1 & 1 & 0 \end{pmatrix}.$$

We can easily check that

$$\langle \mathbf{x_1}, \mathbf{x_2} \rangle = 0$$
$$\langle \mathbf{x_1}, \mathbf{x_3} \rangle = 0.$$

This fact is in agreement with Theorem 10.19 since $\mathbf{x_1} \in V(2)$ and $\mathbf{x_2}, \mathbf{x_2} \in V(-1)$.

On the other hand, it can be observed that

$$\langle \mathbf{x_2}, \mathbf{x_3} \rangle = 1,$$

that is $\mathbf{x_2}$ and $\mathbf{x_3}$ are not orthogonal.

Let us apply the Gram-Schmidt orthonormalization to the column vectors of the matrix $\mathbf{P}$ as shown in Sect. 9.4. The new matrix $\mathbf{P_G}$ are

$$\mathbf{P_G} = \begin{pmatrix} \mathbf{e_1} & \mathbf{e_2} & \mathbf{e_3} \end{pmatrix} = \begin{pmatrix} 0.57735 & -0.70711 & -0.40825 \\ 0.57735 & 0.00000 & 0.81650 \\ 0.57735 & 0.70711 & -0.40825 \end{pmatrix}.$$

We can easily verify that $\mathbf{e_1}, \mathbf{e_2}, \mathbf{e_3}$ are eigenvectors of the mapping. Furthermore, $\mathbf{P_G}$ is an orthogonal matrix, i.e. $\mathbf{P_G^{-1}} = \mathbf{P_G^T}$ and

$$\mathbf{P_G^T A P_G} = \begin{pmatrix} 2 & 0 & 0 \\ 0 & -1 & 0 \\ 0 & 0 & -1 \end{pmatrix} = \mathbf{D}.$$

In other words, the transformation matrix $\mathbf{P}$ (in this case indicated with $\mathbf{P_G}$) is orthogonal. More specifically, in this case there exists an orthogonal transformation matrix that diagonalizes the matrix $\mathbf{A}$.

Definition 10.25. A matrix $\mathbf{A}$ is said *orthogonally diagonalizable* if $\exists$ an orthogonal matrix $\mathbf{P} \ni$ '

$$\mathbf{D} = \mathbf{P^T A P}$$

where $\mathbf{D}$ is a diagonal matrix exhibiting the eigenvalues on the diagonal and $\mathbf{P}$ is an orthogonal matrix whose columns are the eigenvectors.

Theorem 10.20. *If a matrix $\mathbf{A}$ is orthogonally diagonalizable then $\mathbf{A}$ is symmetric.*

Proof. Since $\mathbf{A}$ is orthogonally diagonalizable then $\exists$ an orthogonal matrix $\mathbf{P} \ni$ '

$$\mathbf{D} = \mathbf{P^T A P}.$$

Since $\mathbf{P}$ is orthogonal $\mathbf{P^{-1}} = \mathbf{P^T}$. Thus

$$\mathbf{A} = \mathbf{P D P^T}.$$

Let us calculate the transpose

$$\mathbf{A^T} = \left(\mathbf{P D P^T} \right)^{\mathbf{T}}.$$

For Theorem 2.1 we can write the equation as

$$\mathbf{A}^T = \left(\mathbf{DP}^T\right)^T \mathbf{P}^T = \left(\mathbf{P}^T\right)^T \mathbf{D}^T \mathbf{P}^T = \mathbf{PDP}^T = \mathbf{A},$$

that is $\mathbf{A}$ is symmetric ($\mathbf{A} = \mathbf{A}^T$). $\square$

Example 10.90. Let us consider the following symmetric mapping $f : \mathbb{R}^2 \to \mathbb{R}^2$

$$f(x,y) = (x + 3y, 3x + 3y)$$

whose associate matrix is

$$\mathbf{A} = \begin{pmatrix} 1 & 3 \\ 3 & 3 \end{pmatrix}.$$

The characteristic polynomial is

$$p(\lambda) = (1 - \lambda)(3 - \lambda) - 9 = \lambda^2 - 4\lambda - 6$$

whose roots are

$$\lambda_1 = 2 - \sqrt{10} = -1.1623$$
$$\lambda_2 = 2 + \sqrt{10} = 5.1623.$$

Since the endomorphism has two distinct eigenvalues is diagonalizable and the diagonal matrix is

$$\mathbf{D} = \begin{pmatrix} -1.1623 & 0 \\ 0 & 5.1623 \end{pmatrix}$$

and the transformation matrix $\mathbf{P}$ whose columns are the eigenvectors is

$$\mathbf{P} = \begin{pmatrix} -0.81124 & 0.58471 \\ 0.58471 & 0.81124 \end{pmatrix}.$$

It can be easily verified that the scalar product of the two columns of the matrix $\mathbf{P}$ is null. Thus, $\mathbf{P}$ is orthogonal and since

$$\mathbf{D} = \mathbf{P}^T \mathbf{A} \mathbf{P}$$

the matrix $\mathbf{A}$ is orthogonally diagonalizable.

Theorem 10.21. *Let $f : \mathbb{R}^n \to \mathbb{R}^n$ be a symmetric mapping associated with a matrix $\mathbf{A} \in \mathbb{R}_{n,n}$. Let $\lambda_1, \lambda_2, \ldots, \lambda_r$ be the r eigenvalues of the mapping and*

$$V(\lambda_1), V(\lambda_2), \ldots, V(\lambda_r)$$

be the corresponding eigenspaces.
The vector space sum

$$V(\lambda_1) + V(\lambda_2) + \ldots + V(\lambda_r) = \mathbb{R}^n.$$

Proof. Let us assume by contradiction that

$$V(\lambda_1) + V(\lambda_2) + \ldots + V(\lambda_r) = V \neq \mathbb{R}^n.$$

Let us assume that

$$\dim(V) = n - 1.$$

Since V has $n - 1$ dimensions we may think about another set, W where the nth dimension lies and whose vectors are orthogonal to all the vectors in V. Thus, $\forall \mathbf{x} \in V$ and $\forall \mathbf{w} \in W$ it follows that

$$\langle \mathbf{x}, \mathbf{w} \rangle = 0.$$

Considering that $\mathbf{x}$ is an eigenvector associated with an eigenvalue λ (and thus $\mathbf{Ax} = \lambda \mathbf{x}$) it follows that

$$0 = \langle \mathbf{x}, \mathbf{w} \rangle = \lambda \langle \mathbf{x}, \mathbf{w} \rangle = \langle \lambda \mathbf{x}, \mathbf{w} \rangle = \langle \mathbf{Ax}, \mathbf{w} \rangle = \langle \mathbf{x}, \mathbf{Aw} \rangle. \tag{10.1}$$

Since $\mathbf{x}$ and $\mathbf{Aw}$ are orthogonal and only one direction orthogonal to all of the vectors in V exists, it follows that $\mathbf{Aw} \in W$, that $\mathbf{w}$ and $\mathbf{Aw}$ are parallel. This parallelism can be expressed by saying that $\exists$ a scalar $\lambda_w \ni$ '

$$\mathbf{Aw} = \lambda_w \mathbf{w}.$$

This fact means that λ_w is another eigenvalue of the mapping and $\mathbf{w}$ its corresponding eigenvector. Since $\mathbf{w} \notin V$ where V is the sum set of all the eigenspaces, this is impossible.

Thus, $V = \mathbb{R}^n$. $\quad \square$

Example 10.91. Let us consider again the endomorphism in Example 10.88:

$$f(x, y) = (5x + y, x + 5x)$$

which is associated with the following symmetric matrix

$$\mathbf{A} = \begin{pmatrix} 5 & 1 \\ 1 & 5 \end{pmatrix}.$$

The transformation matrix $\mathbf{P}$ whose columns are the eigenvector is

$$\mathbf{P} = \begin{pmatrix} 1 & 1 \\ -1 & 1 \end{pmatrix}.$$

The eigenvectors are orthogonal and linearly independent. Thus, the columns of $\mathbf{P}$ span $\mathbb{R}^2$.

Example 10.92. Let us consider again the endomorphism from Example 10.89

$$f(x, y) = (y + z, x + z, x + y)$$

which is associated with the following symmetric matrix

$$A = \begin{pmatrix} 0 & 1 & 1 \\ 1 & 0 & 1 \\ 1 & 1 & 0 \end{pmatrix}.$$

The a transformation matrix $\mathbf{P}$ whose columns are eigenvectors is

$$\mathbf{P} = \begin{pmatrix} \mathbf{x_1} & \mathbf{x_2} & \mathbf{x_3} \end{pmatrix} = \begin{pmatrix} 1 & -1 & -1 \\ 1 & 0 & 1 \\ 1 & 1 & 0 \end{pmatrix}.$$

We can easily verify that the eigenvectors are linearly independent and thus span $\mathbb{R}^3$.

Theorem 10.22. *Let* $\mathbf{A} \in \mathbb{R}_{n,n}$. *If* $\mathbf{A}$ *is symmetric then it is diagonalizable.*

Proof. Since $\mathbf{A}$ is symmetric, for Theorem 10.21 the sum of all the eigenspaces $V = \mathbb{R}^n$. Thus $\dim(V) = n$. It follows that there exist n linearly independent eigenvectors. This means that the matrix $\mathbf{P}$ is non-singular and thus invertible. Thus $\mathbf{A}$ is diagonalizable. $\square$

Corollary 10.8. *Let* $f : E \to E$ *be a symmetric mapping associated with a matrix* $\mathbf{A} \in \mathbb{R}_{n,n}$. *The mapping is orthogonally diagonalizable, i.e.* $\exists$ *an orthogonal transformation matrix* $\mathbf{P}$ *such that*

$$D = \mathbf{P}^T \mathbf{A} \mathbf{P}$$

where $\mathbf{D}$ *is a diagonal matrix exhibiting the eigenvalues of* $\mathbf{A}$ *on its diagonal.*

10.7 Power Method

When a linear mapping is highly multivariate, the matrix associated with it is characterized by many columns (as many as the dimensions of the domain). In the case of endomorphism, the number of variables is equal to the order of the associated matrix. Under these conditions, the calculation of the roots of the characteristic polynomial can be computationally onerous. This section describes an iterative method for determining one eigenvalue without calculating the roots of the characteristic polynomial.

Definition 10.26. Let $\lambda_1, \lambda_2, \ldots, \lambda_n$ be the eigenvalues associated with an endomorphism $f : \mathbb{R}^n \to \mathbb{R}^n$. The eigenvalue λ_1 is said *dominant eigenvalue* if

$$|\lambda_1| > |\lambda_i|$$

for $i = 2, \ldots, n$. The eigenvectors associated with the eigenvalue λ_1 are said *dominant eigenvectors*.

The Power Method is an iterative method easily allowing the calculation of the dominant eigenvalue for endomorphisms having a dominant eigenvalue. Let us indicate with $\mathbf{A}$ the matrix associated with the endomorphism. The method processes an initial eigenvector guess $\mathbf{x_0}$ and is described in Algorithm 10.

Algorithm 10 Power Method

$\mathbf{x_1} = \mathbf{A}\mathbf{x_0}$
$\mathbf{x_2} = \mathbf{A}\mathbf{x_1} = \mathbf{A}^2\mathbf{x_0}$
$\dots$

$\mathbf{x_k} = \mathbf{A}\mathbf{x_{k-1}} = \mathbf{A}^k\mathbf{x_0}$

Example 10.93. Let us consider an endomorphism $f : \mathbb{R}^2 \to \mathbb{R}^2$ represented by the following matrix, see [20].

$$\mathbf{A} = \begin{pmatrix} 2 & -12 \\ 1 & -5 \end{pmatrix}.$$

The characteristic polynomial is

$$\det \begin{pmatrix} 2-\lambda & -12 \\ 1 & -5-\lambda \end{pmatrix} = (2-\lambda)(-5-\lambda) + 12 = \lambda^2 + 3\lambda + 2$$

whose roots are $\lambda_1 = -1$ and $\lambda_2 = -2$. Hence, for the definition of dominant eigenvalue λ_2 is the dominant eigenvalue. The corresponding eigenvectors are derived from

$$\det \begin{pmatrix} 4 & -12 \\ 1 & -3 \end{pmatrix} \begin{pmatrix} x_1 \\ x_2 \end{pmatrix} = \begin{pmatrix} 0 \\ 0 \end{pmatrix}$$

whose ∞^1 solutions are proportional to

$$\begin{pmatrix} 3 \\ 1 \end{pmatrix}.$$

Let us reach the same result by means of the Power Method where the initial guess $\mathbf{x_0} = (1,1)$:

$$\mathbf{x_1} = \mathbf{A}\mathbf{x_0} = \begin{pmatrix} 2 & -12 \\ 1 & -5 \end{pmatrix} \begin{pmatrix} 1 \\ 1 \end{pmatrix} = \begin{pmatrix} -10 \\ -4 \end{pmatrix} = -4 \begin{pmatrix} 2.5 \\ 1 \end{pmatrix}$$

$$\mathbf{x_2} = \mathbf{A}\mathbf{x_1} = \begin{pmatrix} 2 & -12 \\ 1 & -5 \end{pmatrix} \begin{pmatrix} -10 \\ -4 \end{pmatrix} = \begin{pmatrix} 28 \\ 10 \end{pmatrix} = 10 \begin{pmatrix} 2.8 \\ 1 \end{pmatrix}$$

$$\mathbf{x_3} = \mathbf{A}\mathbf{x_2} = \begin{pmatrix} 2 & -12 \\ 1 & -5 \end{pmatrix} \begin{pmatrix} 28 \\ 10 \end{pmatrix} = \begin{pmatrix} -64 \\ -22 \end{pmatrix} = -22 \begin{pmatrix} 2.91 \\ 1 \end{pmatrix}$$

$$\mathbf{x_4} = \mathbf{A}\mathbf{x_3} = \begin{pmatrix} 2 & -12 \\ 1 & -5 \end{pmatrix} \begin{pmatrix} -64 \\ -22 \end{pmatrix} = \begin{pmatrix} 136 \\ 46 \end{pmatrix} = 46 \begin{pmatrix} 2.96 \\ 1 \end{pmatrix}$$

$$\mathbf{x_5} = \mathbf{Ax_4} = \begin{pmatrix} 2 & -12 \\ 1 & -5 \end{pmatrix} \begin{pmatrix} 136 \\ 46 \end{pmatrix} = \begin{pmatrix} -280 \\ -94 \end{pmatrix} = -94 \begin{pmatrix} 2.98 \\ 1 \end{pmatrix}$$

$$\mathbf{x_6} = \mathbf{Ax_5} = \begin{pmatrix} 2 & -12 \\ 1 & -5 \end{pmatrix} \begin{pmatrix} -280 \\ -94 \end{pmatrix} = \begin{pmatrix} 568 \\ 190 \end{pmatrix} = 190 \begin{pmatrix} 2.99 \\ 1 \end{pmatrix}.$$

We have found an approximation of the dominant eigenvector. In order to find the corresponding eigenvalue we need to introduce the following theorem.

Theorem 10.23. Rayleigh's Quotient. *Let* $f : \mathbb{R}^n \to \mathbb{R}^n$ *be an endomorphism and* $\mathbf{x}$ *one of its eigenvectors. The corresponding eigenvalue* λ *is given by*

$$\lambda = \frac{\mathbf{x}^T \mathbf{A} \mathbf{x}}{\mathbf{x}^T \mathbf{x}}.$$

Proof. Since $\mathbf{x}$ is an eigenvector then $\mathbf{Ax} = \lambda \mathbf{x}$. Hence,

$$\frac{\mathbf{x}^T \mathbf{A} \mathbf{x}}{\mathbf{x}^T \mathbf{x}} = \frac{\lambda \mathbf{x}^T \mathbf{x}}{\mathbf{x}^T \mathbf{x}} = \lambda. \quad \square$$

Example 10.94. In the example above, the dominant eigenvalue is given by

$$\lambda = \frac{(2.99, 1) \begin{pmatrix} 2 & -12 \\ 1 & -5 \end{pmatrix} \begin{pmatrix} 2.99 \\ 1 \end{pmatrix}}{(2.99, 1) \begin{pmatrix} 2.99 \\ 1 \end{pmatrix}} \approx \frac{-20}{9.94} \approx -2.01.$$

We have found an approximation of the dominant eigenvalue.

Since iterative multiplications can produce large numbers, the solutions are usually normalized with respect to the highest number in the vector. For example, if we consider the matrix

$$\begin{pmatrix} 2 & 4 & 5 \\ 3 & 1 & 2 \\ 2 & 2 & 2 \end{pmatrix}$$

and apply the Power Method with an initial guess $(1,1,1)$, we obtain

$$\mathbf{x_1} = \begin{pmatrix} 11 \\ 6 \\ 6 \end{pmatrix}.$$

Instead of carrying this vector, we can divide the vector elements by 11 and use the modified $\mathbf{x_1'}$ for the following iteration.

$$\mathbf{x_1} = \begin{pmatrix} 1 \\ 0.54 \\ 0.54 \end{pmatrix}.$$

This normalization is named *scaling* and the Power Method that applied the scaling at each iteration is said *Power Method with scaling*.

Let us now give a rigorous proof of the convergence of the Power Method.

Theorem 10.24. *Let $f : \mathbb{R}^n \to \mathbb{R}^n$ be a diagonalizable endomorphism having a dominant eigenvalue and $\mathbf{A}$ the matrix associated with it. It follows that $\exists$ a non-null vector $\mathbf{x_0}$ such that for $k = 1, 2, \ldots$*

$$\mathbf{x_k} = \mathbf{A}^k \mathbf{x_0}$$

approaches the dominant eigenvector.

Proof. Since $\mathbf{A}$ is diagonalizable, for Theorem 10.14 a basis of eigenvectors $\mathbf{x_1}, \mathbf{x_2}, \ldots, \mathbf{x_n}$ of $\mathbb{R}^n$ exists. These eigenvectors are associated with the corresponding eigenvalues $\lambda_1, \lambda_2, \ldots, \lambda_n$. Without a loss of generality, let us assume that λ_1 is the dominant eigenvalue and $\mathbf{x_1}$ the corresponding dominant eigenvector. Since these eigenvectors compose a basis, they are linearly independent. We can choose an initial guess $\mathbf{x_0}$ such that

$$\mathbf{x_0} = c_1 \mathbf{x_1} + c_2 \mathbf{x_2} + \ldots + c_n \mathbf{x_n}$$

where the scalars $c_1, c_2, \ldots, c_n \neq 0, 0, \ldots, 0$ and $c_1 \neq 0$. Let us multiply the terms in this equation by $\mathbf{A}$:

$$\begin{aligned}
\mathbf{A}\mathbf{x_0} &= \mathbf{A}\left(c_1 \mathbf{x_1} + c_2 \mathbf{x_2} + \ldots + c_n \mathbf{x_n}\right) = \\
&= c_1 \mathbf{A}\mathbf{x_1} + c_2 \mathbf{A}\mathbf{x_2} + \ldots + c_n \mathbf{A}\mathbf{x_n} = \\
&= c_1 \lambda_1 \mathbf{x_1} + c_2 \lambda_2 \mathbf{x_2} + \ldots + c_n \lambda_n \mathbf{x_n}.
\end{aligned}$$

If we multiply the terms of the equation k times by the matrix $\mathbf{A}$ we obtain

$$\mathbf{A}^k \mathbf{x_0} = c_1 \lambda_1^k \mathbf{x_1} + c_2 \lambda_2^k \mathbf{x_2} + \ldots + c_n \lambda_n^k \mathbf{x_n}$$

that leads to

$$\begin{aligned}
\mathbf{A}^k \mathbf{x_0} &= \lambda_1^k \left(c_1 \mathbf{x_1} + c_2 \frac{\lambda_2^k}{\lambda_1^k} \mathbf{x_2} + \ldots + c_n \frac{\lambda_n^k}{\lambda_1^k} \mathbf{x_n}\right) = \\
&= \lambda_1^k \left(c_1 \mathbf{x_1} + c_2 \left(\frac{\lambda_2}{\lambda_1}\right)^k \mathbf{x_2} + \ldots + c_n \left(\frac{\lambda_n}{\lambda_1}\right)^k \mathbf{x_n}\right).
\end{aligned}$$

Since λ_1 is the largest eigenvalue in absolute value, the fractions

$$\frac{\lambda_2}{\lambda_1}, \frac{\lambda_3}{\lambda_1}, \ldots, \frac{\lambda_n}{\lambda_1}$$

are all smaller than 1. Hence, the fractions

$$\left(\frac{\lambda_2}{\lambda_1}\right)^k, \left(\frac{\lambda_3}{\lambda_1}\right)^k, \ldots, \left(\frac{\lambda_n}{\lambda_1}\right)^k$$

approach 0 as k approaches infinity. It follows that

$$\mathbf{A}^k \mathbf{x_0} = \lambda_1^k \left(c_1 \mathbf{x_1} + c_2 \left(\frac{\lambda_2}{\lambda_1}\right)^k \mathbf{x_2} + \ldots + c_n \left(\frac{\lambda_n}{\lambda_1}\right)^k \mathbf{x_n}\right) \Rightarrow$$

$$\Rightarrow \mathbf{A}^k \mathbf{x_0} \approx \lambda_1^k c_1 \mathbf{x_1}.$$

with $c_1 \neq 0$. This means that as k grows the method converges to a vector that is proportional to the dominant eigenvector. $\square$

Exercises

10.1. Let the endomorphism $\mathbf{O} : \mathbb{R}^n \to \mathbb{R}^n$ be the null mapping. Determine its rank and nullity. State whether or not this mapping is invertible. Justify the answer.

10.2. Determine kernel, nullity and rank of the following endomorphism $f : \mathbb{R}^3 \to \mathbb{R}^3$

$$f(x,y,z) = (x+y+z, x-y-z, 2x+2y+2z)$$

Verify the validity of the rank-nullity theorem.
Determine whether or not this mapping is (1) injective; (2) surjective.
Determine the Image of the mapping.
Provide a geometrical interpretation of the mapping.

10.3. Let us consider the linear mapping $\mathbb{R}^3 \to \mathbb{R}^3$ defined as

$$f(x,y,z) = (x+2y+z, 3x+6y+3z, 5x+10y+5z).$$

Determine kernel, nullity and rank of the endomorphism. Verify the validity of the rank-nullity theorem.
Determine whether or not this mapping is (1) injective; (2) surjective.
Determine the Image of the mapping.
Provide a geometrical interpretation of the mapping.

10.4. Determine eigenvalues and one eigenvector associated with the following endomorphism:

$$f(x,y,z) = (x+2y, 3y, 2x-4y+2z)$$

Determine whether or not this endomorphism is diagonalizable.

10.5. Let us consider the following endomorphism $f : \mathbb{R}^2 \to \mathbb{R}^2$

$$f(x,y) = (x+y, 2x).$$

Determine, if they exist, the eigenvalues of the mapping. If possible, diagonalize the matrix.

10.6. Let us consider the endomorphism associated with the matrix $\mathbf{A}$. Apply the Power Method to determine the dominant eigenvector and the Rayleigh's Theorem to find the corresponding eigenvalue.

$$\mathbf{A} = \begin{pmatrix} -1 & -6 & 0 \\ 2 & 7 & 0 \\ 1 & 2 & -1 \end{pmatrix}.$$

10.7. Apply the Power Method to the following matrix notwithstanding the fact that this matrix does not have a dominant eigenvalue.

$$\mathbf{A} = \begin{pmatrix} 1 & 1 & 0 \\ 3 & -1 & 0 \\ 0 & 0 & -2 \end{pmatrix}.$$

Comment on the results.

Chapter 11
An Introduction to Computational Complexity

This chapter is not strictly about algebra. However, this chapter offers a set of mathematical and computational instruments that will allow us to introduce several concepts in the following chapters. Moreover, the contents of this chapter are related to algebra as they are ancillary concepts that help (and in some cases allow) the understanding of algebra. More specifically, this chapter gives some basics of complexity theory and discrete mathematics and will attempt to answer to the question: "What is a hard problem?"

We already know that the hardness of a problem strictly depends on the solver. In a human context, the same problem, e.g. learning how to play guitar, can be fairly easy for someone and extremely hard for someone else. In this chapter we will refer to problems that usually humans cannot solve (not in a reasonable time at least). Hence, we will refer to the harness of a problem for a computational machine.

11.1 Complexity of Algorithms and Big-O Notation

A *decision problem* is a question whose answer can be either "yes" or "no". It can be proved that all the problems can be decomposed as a sequence of decision problems. Due to the physics of the machines, computers ultimately can solve only decision problems. Since complex problems can be decomposed into a sequence of decision problems, a computing machine can tackle a complex problem by solving, one by one, each decision problem that compose it.

An algorithm is a finite sequence of instructions that allows to solve a given problem. For example, each cooking recipe is an algorithm as it provides for a number of instructions: ingredients, measures, method, oven temperature, etc. The recipe of a chocolate cake is an algorithm that allows in a finite number of steps to transforms a set of inputs, such as sugar, butter, flour, cocoa, into a delicious dessert. However, not every problem can be solved by an algorithm.

© Springer Nature Switzerland AG 2019

F. Neri, *Linear Algebra for Computational Sciences and Engineering*,
https://doi.org/10.1007/978-3-030-21321-3_11

Definition 11.1. A *Turing Machine* is a conceptual machine able to write and read on a tape of infinite length and to execute a set of definite operations. At most one action is allowed for any given situation. These operations compose the algorithms.

Throughout this chapter, every time we will refer to a machine and to an algorithm running within it we will refer to the Turing Machine, unless otherwise specified.

A problem is said *decidable* (or *computable*) if an algorithm that solves the problem exists or, equivalently, if the problem can be solved in finite (even if enormous) amount of time. This means that for each instance of the input the corresponding problem solution is returned. A problem is said *undecidable* (or *uncomputable*) otherwise.

A example of undecidable problem is the so-called problem "halting problem", formulated by the British mathematician Alan Turing in 1936. The halting problem is the problem of determining, from a description of an arbitrary computer program and an input, whether the program will stop running or continue to run forever.

Theorem 11.1. *The vast majority of the problems are uncomputable.*

Proof. Let *Alg* and *Pro* be the sets of all possible algorithms and problems, respectively.

The elements of the set *Alg* are algorithms. We may think of them as computer programs, i.e. set of instructions that run within a computer. As such, they are represented within a machine as a sequence of binary numbers (binary string). We may think that this sequence is a binary number itself. The number can be converted into base 10 number, which will be a positive integer. Hence every algorithm can be represented as a natural number. In other words,

$$\forall x \in Alg : x \in \mathbb{N}.$$

The elements of the set *Pro* are problems. As stated above, each problem can be represented as a sequence of decision problems. Hence, a problem can be seen as a function p that processes some inputs to ultimately give an answer "yes" or "no". Obviously {"yes", "no" } can be modelled as $\{0, 1\}$. Since an input can be seen as a binary string, it can also be seen as a natural number. The function p is then $p : \mathbb{N} \to \{0, 1\}$.

This function can be expressed by means of a table with infinite columns.

0 1 2 ... 100 ...

0 1 1 ... 0 ...

The second line of this table is an infinite binary string. An infinite binary string can be interpreted as a real number. If we put a decimal (strictly speaking binary) point right before the first digit, the binary string $.011 \ldots 0 \ldots$ can be interpreted as a real number between 0 and 1. In other words, every decision problem can be represented as a real number:

$$\forall y \in Pro : y \in \mathbb{R}.$$

We know that $\mathbb{R}$ is uncountably infinite while $\mathbb{N}$ is countably infinite. Moreover we know that $\mathbb{N} \subset \mathbb{R}$ and that the cardinality of $\mathbb{N}$ is much smaller than the cardinality of $\mathbb{R}$ because $\forall a, b \in \mathbb{N}, \exists$ infinite numbers $c \in \mathbb{R}$ such that $a < c < b$. Hence, the number of the algorithms is much smaller that the number of problems. This means that the vast majority of problems cannot be solved, i.e. are uncomputable. $\quad\square$

Although, in general, the majority of problems is undecidable, in this chapter we are interested in the study of decidable problems and the comparison of several algorithms that solve them. Let us suppose we want to build a house. Let us imagine that as a first action we hire several engineers and architects to perform the initial project design. Let us imagine that a number of projects are then produced. Not all the projects can be taken into account because they may violate some law requirements, e.g. the use of appropriate materials, the safety requirements, the minimum distance from surrounding buildings, etc. Even if the project is done within a perfect respect of laws and regulations, we may not consider it viable because of personal necessities. For example we may exclude a project due to its excessive cost or associated construction time. In order to compare different projects we have to make clear what our preferences are and what would be the most important requirement. A similar consideration can be done for algorithms. Before entering into the details let us consider that the algorithm is automatically performed by a machine. Under these conditions, firstly, an algorithm must be *correct*, i.e. the must produce, at each input, the right output. Secondly, the algorithm must be *efficient*. In order to assess the efficiency of an algorithm the *complexity* of the algorithm must be evaluated. Two types of complexities are taken into account.

- *Space complexity*: amount of memory necessary for the algorithm to return the correct result
- *Time complexity*: number of elementary operations that the processor must perform to return the correct result

Hence, the efficiency of an algorithm can be seen as a function of the length of the input. In this book we will focus on the time complexity. For a given problem there can be many algorithms that solve it. In this case, a natural question will be to assess which algorithm solves the problem in the best way. Obviously this problem can be decomposed as a set of pairwise comparisons: for a given problem and two algorithms solving it how can we assess which algorithm is better and which one is worse? Before explaining this we should clarify what is meant with "better". Here we enter the field of analysis of algorithms and complexity theory. We do not intend to develop these topics in this book as it is about algebra. However, we will limit ourselves to a few simple considerations. Once an algorithms has been developed, it is fundamental to estimate prior to the algorithm's execution on a machine, the time will take to complete the task. This estimation (or algorithm examination) is said analysis of *feasibility* of the algorithm. An algorithm is said feasible if its execution time is acceptable with respect to the necessity of solving a problem. Although this concept is fairly vague in this formulation, it is fairly easy to understand that a very accurate algorithm that requires 100 years to return the solution cannot be used in a real-time industrial process. Probably, a 100 year waiting time would be unacceptable also in the case of a design.

More formally, we can identify a function of the inputs $t(input)$. This function makes correspond, to the inputs, the number of elementary operations needed to solve the problem/return the desired result where the execution time of an elementary operation is assumed to be constant. Obviously, the amount of operations within a machine is directly proportionate to the execution time. Hence, we may think that, for a given problem, each algorithm is characterized by its own function $t(input)$. This piece of information is static, i.e. is related to a specific problem.

Let us consider that the problem is *scalable*, i.e. it can be defined for a growing amount of homogeneous inputs. For example, a distribution network can be associated with a certain problem and algorithm. The variation of the amount of users make the problem vary in size. The time required by the algorithm to solve the problem also varies. The main issue is to assess the impact of the variation in size of the problem on the time required by the machine to solve the problem itself.

Let us consider a simple example.

Example 11.1. Let **a** and **b** be two vectors of size 5. Hence, the two vectors are $\mathbf{a} = (a_1, a_2, a_3, a_4, a_5)$ and $\mathbf{b} = (b_1, b_2, b_3, b_4, b_5)$ respectively. Let us now perform the scalar product between the two vectors

$$\mathbf{ab} = a_1 b_1 + a_2 b_2 + a_3 b_3 + a_4 b_4 + a_5 b_5.$$

The algorithm calculating the scalar products performs the five products $(a_i b_i)$ and four sums. Hence, the complexity of the scalar product solver for vectors composed of five elements is $5 + 4 = 9$. More generally, a scalar product between two vectors of length n requires the calculation of n products and $n - 1$ sums. Hence the time complexity of a scalar product is $2n - 1$ elementary operations.

If we now consider the matrix product between two matrices having size n, we have to perform a scalar product for each element of the product matrix. Thus, we need to compute n^2 scalar products. This means that the time complexity of a product between matrix of size n is

$$n^2 (2n - 1) = 2n^3 - n^2.$$

In other words, for a given n value, if we double the dimensionality we approximately double the corresponding time of scalar product while we make the calculation time of the product between matrices extremely bigger. For example if $n = 5$, the complexity of the scalar product is 9 and of the matrix product 225. If $n = 10$, the complexity of the scalar product becomes 19 while the complexity of the matrix product 1900.

This example shows how algorithms can be characterized by a different complexity. In order to estimate the machine time of an algorithm execution, the exact number of elementary calculations is not too important. The actual interest of the mathematician/computer scientist is to estimate which kind of relation is between dimensionality and time complexity. In the previous example, we have linear and cubic growth, respectively. The following trends of complexity are usually taken into account.

- k constant
- $\log(n)$ logarithmic
- n linear
- n^2 quadratic
- n^3 cubic
- n^t polynomial
- k^n exponential
- $n!$ factorial

These trends are indicated with the symbol $\mathcal{O}(\cdot)$ and named *big-O notation*. For example if an algorithms presents a linear growth of the number of elementary operations in dependence on the problem dimensionality is said to have a $\mathcal{O}(n)$ complexity.

As it can be easily observed the big-O notation gives us an understanding of the order of magnitude of the complexity of the problem. It is not too relevant to distinguish the details of the growth of the complexity while it is fundamental to understand the growth of the complexity when the dimensionality increases. In other words, an increase of the complexity according to n or $30n$ are substantially comparable. Hence, the corresponding problems (and algorithms) belong to the same class. On the contrary, a growth k^n is a completely different problem.

11.2 P, NP, NP-Hard, NP-Complete Problems

Definition 11.2. The class of problems that can be exactly solved by an algorithm within a finite time and such that the time complexity is of the kind $\mathcal{O}(n^k)$ with k finite number $\in \mathbb{R}$ is said to have a *polynomial complexity*. This class of problems composes a set that will be indicated with **P**, see [21].

These problems are important because they are the problems that can surely be solved by a machine in a finite time, see e.g. [22]. Those problems that can be solved by an algorithm within polynomial time are often referred to as *feasible* problems while the solving algorithm is often said to be *efficient*. It must be remarked that although we may use the words feasible and efficient, the solution of the problem can be extremely time-consuming if k is a large number. For example if $k = 10^{252}$ the waiting time in a modern computer would likely be unreasonable even though the problem is feasible and the algorithm is efficient.

Definition 11.3. A *Nondeterministic Turing Machine* is a Turing Machine where the rules can have multiple actions for a given situation (one situation is prescribed into two or more rules by different actions).

Example 11.2. A simple Turing Machine may have the rule "If the light is ON turns right" and no more rules about the light being ON. A Nondeterministic Turing Machine may have both the rules "If the light is ON turns right" and "If the light is ON turns left".

Definition 11.4. An algorithm that runs with a polynomial time complexity on a Nondeterministic Turing Machine is said to have a *Nondeterministic Polynomial* complexity. The corresponding class of algorithms is indicated with **NP**.

An alternative and equivalent way to define and characterize **NP** problems is given by the following definition.

Definition 11.5. Let A be a problem. The problem A is a *Nondeterministic Polynomial* problem, **NP** problem, if and only if a given solution requires at most a polynomial time to be verified.

It can be easily observed that all the problems in **P** are also in **NP**, i.e. **P** $\subset$ **NP**. Obviously many problems in **NP** are not also in **P**. In other words, many problems require a non-polynomial, e.g. exponential time, to be solved but a polynomial time to be verified.

Example 11.3. Let us consider the following discrete set of numbers

$$\{1, 2, 3, \ldots, 781\}.$$

We want to compute the sum of these numbers. This is a very easy task because from basic arithmetic we know that

$$\sum_{i=1, i \in \mathbb{N}}^{n} i = \frac{n(n+1)}{2} = \frac{n^2 + n}{2}.$$

Hence the problem above would simply be $\frac{781 \times 782}{2}$. In general the sum on n natural numbers requires a constant number of operations regardless of n. In other words, this problem is characterized by $\mathcal{O}(k)$.

Let us consider, now, the sum of n generic numbers such as

$$\{-31, 57, 6, -4, 13, 22, 81\}.$$

The sum of these numbers requires $n - 1$ operations, hence the complexity of this problem is linear, i.e. $\mathcal{O}(n)$.

For the same set of numbers, we may want know whether or not there is a subset such that its sum is 24. If we asked this question to a machine, $2^7 - 1$ operations should be performed by checking all the possible grouping. In general, if we consider a set of n numbers, this problem requires $2^n - 1$ calculations. Hence, this is not a **P** problem. More precisely this is an **EXP** problem since the time complexity associated with the problem is exponential. However, if a solution is given, e.g. $\{6, -4, 22\}$, it will take only two operations to verify whether or not their sum is 24. In general, given a candidate solution of m numbers, we will perform $m - 1$ operations to verify it. This means that the search of the solution requires more than a polynomial time while the verification of any solution can be performed in a polynomial time. This problem belongs to **NP**.

Some mathematicians and computer scientists are investigating whether or not **P** and **NP** coincide. The fact that the search of the solution is very different in terms of hardness/complexity is an argument to claim that they are different. On the other hand, the fact that the verification of a solution is equally easy make the look the two sets as the same concept. Whether or not **P** and **NP** are coinciding sets is an open question in computer science that is still a hot topic of discussion and is beyond the scopes of this book.

Several problems can be solved by transforming the original problem into a different problem. Let us consider a generic problem A and let us assume we want to solve it. This problem may be very hard to solve but can be transformed into a mirror problem B that is easier. We could then transform the input from A to B space, solve by an algorithm in B and anti-transform the solution of B back to the A space. In this way we obtain a solution of the our original problem. The set of step just describe are said *reduction* of the problem A to the problem B. This fact is the basic concept behind the following definition.

Definition 11.6. A decision problem H is said **NP**-*hard* when every **NP** problem L can be reduced to H within a polynomial time.

Equivalently, we can give an alternative definition of **NP**-hard problems.

Definition 11.7. A decision problem H is said **NP**-*hard* when is at least as hard as the hardest **NP** problem.

The class of **NP**-hard problems is very large (actually infinite) and includes all the problems that are at least as hard as the hardest **NP** problem. This means that the vast majority of **NP**-hard problems are not in **NP**. For example, undecidable problems, such as the above-mentioned halting problem, are always **NP**-hard. A special role is played by a subset of **NP**-hard that is composed also by **NP** problems.

Definition 11.8. A problem is **NP**-complete if the problem is both **NP**-hard, and in **NP**.

Thus, **NP**-complete problems lay in the intersection between **NP** and **NP**-hard problems. These problems are the hardest **NP** problems that can be found. Many examples can be made, especially from modern games. However, for the purpose of this book, that is to have an initial understanding of computational complexity, the same example shown above will be considered again. Given the set of integer numbers $\{-31, 57, 6, -4, 13, 22, 81\}$ we want to find a non-null subset whose sum is 24. We have intuitively shown that this problem is in **NP**. It can be proved by reduction that problems of this type, namely subset sum problem, are also **NP**-hard. A famous proof of this fact is performed by reduction using an important problem in computational complexity, namely Boolean satisfiability problem. As an outcome, the subset sum problem is **NP**-complete, see e.g. [23] for the proof.

In order to clarify the contents of this section and give an intuitive representation of computational complexity, the following diagram is presented. All the problems of the universe are represented over a line. Let us imagine we ideally sort them from the easiest (on the left) to the most difficult (on the right). The solid part of this line represents the set of decidable problems while the dashed part undecidable problems. The sets of **P** and **NP** problems are highlighted as well as the class of problems **EXP** that can be solved within an exponential time. Also, **NP**-hard and **NP**-complete (grey rectangle) sets are represented.

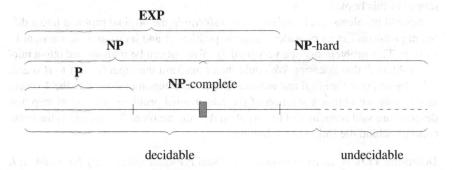

11.3 Representing Information

An important challenge in applied mathematics and theoretical computer science is the efficient representation and manipulation of the information. This challenge, albeit not coinciding with the computational complexity, it is related to it. In this section, a technique to efficiently represent data within a machine, the Huffman coding, and a representation of arithmetic operations to efficiently exploit the architecture of a machine, the polish and reverse polish notations, are given.

11.3.1 Huffman Coding

The Huffman coding is an algorithm to represent data in a compressed way by reserving the shorter length representation for frequent pieces of information and longer for less frequent pieces of information, see [24]. Since the details are outside the scopes of this book, let us explain the algorithmic functioning by means of an example and a graphical representation. Let us consider the sentence "Mississippi river". This sentence contains 17 characters. Considering that each character requires eight bits, the standard representation of this sentence requires 136 bits in total.

Let us see for this example, how Huffman coding can lead to a substantial saving of the memory requirements. In order to do that, let us write the occurrences for each letter appearing in the sentence:

$$M \mapsto 1$$
$$I \mapsto 5$$
$$S \mapsto 4$$
$$P \mapsto 2$$
$$R \mapsto 2$$
$$V \mapsto 1$$
$$E \mapsto 1$$
$$- \mapsto 1$$

where "–" indicates the blank space and $\mapsto$ relates the occurrences to the letter. The first step of the Huffman coding simply consists of sorting the letters from the most frequent to the least frequent.

From this diagram, we now connect the vertices associated with the least occurrence and sum the occurrences:

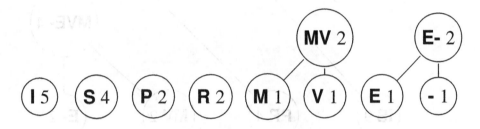

The operation can be iterated to obtain:

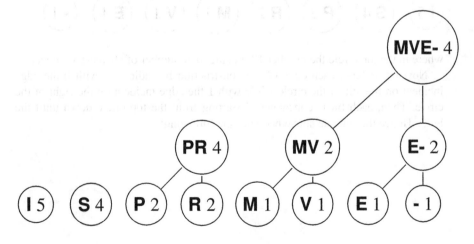

Eventually the complete scheme is the following:

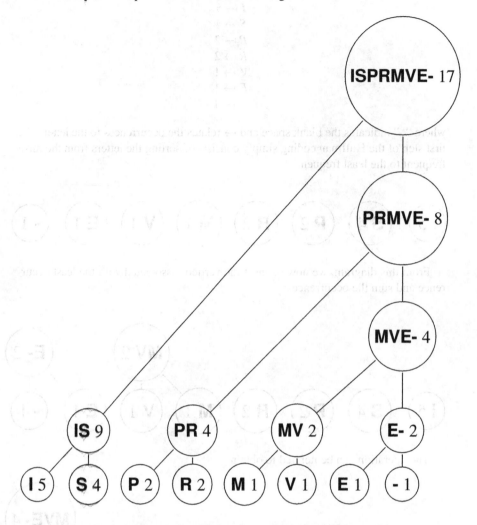

where in the top circle the number 17, i.e. the total number of characters, appears.

Now let us label each edge of the construction by indicating with 0 the edge incident on the left of the circle while with 1 the edge incident on the right of the circle. Then, each bit is concatenated starting from the top circle down until the letter. Hence the scheme above becomes the following:

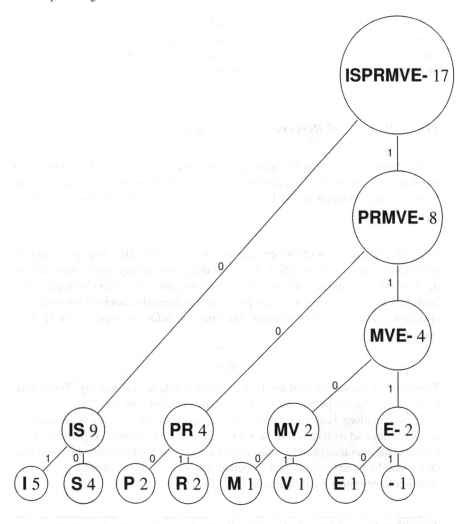

and the Huffman coding of the letters is

$$
\begin{aligned}
I &= 01 \\
S &= 00 \\
P &= 100 \\
R &= 101 \\
M &= 1100 \\
V &= 1101 \\
E &= 1110 \\
- &= 1111.
\end{aligned}
$$

It can be observed that the letters with the highest frequency have the shortest bit representation. On the contrary, the letters with the lowest frequency have the longest bit representation. If we write again the "Mississippi River" sentence by means of the Huffman coding we will need 46 bits instead of the 136 bits for the standard 4-bit binary representation. It must be appreciated that this massive memory saving has been achieved without any loss in the delivered information and only by using an intelligent algorithmic solution.

11.3.2 Polish and Reverse Polish Notation

At the abstract level, a simple arithmetic operation can be interpreted as an operation involving two operands and one operator. For example the operation sum of a and b involves the operands a and b with the operator $+$. Normally, this operation is represented as

$$a+b$$

where the operator is written between the two operands. This way of writing the operation is named *infix notation*. Although this notation may appear naturally understandable for a human, it is indeed not the most efficient for an electronic calculator. The logician Jan Łukasiewicz proposed an alternative notation for arithmetic operations consisting of representing the operator before the operand or after the operands:

$$+ab$$
$$ab+.$$

These two notations are said *prefix* and *postfix* notation, respectively. These notations are also named polish and reverse notation, respectively, see [25].

These notations have essentially three advantages. In order to understand the first advantage let us focus on the $a+b$ example. If this operation is performed in a machine, the operands must be loaded into the memory, the lifted into the computing unit when the operation can then be performed. Hence, the most natural way for a compiler to prioritize the instructions is the following:

Algorithm 11 $a+b$ from the Compiler Perspective

Load memory register R_1 with a
Load memory register R_2 with b
Add what is in R_1 with what is in R_2 and write the answer into the memory register R_3.

Since a machine has to execute the instructions in this order, the most efficient way to pass the information to a machine is by following the same order. In this way, the machine does not require to interpret the notation and can immediately start performing the operations as they are written.

The second advantage is that the polish notation and especially the reverse polish notation work in the same way as a stack memory of a calculator works. A stack memory saves (pushes) and extracts (pops) the items sequentially. The stack memory architecture requires that the first item to be popped is the last that has been pushed. If we consider for example the arithmetic expression (the symbol $*$ indicates the multiplication)

$$a + b * c,$$

in reverse polish notation becomes

$$abc * +,$$

which corresponds exactly to the most efficient way to utilize a stack memory. More specifically, the calculation of the arithmetic expression above is achieved, from the perspective of the stack memory, by the following steps

---

Algorithm 12 $a + b * c$ from the Stack Memory Perspective

---

push a
push b
push c
pop c
pop b
compute $d = b * c$
push d
pop d
pop a
compute $e = a + d$
push e
pop e,

---

where the variables d and e are indicated just for explanation convenience (but are not actual variables). The same set of instructions are graphically represented in the diagram below.

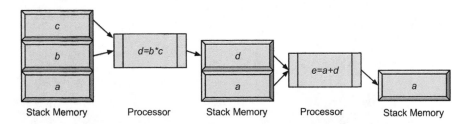

This sequence of instructions is essentially the very same of what represented by the reverse polish notation. In other words, the reverse polish notation explains exactly and in a compact way what the memory stack of a machine supposed to do.

The third advantage is that the reverse polish notation allows to univocally write all the arithmetic expressions without the need to writing parentheses. This was the reason why Łukasiewicz originally introduced it, i.e. to simplify the notation in logical proofs. Let us consider the following expression:

$$(a+b)*c.$$

In this case the parentheses indicate that the sum must be performed before the multiplication. If we removed the parentheses the arithmetic expression would mean something else, that is the multiplication of b by c first and then the sum of the result with a. Thus, when we write with infix notation, the use of parentheses can be necessary to avoid ambiguities. In polish or reverse polish notation, arithmetic expressions can be written without ambiguities without the aid of parentheses. In particular, the expression $(a+b)*c$ in reverse polish notation is:

$$cab+*.$$

The operations of a complex arithmetic expression described in reverse polish notation are performed from the most internal towards the most external.

Example 11.4. Let us consider the following arithmetic expression:

$$5*(4+3)+2*6.$$

In reverse polish notation it can be written as

$$543+*26*+.$$

Chapter 12
Graph Theory

In this chapter we introduce a notion of fundamental importance for modelling in schematic way a large amount of problems. This is the concept of a *graph*. This concept applies not only to computer science and mathematics, but even in fields as diverse as chemistry, biology, physics, civil engineering, mapping, telephone networks, electrical circuits, operational research, sociology, industrial organization, the theory of transport, artificial intelligence.

We will present some concepts of graph theory, those that seem most relevant for our purposes, omitting many others. A complete discussion of *graph theory*, on the other hand, would require more than a chapter. These pages are only an introduction to this fascinating theory.

12.1 Motivation and Basic Concepts

Historically graph theory was born in 1736 with Leonard Euler's when he solved the so called Königsberg bridges problem, see [26]. The problem consisted of the following. The Prussian town of Königsberg (Kaliningrad, in our days) was divided into four parts by the Pregel river, one of these parts being an island in the river. The four regions of the town were connected by seven bridges (Fig. 12.1). On sunny Sundays the people of Königsberg used to go walking along the river and over the bridges. The question was the following: is there a walk using all the bridges once, that brings the pedestrian back to its starting point? The problem was solved by Euler, who showed that such a walk is not possible.

The importance of this result lies above all in the idea that Euler introduced to solve this problem, and that is precisely gave rise to the theory of graphs. Euler realized that to solve the problem it was necessary to identify its essential elements, neglecting accessory or irrelevant items. For this purpose, Euler considered the model in Fig. 12.2.

As it will be shown later, the general result obtained by Euler allows to solve the seven edges problem.

© Springer Nature Switzerland AG 2019
F. Neri, *Linear Algebra for Computational Sciences and Engineering*,
https://doi.org/10.1007/978-3-030-21321-3_12

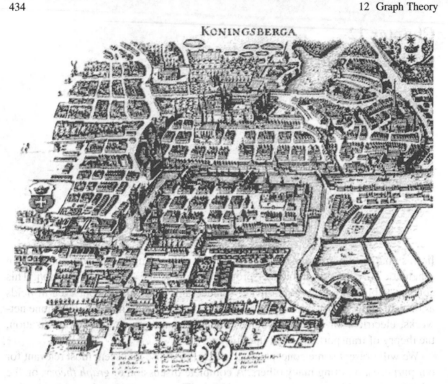

Fig. 12.1 The Königsberg bridges

Definition 12.1. A digraph or directed graph G is a pair of sets (V, E) consisting of a finite set $V \neq \emptyset$ of elements called *vertices* (or *nodes*) and a set $E \subseteq V \times V$ of ordered pairs of distinct vertices called *arcs* or *directed edges*.

Definition 12.2. Let G be a graph composed of the sets (V, E). A graph S_G composed of $(S_V \subset V, S_E \subset E)$ is said *subgraph* of G.

A directed edge represents an element $e = (v, w) \in E$ as an edge oriented from vertex v, called starting point, until vertex w, called end point. The edges describe the links of the vertices.

Definition 12.3. Let G be a graph composed of the sets (V, E). Two vertices w and $v \in V$ are said *adjacent* if $(v, w) \in E$.

If the vertices w and v are adjacent they are also called *neighbours*. The set of neighbours of a vertex v is said its neighbourhood $N(v)$.

Definition 12.4. Let G be a graph composed of the sets (V, E). Two edges are *adjacent edges* if they have a vertex in common.

Example 12.1. The graph $G = (V, E)$ represented in the following figure is a digraph, where $V = \{v_1, v_2, v_3, v_4, v_5, v_6\}$, $E = \{(v_1, v_1), (v_1, v_2), (v_2, v_3), (v_5, v_6)\}$.

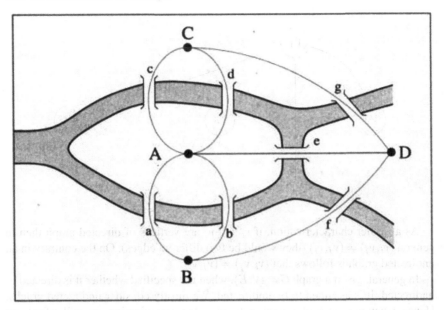

Fig. 12.2 Königsberg bridges graph

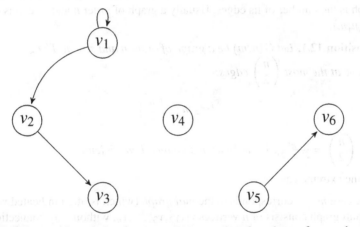

To model a problem, it is often not necessary that the edges of a graph are oriented: for example, to make a map of the streets of a city where every road is passable in both senses enough to indicate the arc between two points of the city and not the direction of travel.

Definition 12.5. An undirected graph, or simply graph G is a pair $G = (V, E)$ which consists of a finite set $V \neq \emptyset$ and a set E of unordered pairs of elements (not necessarily distinct from V).

Example 12.2. An undirected graph is given by $G = (V, E)$ where $V = \{v_1, v_2, v_3, v_4, v_5, v_6\}$ and $E = \{(v_1, v_2), (v_1, v_3), (v_6, v_1)\}$

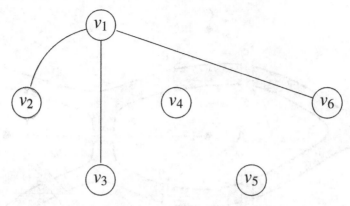

As a further characterization, if v_i and v_j are vertices of directed graph then in general $(v_i, v_j) \neq (v_j, v_i)$ (they would be two different edges). On the contrary in an undirected graph is follows that $(v_i, v_j) = (v_j, v_i)$.

In general, given a graph $G = (V, E)$, when not specified whether it is directed or undirected, it is assumed to be undirected. We mostly consider undirected graphs. Hence, when we speak of a graph without further specification, we will refer to an undirected graph.

Definition 12.6. The *order* of a graph is the number of its vertices, while the *size* of a graph is the number of its edges. Usually a graph of order n and size m is denoted by $G(n, m)$.

Proposition 12.1. *Let* $G(n, m)$ *be a graph of order n and size m. The graph* $G(n, m)$ *can have at the most* $\binom{n}{2}$ *edges:*

$$0 \leq m \leq \binom{n}{2}$$

where $\binom{n}{2} = \frac{n!}{2!(n-2)!}$ *is the Newton's binomial coefficient.*

In the extreme cases:

1. The case $m = 0$ corresponds to the *null graph* (which is often indicated with N_n). A null graph consists of n vertices $v_1, v_2, v_3, \ldots, v_n$, without any connection.

   Example 12.3. An example of null graphs is shown in the following figure.

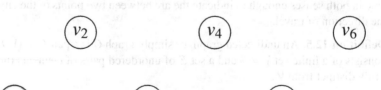

2. The case $m = \binom{n}{2}$ corresponds to the case in which each vertex is connected with all the others.

Definition 12.7. A *complete graph* on n vertices, indicated with K_n, is a graph $G\left(n, \binom{n}{2}\right)$, i.e. a graph having n vertices and an edge for each pair of distinct vertices.

Example 12.4. The chart below shows the graphs $K_1, \ldots, K_4$

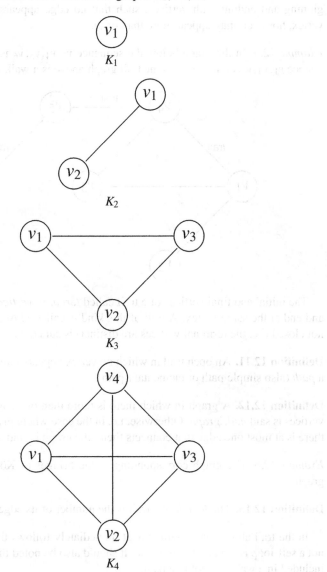

It may be observed that in a directed graph since $E \subseteq V \times V$ the maximum number of edges is n^2, where n is the number of nodes.

Definition 12.8. An edge of the type (v_i, v_i) is called *(self)-loop.*

Definition 12.9. A *walk* of a graph G is a sequence of vertices $v_1, v_2, \ldots, v_n$ where $(v_1, v_2), (v_2, v_3), (v_3, v_4), \ldots, (v_i, v_{i+1}), \ldots, (v_{n-1}, v_n)$ are edges of the graph. Vertices and edges may appear more than once.

Definition 12.10. A *trail* is a finite alternating sequence of vertices and edges, beginning and ending with vertices, such that no edge appears more than once. A vertex, however, may appear more than once.

Example 12.5. In the graph below the sequence v_1, v_4, v_5, v_6 is a trail while the sequence $v_1, v_4, v_3, v_1, v_4, v_5, v_6$ in the trail graph above is a walk (and not a trail).

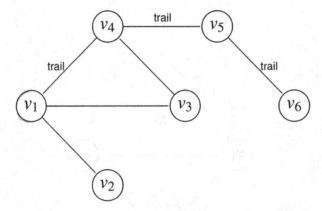

The initial and final vertices of a trail called *terminal vertices.* A trail may begin and end at the same vertex. A trail of this kind is called *closed trail.* A trail that is not closed (i.e. the terminal vertices are distinct) is called *open trail.*

Definition 12.11. An open trail in which no vertex appears more than once is called a *path* (also simple path or elementary path).

Definition 12.12. A graph in which there is more than one edge that connects two vertices is said *multigraph.* Otherwise, i.e. in the case where given any two vertices, there is at most one edge that connects them, the graph is said *simple.*

Example 12.6. The graph corresponding to the bridges of Königsberg is a multigraph.

Definition 12.13. The *length* of a trail is the number of its edges.

In the trail above, the length is 3. It immediately follows that an edge which is not a self-loop is a path of length one. It should also be noted that a self-loop can be included in a walk but not in a path.

Definition 12.14. The distance $d(v_i, v_j)$ between two vertices v_i and v_j of a graph G is the minimal length between all the trails (if they exist) which link them. If the vertexes are not linked $d(v_i, v_j) = \infty$. A trail of minimal length between two vertices of a graph is said to be *geodesic*.

Example 12.7. In the graph above, a geodesic trail is marked out: another trail is for example v_1, v_3, v_4, v_5, v_6. The geodesic trail has length 3 while the latter has length 4.

Proposition 12.2. *The distance in a graph satisfies all the properties of a metric distance. For all vertices u, v, and w:*

- $d(v, w) \geq 0$, with $d(v, w) = 0$ *if and only if* $v = w$
- $d(v, w) = d(w, v)$
- $d(u, w) \leq d(u, v) + d(v, w)$

Definition 12.15. The *diameter* of a graph G is the maximum distance between two vertices of G.

Definition 12.16. A *circuit* or *cycle* in a graph is a closed trail.

A circuit is also called elementary cycle, circular path, and polygon.

Definition 12.17. If a graph G has circuits, the *girth* of G is the length of the shortest cycle contained in G and the *circumference* is the length of the longest cycle contained in G.

It can be easily seen that the length of a circuit $v_1, v_2, \ldots, v_n = v_1$ is $n - 1$.

Definition 12.18. A circuit is said even (odd) if its length is even (odd).

Definition 12.19. A graph is *connected* if we can reach any vertex from any other vertex by travelling along a trail. More formally: a graph G is said to be connected if there is at least one trail between every pair of vertices in G. Otherwise, G is *disconnected*.

It is easy to see that one may divide each graph in connected subgraphs.

Definition 12.20. A connected subgraph containing the maximum number of (connected) edges is said *connected component* (or simply component) of the graph G.

Example 12.8. In the graph below, the subgraph v_1, v_2, v_3 is connected but is not a connected component. The subgraph v_1, v_2, v_3, v_4 is a connected component.

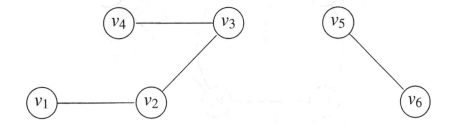

A null graph of more than one vertex is disconnected.

Definition 12.21. A vertex that is not connected to any part of a graph is said to be *isolated*.

Obviously, the concept of connected component of a graph must not be confused with the concept of vector component seen in the previous chapters.

Definition 12.22. The *rank of a graph* ρ is equal to the number of vertices n minus the number of components c:

$$\rho = n - c.$$

Definition 12.23. The *nullity of a graph* v is equal to number of edges m minus the rank ρ:

$$v = m - \rho.$$

If we combine the two definitions above it follows that $v = m - n + c$. Moreover, if the graph is connected then $\rho = n - 1$ and $v = m - n + 1$.

Furthermore, the equation $m = \rho + v$ can be interpreted rank-nullity theorem for graphs, see Theorem 10.7. Let us consider a mapping f that consists of connecting n fixed nodes by means m edges. The number of edges m is the dimension of a vector space, the rank ρ is the dimension of the image, and the nullity is the dimension of the kernel.

Proposition 12.3. *Let G be a graph. If G is not connected then its rank is equal to the sum of the ranks of each of its connected component. Its nullity is equal to the sum of the nullities of each of its connected component.*

Graphs are mathematical entities like sets, matrices and vectors. As such, for a graph G, a set operations can be defined for graphs. Among all the possible operations, some are listed in the following.

1. *Vertex Removal.* A vertex v_i can be removed from the graph G. Unless v_i is an isolated vertex, we obtain something that is not a graph, because there will be edges that have only one end. Hence, we are then forced to remove, along with v_i, all the edges that pass through v_i. Let us indicate with the notation $G - v_i$ the subgraph obtained from G by removing v_i with all the edges passing through v_i. So $G - v_i$ is the largest subgraph of G that does not contain the vertex v_i.

Example 12.9. If we consider the following graph

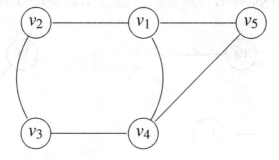

and we remove v_5, we obtain

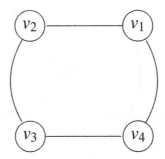

2. *Edge Removal.* An edge can be removed from a graph. Unlike for the vertex removal, the edge removal results into the removal of the edge but not the vertices.

   Example 12.10. If from the graph in the example above the edge v_4, v_5 is removed the following graph is obtained.

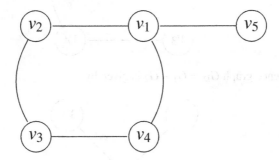

3. *Union of Graphs.* Let $G_1(V_1, E_1)$ and $G_2(V_2, E_2)$ be two graphs. The union graph $G_U(V_U, E_U) = G_1 \cup G_2$ is a graph such that $V_U = V_1 \cup V_2$ and $E_U = E_1 \cup E_2$. In other words, the union graph contains all vertices and edges of both G_1 and G_2.
4. *Intersection of Graphs.* Let $G_1(V_1, E_1)$ and $G_2(V_2, E_2)$ be two graphs. The intersection graph $G_I(V_I, E_I) = G_1 \cap G_2$ is a graph such that $V_I = V_1 \cap V_2$ and $E_I = E_1 \cap E_2$. In other words, the intersection graph contains vertices and edges belonging to G_1 and at the same time to G_2.
5. *Difference Between Graphs.* Let $G_1(V_1, E_1)$ and $G_2(V_2, E_2)$ be two graphs. The difference graph $G_D(V_D, E_D) = G_1 - G_2$ is a graph containing the edges belonging to G_1 but not to G_2 and all the vertices associated with the edges E_D.

Example 12.11. Let us name the following graph G_1

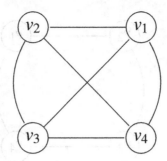

and the following G_2

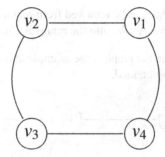

The difference graph $G_D = G_1 - G_2$ is given by

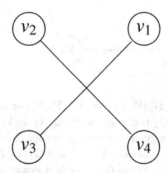

6. *Ring Sum.* Let $G_1(V_1, E_1)$ and $G_2(V_2, E_2)$ be two graphs. The ring sum graph is
   $G_R(V_R, E_R) = (G_1 \cup G_2) - (G_1 \cap G_2)$.

Definition 12.24. Let G be a graph and v one of its vertices. The vertex v is a *cut-point* if its removal causes the increase of amount of connected components of the graph.

Proposition 12.4. *Let G be a graph and v_c one of its cut-points. There exist at least a pair of vertices v_1 and v_2, such that the trail connecting these two vertices passes through v_c.*

Example 12.12. Let us consider the following graph

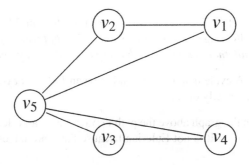

The node v_5 is a cut-point. We may notice that its removal causes an increase in the number of components:

Furthermore, v_1 and v_4 are connected by means of a trail passing through v_5. This trail is v_1, v_2, v_5, v_3, v_4.

Definition 12.25. A graph G is said *non-separable* is it does not contain cut-points.

Example 12.13. The following graph would be non-separable since the removal of any node would not result into an increase in the number of components.

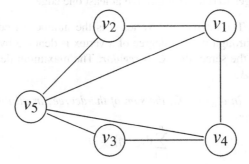

12.2 Eulerian and Hamiltonian Graphs

In the same paper where Euler solved the "seven bridges problem", he posed (and then solved) the following more general problem: *in what type of graph is it possible to find a closed walk that passes through every edge of G only ones?*

Definition 12.26. A cycle in a graph is said to be an *Eulerian cycle* if it contains all edges of the graph exactly once.

Example 12.14. In the graph above the path v_1, v_4, v_3, v_1 is a cycle; this circuit is odd because its length is 3. It is not Eulerian because there are edges not belonging to this cycle.

Definition 12.27. Euler's Graph. A graph which contains an Eulerian cycle is called *Eulerian graph*.

Example 12.15. The following graph is Eulerian since the cycle $v_1, v_2, v_3, v_5, v_4, v_1, v_2, v_4, v_3, v_1$ is Eulerian.

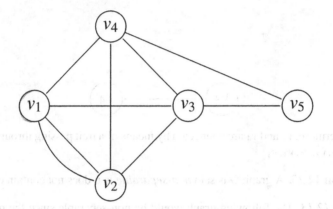

On the contrary, the complete graph K_4 is not Eulerian since a circuit including all the edges obliges to run twice through at least one edge.

Definition 12.28. The *degree* of a vertex v is the number of edges incident to v (i.e. which pass through v). The degree of a vertex is denoted by $\deg(v)$. If all the degrees of G are the same, then G is *regular*. The maximum degree in a graph is often denoted by Δ.

Proposition 12.5. *In a graph G, the sum of the degrees in all the vertices is twice the number of edges:*

$$\sum_{i=1}^{n} \deg(v_i) = 2m.$$

For example, In the graph of Fig. 12.2 associated with the Königsberg bridges all vertices have degree 3, except that A has degree 5.

The following important theorem characterizes Eulerian graphs.

Theorem 12.1. Euler's Theorem. *A graph G without any isolated vertices is Eulerian if and only if*

(a) *is a connected graph*
(b) *all vertices have even degree*

Proof. Let us prove that Eulerian graphs are connected. Let v and w be two arbitrary, distinct vertices of G. Since the graph has no isolated vertices, v e w will belong to an edge. As the graph is Eulerian, it will contain an Eulerian cycle, passing through all edges of the graph and thus containing the two vertices. Thus, G is connected. □

Now we prove that all vertices have even degree. Let v be a vertex. The vertex v belongs to an Eulerian cycle, which surely exists. If we start from v and follow the Eulerian cycle, at the end we will come back to v. If we go through v multiple times, we will go out from v (i.e. the times in which v is a starting point of an edge) the same number of times in which we will come back in v (i.e. the times in which v appears as end point of an edge). Hence, the edges which go in or go out each vertex is an even number. □

Let us prove that a connected graph with vertices having all even degree is Eulerian. We choose an arbitrary initial vertex, say v, and we walk along the edges without walking over the same edge more than once. We continue until we get stuck to a vertex w: all outgoing edges from w have already been covered. There are two possibilities:

- $w \neq v$. This case cannot occur, because the degree of w would be an odd number, and this contradicts the hypothesis;
- $w = v$. This means that if we get stuck, we will be at the starting point. Thus, the arcs walked until they form a circuit is $C = v_1, v_2, \ldots, v_n$ where $v_1 = v = w = v_n$.

It may happen that there is another edge outgoing from some vertex v_i of the circuit C that there was not followed by our walk. We will call this situation a *leak* from the vertex v_i. In this case, we replace the circuit C with the following path P: $v_i, v_{i+1}, \ldots, v_n, v_1, v_2, \ldots, v_i, u$ where u is the vertex of the edge outgoing from v_i. This path contains the circuit C within it (since $C = v_1, v_2, \ldots, v_n$).

From the newly constructed path P having last vertex u, we continue as previously by adding edges and nodes to the path until a new circuit is identified. If a leak is identified, the algorithm is repeated. Since, each node has an even degree there is always a returning path to a node without having to walk on the same edge more than once. Eventually a circuit C' without any leak is identified.

Let us prove that C' contains all the edges of the graph. Let α be any edge in the graph G, with terminating vertices x and y. Since G is connected there will certainly be a trail that connects the initially chosen vertex v with x. Let us name this trial $v, w_1, w_2, \ldots, x$. The edge v, w_1 must be on the cycle C' (otherwise there would be a leak from v, which we excluded). Then $w_1 \in C'$ and also the edge w_1, w_2 must stay on the cycle C' for the same reason. Iterating this process, $w_3, \ldots, x$ must belong to the cycle. This means that the edge α connecting x and y must also belong to the same cycle.

Since the cycle C' contains all the edges and never counts the same edge more than once, the cycle C' is Eulerian. This means that the graph is Eulerian. □

Thus, the Euler's Theorem for Eulerian graphs allows to find out if a graph is Eulerian simply by observing the degree of its vertices: if even only one vertex has odd degree, the graph is not Eulerian. Notice that the above not only proves the theorem, but also describes an algorithm to find an Eulerian cycle. At this point we have all the elements on hand to give an answer to the problem of the bridges of Königsberg. The problem is clearly equivalent to determine if the associated with the bridges of Königsberg is an Eulerian graph. To find that it is not Eulerian just check if it has a vertex which has odd degree, we have seen that all the vertices of the graph have odd degree, hence it is not an Eulerian graph. Thus, it is not possible to walk over each of the seven bridges exactly once and return to the starting point.

In order to be able to formulate the Euler's Theorem for directed graphs we need to introduce the concept of *vertex-balance*.

Definition 12.29. A vertex of a directed graph is *balanced* if the number of incoming edges equals the number of outgoing edges.

Proposition 12.6. *Let G be a directed graph with no isolated vertices. The directed graph G is Eulerian if and only if*

1. the graph is connected
2. each vertex is balanced

Definition 12.30. An *Eulerian trail* is a trail that passes through all the edges and is such that initial and final vertices are not the same (unlike in circuits).

We will now analyse the case of a graph that contains an Eulerian trail.

Theorem 12.2. *Let G be a graph with no isolated vertices. The graph G contains an Eulerian trail if and only if*

(a) is connected
(b) each vertex of G has even degree except exactly two vertices

We have briefly discussed the problem of when a graph contains an Eulerian cycle, a closed walk traversing every edge exactly once. Let us now briefly introduce the analogous requirement for vertices: a cycle of a graph G that contains all the vertices of G exactly once.

Definition 12.31. Let G be a graph containing at least three vertices. Every walk that includes all the vertices on G exactly once is named *Hamiltonian cycle* of G.

Definition 12.32. Let G be a graph. If G contains a Hamiltonian cycle, it is called *Hamiltonian graph*.

Obviously, for the walk to be a cycle, it is required that G contains at least three vertices. If this condition is not satisfied the walk is not closed.

Definition 12.33. A path in G containing every vertex of G is an *Hamiltonian path* and a graph containing a Hamiltonian path is said to be *traceable*.

Example 12.16. An example of Hamiltonian graph is given in the following. As it can be easily checked, the cycle $v_1, v_2, v_3, v_5, v_6, v_4$ includes all the vertices exactly once.

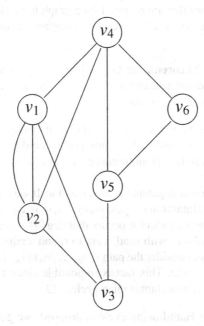

To determine whether or not a given graph is Hamiltonian is much harder than determining whether or not it is Eulerian.

Hamiltonian graphs are named after Sir William Hamilton, an Irish Mathematician (1805–1865), who invented a puzzle, called the Icosian game, which he sold for 25 guineas to a game manufacturer in Dublin. The puzzle involved a dodecahedron on which each of the 20 vertices was labelled by the name of some capital town in the world. The aim of the game was to construct, using the edges of the dodecahedron a closed walk which traversed each town exactly once. In other words, one had essentially to form a Hamiltonian cycle in the graph corresponding to the dodecahedron.

Definition 12.34. A simple graph G is called *maximal non-Hamiltonian* if it is not Hamiltonian and the addition of an edge between any two non-adjacent vertices of it forms a Hamiltonian graph.

Proposition 12.7. *A complete graph is always Hamiltonian.*

Let us introduce the following theorem in order to inspect a graph G and check the existence of a Hamiltonian cycle within it.

Theorem 12.3. Dirac's Theorem. *A graph $G(n,m)$ with $n \geq 3$ vertices and in which each vertex has degree at least $\frac{n}{2}$ has, a Hamiltonian cycle.*

Dirac's theorem states that a specific class of graphs must have a Hamiltonian cycle. This is only a sufficient condition, other types of graphs can still be Hamiltonian. Historically, Dirac's theorem represented the starting point for the investigation about the conditions that are required for a graph to be Hamiltonian. We report here the Ore's Theorem that is a later result encompassing the previous research on Hamiltonian cycles.

Theorem 12.4. Ore's Theorem. *Let $G(n,m)$ be a graph containing n vertices. If for every two non-adjacent vertices u and w it occurs that $\deg(u) + \deg(w) \geq n$, then G contains a Hamiltonian cycle.*

Proof. Let us suppose, by contradiction, that G contains no Hamiltonian cycles. If we added edges to G, eventually a Hamilton cycle would be produced. Thus, let us add edges avoiding to produce Hamiltonian cycles until a maximal non-Hamiltonian G_0 is obtained.

Let us consider two non-adjacent vertices u and w. If we added the edge (u,w) to G_0, we would obtain a Hamiltonian cycle $v_1, v_2, \ldots, v_n$ in G_0 with $u = v_1$ and $w = v_n$.

Since u and w are non adjacent it occurs that $\deg(u) + \deg(w) \geq n$. Thus, there exist two vertices v_i and v_{i+1} with w adjacent to v_i and u adjacent to v_{i+1}.

This means that if we consider the path $u, v_2, \ldots, v_i, w, v_{n-1}, \ldots, v_{i+1}, u$ in G_0, this path is a Hamiltonian cycle. This fact is impossible since G_0 is a maximal non-Hamiltonian. Hence, G has a Hamiltonian cycle. $\square$

If the last edge of a Hamiltonian cycle is dropped, we get a Hamiltonian path. However, a non-Hamiltonian graph can have a Hamiltonian path.

Example 12.17. In the following graphs, G_1 has no Hamiltonian path as well as no Hamiltonian cycle; G_2 has the Hamiltonian path v_1, v_2, v_3, v_4 but no Hamiltonian cycle; G_3 has the Hamiltonian cycle v_1, v_2, v_3, v_4, v_1.

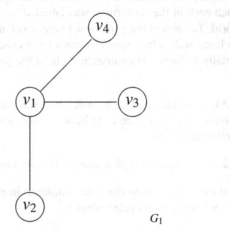

G_1

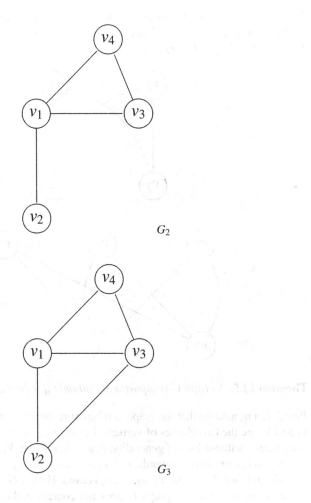

12.3 Bipartite Graphs

Definition 12.35. A graph is said *bipartite* if its vertices set V can be written as the union of two disjoint sets $V = V_1 \cup V_2$ (with $V_1 \cap V_2 = \emptyset$) such that each vertex of V_1 is connected to at least one vertex of V_2 while there are no connections within V_1 nor V_2. In other words, if the edges connecting the vertices of the vertices of V_1 to those of V_2 are removed, the two sets V_1 and V_2 are composed of isolated vertices only. V_1 and V_2 are said classes of vertices.

Definition 12.36. Let G be a bipartite graph composed of the classes of vertices V_1 and V_2. Let V_1 be composed of n elements and V_2 of m elements. If there is an edge connecting all the vertices of V_1 to all the vertices of V_2 the graph is said *bipartite complete* and is indicated by the symbol $K_{m,n}$.

Example 12.18. Examples of graphs $K_{2,2}$ and $K_{3,2}$ are shown in the following. In the first example $V_1 = \{v_1, v_2\}$ and $V_2 = \{v_3, v_4\}$ while in the second example $V_1 = \{v_1, v_2, v_5\}$ and $V_2 = \{v_3, v_4\}$.

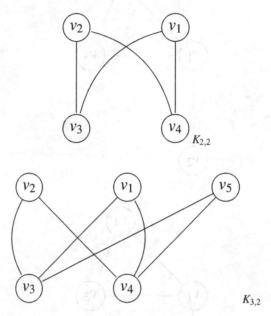

Theorem 12.5. *A graph G is bipartite if and only if it contains no odd circuits.*

Proof. Let us assume that the graph G is bipartite. Suppose the bipartite graph G and V_1 and V_2 are the two classes of vertices. Let $v_1, v_2 \ldots, v_n = v_1$ be a circuit in G. We can assume, without loss of generality, that v_1 belongs to V_1. Then v_2 belongs to V_2, $v_3 \in V_1$, and so on. In other words, $v_i \in V_1, \forall i$ odd. Since $v_n = v_1$ is in V_1 we have that n is odd and then the circuit $v_1, v_2, \ldots, v_n$ is even. Hence, G contains no odd circuits.

Let us assume that the graph G does not contain odd circuits. We can assume, without loss of generality, that the graph is connected: otherwise, you can consider separately the connected components (that are also bipartite). Let v_1 be any vertex of G and V_1 be the set containing v_1 and vertices that have even distance from v_1. V_2 is the complement of V_1. In order to show that G is bipartite, it is enough to prove that each edge of G joins each vertex of V_1 with a vertex in V_2. Let us suppose by contradiction that there is an edge that connects two vertices x, y of V_1. Under these conditions, the union of all the geodesics from v_1 to x, all the geodesic from v_1 to y and of the edge x, y contains an odd circuit, which contradicts the hypothesis. $\square$

12.4 Planar Graphs

In graph theory an edge is identified solely by a pair of vertices. Length and shape of the edges are not important.

Example 12.19. The following two figures are two representations of the same graph, that is K_4.

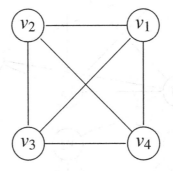

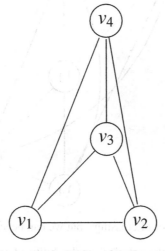

It is important to note how in the first representation two edges intersect, although their intersection is not a vertex. Such a situation takes the name of *crossing*. In the second representation there are no crossings.

Definition 12.37. A graph G is said to be *planar* if it can be drawn on a plane without crossings.

In other words, in a planar graph arcs intersect only at the vertices that have in common. As we have just seen, the complete graph K_4 is planar.

Example 12.20. Let us consider the complete graph K_5. Two of its representations are given in the following.

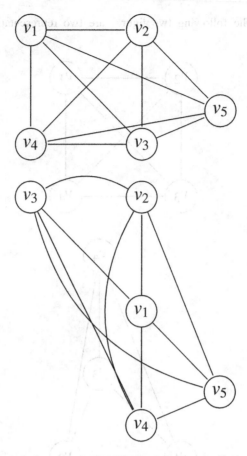

We can vary the number of crossings, but we can never eliminate them.

In order to visualize the practical applicability of the concept of planar/non-planar graphs, let A, B and C, be three houses, and E, F and G, the electric power, gas and water, respectively. One wonders if it is possible to connect each of the three houses with each of the stations so that the pipes do not intersect. The graph that models this situation is clearly the complete bipartite graph on six vertices $K_{3,3}$ that is not planar. So there is no way to build pipelines that connect the three units at three power plants without overlap.

Many algorithms to assess whether or not a graph is planar, or to understand what is the minimum number of crossings in the graph have been designed. Although an exhaustive analysis of these algorithms falls beyond the scopes of this book, a characterization of non-planar graphs, provided by the Polish mathematician Kuratowski is reported here.

Definition 12.38. Let G be a finite graph. Let u and v be two vertices of G connected by an edge. It is said that the edge joining u and v has been *contracted* if the edge has been eliminated and the vertices u and v have been merged.

An example of contraction is shown in the following.

Example 12.21. Let us consider the second representation of K_4 and remove the edge between v_3 and v_4. Let us name v_{col} the collapsed vertex replacing the other two. The graph before and after the contraction are given in the following.

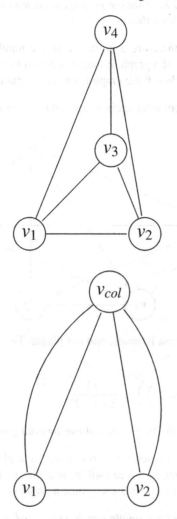

Definition 12.39. Let G be a graph. If the contraction operation is possible, the graph G is said to be *contractible* for edges to the graph G'.

Theorem 12.6. *(Kuratowski's Theorem) A finite graph G is planar if and only if it contains no subgraph contractible for edges to K_5 or to $K_{3,3}$.*

Since the detection of these contractible subgraphs can be difficult, Kuratowski's criterion is not very convenient to check the planarity (or non-planarity) of a graph. Many efficient algorithms in time $\mathcal{O}(n)$, i.e. linear complexity with respect to the number of vertices, to decide if and only if a graph is planar or not. The study of these algorithms falls beyond the objectives of this book.

Nonetheless, it is worthwhile to mention in this chapter the following criteria that help to easily recognize non-planar graphs.

Proposition 12.8. *Let G be a simple graph, connected and planar with $n \geq 3$ vertices and m edges. It follows that $m \leq 3n - 6$.*

Planar graphs have therefore a limitation in the number of edges. In order to detect the non-planarity of a graph, it may be enough to count the number of edges in the sense that if $m > 3n - 6$ the graph is surely not planar.

Example 12.22. Let us consider again a representation of K_5.

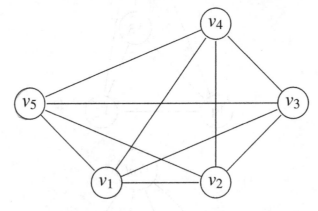

We know that this graph is simple and not planar. The number of nodes is $n = 5$ and the number of edges is

$$m = \binom{n}{2} = \frac{n!}{2!\,(n-2)!} = \frac{n^2 - n}{2} = 10.$$

We can easily verify the Proposition above by noting that $m > 3n - 6 = 9$.

Although a graph with at least $3n - 6$ is always not planar, the vice-versa is not true: a graph with less than $3n - 6$ can still be non-planar. The following proposition gives another condition that occurs every time a graph is planar.

Proposition 12.9. *Let G be a simple graph, connected and planar with $n > 3$ vertices and m edges. Let G be such that there are no cycles of length 3 (all the cycles are of length at least 4). It follows that $m \leq 2n - 4$.*

This proposition says that if the graph has no short cycles (of length 3) then $2n - 3$ edges are enough to conclude that it is not planar. The following example clarifies this fact.

Example 12.23. The graph $K_{3,3}$ is simple, connected, has $n = 6$ vertices and $m = 9$ edges. By checking the criterion in Proposition 12.8, $m \leq 3n - 6 = 18 - 6 = 12$. Thus, we cannot make conclusions about the planarity. However, the graph $K_{3,3}$ has no cycles of length 3 (it has cycles of length at least 4). Thus, if it has more than $2n - 4$ edges, it surely is not planar. In this case $2n - 4 = 8$. We know that in $K_{3,3}$ there are $m = 9$. Thus, we can conclude $K_{3,3}$ is not planar.

Definition 12.40. Let G be a planar graph. An *internal region* (or internal face) of a graph is a part of plane surrounded by edges. The part of plane not surrounded by edges is called *external region* of a graph.

In the following sections of this chapter when we refer to the *regions of a graph* or simply regions we generically mean either the internal regions or the external one. A graph can thus be divided into regions, one being external and all the others being internal.

Definition 12.41. Let G_1 and G_2 be two planar graphs. The graph G_2 is constructed starting from G_1. Each vertex of G_2 is placed in one region of G_1 (only one vertex per region). Then, the edges of G_2 are placed in that way such that each of G_2 crosses each edge of G_1 only once. The graph G_2 is said *geometric dual* (or simply *dual*) of the graph G_1.

Example 12.24. The graph below shows the graph G_1 composed of it vertices v_i and edges pictured in solid lines and the graph G_2 composed of the vertices w_i and edges pictured in dashed lines.

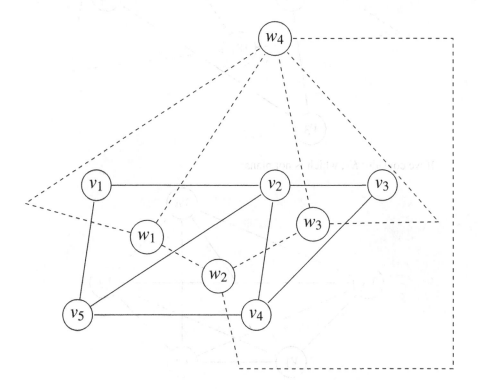

Proposition 12.10. *Let G_1 be a planar graph and G_2 be its geometric dual. Then, the geometric dual of G_2 is G_1.*

The existence of a dual is a very important concept in graph theory as it is a check for the planarity of a graph. The following Proposition explicitly states this fact.

Proposition 12.11. *Let G be a planar graph. The graph G is planar if and only if has a dual.*

Example 12.25. The following graph (solid edges) is planar and has a dual (dashed edges).

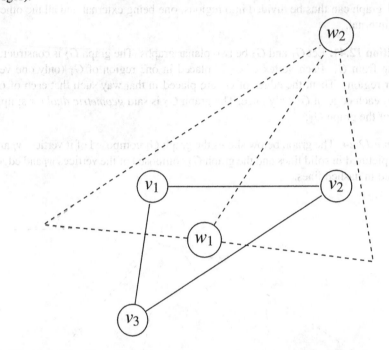

If we consider K_5, which is not planar

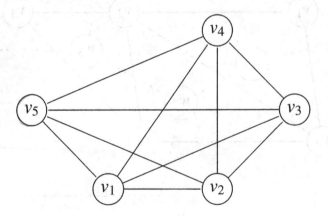

we can represent it as

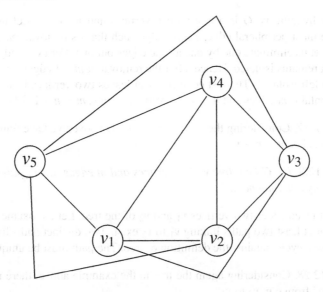

Regardless of how we represent K_5, still at least one crossing will appear. Hence, there will be no faces for constructing the dual.

Proposition 12.12. *Let G be a planar graph. If the graph G is non-separable its dual is non-separable.*

12.4.1 Trees and Cotrees

Definition 12.42. A *tree* is a connected graph which contains no cycles.

Example 12.26. An example of tree is given in the following.

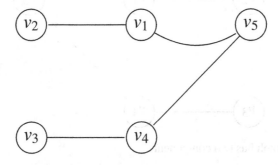

A graph composed of disconnected trees is called *forest*. So a forest is the disjoint union of trees. In other words, each connected component of a forest is a tree. The following theorem gives a characterizations of a tree.

Theorem 12.7. *Let G be a tree with n vertices and m edges. It follows that $m = n - 1$.*

Proof. By hypothesis, G is a tree with n vertices and m edges. Let us eliminate one of the most peripheral edge, i.e. an edge such that its removal leaves a vertex isolated. Let us eliminate one by one all the edges but one. For every edge removal, one vertex remains isolated. Hence, after the removal of $m - 1$ edges, $m - 1$ vertices have been left isolated. The last edge removal leaves two vertices isolated. Hence, the tree contains m edges and $n = m + 1$ vertices. Thus, $m = n - 1$. □

Example 12.27. Considering the tree in the example above, we have four edges and five vertices, i.e. $m = n - 1$.

Theorem 12.8. *Let G be a tree with n vertices and m edges. Every pair of vertices are joined by a unique path.*

Proof. Let us consider two vertices v_1 and v_2 of the tree. Let us assume by contradiction that at least two paths linking v_1 to v_2 exist. Two distinct paths linking v_1 to v_2 compose a cycle, against the definition of tree. The path must be unique. □

Example 12.28. Considering again the tree in the example above, there is only one way to "go" from e.g. v_2 to v_4.

Theorem 12.9. *Let G be a tree with n vertices and m edges. If an arbitrary edge is removed, the tree is then divided into two components.*

Proof. Let us consider two adjacent vertices v_1 and v_2 of the tree. Let us remove the edge linking v_1 to v_2. Since a tree does not contain cycles, there are no alternative paths to connect v_1 to v_2. The removal of the edge disconnects v_1 from v_2. This means that the tree is divided into two components. □

Example 12.29. If we consider again the tree above and remove an arbitrary edge, e.g. (v_1, v_5)

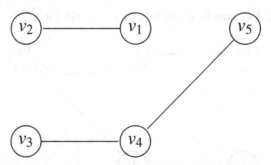

the resulting graph has two components.

The last theorem can also be equivalently expressed by the following proposition.

Proposition 12.13. *Every tree is a bipartite graph.*

Proof. A tree contains no cycles. Then, if an arbitrary edge connecting the arbitrary vertices v and v' is removed the two vertices will be disconnected. Hence the graph will be composed of two components. $\square$

Definition 12.43. Let G be a graph having n vertices and m edges. A *spanning tree* of the graph G is a tree containing all the n vertices of G. The edges of a spanning tree are said *branches*.

Example 12.30. Let us consider the following graph

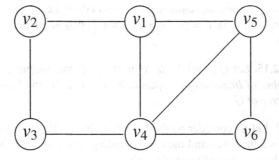

A spanning tree is

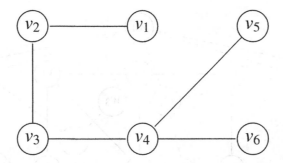

Definition 12.44. Let G be a graph and T one of its spanning trees. The *cotree* C of the tree T is a graph composed of the n vertices of T (and G) and those edges belonging to G but not to T. The edges of the cotree are said *chords*.

Example 12.31. The cotree corresponding to the spanning tree above is

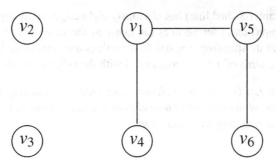

Proposition 12.14. *Let $G(n,m)$ be a connected, planar, simple graph composed of n vertices and m edges. Let ρ and ν be rank and nullity of $G(n,m)$, respectively. For every arbitrary spanning tree, the rank ρ is equal to the number of branches while the nullity ν is equal to the number of chords of the corresponding cotree.*

Proof. Any arbitrary spanning tree of the graph $G(n,m)$ contains all the n nodes of the graph. We know from Theorem 12.7 that the spanning tree contains $n-1$ branches. The rank of the graph is $\rho = n - c$ with c number of components. Since the graph is connected, it follows that $\rho = n - 1$, that is the number of branches.

The cotree will contain the remaining $m - n + 1$ edges, i.e. $m - n + 1$ chords. Since by definition the nullity is $\nu = m - \rho = m - n + 1$, that is the number of chords of the cotree. $\square$

Proposition 12.15. *Let $G(n,m)$ be a connected, planar, simple graph and G' its dual. The number of branches of a spanning tree of G' is equal to the numbers of chords of a cotree of G.*

Example 12.32. Let us consider again the graph above. We know that its spanning tree has $n - 1 = 5$ branches and the corresponding cotree $m - 5 = 3$ chords. Let us now consider the corresponding dual graph.

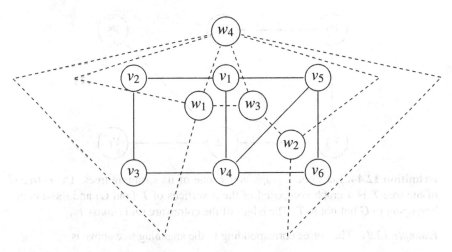

The dual graph (dashed line) has obviously eight edges (as many as the edges of the original graph) and four vertices (as many as the faces as the original graph). This means that its spanning tree has four vertices and thus three branches, that is the number of chords of a cotree associated with the original graph.

Corollary 12.1. *Let $G(n,m)$ be a planar connected graph having m edges and G' its dual. The sum of the number of branches of a spanning tree of G and the number of branches of a spanning tree of G' is m.*

12.4.1.1 Analogy Between Graphs and Vector Spaces

We can intuitively observe that there is an analogy between graphs and vector spaces. In particular, we can define a vector space on the set of edges E. The basis of vectors is replaced, in graph theory, by the spanning tree. Similar to vectors that must be n to span an n-dimensional vector space, a spanning tree must contain (and connect) all the n nodes. The concept of linear independence is replaced by the absence of cycles within a spanning tree. The presence of a cycle can be interpreted as a redundancy since it corresponds to multiple paths to connect a pair of nodes.

Consequently, the rank of a graph can be interpreted as the number of edges (branches) of its spanning tree. This is analogous to the rank of a vector space that is the number of vectors composing its basis. By recalling the rank-nullity theorem we may think that the nullity is the dimension of another vector space such that, when it is summed up to the rank, the dimension of the entire set (of edges) is obtained. In this sense, the cotree can also be interpreted as the basis of a vector space whose dimension is its number of edges (chords), that is the nullity. The sum of rank and nullity is the total number of edges of the graph.

The subject that studies this analogy falls outside the scopes of this book and is named *Matroids*.

12.4.2 Euler's Formula

A very important property for planar graphs is the Euler's formula.

Theorem 12.10. Euler's Formula. *Let $G(n,m)$ be a planar connected graph with n vertices and m edges. Let f be the number of faces. It follows that*

$$n - m + f = 2.$$

Proof. Let us consider a spanning tree T of the graph $G(n,m)$. We know from Proposition 12.14 that T has n vertices and $n-1$ branches.

The dual graph G' of $G(n,m)$ has one vertex in each face of G. Hence, it has f vertices. A spanning tree of the dual graph G' has $f-1$ branches.

From Corollary 12.1 it follows that

$$n - 1 + f - 1 = m \Rightarrow n - m + f = 2. \square$$

Example 12.33. Let us consider the following graph.

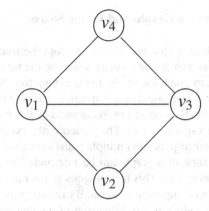

We have $n = 4$, $m = 5$, $f = 3$, and $n - m + f = 4 - 5 + 3$.

It must be remarked that Euler's formula was originally derived for polyhedra (i.e. a solid bounded by polygons). For example, polyhedra are the so-called Platonic solids: tetrahedron, cube, octahedron, dodecahedron, and icosahedron.

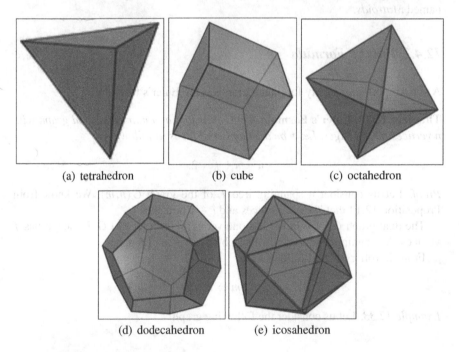

| (a) tetrahedron | (b) cube | (c) octahedron |

| (d) dodecahedron | (e) icosahedron |

Euler found the relation between the number of faces F, vertices V, and edges E of any simple (i.e. without holes) polyhedron:

$$V - E + F = 2.$$

This formula is identical to that seen for graphs because a polyhedron can be can be transformed into a simple, connected, planar, graph, taking the vertices of the polyhedron as vertices of the graph, the sides of the polyhedron as arcs of the graph, the faces of the polyhedron in this way correspond to the faces of the graph.

Example 12.34. The following two graphs show a planar representation of tetrahedron and cube, respectively.

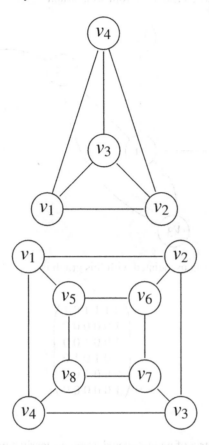

12.5 Graph Matrices

A useful way to represent a graph is through a matrix. This representation is conceptually very important as it links graph theory and matrix algebra, showing once again how mathematics is composed of interconnected concepts that are subject to examination from various perspectives.

12.5.1 Adjacency Matrices

Definition 12.45. Given a simple graph G with n vertices, its *adjacency matrix* is the $n \times n$ matrix which has 1 in position (i, j) if the vertex v_i and the vertex v_j are connected by an edge, 0 otherwise.

Example 12.35. Let us consider the following graph.

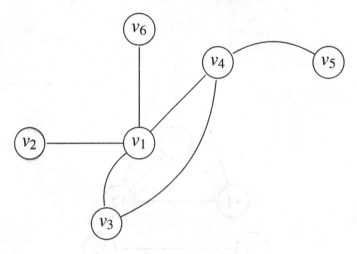

The adjacency matrix associated with this graph is

$$\mathbf{A} = \begin{pmatrix} 0 & 1 & 1 & 1 & 0 & 1 \\ 1 & 0 & 0 & 0 & 0 & 0 \\ 1 & 0 & 0 & 1 & 0 & 0 \\ 1 & 0 & 1 & 0 & 1 & 0 \\ 0 & 0 & 0 & 1 & 0 & 0 \\ 1 & 0 & 0 & 0 & 0 & 0 \end{pmatrix}.$$

The adjacency matrix of an undirected graph is clearly symmetric. In the case of an oriented graph, the associated adjacency matrix exhibits a 1 in position (i, j) if there exists an edge from vertex v_i to vertex v_j. On the contrary, in position (j, i) the will be 0. Hence, the adjacency matrix of an oriented graph is, in general, not symmetric.

Example 12.36. Let us consider the following oriented graph.

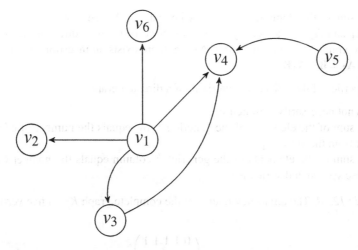

The adjacency matrix associated with this graph is:

$$\mathbf{A} = \begin{pmatrix} 0\ 1\ 1\ 1\ 0\ 1 \\ 0\ 0\ 0\ 0\ 0\ 0 \\ 0\ 0\ 0\ 1\ 0\ 0 \\ 0\ 0\ 0\ 0\ 0\ 0 \\ 0\ 0\ 0\ 1\ 0\ 0 \\ 0\ 0\ 0\ 0\ 0\ 0 \end{pmatrix}.$$

In the case of non-oriented multigraphs, if there are multiple edges (let us say s) connecting the vertex v_i with the vertex v_j, the matrix exhibits the integer s in the position (i, j) and in the position (j, i). In the case of oriented multigraphs, if there are s edges which go from v_i to v_j, the matrix exhibits s in the position (i, j) and if there are t edges that start at v_j and go in v_i the matrix exhibits t in the position (j, i), and 0 if there is no edge from v_j to v_i.

Example 12.37. The adjacency matrix of the graph of Konigsberg bridges is:

$$\begin{pmatrix} 0\ 2\ 2\ 1 \\ 2\ 0\ 0\ 1 \\ 2\ 0\ 0\ 1 \\ 1\ 1\ 1\ 0 \end{pmatrix}.$$

Definition 12.46. A graph is *labelled* if its vertices are distinguishable from one to another because a name was given to them.

The encoding of a graph by means of its associated adjacency matrix is the most suitable method for communicating the structure of a graph on a computer.

Properties of the adjacency matrix $\mathbf{A}$ of an undirected graph:

- $\mathbf{A}$ is symmetric

- The sum of the elements of each row i equals the degree of v_i
- If $\mathbf{A_1}$ and $\mathbf{A_2}$ are adjacency matrices that correspond to different labels of the same graph, $\mathbf{A_1}$ is conjugated to $\mathbf{A_2}$, i.e. there exists an invertible matrix $\mathbf{B}$ such that $\mathbf{A_2} = \mathbf{B}^{-1}\mathbf{A_1}\mathbf{B}$

Properties of the adjacency matrix $\mathbf{A}$ of a directed graph

- $\mathbf{A}$ is not necessarily symmetric
- The sum of the elements of the generic i^{th} row equals the number of edges that start from the vertex v_i
- The sum of the elements of the generic i^{th} column equals the number of edges whose second end vertex is v_i

Example 12.38. The adjacency matrix of the complete graph K_5 on five vertices.

$$\mathbf{A} = \begin{pmatrix} 0\ 1\ 1\ 1\ 1 \\ 1\ 0\ 1\ 1\ 1 \\ 1\ 1\ 0\ 1\ 1 \\ 1\ 1\ 1\ 0\ 1 \\ 1\ 1\ 1\ 1\ 0 \end{pmatrix}$$

Note that it is a symmetric matrix, corresponding to an undirected graph, that has all zeros on the main diagonal, given that there are no cycles.

Example 12.39. The following matrix

$$\begin{pmatrix} 0\ 0\ 2\ 2 \\ 1\ 0\ 2\ 0 \\ 3\ 0\ 1\ 1 \\ 2\ 1\ 0\ 0 \end{pmatrix}$$

is the adjacency matrix of the following graph.

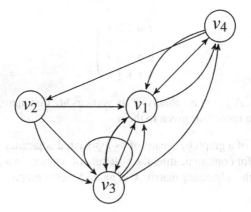

The adjacency matrix of a graph is not only an important tool to describe a graph but it is also a coherent representation that allows graph manipulations by operating within the algebraic space of matrices.

The definition of sum and multiplication of square matrices are valid also for adjacency matrices and have a meaning also on the resulting graphs. The meaning is analogous to that of union and Cartesian product between sets.

We know that, if A is the adjacency matrix of a (multi)graph G, the number that is in the position i, j represents the number of edges connecting the vertex v_i with the vertex v_j. The study of the powers of the matrix A gives us an important piece of information about the graph. Specifically, as stated by the following proposition, the powers of the matrix A give information on the number of walks (see Definition 12.10) of the graph.

Proposition 12.16. *If A is the adjacency matrix of a graph G, then the number of walks in G from the vertex v_i to the vertex v_j of length k ($k \geq 1$) is given by the element i, j of the matrix A^k.*

The proof of this proposition can be performed by induction on k.

Example 12.40. In order to calculate the amount of walks of length 2 in a graph, we have to calculate the power A^2 where A is the adjacency matrix of the graph. Let us consider the following adjacency matrix.

$$A = \begin{pmatrix} 0 & 1 & 1 & 1 & 0 & 1 \\ 1 & 0 & 0 & 0 & 0 & 0 \\ 1 & 0 & 0 & 1 & 0 & 0 \\ 1 & 0 & 1 & 0 & 1 & 0 \\ 0 & 0 & 0 & 1 & 0 & 0 \\ 1 & 0 & 0 & 0 & 0 & 0 \end{pmatrix}$$

The matrix A^2 is given by

$$A^2 = \begin{pmatrix} 0 & 1 & 1 & 1 & 0 & 1 \\ 1 & 0 & 0 & 0 & 0 & 0 \\ 1 & 0 & 0 & 1 & 0 & 0 \\ 1 & 0 & 1 & 0 & 1 & 0 \\ 0 & 0 & 0 & 1 & 0 & 0 \\ 1 & 0 & 0 & 0 & 0 & 0 \end{pmatrix} \begin{pmatrix} 0 & 1 & 1 & 1 & 0 & 1 \\ 1 & 0 & 0 & 0 & 0 & 0 \\ 1 & 0 & 0 & 1 & 0 & 0 \\ 1 & 0 & 1 & 0 & 1 & 0 \\ 0 & 0 & 0 & 1 & 0 & 0 \\ 1 & 0 & 0 & 0 & 0 & 0 \end{pmatrix} = \begin{pmatrix} 4 & 0 & 1 & 1 & 1 & 0 \\ 0 & 1 & 1 & 1 & 0 & 1 \\ 1 & 1 & 2 & 1 & 1 & 1 \\ 1 & 1 & 1 & 3 & 0 & 1 \\ 1 & 0 & 1 & 0 & 1 & 0 \\ 0 & 1 & 1 & 1 & 0 & 1 \end{pmatrix}$$

The walks from v_1 to v_1 of length 2 are 4, i.e. v_1, v_2, v_1; v_1, v_6, v_1; v_1, v_3, v_1; v_1, v_4, v_1 (remember that they are walks, not trails, then the edges can retrace). There is no walk from v_1 to v_2 of length 2 ($a_{1,2} = 0$). There are three walks from v_4 to v_4 of length 2 ($a_{4,4} = 3$) and so on.

The powers of the adjacency matrix of a graph also give us information on the connection of the graph. The following proposition formalizes this statement.

Proposition 12.17. *Let* $\mathbf{A}$ *be the adjacency matrix of a graph G with n vertices. Then*

1. *G is connected if and only if $\mathbf{I}+\mathbf{A}+\mathbf{A}^2+\ldots+\mathbf{A}^{n-1}$ contains only strictly positive integers.*
2. *G is connected if and only if $(\mathbf{I}+\mathbf{A})^{n-1}$ contains only strictly positive integers.*

12.5.2 Incidence Matrices

Definition 12.47. Let G be a graph containing n vertices and m edges and without self-loops. The *vertex-edge incidence matrix* (or simply incidence matrix) is that $n \times m$ matrix displaying in position (i, j) a 1 if the edge e_j terminates into the vertex v_i, and a 0 otherwise.

Example 12.41. In order to clarify this concept, let us consider the graph used to introduce the adjacency matrix but with labelled edges.

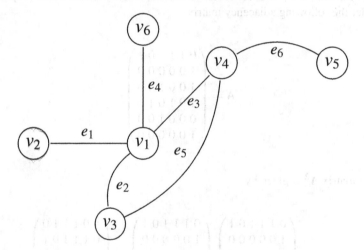

The incidence matrix associated with this graph is

$$
\mathbf{A_a} = \begin{pmatrix}
1 & 1 & 1 & 1 & 0 & 0 \\
1 & 0 & 0 & 0 & 0 & 0 \\
0 & 1 & 0 & 0 & 1 & 0 \\
0 & 0 & 1 & 0 & 1 & 1 \\
0 & 0 & 0 & 0 & 0 & 1 \\
0 & 0 & 0 & 1 & 0 & 0
\end{pmatrix}.
$$

It may be observed that the incidence matrix, in general, are not symmetric and not square. However, incidence matrices have other features. At first, as it is obvious from its definition, each column of a incidence matrix contains two 1 values. This fact be easily explained considering that the jth column displays a 1 for every intersection between the edge e_j and the vertices. Since each edge intersects always two vertices, then the jth column displays two 1 values. Let us now introduce the sum operation on the elements of incidence matrices.

$$0 + 0 = 0$$
$$1 + 0 = 1$$
$$0 + 1 = 1$$
$$1 + 1 = 0$$

An important property of incidence matrices is expressed in the following proposition.

Proposition 12.18. *Let G be a graph containing n vertices and m edges. Let* $\mathbf{A_a}$ *be the incidence matrix associated with the graph G. Let us indicate with* $\mathbf{a_i}$ *the generic* i^{th} *column vector of the matrix* $\mathbf{A_a}$. *It follows that*

$$\sum_{i=1}^{n} \mathbf{a_i} = \mathbf{o}.$$

Proof. For an arbitrary j value, let us calculate $\sum_{i=1}^{n} a_{i,j}$. This is the sum of a column vector which displays two 1 values. Hence, the sum is equal to 0. We can repeat the same calculation $\forall j = 1, \ldots, m$ and we will obtain 0 every time. Hence,

$$\sum_{i=1}^{n} \mathbf{a_i} = \left(\sum_{i=1}^{n} a_{i,1}, \ldots, \sum_{i=1}^{n} a_{i,j}, \ldots, \sum_{i=1}^{n} a_{i,m} \right) = (0, \ldots, 0, \ldots, 0) = \mathbf{o}. \quad \square$$

Theorem 12.11. *Let G be a connected graph containing n vertices and m edges. Let* $\mathbf{A_a}$ *be the incidence matrix associated with the graph G. It follows that the rank of G and the rank of the matrix* $\mathbf{A_a}$ *are the same.*

Proof. If G is a connected graph then its rank is $n - c = n - 1$. The matrix $\mathbf{A_a}$ has size $n \times m$. Since $\sum_{i=1}^{n} \mathbf{a_i} = \mathbf{o}$, then the rank of the matrix $\mathbf{A_a}$ is at most $n - 1$.

In order to prove that graph and matrix have the same rank, we need to prove that the rank of the matrix $\mathbf{A_a}$ is at least $n - 1$. This means that we need to prove that the sum of $k < n$ row vectors is $\neq \mathbf{o}$. Let us assume, by contradiction, that there exist k row vectors such that $\sum_{i=1}^{k} \mathbf{a_i} = \mathbf{o}$.

Let us permute the rows of the matrix so that these k rows appear on the top of the matrix. The columns of these k rows either contain two 1 values or all 0 values. The columns can be permuted in order to have the first l columns displaying two 1 values and the remaining $m - l$ columns composed of all 0 values. As a consequence of these column swaps, the remaining $n - k$ rows are arranged to display all 0 values in the first l columns and two 1 values per column for the remaining $m - l$ columns. The matrix appears mapped in the following way:

$$\mathbf{A_a} = \left(\begin{array}{c|c} \mathbf{A_{1,1}} & \mathbf{0} \\ \hline \mathbf{0} & \mathbf{A_{2,2}} \end{array} \right).$$

This would be an incidence matrix associated with a graph composed of two components, $\mathbf{A_{1,1}}$ and $\mathbf{A_{2,2}}$ respectively. The two components are clearly disconnected because there is no edge connecting them (there is no column with only a 1 in $\mathbf{A_{1,1}}$ and only a 1 in $\mathbf{A_{2,2}}$). This is impossible as, by hypothesis, the graph is connected. Hence, the rank of the matrix is at least $n - 1$ and, for what seen above, exactly $n - 1$ just like the rank of the graph. $\square$

Example 12.42. Let us consider again the graph of the example above and its associated matrix $\mathbf{A_a}$. The graph is composed of six nodes and one component. Hence, its rank is 5. The associated incidence matrix has $n = 6$ rows and $m = 6$ columns. The matrix is singular and has rank 5.

Corollary 12.2. *Let G be a graph containing n vertices and m edges and composed of c components. Let $\mathbf{A_a}$ be the incidence matrix associated with the graph G. The rank of $\mathbf{A_a}$ is $n - c$.*

Definition 12.48. Let G be a graph containing n vertices and m edges. Let $\mathbf{A_a}$ be the incidence matrix associated with the graph G. A *reduced incidence matrix* $\mathbf{A_f}$ is any $((n-1) \times m)$ matrix obtained from $\mathbf{A_a}$ after the cancellation of a row. The vertex corresponding to the deleted row in $\mathbf{A_f}$ is said *reference vertex*.

Obviously, since the rank of an incidence matrix associated with a connected graph is $n - 1$ the row vectors of a reduced incidence matrix are linearly independent.

Corollary 12.3. *Let G be a graph containing n vertices. The reduced incidence matrix of G is non-singular if and only if G is a tree.*

Proof. If G is a tree with n vertices, it is connected and contains $n - 1$ edges. The incidence matrix associated with a tree has size $n \times ((n-1))$ and rank $n - 1$. If we remove one row, the reduced incidence matrix associated with this tree is a square matrix having size $((n-1) \times (n-1))$ and rank $n - 1$ for the Theorem 12.11. Since the size of the matrix is equal to its rank, the matrix is non-singular. $\square$

If the reduced incidence matrix is non-singular, it is obviously square and, more specifically, for the hypothesis is $((n-1) \times (n-1))$. This means that the rank of the matrix is equal to its size, i.e. $n - 1$. The graph must be connected because, otherwise, its rank would be less than $n - 1$ (if the graph had more than $n - 1$ edges the matrix would not be square and we could not speak about singularity). Moreover, the graph associated with this reduced incidence matrix has n vertices and $n - 1$ edges. Hence, the graph G is a tree. $\square$

Example 12.43. Essentially, this theorem says that the reduced incidence matrix is square when the associated graph is a tree. If the graph contained circuits, they will result into extra columns in the incidence matrix. If the graph is disconnected, the rank of the associated reduced incidence matrix would be $< n - 1$, i.e. the reduced incidence matrix would either be rectangular (an edge removal corresponds to a column removal) or a singular matrix (disconnected graph with at least one circuit).

Probably, this fact clarifies the analogy between spanning trees and vector bases and why the presence of a cycle is a redundancy.

Let us consider a tree and its incidence matrix.

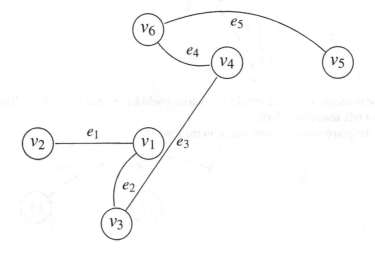

$$\mathbf{A_a} = \begin{pmatrix} 1 & 1 & 0 & 0 & 0 \\ 1 & 0 & 0 & 0 & 0 \\ 0 & 1 & 1 & 0 & 0 \\ 0 & 0 & 1 & 1 & 0 \\ 0 & 0 & 0 & 0 & 1 \\ 0 & 0 & 0 & 1 & 1 \end{pmatrix}$$

Let us take v_6 as the reference vertex and write the reduced incidence matrix:

$$\mathbf{A_f} = \begin{pmatrix} 1 & 1 & 0 & 0 & 0 \\ 1 & 0 & 0 & 0 & 0 \\ 0 & 1 & 1 & 0 & 0 \\ 0 & 0 & 1 & 1 & 0 \\ 0 & 0 & 0 & 0 & 1 \end{pmatrix}.$$

It can be verified that $\mathbf{A_f}$ is non-singular.

If we added an edge we would have

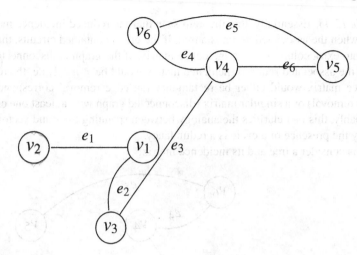

whose associated reduced incidence matrix would have dimensions 5×6. Thus, we cannot talk about singularity.

If the graph was disconnected as in the case of

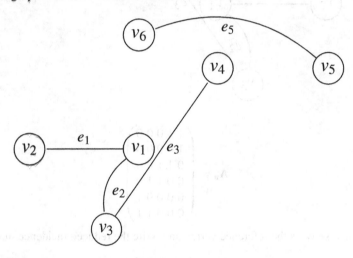

the resulting reduced incidence matrix would be of size 5×4.

If the graph contains cycles and is disconnected the corresponding reduced incidence matrix can be square.

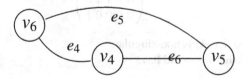

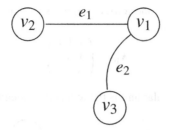

The associated matrix would have rank $n - c = 6 - 2 = 4$. Thus, the reduced incidence matrix would be singular.

Corollary 12.4. *Let G be a connected graph containing n vertices and m edges. Let $\mathbf{A_a}$ be the incidence matrix associated with the graph G. Let $\mathbf{A_b}$ be a $(n-1) \times (n-1)$ submatrix extracted from $\mathbf{A_a}$. The matrix $\mathbf{A_b}$ is non-singular if and only if the matrix $\mathbf{A_b}$ is the reduced incidence matrix associated with a spanning tree of G.*

Proof. For the hypothesis G is a connected graph with n vertices and m edges. Hence, $m \geq n - 1$. The incidence matrix $\mathbf{A_a}$ is composed of n rows and $m \geq n - 1$ columns. A $(n-1) \times (n-1)$ submatrix $\mathbf{A_b}$ extracted from $\mathbf{A_a}$ is the reduced incidence matrix associated with a subgraph of $G_b \subset G$ containing n vertices and $n - 1$ edges.

Let us assume that $\mathbf{A_b}$ is non-singular. The submatrix $\mathbf{A_b}$ is non-singular if and only if its associated graph G_b is a tree. Since G_b contains n vertices and $n-1$ edges, it is a spanning tree. □

Let us assume that G_b is a spanning tree. Hence, it has n vertices and $n - 1$ edges. The associated reduced incidence matrix $\mathbf{A_b}$ is square and has size $n - 1$. For Corollary 12.3 the matrix of a tree is non-singular. □

Example 12.44. Let us consider the following graph.

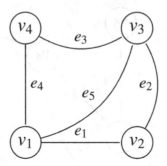

The corresponding incidence matrix is

$$\mathbf{A_a} = \begin{pmatrix} 1 & 0 & 0 & 1 & 1 \\ 1 & 1 & 0 & 0 & 0 \\ 0 & 1 & 1 & 0 & 1 \\ 0 & 0 & 1 & 1 & 0 \end{pmatrix}.$$

Let us cancel the second and third columns as well as the fourth row. Let us call $\mathbf{A_b}$ the resulting sub-matrix

$$\mathbf{A_b} = \begin{pmatrix} 1 & 1 & 1 \\ 1 & 0 & 0 \\ 0 & 0 & 1 \end{pmatrix}.$$

This matrix is non-singular and is the reduced incidence matrix of

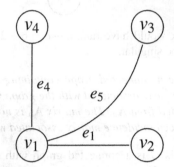

that is a spanning tree of the graph.

12.5.3 Cycle Matrices

Definition 12.49. Let G be a graph containing n vertices and m edges. Let us assume that the graph contains q cycles. The *cycle matrix* $\mathbf{B_a}$ is that $q \times m$ matrix displaying in position (i, j) a 1 if the edge e_j belongs to the cycle z_i and a 0 otherwise.

Example 12.45. Let us consider again the graph

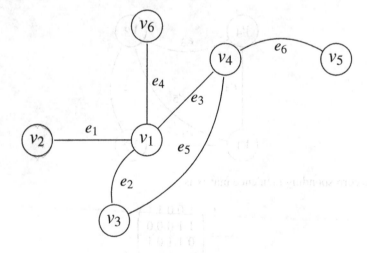

The only cycle appearing in this graph is $z_1 = (e_2, e_3, e_5)$. Hence, in this case the cycle matrix is trivial and given by

$$\mathbf{B_a} = \begin{pmatrix} 0 & 1 & 1 & 0 & 1 & 0 \end{pmatrix}.$$

Theorem 12.12. *Let G be a graph containing n vertices and m edges and no self-loops. Let $\mathbf{A_a}$ and $\mathbf{B_a}$ be the incidence and cycle matrices associated with the graph, respectively. Let the columns of the two matrices arranged in the same way, i.e. same labelling of the edges. It follows that $\mathbf{A_a B_a^T} = \mathbf{A_a^T B_a} = \mathbf{O}$.*

Proof. Let v_i be a vertex of the graph G and z_j one of its cycles. If the vertex v_i does not belong to the cycle z_j, then there are no edged being incident the vertex v_i and belonging to the cycle. If v_i belongs to z_j, then exactly two edges belong to the cycle and are incident on v_i.

Let us consider now the ith row of the matrix $\mathbf{A_a}$ and j^{th} row $\mathbf{B_a}$ (which is the jth column). Let us perform the scalar product between the two vectors: $\sum_{r=1}^{m} a_{i,r} b_{j,r}$. If we consider the edge e_r the following scenarios are possible.

1. e_r is not incident on v_i and does not belong to z_j. Hence, $a_{i,r} = 0$, $b_{j,r} = 0$ and $a_{i,r} b_{j,r} = 0$
2. e_r is not incident on v_i and belongs to z_j. Hence, $a_{i,r} = 0$, $b_{j,r} = 1$ and $a_{i,r} b_{j,r} = 0$
3. e_r is incident on v_i and does not belong to z_j. Hence, $a_{i,r} = 1$, $b_{j,r} = 0$ and $a_{i,r} b_{j,r} = 0$
4. e_r is incident on v_i and belongs to z_j. Hence, $a_{i,r} = 1$, $b_{j,r} = 1$ and $a_{i,r} b_{j,r} = 1$.

The product $a_{i,r} b_{j,r} = 1$ occurs exactly twice, for those two edges, $r1$ and $r2$, that are incident on v_i and belong to the cycle. Thus, the scalar product is the sum of null terms except two: $\sum_{r=1}^{m} a_{i,r} b_{j,r} = 0 + 0 + \ldots + a_{i,r1} b_{j,r1} + \ldots + a_{i,r2} b_{j,r2} + \ldots + 0 = a_{i,r1} b_{j,r1} + a_{i,r2} b_{j,r2} = 1 + 1 = 0$. Each scalar product is 0 and then $\mathbf{A_a B_a^T} = \mathbf{O}$.

If we consider the product $\mathbf{A_a^T B_a}$, we have scalar products between the columns of the matrix $\mathbf{A_a}$ and the columns of the matrix $\mathbf{B_a}$. Each column of $\mathbf{A_a}$ contains only two 1 values. If the edge incident on the vertex also belongs to the cycle, the situation $1 + 1 = 0$ described above is verified. All the other products return 0. Hence $\mathbf{A_a^T B_a} = \mathbf{O}$.

In general it is true that $\mathbf{A_a B_a^T} = \mathbf{A_a^T B_a} = \mathbf{O}$. $\square$

Example 12.46. In order to understand Theorem 12.12, let us consider the following graph.

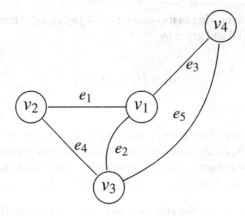

The associated incidence matrix is

$$
\mathbf{A_a} =
\begin{pmatrix}
1 & 1 & 1 & 0 & 0 \\
1 & 0 & 0 & 1 & 0 \\
0 & 1 & 0 & 1 & 1 \\
0 & 0 & 1 & 0 & 1
\end{pmatrix}.
$$

Totally, three cycles can be identified, i.e. (e_1,e_2,e_4), (e_2,e_3,e_5), and (e_1,e_3,e_5, e_4) The cycle matrix $\mathbf{B_a}$ is the following.

$$
\mathbf{B_a} =
\begin{pmatrix}
1 & 1 & 0 & 1 & 0 \\
0 & 1 & 1 & 0 & 1 \\
1 & 0 & 1 & 1 & 1
\end{pmatrix}.
$$

Let us compute

$$
\mathbf{A_a B_a^T} =
\begin{pmatrix}
1 & 1 & 1 & 0 & 0 \\
1 & 0 & 0 & 1 & 0 \\
0 & 1 & 0 & 1 & 1 \\
0 & 0 & 1 & 0 & 1
\end{pmatrix}
\begin{pmatrix}
1 & 0 & 1 \\
1 & 1 & 0 \\
0 & 1 & 1 \\
1 & 0 & 1 \\
0 & 1 & 1
\end{pmatrix}
=
\begin{pmatrix}
1+1 & 1+1 & 1+1 \\
1+1 & 0 & 1+1 \\
1+1 & 1+1 & 1+1 \\
0 & 1+1 & 1+1
\end{pmatrix}
= \mathbf{O}.
$$

The first element of the product matrix is given by the scalar product

$$
\begin{pmatrix} 1 & 1 & 1 & 0 & 0 \end{pmatrix}
\begin{pmatrix} 1 \\ 1 \\ 0 \\ 1 \\ 0 \end{pmatrix} = 1+1+0+0+0.
$$

The first 1 corresponds to the edge e_1 inciding on the vertex v_1 and belonging to the circuit (e_1,e_2,e_4). The second 1 corresponds to the edge e_2 inciding on the vertex v_1 and belonging to the circuit (e_1,e_2,e_4). The matrix product is necessarily zero because we have always exactly two edges belonging to the same circuit and inciding to the same node. A third edge belonging to the same circuit (such as e_4) would not incide on the same node. Analogously, a third edge inciding on the same node (such as e_3) would not be part of the same circuit.

Definition 12.50. Let G be a connected graph having n vertices and m edges. Let T be a spanning tree of the graph G. The spanning tree univocally determines a cotree C. If one arbitrary chord from C is added to T one cycle is identified. This cycle is said *fundamental cycle* (or fundamental circuit).

Proposition 12.19. *Let G be a connected graph having n vertices and m edges. Regardless of the choice of the spanning tree T, the spanning tree T has $n-1$ edges and the graph G has $m-n+1$ fundamental cycles.*

Proof. Since the graph G has n vertices, for Theorem 12.7, $n-1$ edges of the total m edges belong to a spanning tree (branches). It follows that the remaining $m-n+1$ edges belong to the cotree (chords). Every time one chord is added to the spanning tree a fundamental cycle is identified. Thus, each chord of the cotree corresponds to a fundamental cycle. Since there exist $m-n+1$ chords, then there exist $m-n+1$ fundamental cycles. $\square$

Definition 12.51. Let G be a connected graph having n vertices and m edges. The matrix $\mathbf{B_f}$ having size $(m-n+1) \times m$ and containing the fundamental cycles associated with an arbitrary spanning tree is said *fundamental cycle matrix*.

Obviously when we speak about spanning tree we refer to connected graph since the spanning tree must connect all the vertices of the graph.

Example 12.47. Let us consider the following graph.

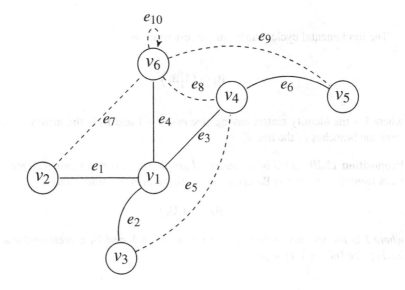

The graph contains $10 - 6 + 1 = 5$ fundamental cycles. If we consider the spanning tree identified by the edges $(e_1, e_2, e_3, e_4, e_6)$, the fundamental cycles are

- e_2, e_3, e_5
- e_1, e_4, e_7
- e_3, e_4, e_8
- e_3, e_4, e_6, e_9
- e_{10}

The cycle matrix associated with these fundamental cycles is the following.

$$
\mathbf{B_f} = \begin{pmatrix}
0 & 1 & 1 & 0 & 1 & 0 & 0 & 0 & 0 & 0 \\
1 & 0 & 0 & 1 & 0 & 0 & 1 & 0 & 0 & 0 \\
0 & 0 & 1 & 1 & 0 & 0 & 0 & 1 & 0 & 0 \\
0 & 0 & 1 & 1 & 0 & 1 & 0 & 0 & 1 & 0 \\
0 & 0 & 0 & 0 & 0 & 0 & 0 & 0 & 0 & 1
\end{pmatrix}.
$$

Let us now rearrange the columns to have the cotree's chords first and the tree's branches then,

$$
\mathbf{B_f} = \left(\begin{array}{ccccc|ccccc}
e_5 & e_7 & e_8 & e_9 & e_{10} & e_1 & e_2 & e_3 & e_4 & e_6 \\
\hline
1 & 0 & 0 & 0 & 0 & 0 & 1 & 1 & 0 & 0 \\
0 & 1 & 0 & 0 & 0 & 1 & 0 & 0 & 0 & 0 \\
0 & 0 & 1 & 0 & 0 & 0 & 0 & 1 & 1 & 0 \\
0 & 0 & 0 & 1 & 0 & 0 & 0 & 1 & 1 & 1 \\
0 & 0 & 0 & 0 & 1 & 0 & 0 & 0 & 0 & 0
\end{array}\right).
$$

The fundamental cycle matrix can be re-written as

$$
\mathbf{B_f} = \left(\mathbf{I} | \mathbf{B_t}\right).
$$

where $\mathbf{I}$ is the identity matrix having size $m - n + 1$ and $\mathbf{B_t}$ is the matrix characterizing the branches of the tree T.

Proposition 12.20. *Let G be a connected graph with n vertices and m edges. Every fundamental cycle matrix $\mathbf{B_f}$ associated with it can be partitioned as*

$$
\mathbf{B_f} = \left(\mathbf{I} | \mathbf{B_t}\right)
$$

where $\mathbf{I}$ is the identity matrix having size $m - n + 1$ and $\mathbf{B_t}$ a rectangular matrix having size $(m - n + 1) \times (n - 1)$.

Proof. Each circuit represented in $\mathbf{B_f}$ (fundamental circuit) is composed of some branches of spanning tree and only one chord of cotree. Thus, if we arrange the columns (edges) of $\mathbf{B_f}$ so that the chords appear in the first columns and while in the remaining columns the branches of the tree appear:

$$\mathbf{B_f} = \left(\mathbf{B_{c1}} \big| \mathbf{B_{t1}} \right)$$

where $\mathbf{B_c}$ represents the cotree sub-matrix and $\mathbf{B_t}$ the tree sub-matrix.

We know that a spanning tree contains $n - 1$ branches and that the graph contains m edges in total. Thus, there are $m - n + 1$ cotree's chords. This means that the sub-matrix $\mathbf{B_{c1}}$ has $m - n + 1$ columns. Since each cotree corresponds to a fundamental circuit, the matrix $\mathbf{B_f}$ (and then also the sub-matrix $\mathbf{B_{c1}}$) has $m - n + 1$ rows. It follows that the sub-matrix $\mathbf{B_{c1}}$ is square, displays only one 1 per row while all the other row elements are 0.

We can then swap the rows of the matrix and have the matrix $\mathbf{B_f}$ partitioned as

$$\mathbf{B_f} = \left(\mathbf{I} \big| \mathbf{B_t} \right)$$

where $\mathbf{I}$ is the identity matrix having size $m - n + 1$ and $\mathbf{B_t}$ a $(m - n + 1) \times (n - 1)$ rectangular matrix. $\square$

We can also verify that the relation $\mathbf{A_a B_a^T} = \mathbf{A_a^T B_a} = \mathbf{O}$ is valid also for fundamental matrices.

Example 12.48. Let us consider the following graph.

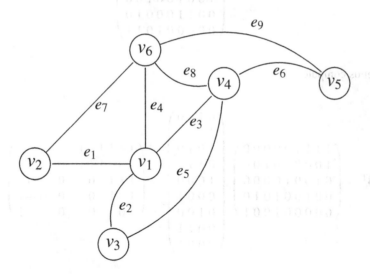

The associated incidence matrix is

$$
\mathbf{A_a} = \begin{pmatrix}
1 & 1 & 1 & 1 & 0 & 0 & 0 & 0 & 0 \\
1 & 0 & 0 & 0 & 0 & 0 & 1 & 0 & 0 \\
0 & 1 & 0 & 0 & 1 & 0 & 0 & 0 & 0 \\
0 & 0 & 1 & 0 & 0 & 1 & 0 & 1 & 0 \\
0 & 0 & 0 & 0 & 0 & 1 & 0 & 0 & 1 \\
0 & 0 & 0 & 1 & 0 & 0 & 1 & 1 & 1
\end{pmatrix}.
$$

Let us remove the last row and we obtain

$$
\mathbf{A_f} = \begin{pmatrix}
1 & 1 & 1 & 1 & 0 & 0 & 0 & 0 & 0 \\
1 & 0 & 0 & 0 & 0 & 0 & 1 & 0 & 0 \\
0 & 1 & 0 & 0 & 1 & 0 & 0 & 0 & 0 \\
0 & 0 & 1 & 0 & 0 & 1 & 0 & 1 & 0 \\
0 & 0 & 0 & 0 & 0 & 1 & 0 & 0 & 1
\end{pmatrix}.
$$

A fundamental cycle matrix $\mathbf{B_f}$ is the following.

$$
\mathbf{B_f} = \begin{pmatrix}
0 & 1 & 1 & 0 & 1 & 0 & 0 & 0 & 0 \\
1 & 0 & 0 & 1 & 0 & 0 & 1 & 0 & 0 \\
0 & 0 & 1 & 1 & 0 & 0 & 0 & 1 & 0 \\
0 & 0 & 0 & 0 & 0 & 1 & 0 & 1 & 1
\end{pmatrix}.
$$

Let us compute

$$
\mathbf{A_f B_f^T} = \begin{pmatrix}
1 & 1 & 1 & 1 & 0 & 0 & 0 & 0 & 0 \\
1 & 0 & 0 & 0 & 0 & 0 & 1 & 0 & 0 \\
0 & 1 & 0 & 0 & 1 & 0 & 0 & 0 & 0 \\
0 & 0 & 1 & 0 & 0 & 1 & 0 & 1 & 0 \\
0 & 0 & 0 & 0 & 0 & 1 & 0 & 0 & 1
\end{pmatrix}
\begin{pmatrix}
0 & 1 & 0 & 0 \\
1 & 0 & 0 & 0 \\
1 & 0 & 1 & 0 \\
0 & 1 & 1 & 0 \\
1 & 0 & 0 & 0 \\
0 & 0 & 0 & 1 \\
0 & 1 & 0 & 0 \\
0 & 0 & 1 & 1 \\
0 & 0 & 0 & 1
\end{pmatrix}
=
\begin{pmatrix}
1+1 & 1+1 & 1+1 & 0 \\
0 & 1+1 & 0 & 0 \\
1+1 & 0 & 0 & 0 \\
1+1 & 0 & 0 & 1+1 \\
0 & 0 & 0 & 1+1
\end{pmatrix} = \mathbf{O}.
$$

Theorem 12.13. *Let G be a connected graph having n vertices and m edges. Let $\mathbf{B_a}$ be the cycle matrix associated with the graph. The rank of $\mathbf{B_a}$ indicated with $\rho_{\mathbf{B_a}}$, is $m - n + 1$.*

Proof. Since $\mathbf{B_f}$ contains an identity matrix having order $m - n - 1$ also $\mathbf{B_a}$ contains the same matrix. Hence, the rank of $\mathbf{B_a}$, $\rho_{\mathbf{B_a}}$ is at least $m - n + 1$:

$$\rho_{\mathbf{B_a}} \geq m - n + 1.$$

We know that $\mathbf{A_a}\mathbf{B_a^T} = \mathbf{O}$. For the Theorem 2.12, (Weak Sylvester's Law of Nullity)

$$\rho_{\mathbf{A_a}} + \rho_{\mathbf{B_a}} \leq m$$

with $\rho_{\mathbf{A_a}}$ rank of the incidence matrix $\mathbf{A_a}$.

Since $\rho_{\mathbf{A_a}} = n - 1$, we obtain

$$\rho_{\mathbf{B_a}} \leq m - n + 1.$$

Thus, we can conclude that

$$\rho_{\mathbf{B_a}} = m - n + 1. \square$$

Example 12.49. Let us consider the following graph

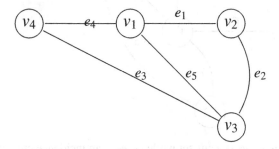

The cycles of the graph are

$$e_1, e_2, e_5$$
$$e_3, e_4, e_5$$
$$e_1, e_2, e_3, e_4$$

and the cycle matrix is

$$\mathbf{B_a} = \begin{pmatrix} 1 & 1 & 0 & 0 & 1 \\ 0 & 0 & 1 & 1 & 1 \\ 1 & 1 & 1 & 1 & 0 \end{pmatrix}.$$

We can easily check that the determinant of all the order 3 sub-matrices is null, while we can find non-singular order 2 sub-matrices. In other words $\rho_{\mathbf{B_a}} = 2$. Considering that $m = 5$ and $n = 4$, it follow that $\rho_{\mathbf{B_a}} = m - n + 1 = 5 - 4 + 1 = 2$.

Theorem 12.14. *Let G be a graph having n vertices, m edges, and c components. Let $\mathbf{B_a}$ be the cycle matrix associated with the graph. The rank of $\mathbf{B_a}$ indicated with $\rho_{\mathbf{B_a}}$, is $m - n + c$.*

12.5.4 Cut-Set Matrices

Definition 12.52. Let G be a graph containing n vertices and m edges. A *cut-set* is a set composed of the minimum number of edges whose removal cause the increase of components by exactly one.

Definition 12.53. An *incidence cut-set* is a cut-set such that one of the two components resulting from the edge removals is an isolated vertex.

Example 12.50. Let us consider the following graph again.

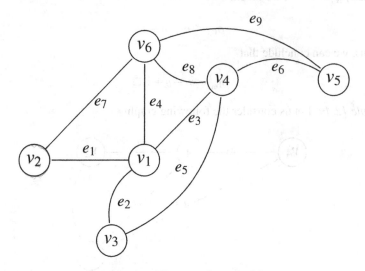

The set e_3, e_4, e_5, e_7 is a cut-set while e_6, e_9 is an incidence cut-set.

Definition 12.54. Let G be a graph containing n vertices and m edges. Let us assume that the graph contains p cut-sets. The *cut-set matrix* $\mathbf{Q_a}$ is that $p \times m$ matrix displaying in position (i, j) a 1 if the edge e_j belongs to the cut-set cs_i and a 0 otherwise.

Proposition 12.21. *Let $G(n, m)$ be a connected graph having n nodes and m edges. Let $\mathbf{Q_a}$ be the cut-set matrix of the graph. It follows that*

$$\rho_{\mathbf{Q_a}} \geq n - 1$$

where $\rho_{\mathbf{Q_a}}$ is the rank of the matrix $\mathbf{Q_a}$.

Proof. In non-separable graphs, since every set of edges incident on a vertex is a cut-set, every row of the incidence matrix $\mathbf{A_a}$ is also a row of the cut-set matrix $\mathbf{Q_a}$. Hence, the incidence matrix and the cut-set matrix coincide: $\mathbf{A_a} = \mathbf{Q_a}$. Let us name $\rho_{\mathbf{A_a}}$ the rank of the matrix $\mathbf{A_a}$. Since $\rho_{\mathbf{A_a}} = n - 1$, also $\rho_{\mathbf{Q_a}} = n - 1$.

In separable graphs, the incidence matrix is "contained" in the cut-set matrix. In other words, the incidence matrix $\mathbf{A_a}$ is a submatrix of the cut-set matrix $\mathbf{Q_a}$.

If follows that in general $\rho_{\mathbf{Q_a}} \geq \rho_{\mathbf{A_a}}$. Considering that, $\rho_{\mathbf{A_a}} = n - 1$, it follows that

$$\rho_{\mathbf{Q_a}} \geq n - 1. \quad \square$$

Theorem 12.15. *Let G be a graph containing n vertices and m edges and no self-loops. Let $\mathbf{Q_a}$ and $\mathbf{B_a}$ be the cut-set and cycle matrices associated with the graph, respectively. Let the columns of the two matrices arranged in the same way, i.e. same labelling of the edges. It follows that $\mathbf{Q_a B_a^T} = \mathbf{Q_a^T B_a} = \mathbf{O}$.*

Proof. Let us perform the product $\mathbf{Q_a B_a^T}$. Let us consider now the ith row of the matrix $\mathbf{Q_a}$ and j^{th} row $\mathbf{B_a}$ (which is the jth column). Let us perform the scalar product between the two vectors: $\sum_{r=1}^{m} q_{i,r} b_{j,r}$. There are two options. Either there are no edges belonging to the cycle and the cut-set or the number of edges belonging to both cycle and cut-set is even. In both cases the scalar product would be 0. Since all the scalar products are 0, then $\mathbf{Q_a B_a^T} = \mathbf{O}$. Since, for the transpose of the product, $\mathbf{Q_a^T B_a} = \left(\mathbf{Q_a B_a^T} \right)^{\mathbf{T}} = \mathbf{O^T}$, we have proved that $\mathbf{Q_a B_a^T} = \mathbf{Q_a^T B_a} = \mathbf{O}$. $\quad \square$

Theorem 12.16. *Let G be a connected graph containing n vertices and m edges. Let $\mathbf{A_a}$ and $\mathbf{Q_a}$ be the incidence and cut-set matrices, respectively, associated with this graph. It follows that the two matrices have the same rank, which is also equal to the rank of the graph G.*

Proof. Since the graph is connected we know that

$$\rho_{\mathbf{Q_a}} \geq n - 1.$$

Let us indicate with $\mathbf{B_a}$ the cycle matrix. Since cut-set and cycle have always an even number of edges in common, the rows of $\mathbf{Q_a}$ and $\mathbf{B_a}$ are orthogonal if the edges of two matrices are arranged (labelled) in the same way. From Theorem 12.15 we know that $\mathbf{Q_a B_a^T} = \mathbf{B_a Q_a^T} = \mathbf{O}$.

For the Theorem 2.12 (Sylvester's Weak Sylvester's Law of Nullity):

$$\rho_{\mathbf{Q_a}} + \rho_{\mathbf{B_a}} \leq m.$$

Since the rank of $\mathbf{B_a}$ is $m - n + 1$ then

$$\rho_{\mathbf{Q_a}} \leq m - \rho_{\mathbf{B_a}} = n - 1$$

Hence, $\rho_{\mathbf{Q_a}} = n - 1$ that is the rank of $\mathbf{A_a}$ and, by definition of rank of graph, the rank of G. $\quad \square$

Definition 12.55. Let G be a connected graph containing n vertices and m edges and T be a spanning tree associated with the graph. A *fundamental cut-set* with respect to the tree T is a cut-set that contains only one branch of spanning tree. Totally, $n - 1$ fundamental cut-sets exist.

Definition 12.56. Let G be a connected graph containing n vertices and m edges. The graph contains $n - 1$ cut-sets with respect to a spanning tree T. The *fundamental* *cut-set matrix* $\mathbf{Q_f}$ is that $(n-1) \times m$ matrix displaying in position (i, j) a 1 if the edge e_j belongs to the fundamental cut-set cs_i and a 0 otherwise. A *fundamental cut-set matrix*

Example 12.51. In order to understand the concept of fundamental cut-set let us consider again the following graph where with a solid line the spanning tree T is indicated while with a dashed line the cotree chords.

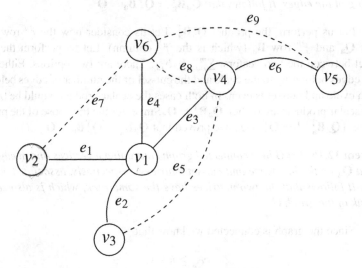

The cut-sets with respect to the spanning tree T are

- e_2, e_5
- e_1, e_7
- e_6, e_9
- e_3, e_5, e_8, e_9
- e_4, e_7, e_8, e_9

The $n-1$ cut-set above are the fundamental cut-set and can be represented in the following matrix

$$
\mathbf{Q_f} = \begin{pmatrix}
0 & 1 & 0 & 0 & 1 & 0 & 0 & 0 & 0 \\
1 & 0 & 0 & 0 & 0 & 0 & 1 & 0 & 0 \\
0 & 0 & 0 & 0 & 0 & 1 & 0 & 0 & 1 \\
0 & 0 & 1 & 0 & 1 & 0 & 0 & 1 & 1 \\
0 & 0 & 0 & 1 & 0 & 0 & 1 & 1 & 1
\end{pmatrix}.
$$

Let us now rearrange the columns to have the cotree's chords first and the tree's branches then,

$$\mathbf{Q_f} = \begin{pmatrix} e_5 & e_7 & e_8 & e_9 & e_1 & e_2 & e_3 & e_4 & e_6 \\ \hline 1 & 0 & 0 & 0 & 0 & 1 & 0 & 0 & 0 \\ 0 & 1 & 0 & 0 & 1 & 0 & 0 & 0 & 0 \\ 0 & 0 & 0 & 1 & 0 & 0 & 0 & 0 & 1 \\ 1 & 0 & 1 & 1 & 0 & 0 & 1 & 0 & 0 \\ 0 & 1 & 1 & 1 & 0 & 0 & 0 & 1 & 0 \end{pmatrix}.$$

If we rearrange the rows of the matrix we obtain

$$\mathbf{Q_f} = \begin{pmatrix} e_5 & e_7 & e_8 & e_9 & e_1 & e_2 & e_3 & e_4 & e_6 \\ \hline 0 & 1 & 0 & 0 & 1 & 0 & 0 & 0 & 0 \\ 1 & 0 & 0 & 0 & 0 & 1 & 0 & 0 & 0 \\ 1 & 0 & 1 & 1 & 0 & 0 & 1 & 0 & 0 \\ 0 & 1 & 1 & 1 & 0 & 0 & 0 & 1 & 0 \\ 0 & 0 & 0 & 1 & 0 & 0 & 0 & 0 & 1 \end{pmatrix} = \left(\mathbf{Q_c} \, \mathbf{I} \right).$$

This matrix partitioning can be always done over a fundamental cut-set matrix.

Proposition 12.22. *Every fundamental cut-set matrix $\mathbf{Q_f}$ associated with a connected graph G can be partitioned as $\left(\mathbf{Q_c} | \mathbf{I} \right)$ where $\mathbf{I}$ is the identity matrix of size $(n-1)$ representing the branches of the spanning tree while $\mathbf{Q_c}$ is a rectangular matrix having size $(n-1) \times (m-n+1)$ and represents the chords of the associated cotree.*

12.5.5 Relation Among Fundamental Matrices

Let G be a graph having n vertices and m edges. A reduced incidence matrix $\mathbf{A_f}$ can be easily constructed. We can identify a spanning tree T. With respect to this spanning tree we can construct the fundamental cycle matrix $\mathbf{B_f}$ and the fundamental cute-set matrix $\mathbf{Q_f}$.

We know that $\mathbf{B_f} = \left(\mathbf{I} | \mathbf{B_b} \right)$ and $\left(\mathbf{Q_c} | \mathbf{I} \right)$.

Let us consider the reduced incidence matrix $\mathbf{A_f}$. The matrix has size $(n-1) \times m$ and can be partitioned as $\mathbf{A_f} = \left(\mathbf{A_c} | \mathbf{A_b} \right)$ where $\mathbf{A_c}$ contains the incidence cotree's chords while $\mathbf{A_b}$ contains the incidence of the tree's branches. Let us now perform the following calculation

$$\mathbf{A_f B_f^T} = \left(\mathbf{A_c} | \mathbf{A_b} \right) \left(\frac{\mathbf{I}}{\mathbf{B_b}} \right) = \mathbf{A_c} + \mathbf{A_b B_b} = \mathbf{O}.$$

Hence, considering that in this binary arithmetic $-1 = 1$ and that $\mathbf{A_b}$ is nonsingular for the Theorem 12.4, we can write

$$\mathbf{A_c} = -\mathbf{A_b B_b^T} \Rightarrow \mathbf{A_b^{-1} A_c} = \mathbf{B_b^T}.$$

Now let us arrange the edges in the same way for the matrices $\mathbf{B_f}$ and $\mathbf{Q_f}$. We know that $\mathbf{B_f^T} = \left(\mathbf{A_c} | \mathbf{A_b} \right)$ and $\mathbf{Q_f} = \left(\mathbf{Q_c} | \mathbf{I} \right)$. Hence,

$$\mathbf{Q_f} \mathbf{B_f^T} = \left(\mathbf{Q_c} | \mathbf{I} \right) \left(\frac{\mathbf{I}}{\mathbf{B_b^T}} \right) = \mathbf{Q_c} + \mathbf{B_b^T} = \mathbf{O} \Rightarrow \mathbf{Q_c} = \mathbf{B_b^T}.$$

Putting together the equations we see that $\mathbf{Q_c} = \mathbf{A_b^{-1}} \mathbf{A_c}$.
This relation has the following consequences.

1. If $\mathbf{A_a}$ and then $\mathbf{A_f}$ are available, immediately $\mathbf{B_f}$ and $\mathbf{Q_f}$ can be calculated

2. If either $\mathbf{B_f}$ or $\mathbf{Q_f}$ is given, the other can be determined

3. Even if $\mathbf{B_f}$ and $\mathbf{Q_f}$ are both given, the matrix $\mathbf{A_f}$ cannot, in general, be fully determined

Example 12.52. In order to understand these findings, let us consider the following graph where a spanning tree is marked with solid line and cotree with dashed line.

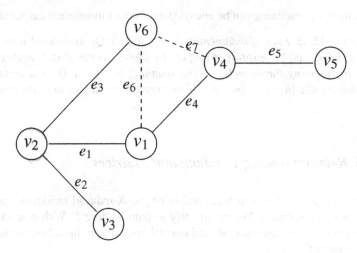

The incidence matrix is

$$\mathbf{A_a} = \begin{pmatrix} & e_1 & e_2 & e_3 & e_4 & e_5 & e_6 & e_7 \\ \hline v_1 & 1 & 0 & 0 & 1 & 0 & 1 & 0 \\ v_2 & 1 & 1 & 1 & 0 & 0 & 0 & 0 \\ v_3 & 0 & 1 & 0 & 0 & 0 & 0 & 0 \\ v_4 & 0 & 0 & 0 & 1 & 1 & 0 & 1 \\ v_5 & 0 & 0 & 0 & 0 & 1 & 0 & 0 \\ v_6 & 0 & 0 & 1 & 0 & 0 & 1 & 1 \end{pmatrix}.$$

The reduced incidence matrix can be easily obtained by cancelling one row (e.g. the last one). The resulting matrix can be rearranged as $\mathbf{A_f} = \left(\mathbf{A_c} | \mathbf{A_b} \right)$ where

$$\mathbf{A_c} = \begin{pmatrix} & e_6 & e_7 \\ \hline v_1 & 1 & 0 \\ v_2 & 0 & 0 \\ v_3 & 0 & 0 \\ v_4 & 0 & 1 \\ v_5 & 0 & 0 \end{pmatrix}$$

and

$$\mathbf{A_b} = \begin{pmatrix} & e_1 & e_2 & e_3 & e_4 & e_5 \\ \hline v_1 & 1 & 0 & 0 & 1 & 0 \\ v_2 & 1 & 1 & 1 & 0 & 0 \\ v_3 & 0 & 1 & 0 & 0 & 0 \\ v_4 & 0 & 0 & 0 & 1 & 1 \\ v_5 & 0 & 0 & 0 & 0 & 1 \end{pmatrix}.$$

The fundamental cycle matrix is

$$\mathbf{B_f} = \begin{pmatrix} & e_1 & e_2 & e_3 & e_4 & e_5 & e_6 & e_7 \\ \hline z_1 & 1 & 0 & 1 & 0 & 0 & 1 & 0 \\ z_2 & 1 & 0 & 1 & 1 & 0 & 0 & 1 \end{pmatrix}.$$

Also this matrix can be rearranged as $\mathbf{B_f} = \left(\mathbf{I} | \mathbf{B_b} \right)$ where

$$\mathbf{I} = \begin{pmatrix} & e_6 & e_7 \\ \hline z_1 & 1 & 0 \\ z_2 & 0 & 1 \end{pmatrix}$$

and

$$\mathbf{B_b} = \begin{pmatrix} & e_1 & e_2 & e_3 & e_4 & e_5 \\ \hline z_1 & 1 & 0 & 1 & 0 & 0 \\ z_2 & 1 & 0 & 1 & 1 & 0 \end{pmatrix}.$$

The fundamental cut-sets are

- e_2
- e_5
- e_3, e_6, e_7
- e_1, e_6, e_7
- e_4, e_7

The fundamental cut-set matrix is

$$
\mathbf{Q_f} =
\begin{pmatrix}
 & e_1 & e_2 & e_3 & e_4 & e_5 & e_6 & e_7 \\
\hline
cs_1 & 0 & 1 & 0 & 0 & 0 & 0 & 0 \\
cs_2 & 0 & 0 & 0 & 0 & 1 & 0 & 0 \\
cs_3 & 0 & 0 & 1 & 0 & 0 & 1 & 1 \\
cs_4 & 1 & 0 & 0 & 0 & 0 & 1 & 1 \\
cs_5 & 0 & 0 & 0 & 1 & 0 & 0 & 1
\end{pmatrix} .
$$

Rearranging the rows we obtain

$$
\mathbf{Q_f} =
\begin{pmatrix}
 & e_1 & e_2 & e_3 & e_4 & e_5 & e_6 & e_7 \\
\hline
cs_4 & 1 & 0 & 0 & 0 & 0 & 1 & 1 \\
cs_1 & 0 & 1 & 0 & 0 & 0 & 0 & 0 \\
cs_3 & 0 & 0 & 1 & 0 & 0 & 1 & 1 \\
cs_5 & 0 & 0 & 0 & 1 & 0 & 0 & 1 \\
cs_2 & 0 & 0 & 0 & 0 & 1 & 0 & 0
\end{pmatrix} .
$$

The matrix can be re-written as $\mathbf{Q_f} = \left(\mathbf{Q_c} | \mathbf{I} \right)$ where

$$
\mathbf{Q_c} =
\begin{pmatrix}
 & e_6 & e_7 \\
\hline
cs_4 & 1 & 1 \\
cs_1 & 0 & 0 \\
cs_3 & 1 & 1 \\
cs_5 & 0 & 1 \\
cs_2 & 0 & 0
\end{pmatrix}
$$

$$
\mathbf{I} =
\begin{pmatrix}
 & e_1 & e_2 & e_3 & e_4 & e_5 \\
\hline
cs_4 & 1 & 0 & 0 & 0 & 0 \\
cs_1 & 0 & 1 & 0 & 0 & 0 \\
cs_3 & 0 & 0 & 1 & 0 & 0 \\
cs_5 & 0 & 0 & 0 & 1 & 0 \\
cs_2 & 0 & 0 & 0 & 0 & 1
\end{pmatrix} .
$$

It is clear that $\mathbf{Q_c} = \mathbf{B_b^T}$. Moreover it can be easily verified that $\mathbf{A_b}$ is non-singular and the inverse matrix $\mathbf{A_b^{-1}}$ is

$$
\mathbf{A_b^{-1}} = \left(\begin{array}{c|ccccc}
 & e_1 & e_2 & e_3 & e_4 & e_5 \\
\hline
v_1 & 1 & 0 & 0 & 1 & 1 \\
v_2 & 0 & 0 & 1 & 0 & 0 \\
v_3 & 1 & 1 & 1 & 1 & 1 \\
v_4 & 0 & 0 & 0 & 1 & 1 \\
v_5 & 0 & 0 & 0 & 0 & 1
\end{array}\right).
$$

If we perform the multiplication

$$
\mathbf{A_b^{-1}A_c} = \begin{pmatrix}
1 & 0 & 0 & 1 & 1 \\
0 & 0 & 1 & 0 & 0 \\
1 & 1 & 1 & 1 & 1 \\
0 & 0 & 0 & 1 & 1 \\
0 & 0 & 0 & 0 & 1
\end{pmatrix}
\begin{pmatrix}
1 & 0 \\
0 & 0 \\
0 & 0 \\
0 & 1 \\
0 & 0
\end{pmatrix}
=
\begin{pmatrix}
1 & 1 \\
0 & 0 \\
1 & 1 \\
0 & 1 \\
0 & 0
\end{pmatrix}
= \mathbf{Q_c} = \mathbf{B_b^T}.
$$

12.5.5.1 Graph Matrices and Vector Spaces

We have extensively analysed the relation between graphs and matrices and, previously, the relation between matrices and vector spaces. Now we will further show how graphs are linear algebra objects and how graphs, matrices, and vector spaces can be seen as the same concept represented from different perspectives.

Let $G(V,E)$ be a graph composed of n vertices and m edges. We can consider every edge as an elementary graph. Hence, the original graph can be seen as the composition of m elementary graphs. Looking at this fact from a different perspective, each edge can be seen as the component of a vector space in m dimensions. This vector space can be divided into two subspaces, the first, having dimension $n-1$, corresponding to the branches of the spanning tree and the second, having dimension $n-m+1$, to the chords of cotree. These two subspaces are then spanned by the rows of the matrices $\mathbf{B_f}$ and $\mathbf{Q_f}$, respectively.

12.6 Graph Isomorphisms and Automorphisms

Definition 12.57. Two graphs $G = (V,E)$ and $G' = (V',E')$ are said to be *isomorphic*, and we write $G = G'$ if there exists a bijection $f : V \to V'$ such that $\forall$ vertices v, w, if $(v,w) \in E$, then $(f(v), f(w)) \in E'$. In other words, two vertices are adjacent in G if and only if their images by f are adjacent in G'.

The isomorphism is then a graph transformation that may change the shape of the graph while preserving its structure, that is the adjacency between each pair of vertices.

Example 12.53. The following two graphs are isomorphic where $f(v_1) = v_a$, $f(v_2) = v_b$, $f(v_3) = v_c$, and $f(v_4) = v_d$.

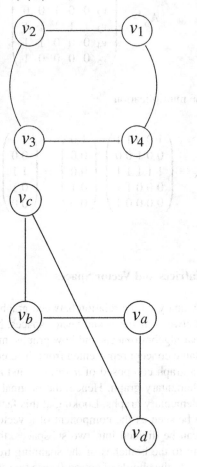

Note that two (or more) isomorphic graphs always have the same number of vertices, the same number of edges. However, these conditions are not enough to ensure the isomorphism of the two graphs.

Example 12.54. If we consider the following two graphs, although they have the same number of vertices and the same number of edges, they are not isomorphic. The reason is that the two graphs have a different structure: the adjacency among vertices is not the same.

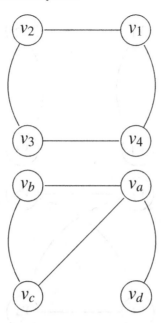

About isomorphisms between graphs there is a famous conjecture by Ulam.

Ulam's Conjecture. *Let G be a graph with n vertices $v_1, v_2, v_3, \ldots, v_n$ and let G' be a graph with n vertices $v'_1, v'_2, v'_3, \ldots, v'_n$. If $\forall i = 1, 2, \ldots, n$, the subgraphs $G - v_i$ and $G' - v'_i$ are isomorphic, then the graphs G and G' are isomorphic too.*

Definition 12.58. An *automorphism* of a graph G is an isomorphism of G to itself. Thus, an automorphism is a bijection $\alpha : V \to V$ such that $\forall$ pair of vertices v_1, v_2, it follows that $\alpha(v_1), \alpha(v_2)$ is an edge if and only if v_1, v_2 is an edge.

From the definition above, the automorphism is a graph transformation that, like isomorphism, preserves the adjacency of the vertices but imposes a stronger condition with respect to isomorphisms: the vertices in the transformed graph are the same as those in the original graph. For all graphs a *trivial automorphism*, namely identity automorphism can be defined. This transformation makes correspond a graph G to itself. In general, automorphism can be obtained by "stretching" the edges of a graph and rearranging the graph in the plane.

Example 12.55. An example of two automorphic graphs are given in the following.

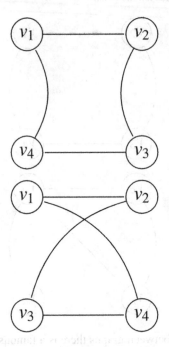

12.7 Some Applications of Graph Theory

In this section we will mention some applications of graph theory, stating and discussing some problems, the most varied nature, which can be set and possibly solved using graph theory. Some applications will be simple exercises, others are more complex or even unresolved.

12.7.1 *The Social Network Problem*

Let us consider a meeting involving six individuals and let us pose the following question: "How many people know each other?" The situation can be represented by a graph G with six vertices representing the six people under consideration. The adjacency of vertices represents whether or not the individuals know each other.

Example 12.56. The graph describing the situation above is the following. Let us suppose a scenario and represent it in the following graph.

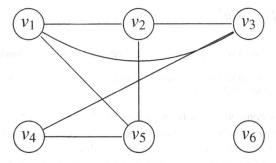

Before tackling the problem, let us introduce the notion of complementary graph of a graph G.

Definition 12.59. Let G be a graph. The *complementary* graph $\tilde{G}$ of the graph G is graph containing the same vertices of the graph G, and in which two vertices are adjacent (non-adjacent) in G if and only if they are adjacent (non-adjacent) in $\tilde{G}$.

Example 12.57. The complementary graph $\tilde{G}$ to the graph above G is given in the following.

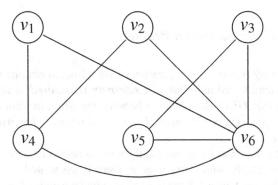

It may be observed that regardless of the initial choice of G, either in G or in $\tilde{G}$ at least one triangle occurs. In other words, in every meeting including six individuals will either have at least a triangle of people that do not each other or a triangle of people that have already met. More formally the following proposition proves what has been stated at the intuitive level.

Proposition 12.23. *For every graph G with six vertices, G or its complement $\tilde{G}$ contains a triangle.*

Proof. Let v be a vertex of G. Each vertex different from v is either connected to another vertex or not connected. Hence, v is adjacent either in G or in $\tilde{G}$ to the other 5 vertices. Thus, one can assume, without loss of generality, that there are three vertices v_1, v_2, v_3 in G adjacent to v. If between these there are two mutually adjacent, then these two are the vertices of a triangle in G whose third side is v. Otherwise (i.e. if none of v_1, v_2, and v_3 is adjacent in G), then v_1, v_2, and v_3 form a triangle in $\tilde{G}$.

12.7.2 The Four Colour Problem

One of the most famous conjectures in mathematics was the following.

Any map on a flat surface (or on a sphere) can be coloured with at most four colors, so that adjacent regions have different colors.

In 1977, this conjecture has been proved by K. Appel and W. Hagen, see [27]. They have proved with the help of the computer that four colors are sufficient. Precisely for this reason (that the proof was done with the help of the computer) some mathematicians do not accept it: despite the accusation of lack of elegance, no errors were found. Some improvements have been made after the theorem.

The four-color problem can be expressed in the language of graph theory. Each region is represented by a vertex of a graph, while each edge is a boundary segment separating two regions. In this formulation the problem becomes: given a planar graph, its vertices can be coloured with a maximum of four colors in such a way that two adjacent vertices never have the same color. This concept can be expressed by saying that every planar graph is *4-colorable*.

12.7.3 Travelling Salesman Problem

A problem closely related to the question of Hamiltonian circuits is the traveling-salesman problem, stated as follows: *a salesman is required to visit a number of cities during a trip. Given the distances between the cities, in what order should he travel so a stop visit every city precisely once and return home, with the minimum mileage traveled?*

Representing the cities by vertices and the roads between them by edges, we get a graph. In this graph, with every edge e_i there is associated a real number (the distance in miles, say), $w(e_i)$. Such a graph is called a weighted graph; $w(e_i)$ being the weight of edge e_i.

In our problem, if each of the cities has a road to every other city, we have a complete weighted graph. This graph has numerous Hamiltonian circuits, and we are to pick the one that has the smallest sum of distances (or weights). The total number of different (not edge disjoint, of course) Hamiltonian circuits in a complete graph of n vertices can be shown to be $\frac{(n-1)!}{2}$. This follows from the fact that starting from any vertex we have $n-1$ edges to choose from the first vertex, $n-2$ from the second, $n-3$ from the third, and so on. These being independent choices, we get $(n-1)!$ possible number of choices. This number is, however, divided by 2, because each Hamiltonian circuit has been counted twice.

Theoretically, the problem of the travelling salesman can always be solved by enumerating all $\frac{(n-1)!}{2}$ Hamiltonian circuits, calculating the distance travelled in each, and then picking the shortest one. However, for a large value of n, the labour involved is too great even for a digital computer.

The problem is to prescribe a manageable algorithm for finding the shortest route. No efficient algorithm for problems of arbitrary size has yet been found, although many attempts have been made. Since this problem has applications in operations research, some specific large-scale examples have been worked out. There are also available several heuristic methods of solution that give a route very close to the shortest one, but do not guarantee the shortest. It is obvious the importance of this kind of problems in the theory of transport in electric circuits, etc.

12.7.4 The Chinese Postman Problem

This problem is so called because it has been discussed by the Chinese mathematician Mei-Ko Kwan in 1962 [28]. A postman wants to deliver all the letters in such a way as to minimize the path and then return to the starting point. It must of course go through all the streets of the path that was entrusted to him at least once, but he would avoid covering too many roads more than once.

We observe that the Chinese postman problem differs from the traveling salesman problem, since the latter only has to visit a number of towns and can choose the most convenient roads to reach them. Now if one represents a map of the city, in which the postman has to make deliveries, by a graph, the problem is equivalent to the determination of the minimum length of a cycle that passes through each edge at least once. If the graph is Eulerian, it is clear that a solution to the problem is given by an Eulerian cycle (which crosses every edge exactly once). However, it is highly unlikely that the road network of the mailman meets the conditions required by Euler's theorem for owning a cycle Eulerian. It can be shown that this cycle of minimum length that the postman look never goes to no arc for more than two times. So for every road network optimal paths exist: you can build by adding to the graph of the road network adequate arc in order to make it Eulerian.

12.7.5 Applications to Sociology or to the Spread of Epidemics

Suppose that by some psychological studies will be able to tell when a person in a group can influence the way of thinking of other members of the group. We can then construct a graph with a vertex v_i for each person in the group and a directed edge (v_i, v_j) every time the person v_i will influence the person v_j. One may ask what is the minimum number of people that can spread an idea to the entire group either directly or by influencing someone who in turn can affect someone else, and so on.

A similar mechanism is typical of epidemics: what is the minimum number of patients who can create an epidemic in a population?

Exercises

12.1. Determine whether or not the graph with vertices $v_1, v_2, v_3, v_4, v_5, v_6$ and edges (v_2, v_3), (v_1, v_4), (v_3, v_1), (v_3, v_4), (v_6, v_4) is connected. If not, determine the connected components.

12.2. Determine the degree of each the graph below.

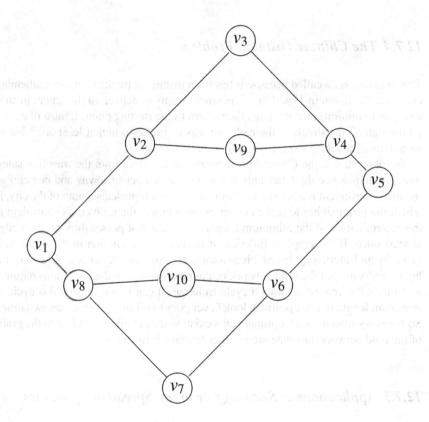

12.3. Consider the graph below. Does this graph contain an Eulerian cycle? If yes, indicate one.

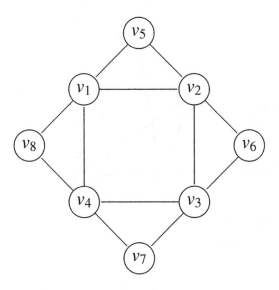

12.4. For the graph below

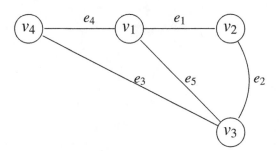

1. Determine the reduced incidence matrix where v_3 is the reference vertex
2. Indicate one spanning tree
3. On the basis of the chosen spanning tree determine the corresponding fundamental cycle matrix

12.5. Assess whether or not the following graph is planar. Verify, if possible, Euler's formula.

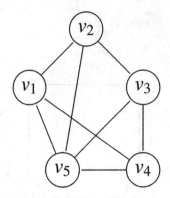

Chapter 13
Applied Linear Algebra: Electrical Networks

13.1 Basic Concepts

This chapter shows how mathematical theory is not an abstract subject which has no connection with the real world. On the contrary, this entire book is written by stating that mathematics in general, and algebra in this case, is an integrating part of every-day real life and that the professional life of computational scientists and engineers requires a solid mathematical background. In order to show how the contents of the previous chapters have an immediate technical application, the last chapter of this book describes a core engineering subject, i.e. electrical networks, as an algebraic exercise. Furthermore, this chapter shows how the combination of the algebraic top-ics give a natural representation of a set of interacting physical phenomena.

Definition 13.1. A *bi-pole* is a generic electrical device having two connections. A bi-pole is always associated with two physical entities namely *voltage v* and *current i*. Both, voltage and current are time-variant entities and thus functions of the time t with codomain $\mathbb{R}$. The unit for voltage is Volt V while the unit for current is Ampere A.

A bi-pole is graphically represented in the following way

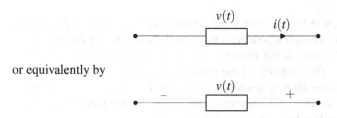

or equivalently by

if we assume that the current always flows from the negative pole (indicated with "−") to the positive pole (indicated with "+").

Obviously electromagnetic phenomena are complex and would require a detailed dissertation which is outside the scopes of this book. From the point of view of

© Springer Nature Switzerland AG 2019
F. Neri, *Linear Algebra for Computational Sciences and Engineering*,
https://doi.org/10.1007/978-3-030-21321-3_13

algebra, current and voltage are two functions of the time and associated with a direction that characterize bi-pole. Furthermore, the *electrical power* $p(t)$ is here defined as

$$p(t) = v(t)i(t)$$

and is measured in Watt W. The power within a time frame is the *electrical energy* and is measured in W per hour h, i.e. Wh.

13.2 Bi-poles

The bi-pole itself is a model of electrical phenomena. While electrical phenomena occur between the entire extension of physical object, the modelling of these phenomena assumes that they can be concentrated in a point. Under this hypothesis, bi-poles belong to two macro-categories, namely *passive* and *active* bi-poles, respectively. The bi-poles belonging to the first category absorb energy while the bi-poles belonging to the second category generate energy. Before analysing the various types of bi-poles, it is worth commenting that each bi-pole is an object characterised by the relation between voltage and current.

13.2.1 Passive Bi-poles

An *electrical conductor* is a material whose electrons can freely flow within it and the current $i(t)$ can (almost rigorously) be interpreted as the flow of electrons within the material, see [29]. A current flowing through a conductor is associated with various concurrent and simultaneous physical phenomena. In physics, these phenomena are divided into three categories that correspond to the three passive bi-poles taken into consideration in this book and presented in the forthcoming subsections.

Resistor

The first phenomenon to be analysed is the *electro-dynamic* effect of the current. When a current flows through a conductor the material contrasts the flowing. As an effect, the electrical energy is not entirely transferred. On the contrary part of the energy is dissipated. The property of the material of contrasting the current flow is said *resistance*. Materials may contrast the current flow with diverse intensities which correspond to diverse resistance values measured in Ohm (Ω). This property is modelled by a bi-pole called *resistor* and indicated as

The relation between voltage and current of a resistor is given by the Ohm's law:

$$v(t) = Ri(t)$$

where R is the resistance of the resistor.

The two extreme conditions $R = 0$ and $R = \infty$ correspond to two special resistors, namely *short circuit* and *open circuit*, respectively. A short circuit is indicated as

and is characterized by the fact that, regardless of the current value, the voltage is always null: $\forall i(t) : v(t) = 0$.

The opposite situation, the open circuit, occurs when regardless of the voltage value the current is null: $\forall v(t) : i(t) = 0$. An open circuit is indicated as

Inductor

The second phenomenon to be analysed is the *electro-magnetic* effect of the current. When a current flows through a conductor other conductors nearby and the conductor itself, secondary voltages (and secondary currents) are created within the conductors. These secondary voltages are said to be *induced* by the main current. The physical property of material of inducing secondary voltages is said *inductance* and is measured in Henry (H). The corresponding bi-pole is said *inductor* and indicated as

The relation between voltage and current of an inductor is given by:

$$v(t) = L\frac{di(t)}{dt}$$

where L is the inductance of the inductor and $\dfrac{di(t)}{dt}$ is the derivative of the current with respect to the time t. A dissertation about derivatives can be found in a calculus book, see e.g. [18].

Capacitor

The third phenomenon to be analysed is the *electro-static* effect of the current. Voltages and currents are influenced also by static electric charges. The property of the material to store the charges (electrons) is said *capacitance* and is measured in Farads (F). The corresponding bi-pole is said *capacitor* and indicated as

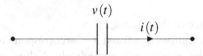

The relation between voltage and current of a resistor is given by:

$$v(t) = \frac{1}{C} \int_0^t i(\tau) \, d\tau$$

where C is the capacitance of the capacitor and $\int_0^t i(\tau) \, d\tau$ is the integral of the current. A dissertation about integrals can be found in a calculus book, see e.g. [18].

13.2.2 Active Bi-poles

Electrical energy can be generated by converting energies of a different nature. For example a battery is a device that transforms chemical energy into electrical energy. Electrical energy can be also converted by other types of energy such as combustion, i.e. thermal energy, or mechanical energy, like the dynamo of a bicycle. Devices of this kind are known as active bi-poles or *sources*. Electrical sources can be divided into *voltage sources* and *current sources*, respectively. Sources impose the values of voltage (or current) measured at its connectors. Two kind of values are the most popular in engineering and are shortly presented in this chapter Direct Current (DC) and Alternating Current (AC) sources, respectively.

DC sources

The most straightforward voltage source is a bi-pole that regardless of the time and the current value takes a constant voltage value E. In other words, the *DC voltage source* is indicated by the symbol

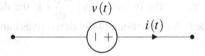

and is characterized by the equation

$$v(t) = E$$

where E is a constant.

Analogously, a *DC current source* is a bi-pole that regardless of the time and the voltage value takes a constant current value I. The symbol of DC current source is

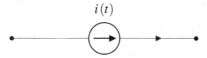

$$i(t)$$

and its equation is

$$i(t) = I$$

where I is a constant.

AC Sources

While DC currents are mostly used in electronics, the most common electrical sources in power distribution are AC. In AC systems, the current, i.e. the flow of the charges periodically reverse its direction. The AC sources are bi-poles characterized by a (sometimes complex) periodical function of the time t. In this book, we will focus only on voltages and currents that depend sinusoidally on the time t, which indeed is an example of AC and the most commonly used approximation of AC in engineering. More specifically, when we refer to *AC voltage source* we mean a bi-pole characterized by the equation

$$v(t) = V_M \cos(\omega t + \theta)$$

where V_M is a constant called *amplitude* and ω is said *angular frequency* of the voltage. An AC voltage source is indicated as

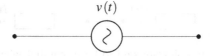

$$v(t)$$

Analogously, an AC current source is characterized by the equation

$$i(t) = I_M \cos(\omega t + \phi)$$

and is represented by

$$i(t)$$

The reason why AC is commonly used is fairly complex, related to energy transmission and distribution, and out of scope for the present book.

13.3 Electrical Networks and Circuits

Definition 13.2. An *electrical network* is a set of interconnected bi-poles.

Example 13.1. The following is an electrical network

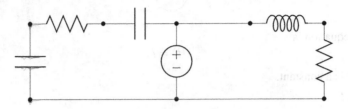

Definition 13.3. An *electrical circuit* is a network consisting of a closed loop, thus giving a return path for the current.

13.3.1 Bi-poles in Series and Parallel

Since a network is composed of interconnected bi-poles, it is fundamental to study, not only each bi-pole but also their connections within the network. Although an in depth explanation of the meaning of current and voltage is not given in this chapter, a current can be seen as a charge flow through a bi-pole while a voltage is a quantity measured across a bi-pole.

Definition 13.4. Two bi-poles are said to be in series when the same current flows through both of them. The topological configuration of two bi-poles in series is

Definition 13.5. Two bi-poles are said to be in parallel when they are associated with the same voltage. The topological configuration of two bi-poles in parallel is

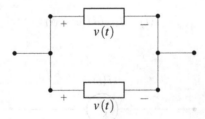

Obviously, more than two bi-poles can be in series and parallel.

13.3.2 Kirchoff's Laws

When multiple bi-poles are connected to compose a network, two energy conservation laws for currents and voltages, respectively are valid. These conservation laws are fundamental to study a network as they describe the network structure. Before stating these laws in details the following definitions must be reported.

Definition 13.6. Given an electrical network, a point that connects three of more bi-poles is said *electrical node* (or simply node) of the network.

Example 13.2. The network

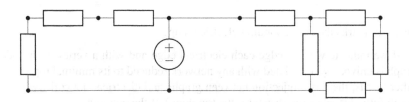

has four nodes that are highlighted in the following by empty circles which simple connectors are indicated with full circles.

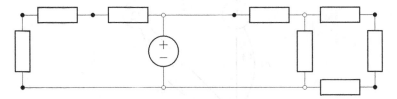

We know that two or more bi-poles in series when the same current flows through all of them. Keeping this in mind, we can write the following definition.

Definition 13.7. An *electrical edge* is a bi-pole or a series of bi-poles such that both current and voltage are non-null. An electrical edge is indicated as

Example 13.3. The network in Example 13.2 has four electrical nodes and five electrical edges. Hence, the network can be depicted as

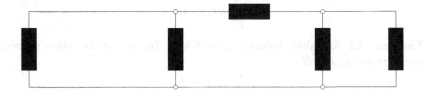

Let us collapse the nodes connecting short circuits into one single node. We obtain a network of interconnected electrical edges. This network is said to be *reduced to its minimal topology*.

Example 13.4. The network above when reduced to its minimal topology has three electrical nodes and five electrical edges and is

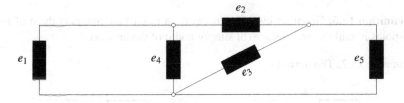

where e_k marks the corresponding electrical edge.

If we indicate with an edge each electrical edge and with a vertex each node, a graph is univocally associated with any network reduced to its minimal topology. In other words, there is a bijection between graphs and electrical networks where the graph fully describes the structure (the topology) of the network.

Example 13.5. The network in Example 13.2 is associated with the following graph

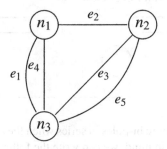

Definition 13.8. Given an electrical network, a *mesh* is a cycle (or circuit) in the associated graph.

This network characterization allows us to state the following experimental principles (thus without a mathematical proof) which connect currents and voltages to the network structures.

Theorem 13.1. Kirchhoff Currents Law (KCL). *The sum of the currents incident into a node is null:*

$$\sum_{k=1}^{n} i_k(t) = 0.$$

Theorem 13.2. Kirchhoff Voltages Law (KVL). *The sum of the voltages associated with a mesh is null:*

$$\sum_{k=1}^{m} v_k(t) = 0.$$

13.3.3 Phasorial Representation of Electrical Quantities

Let us observe that a constant function $f(x) = k$ can be interpreted as a special sinusoidal (or cosinusoidal) function

$$f(x) = A\cos(\omega x + \theta)$$

with A, ω, and θ constants when $\omega = 0$.

In a similar way, a DC voltage $v(t) = E$ can be considered as a special case of AC voltage with null angular frequency. Hence, we may theoretically think that all the possible voltages and currents are AC and can be represented by a certain sinusoidal function.

As a further premise, let us remind that a cosinusoidal function is equivalent to a sinusoidal with a phase lag of $90°$:

$$\cos(x) = \sin(x + 90°).$$

Let us focus on a sinusoidal voltage generated by voltage source. It can be easily verified that the voltages associated with each bi-pole is sinusoidal and has the same angular frequency ω at that of the voltage source. This can be checked by applying KLC and KLV and then calculating the equations characterizing the bi-poles. A similar consideration can be done about the current. It follows that if a sinusoidal voltage is generated with angular frequency ω, within a network, all the electrical quantities associated with the bi-poles are characterized by sinusoidal voltages and currents with the same angular frequency ω.

Since the value of ω is fixed for the entire network, the voltage value $V_M \cos(\omega t + \theta)$ associated with a bi-pole of the network is identified only by the value V_M and the angle θ. Thus, each voltage and each current in the network is univocally identified by a couple of numbers of these type. The pair of numbers (V_M, θ) can be interpreted as a complex number in its polar form, i.e. $(V_M, \angle\theta)$. Equivalently, we may state that, if ω is fixed, there is a bijection between the set of sinusoidal voltages (currents) of an electrical network and the set of complex numbers.

An electrical quantity in the form $(V_M, \angle\theta)$ is said *phasor* which is the combination of the word "phase", i.e. the angle θ, and the and word "vector" to highlight that a voltage (current) can be seen as a vector having modulus V_M and direction determined by θ.

Obviously, since the phasor is a complex number, it can be expressed into its rectangular coordinates by using the transformations described in Chap. 5.

Let us now think about a generic current $I_M \cos(\omega t + \phi) = (I_M, \angle\phi)$ in a network. If we assume that the current is known let us find the voltage associated with each passive bi-pole.

Proposition 13.1. *Let $I_M \cos(\omega t + \phi) = (I_M, \angle\phi)$ be the current flowing through a resistor R. The associated voltage is*

$$V_R = R(I_M, \angle\phi).$$

Proof. For the Ohm's law, obviously

$$v_R(t) = Ri(t) = RI_M \cos(\omega t + \phi) = R(I_M, \angle \phi). \quad \square \qquad (13.1)$$

Theorem 13.3. *Let* $i(t) = I_M \cos(\omega t + \phi) = (I_M, \angle \phi)$ *be the current flowing through an inductor of inductance L. The associated voltage is*

$$V_L = j\omega L(I_M, \angle \phi).$$

Proof. The voltage across the inductor can be determined by applying simple differentiation rules, see [18], and basic trigonometry:

$$
\begin{aligned}
v_L(t) &= L\frac{di(t)}{dt} = L\frac{dI_M \cos(\omega t + \phi)}{dt} = \\
&= -\omega L I_M \sin(\omega t + \phi) = \omega L I_M \sin(-\omega t - \phi) = \\
&= \omega L I_M \cos(90° - \omega t - \phi) = \omega L I_M \cos(90° + \omega t + \phi).
\end{aligned}
$$

For Proposition 5.3, if $\omega L I_M \cos(90° + \omega t + \phi)$ is seen as a complex number (composed of only its real part) it is occurs that

$$\omega L I_M \cos(90° + \omega t + \phi) = j\omega L I_M \cos(\omega t + \phi).$$

Hence,

$$v_L(t) = j\omega L I_M \cos(\omega t + \phi) = j\omega L(I_M, \angle \phi). \quad \square$$

The same result can be achieved in a trigonometrical easier way by means of the following proof.

Proof. Let us pose $\alpha = \phi + 90°$. Then,

$$I_M \cos(\omega t + \phi) = I_M \sin(\omega t + \phi + 90°).$$

It follows that

$$
\begin{aligned}
v_L(t) &= L\frac{di(t)}{dt} = L\frac{dI_M \sin(\omega t + \alpha)}{dt} = \\
&= \omega L I_M \cos(\omega t + \alpha) = \omega L I_M \sin(\omega t + \alpha + 90°) = \\
&= j\omega L I_M \sin(\omega t + \alpha) = j\omega L(I_M, \angle \alpha). \quad \square
\end{aligned}
$$

The proof regarding the voltage across a capacitor is analogous.

Theorem 13.4. *Let* $i(t) = I_M \cos(\omega t + \phi) = (I_M, \angle \phi)$ *be the current flowing through a capacitor of capacitance C. The associated voltage is*

$$V_C = \frac{1}{j\omega C}(I_M, \angle \phi) = \frac{-j}{\omega C}(I_M, \angle \phi).$$

13.3.4 Impedance

We know that an electrical edge is a series of bi-poles, thus resistors, inductors, and capacitors. Moreover, since these bi-poles are in series, the same current flows through them. Let $(I_M, \angle\phi)$ be this current. Let us name the resistors, inductors, and capacitors with the corresponding values of resistance inductance, and capacity and let us indicate as:

$$R_1, R_2, \ldots, R_p$$
$$L_1, L_2, \ldots, L_q$$
$$C_1, C_2, \ldots C_s.$$

Since $\forall k$ the voltages v_{Rk} across R_k, v_{Lk} across L_k and v_{Ck} across C_k are given by

$$v_{Rk} = R_k (I_M, \angle\phi)$$
$$v_{Lk} = j\omega L_k (I_M, \angle\phi)$$
$$v_{Ck} = -\frac{j}{\omega C_k} (I_M, \angle\phi)$$

it follows, by associativity of the sum for complex numbers that the total voltages due to the resistive, inductive and capacitive contributions are, respectively,

$$v_R = R (I_M, \angle\phi)$$
$$v_L = j\omega L (I_M, \angle\phi)$$
$$v_C = -\frac{j}{\omega C} (I_M, \angle\phi)$$

where

$$R = \Sigma_{k=1}^{p} R_k$$

$$L = \Sigma_{k=1}^{q} L_k$$

$$\frac{1}{C} = \Sigma_{k=1}^{s} \frac{1}{C_k}.$$

By applying again associativity of the sum for complex numbers, the voltage across the entire electrical edge is

$$v(t) = V_m \cos(\omega t + \theta) = (V_M \angle\theta) =$$
$$= \left(R + j\left(\omega L - \frac{1}{\omega C}\right)\right)(I_M, \angle\phi) = (R + jX)(I_M, \angle\phi)$$

where

$$X = \left(\omega L - \frac{1}{\omega C}\right)$$

is said *conductance*, and the complex number indicated as

$$\dot{z} = \left(R + j\left(\omega L - \frac{1}{\omega C}\right)\right)$$

is said *electrical impedance* or simply *impedance*. It obviously follows that an electrical bi-pole is mathematically a complex number whose real part is the resistive contribution while its imaginary part is a combination of inductive and capacitive contributions.

Example 13.6. Let us consider the following electrical edge

associated with a current $1\angle 30°A$ and angular frequency $\omega = 1$ rad/s. To find the voltage at the connectors of the electrical edge the impedance of the electrical edge must be found:

$$3 + 4 + j \left(1 \left((15 + 30) \times 10^{-3} \right) - \frac{1}{150 \times 10^{-6}} \right) = 7 + j (0.045 - 6666.666) \approx 7 - j6666\Omega.$$

The voltage is given by

$$(1\angle 30°)(7 - j6666) = (1\angle 30°)(6666\angle - 89.9°) = (6666\angle - 59.9°)V$$

that is

$$6666 \cos (t - 59.9°)V.$$

The Ohm's law in AC is then the product of two complex numbers. The DC case can be seen as a special case of AS where $\omega = 0$. Under such hypotheses it occurs that he current $I_M \cos (\omega t + \phi)$ is a constant (is time-invariant). The voltage is consequently also time-invariant. Regarding the impedance it occurs that

- its resistive contribution R stays unmodified
- its inductive contribution is $\omega L = 0$: the voltage is null regardless of the current value, i.e. the inductor behaves like a short circuit
- its capacitive contribution is $\frac{1}{\omega C} = \infty$: the current is null regardless of the voltage value, i.e. the capacitor behaves like an open circuit

Definition 13.9. The equation expressing the connection between current and voltage at the connector of an electrical edge is said *edge equation*.

Obviously the Ohm's law is the edge equation for passive bi-poles. This fact can be rephrased in the following proposition.

Proposition 13.2. *The electrical impedance $\dot{z}$ of two passive bi-poles in series with impedance $\dot{z}_1$ and $\dot{z}_2$ is the sum of the impedance:*

$$\dot{z} = \dot{z}_1 + \dot{z}_2.$$

Analogously, we may calculate the electrical impedance equivalent to two passive bi-poles in parallel.

Proposition 13.3. *The electrical impedance $\dot{z}$ of two passive bi-poles in parallel with impedance $\dot{z}_1$ and $\dot{z}_2$ is*

$$\dot{z} = \frac{\dot{z}_1 \dot{z}_2}{\dot{z}_1 + \dot{z}_2}.$$

Proof. Let us consider two bi-poles in parallel

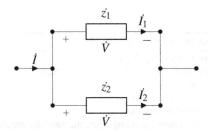

By applying the KCL and Ohm's law we know that

$$\begin{cases} \dot{I} = \dot{I}_1 + \dot{I}_2 \\ \dot{V} = \dot{z}_1 \dot{I}_1 \\ \dot{V} = \dot{z}_2 \dot{I}_2 \end{cases}$$

By combining the equations we may write that

$$\dot{I} = \frac{\dot{V}}{\dot{z}_1} + \frac{\dot{V}}{\dot{z}_2}.$$

We want to find the equivalent impedance, that is that electrical impedance $\dot{z}$ such that

$$\dot{V} = \dot{z}\dot{I} \Rightarrow \dot{I} = \frac{\dot{V}}{\dot{z}}.$$

Hence, by substituting

$$\frac{\dot{V}}{\dot{z}} = \frac{\dot{V}}{\dot{z}_1} + \frac{\dot{V}}{\dot{z}_2}.$$

We can then determine $\dot{z}$:

$$\frac{1}{\dot{z}} = \frac{1}{\dot{z}_1} + \frac{\dot{V}}{\dot{z}_2} = \frac{\dot{z}_1 + \dot{z}_2}{\dot{z}_1 \dot{z}_2} \Rightarrow \dot{z} = \frac{\dot{z}_1 \dot{z}_2}{\dot{z}_1 + \dot{z}_2}. \square$$

13.4 Solving Electrical Networks

The main task of the study/analysis of an electrical network is its solution, that is the identification of all the voltages and currents associated with the passive bi-poles when the voltages (or currents) associated with the sources are supposed to be known.

In general, the solution of an electrical network is a difficult task which requires sophisticated techniques to model the problem and a high computational effort to be solved. In this book, a simplified case involving only the linear bi-poles listed above and neglecting the transient phenomena is taken into consideration. Under these hypotheses, the solution of an electrical network is achieved by applying the following algorithm.

Algorithm 13 Solution of an Electrical Network

---

reduce the network to its minimal topology
assign a tentative voltage orientation to all the electrical edges
write all the edge equations
write the KCL equation for all the nodes
write the KVL equation for all the meshes
solve the resulting system of linear equations where the variables are all the currents and voltages
of the network

While the edge equations can be straightforwardly written, the writing of KCL and KVL equations requires a comment. Each electrical edge of a network reduced to its minimal topology is univocally associated with one current value and one voltage value. This concept can be expressed as: there is a bijection between the set of electrical edges and the set of currents as well as between the set of electrical edges and voltages. Furthermore, as shown above, there is a bijection between the electrical edges of a network and the edges of a graph as well as between the nodes of an electrical network and the vertices of a graph.

In this light, KCL when applied to all the nodes identifies a homogeneous system of linear equations. The associated incomplete matrix is the incidence matrix of the corresponding graph where each edge is oriented (it can take 1 or -1 values). We know from Corollary 12.2 that the rank of an incidence matrix is $n - k$ with n number of vertices and k number of components. In this case, $k = 1$. Hence, out of the n KCL possible equations of a network only $n - 1$ are linearly independent; one equation is surely redundant. Equivalently, we can interpret the nodes of an electrical network as vectors whose components are the currents flowing into them. In this case, the n nodes of a network span a vector space having dimension $n - 1$.

Proposition 13.4. *Given an electrical network, the number of linearly independent KCL equations is $n - 1$.*

Regarding the voltages, KVL applied to all the meshes also identifies a homogeneous system of linear equations. The associated incomplete matrix is the cycle matrix of the corresponding graph where each edge is oriented (it can take 1 or -1 values). From Theorem 12.13, the rank of the cycle matrix is $m - n + 1$ where n is the number of vertices and m is the number of edges. It follows that only $m - n + 1$ KVL equations are linearly independent. Analogous to the KCL case, the meshes of an electrical network can be interpreted as vectors whose components are the voltages. These vectors span a vector space whose dimension $m - n + 1$.

Proposition 13.5. *Given an electrical network, the number of linearly independent KVL equations is $m - n + 1$.*

In order to identify $m - n + 1$ meshes leading to linearly independent equations, we can apply Proposition 12.19, i.e. we identify a spanning tree and obtaining the meshes, one by one, when the cotree chords are included. A spanning tree has $n - 1$ branches. Every time a cotree chord is added a mesh is identified. The meshes obtained in this way are a basis spanning a vector space. This explains the adjective spanning next to this special tree. This tree unequivocally spans a vector space.

Example 13.7. Let us solve the network from the example above with $\omega = 314 \, \text{rad/s}$ and the corresponding values of bi-poles indicated in figure

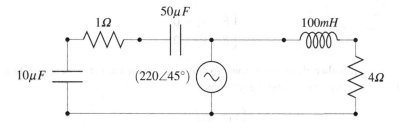

The corresponding network to minimal topology can represented as

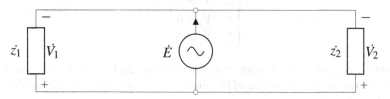

where

$$\dot{z}_1 = 1 - j\frac{6}{\omega 50}10^6 \approx 1 - j382\Omega$$
$$\dot{z}_2 = 4 + j41.4\Omega \, \dot{E} = (220\angle 45°).$$

which lead to the edge equations

$$\begin{cases} \dot{V}_1 = \dot{z}_1 \dot{I}_1 \\ \dot{V}_2 = \dot{z}_2 \dot{I}_2 \end{cases}$$

that is a system of two linear equations in four complex variables, $\dot{I}_1, \dot{I}_2, \dot{V}_1, \dot{V}_2$. Furthermore,

$$\dot{E} = (220\angle 45°) \, \forall \dot{I}_E$$

where $\dot{I}_E$ is the current through the voltage source.

We know that only $n - 1 = 1$ KCL equation is linearly independent, which is for the upper node

$$\dot{I}_E - \dot{I}_1 - \dot{I}_2 = 0$$

where the orientation of the current flowing towards a node is taken positive and the current moving out of the node is taken negative. The equation of the lower node would have been

$$-\dot{I}_E + \dot{I}_1 + \dot{I}_2 = 0$$

which, is obviously redundant.

Finally, $m - n + 1 = 3 - 2 + 1 = 2$ KVL independent equations can be written. We choose the following ones

$$\begin{cases} \dot{E} - \dot{V}_1 = 0 \\ \dot{E} - \dot{V}_2 = 0. \end{cases}$$

Putting together all the equations, to solve this electrical network means to solve the following system of linear equations:

$$\begin{cases} \dot{V}_1 = \dot{z}_1 \dot{I}_1 \\ \dot{V}_2 = \dot{z}_2 \dot{I}_2 \\ \dot{I}_E - \dot{I}_1 - \dot{I}_2 = 0 \\ \dot{E} - \dot{V}_1 = 0 \\ \dot{E} - \dot{V}_2 = 0. \end{cases}$$

where $\dot{E}$, $\dot{z}_1$, and $\dot{z}_2$ are known complex constants and $\dot{I}_1$, $\dot{I}_2$, $\dot{I}_E$, $\dot{V}_1$, and $\dot{V}_2$ are complex variables. This system of five linear equations in five variables can be easily simplified by substitution:

$$\begin{cases} \dot{I}_E - \dot{I}_1 - \dot{I}_2 = 0 \\ \dot{E} - \dot{z}_1 \dot{I}_1 = 0 \\ \dot{E} - \dot{z}_2 \dot{I}_2 = 0. \end{cases}$$

The equations are directly solved by calculating

$$\begin{cases} \dot{I}_1 = \frac{\dot{E}}{\dot{z}_1} = -0.4062 + j0.4083A \\ \dot{I}_2 = \frac{\dot{E}}{\dot{z}_2} = 4.083 - j3.363A \\ \dot{I}_E = \dot{I}_1 + \dot{I}_2 = -0.4062 + j0.4083 + 4.083 - j3.363 = 3.676 - j2.955A \end{cases}$$

In this case, $\dot{E} = \dot{V}_1$ and $\dot{E} = \dot{V}_2$ which completes the solution of the network.

The network above was especially easy to solve because the equations were uncoupled and then there was no need to solve the system of linear equations. In the following example this does not happen and a more clear feeling of what is the solution of an electrical network is given.

Example 13.8. The following network ($\omega = 314\,\text{rad/s}$)

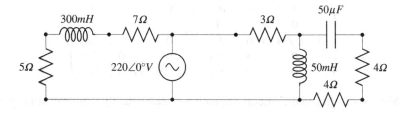

can be represented as

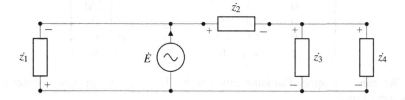

where

$$\dot{z}_1 = 12 + j94.2\,\Omega$$
$$\dot{z}_2 = 3\,\Omega$$
$$\dot{z}_3 = j15.7\,\Omega$$
$$\dot{z}_4 \approx 8 - j63.7\,\Omega$$
$$\dot{E} = 220\angle 0°\,V.$$

We can now write the system of linear equations solving the network:

$$\begin{cases} -\dot{I}_1 + \dot{I}_2 + \dot{I}_E = 0 \\ -\dot{I}_2 + \dot{I}_3 + \dot{I}_4 = 0 \\ \dot{z}_1 \dot{I}_1 = -\dot{E} \\ \dot{z}_2 \dot{I}_2 + \dot{z}_3 \dot{I}_3 = \dot{E} \\ \dot{z}_3 \dot{I}_3 - \dot{z}_4 \dot{I}_4 = 0 \end{cases}$$

which can be written as the following matrix equation:

$$\begin{pmatrix} -1 & 1 & 0 & 0 & 1 \\ 0 & -1 & 1 & 1 & 0 \\ \dot{z}_1 & 0 & 0 & 0 & 0 \\ 0 & \dot{z}_2 & \dot{z}_3 & 0 & 0 \\ 0 & 0 & \dot{z}_3 & -\dot{z}_4 & 0 \end{pmatrix} \begin{pmatrix} \dot{I}_1 \\ \dot{I}_2 \\ \dot{I}_3 \\ \dot{I}_4 \\ \dot{I}_E \end{pmatrix} = \begin{pmatrix} 0 \\ 0 \\ -\dot{E} \\ \dot{E} \\ 0 \end{pmatrix}.$$

This system of linear equations requires a certain effort to be solved manually by Cramer's method. Let us solve it by Gaussian elimination. At first, let us write the complete matrix (and let us multiply the first row by -1):

$$\begin{pmatrix} 1 & -1 & 0 & 0 & -1 & | & 0 \\ 0 & 1 & -1 & -1 & 0 & | & 0 \\ (12+j94.2) & 0 & 0 & 0 & 0 & | & -220 \\ 0 & 3 & j15.7 & 0 & 0 & | & 220 \\ 0 & 0 & j15.7 & -(8-j63.7) & 0 & | & 0 \end{pmatrix}.$$

Since some diagonal elements are null, we need to apply a pivotal strategy. Let us swap the rows to have diagonal elements always non-null:

$$\begin{pmatrix} (12+j94.2) & 0 & 0 & 0 & 0 & | & -220 \\ 0 & 3 & j15.7 & 0 & 0 & | & 220 \\ 0 & 0 & j15.7 & -(8-j63.7) & 0 & | & 0 \\ 0 & 1 & -1 & -1 & 0 & | & 0 \\ 1 & -1 & 0 & 0 & -1 & | & 0 \end{pmatrix}.$$

We can now apply Gaussian elimination. To eliminate the first column it is enough to apply

$$\mathbf{r}_5 = \mathbf{r}_5 - \frac{1}{(12+j94.2)}\mathbf{r}_1.$$

The resulting matrix is

$$\begin{pmatrix} (12+j94.2) & 0 & 0 & 0 & 0 & | & -220 \\ 0 & 3 & j15.7 & 0 & 0 & | & 220 \\ 0 & 0 & j15.7 & -(8-j63.7) & 0 & | & 0 \\ 0 & 1 & -1 & -1 & 0 & | & 0 \\ 0 & -1 & 0 & 0 & -1 & | & 0.29-j2.29 \end{pmatrix}.$$

In order to eliminate the second column we need to perform the following transformation:

$$\mathbf{r}_4 = \mathbf{r}_4 - \tfrac{1}{3}\mathbf{r}_2$$
$$\mathbf{r}_5 = \mathbf{r}_5 + \tfrac{1}{3}\mathbf{r}_2.$$

The resulting matrix is

$$\begin{pmatrix} (12+j94.2) & 0 & 0 & 0 & 0 & | & -220 \\ 0 & 3 & j15.7 & 0 & 0 & | & 220 \\ 0 & 0 & j15.7 & -(8-j63.7) & 0 & | & 0 \\ 0 & 0 & -1-j5.2 & -1 & 0 & | & -73.3 \\ 0 & 0 & -j5.2 & 0 & -1 & | & 70-j2.29 \end{pmatrix}.$$

Let us eliminate the third column by performing the following row transformations:

$$\mathbf{r_4} = \mathbf{r_4} - \frac{-1 - j5.2}{j15.7}\mathbf{r_3}$$
$$\mathbf{r_5} = \mathbf{r_5} + \frac{j5.2}{j15.7}\mathbf{r_3}.$$

The resulting matrix is

$$\begin{pmatrix} (12 + j94.2) & 0 & 0 & 0 & 0 & -220 \\ 0 & 3 & j15.7 & 0 & 0 & 220 \\ 0 & 0 & j15.7 & -(8 - j63.7) & 0 & 0 \\ 0 & 0 & 0 & 0.4 + j21.6 & 0 & -73.3 \\ 0 & 0 & 0 & -2.6 + j21.2 & -1 & 70 - j2.29 \end{pmatrix}.$$

Finally, the cancellation of the fourth column is obtained by applying

$$\mathbf{r_5} = \mathbf{r_5} + \frac{-2.6 + j21.2}{0.4 + j21.6}\mathbf{r_4}.$$

The final triangular matrix is

$$\begin{pmatrix} (12 + j94.2) & 0 & 0 & 0 & 0 & -220 \\ 0 & 3 & j15.7 & 0 & 0 & 220 \\ 0 & 0 & j15.7 & -(8 - j63.7) & 0 & 0 \\ 0 & 0 & 0 & 0.4 + j21.6 & 0 & -73.3 \\ 0 & 0 & 0 & 0 & -1 & -1.75 - j12.44 \end{pmatrix}.$$

By solving the triangular system we obtain:

$$\dot{I}_1 \approx -0.2940 + j2.3029A$$
$$\dot{I}_2 \approx 1.4698 - j10.3802A$$
$$\dot{I}_3 \approx 1.9835 - j13.7319A$$
$$\dot{I}_4 \approx -0.06282 + j3.39235A$$
$$\dot{I}_E \approx 1.75 + j12.44A$$

which is essentially the solution of the network.

13.5 Remark

This chapter shows how mathematics (and algebra in this case) has an immediate application in real-world. Conversely, engineering and computational sciences are built up on mathematics. Therefore a computer scientists or an engineer would greatly benefit of an understanding of basic mathematics every time they are supposed to propose a novel technological solution.

This chapter is an example of this fact. The study of electrical network is a major task in electrical and electronic engineering. Still, to solve a network in a simplified manner almost all topics studied in the previous chapters have been invoked:

matrices, systems of linear equations, vector spaces, linear mappings, graph theory, complex numbers and polynomials etc. Essentially, the solution of an electrical network requires the entire algebra as a theoretical foundation. If some simplifications are removed, for example if we take into account the transient phenomena, algebra on its own would not be enough and a substantial portion of calculus would be needed. A similar conclusion could be achieved by studying a mechanical dynamic system (mass, spring, damper) or a static phenomenon in building science.

Exercises

13.1. Solve the following electrical network for $\omega = 324$ rad/s:

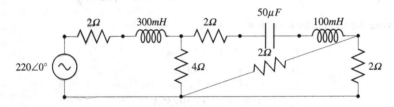

13.2. Solve the following electrical network for $\omega = 324$ rad/s:

Appendix A
A Non-linear Algebra: An Introduction to Boolean Algebra

A.1 Basic Logical Gates

This chapter is, strictly speaking, not about linear algebra. Nonetheless, the topic of this chapter, i.e. Boolean algebra, is related to linear algebra and it has been crucially important, over the last decades, for the progress of electronics. More specifically, while linear algebra deals with numbers, vectors, and matrices, Boolean algebra deals with binary states 0 and 1. In addition, the basic operators of Boolean algebra are non-linear.

Furthermore, it must be remarked that Boolean algebra is related to linear algebra as it allows its implementation within digital circuits. Thus, Boolean algebra can be seen as the "trait d'union" between abstract algebra and computational science.

In order to introduce Boolean algebra, let us consider an object x and a set A. A *membership function $m_f(x)$* scores 0 if $x \notin A$ and 1 if $x \in A$. At an abstract level, if we consider all the possible objects and sets of the universe, we can associate a membership relationship between each object and set. Each relationship will score 0 when the belonging relationship is not verified (the statement is false) and 1 when the belonging relationship is verified (the statement is true).

We can think about an image space where only the *true* and *false* (or 1 and 0, respectively) are allowed. In this space, the variables, namely *binary variables* can be combined to generate a *binary algebra*, namely *Boolean algebra*. Since the latter has to obey to logical rules, the same subject can be seen from a different perspective and named *Boolean logic*. This name is due to George Boole, the English mathematician that in 1853 described this logic in his book, "An Investigation of the Laws of Thought", see [30].

As mentioned above, in Boolean algebra, a variable x can take either the value 0 or 1. Thus, x is a generic variable of a binary set $B = \{0, 1\}$. Three elementary operators (or basic logical gates), one of them unary (that is applied to only one variable)

© Springer Nature Switzerland AG 2019
F. Neri, *Linear Algebra for Computational Sciences and Engineering*,
https://doi.org/10.1007/978-3-030-21321-3

and two of them binary (that is applied to two variables) are here introduced. The first operator, namely *NOT*, is defined in the following way:

$$if\ x = 1\ then\ \bar{x} = NOT\ (x) = 0$$
$$if\ x = 0\ then\ \bar{x} = NOT\ (x) = 1$$

The *NOT* operators is also graphically represented by the following symbol.

The second operator, namely *AND* or logical multiplication, processes two inputs, x and y respectively, and returns the values $x \wedge y$ according to the following rules:

x	y	$x \wedge y$
0	0	0
1	0	0
0	1	0
1	1	1

A graphical representation of the *AND* operator is given in the following way.

$$x - \!\!\!\!\!\!\!\!\!\!\!\!\!\! \rangle\!\!- x \wedge y$$

The third operator, namely *OR* or logical sum, processes two inputs, x and y respectively, and returns the values $x \vee y$ according to the following rules:

x	y	$x \vee y$
0	0	0
1	0	1
0	1	1
1	1	1

A graphical representation of the *OR* operator is given in the following way.

$$x - \!\!\!\!\!\!\!\!\!\!\!\!\!\! \rangle\!\!- x \vee y$$

A.2 Properties of Boolean Algebra

As for numerical linear algebra, the Boolean operators are characterized by basic properties. These properties are here listed.

- neutral element for *AND*: $\forall x$ binary number, $x \wedge 1 = x$
- absorbing element for *AND*: $\forall x$ binary number, $x \wedge 0 = 0$
- neutral element for *OR*: $\forall x$ binary number, $x \vee 0 = x$
- absorbing element for *OR*: $\forall x$ binary number, $x \vee 1 = 1$
- commutativity with respect to *AND*: $x \wedge y = y \wedge x$
- commutativity with respect to *OR*: $x \vee y = y \vee x$
- distributivity 1: $x \wedge (y \vee z) = (x \wedge y) \vee (x \wedge z)$
- distributivity 2: $x \vee (y \wedge z) = (x \vee y) \wedge (x \vee z)$
- identity property 1: $x \wedge x = x$
- identity property 2: $x \vee x = x$
- negative property 1: $x \wedge \bar{x} = 0$
- negative property 2: $x \vee \bar{x} = 1$

These basic properties, when combined, allow us to detect more complex relationships, as shown in the following example.

Theorem A.1. $x \wedge (x \vee y) = x$

Proof. $x \wedge (x \vee y) = (x \wedge x) \vee (x \wedge y) = x \vee (x \wedge y) = x \wedge (1 + y) = x \wedge 1 = x.$ □

Important properties of Boolean algebra are the so-called De Morgan's laws.

Theorem A.2. First De Morgan's Law. *The negation of a disjunction is the conjunction of the negations.* $\overline{(x \vee y)} = \bar{x} \wedge \bar{y}$

Proof. In order to prove the first De Morgan's law, we consider that $\overline{(x \vee y)} = \bar{x} \wedge \bar{y}$ is equivalent to write that $(x \vee y) \wedge (\bar{x} \wedge \bar{y}) = 0$ (for the negative property 1).

The latter equation can be written as $(x \vee y) \wedge (\bar{x} \wedge \bar{y}) = ((x \vee y) \wedge \bar{x}) \wedge \bar{y}$
$= (0 \vee (\bar{x} \wedge y)) \wedge \bar{y} = \bar{x} \wedge y \wedge \bar{y} = \bar{x} \wedge 0 = 0$ □

Theorem A.3. Second De Morgan's Law. *The negation of a conjunction is the disjunction of the negations.* $\overline{(x \wedge y)} = \bar{x} \vee \bar{y}$

The second De Morgan's Law can be proved in an analogous way.

Example A.1. Let us consider the following expression $(x \vee y) \wedge (x \vee \bar{y})$. This expression can be simplified. Let us re-write it:

$$(x \vee y) \wedge (x \vee \bar{y}) = x \vee x \wedge \bar{y} \vee x \wedge y \vee 0 =$$
$$= x \wedge (1 \vee y \vee \bar{y}) = x \wedge (1 \vee y) = x.$$

A.3 Boolean Algebra in Algebraic Structures

The abstract aspects of Boolean Algebra are very complex and deserves possibly a separate book. However, with the purpose of linking Boolean Algebra to the other chapters of this book, especially to place it within algebraic structures, this section introduces some other concepts of abstract algebra.

Definition A.1. The algebraic structure composed of a lattice set L endowed with two binary operators $\vee$ and $\wedge$, respectively, is said *lattice* and indicated with $(L, \vee, \wedge)$. For lattices and $x, y, z \in L$, the following properties hold:

- commutativity 1: $x \vee y = y \vee x$
- commutativity 2: $x \wedge y = y \wedge x$
- associativity 1: $(x \vee y) \vee z = x \vee (y \vee z)$
- associativity 2: $(x \wedge y) \wedge z = x \wedge (y \wedge z)$
- absorption 1: $x \vee (x \wedge y)$
- absorption 2: $x \wedge (x \vee y)$
- idempotence 1: $a \vee a = a$
- idempotence 2: $a \wedge a = a$

Definition A.2. A lattice $(L, \vee, land)$ is said to be *bounded* when 0 is the neutral element for $\vee$ ($x \vee 0 = x$) and 1 is the neutral element for $\wedge$ ($x \wedge 1 = x$).

It can be observed that a lattice can be seen as the combination of two semi-groups: $(L, \vee)$ and $(L, \wedge)$ respectively. In the cases of bounded lattice, it is the combination of two monoids.

Definition A.3. A lattice $(L, \vee, land)$ is said to be *complemented* when it is bounded (with infimum equal to 0 and supremum equal to 1) and for all $x \in L$, there exists an element y such that
$$x \wedge y = 0$$
and
$$x \vee y = 1.$$

Definition A.4. A lattice $(L, \vee, \wedge)$ is said to be *distributive* when the following equality holds
$$x \wedge (y \vee z) = (x \wedge y) \vee (x \wedge z).$$

The latter equality can be proved to be equivalent to
$$x \vee (y \wedge z) = (x \vee y) \wedge (x \vee z).$$

It can be easily seen that Boolean algebra is a complemented distributive lattice.

A.4 Composed Boolean Gates

From the three basic gates several composed gates can be generated. An important example is the *NAND* operator composed of an *AND* and a *NOT* and here represented.

$$x \quad \overline{(x \wedge y)}$$
$$y$$

A more complex example is the *XOR* operator. This operator processes two inputs and returns 0 when the inputs are the same and 1 when they are different.

x	y	$x \vee y$
0	0	0
1	0	1
0	1	1
1	1	0

This operator is composed as $x \wedge \bar{y} \vee \bar{x} \wedge y$. This expression is graphically represented in the following way,

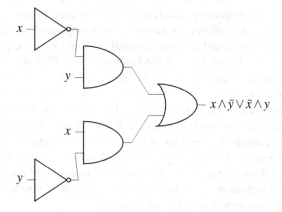

$$x \wedge \bar{y} \vee \bar{x} \wedge y$$

or, more compactly as

$$x \quad x \wedge \bar{y} \vee \bar{x} \wedge y$$
$$y$$

In order to appreciate the practical implications of Boolean logic and it composed gates, let us consider a so-called half adder circuit. The latter is a logic structure, obtained by a straightforward combination of *XOR* and *AND* gates, that performs sum. The scheme is given by:

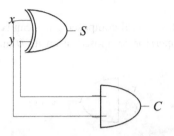

If we consider that S stands for sum while C stands for carry, the functioning of the half adder can be summarized in the following way. A more complex example is the XOR operator. This operator processes two inputs and returns 0 when the inputs are the same and 1 when they are different.

x	y	C	S
0	0	0	0
1	0	0	1
0	1	0	1
1	1	1	0

As shown the result of the operation $1 + 1 = 10$ (where 10 is 2 in binary). Thus, the half adder is an elementary structure that can be used to perform sums. More generally, Boolean logic allows, by means of binary operator, to define a complex logic. This feature is relevant in computational devices where the physics imposes the employment of a binary logic at the hardware level. Without entering into the details of a computer hardware, it can be easily seen that it is easier to measure whether or not an amperage flows through a conductor rather than measuring it intensity and associate a semantic value to it. In other words, in order to be reliable a computer hardware must be kept simple at the low level. Then these simple gates can be logically combined in billions of way in order to build a complex logic.

As a further remark, Boolean logic was not defined to satisfy the necessities of a computational device since it was defined about one century earlier that the first computers. This statement is to highlight that research in mathematics normally precedes applied research of years if not centuries.

A.5 Crisp and Fuzzy sets

As stated above, Boolean logic is derived from the concept that an object x either belongs or does not belong to a set A. The idea of membership of an object to a set is formulated as a membership function m_F that can take only two values, i.e. 0 and 1. In this case, the set A is said to be *crisp*.

However, an object x can belong to a set A with a certain degree of membership. For example if person is considered "quite tall", he/she is not fully a member of the set of tall people. More mathematically this person is said to belong to the set of tall people with a certain degree or, this person is associated with a membership function value between 0 and 1. In general we can associate to each set A a continuous membership function m_F that makes correspond to each object x its degree of membership to the set. For example, we can say that with respect to the set A, $m_F = 0.8$. In this case, the set A is said to be a *fuzzy set*, see [31].

Exercises

A.1. Proof of the Equivalence
Select the expression equivalent to $x \wedge y \vee x \wedge y \wedge z$

(a) $x \wedge y$
(b) $\bar{y}$
(c) $x \wedge z$
(d) $x \wedge y \wedge z$

A.2. Determine the values of x, y, z, v such that $\bar{x} \vee y \vee \bar{z} \vee v = 0$

A.3. Derive the Boolean expression from the following logical circuit

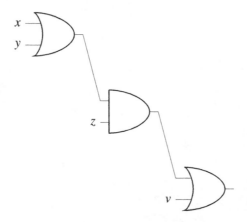

(a) $x \vee y \wedge z \vee v$
(b) $z(x \vee y) \vee v$
(c) $(z \vee v) \wedge (x \vee y)$
(d) $(x \wedge y) \vee z \vee v$

A.4. Considering the result of the Boolean expression (truth table)

x	y	z	Out
0	0	0	0
0	0	1	1
0	1	0	0
0	1	1	1
1	0	0	0
1	0	1	0
1	1	0	1
1	1	1	0

extract the corresponding Boolean expression.

(a) $x \wedge y \wedge z \vee x \wedge y \wedge z \vee x \wedge y \wedge z$
(b) $x \wedge y \wedge z \vee x \wedge y \wedge z \vee \bar{x} \wedge y \wedge \bar{z}$
(c) $\bar{x} \wedge \bar{y} \wedge z \vee \bar{x} \wedge y \wedge z \vee x \wedge y \wedge \bar{z}$
(d) $x \wedge \bar{y} \wedge \bar{z} \vee \bar{x} \wedge y \wedge \bar{z} \vee x \wedge \bar{y} \wedge z$

Appendix B
Proofs of Theorems That Require Further Knowledge of Mathematics

Theorem B.1. Rouchè-Capelli Theorem (Kronecker-Capelli Theorem). *A system of m linear equations in n variables* $\mathbf{Ax} = \mathbf{b}$ *is compatible if and only if both the incomplete and complete matrices ($\mathbf{A}$ and $\mathbf{A^c}$ respectively) are characterised by the same rank* $\rho_{\mathbf{A}} = \rho_{\mathbf{A^c}} = \rho$ *named rank of the system.*

Proof. The system of linear equations $\mathbf{Ax} = \mathbf{b}$ can be interpreted as a linear mapping $f : \mathbb{R}^n \to \mathbb{R}^m$

$$f(\mathbf{x}) = \mathbf{Ax}.$$

This system is determined if one solution exists, i.e. if $\exists \mathbf{x_0}$ such that $f(\mathbf{x}) = \mathbf{b}$. This means that the system is determined if $\mathbf{b} \in \mathrm{Im}(f)$.

The basis spanning the image vector space $(\mathrm{Im}(f), +, \cdot)$ is composed of the column vectors of the matrix $\mathbf{A}$:

$$B_{\mathrm{Im}(f)} = \{\mathbf{I_1}, \mathbf{I_2}, \ldots, \mathbf{I_n}\}$$
$$\mathbf{A} = \left(\mathbf{I_1}\ \mathbf{I_2}\ \ldots\ \mathbf{I_n}\right).$$

Thus, the fact that $\mathbf{b} \in \mathrm{Im}(f)$ is equivalent to the fact that $\mathbf{b}$ belongs to the span of the column vectors of the matrix $\mathbf{A}$:

$$\mathbf{b} = L(\mathbf{I_1}, \mathbf{I_2}, \ldots, \mathbf{I_n}).$$

This is equivalent to say that the rank of

$$\mathbf{A} = \left(\mathbf{I_1}\ \mathbf{I_2}\ \ldots\ \mathbf{I_n}\right)$$

and

$$\mathbf{A^c} = \left(\mathbf{I_1}\ \mathbf{I_2}\ \ldots\ \mathbf{I_n}\ \mathbf{b}\right)$$

have the same rank.

Thus, the system is compatible if $\rho_{\mathbf{A}} = \rho_{\mathbf{A^c}} = \rho$. $\square$

© Springer Nature Switzerland AG 2019
F. Neri, *Linear Algebra for Computational Sciences and Engineering*,
https://doi.org/10.1007/978-3-030-21321-3

Theorem B.2. Euler's Formula. *For every real number* $x \in \mathbb{R}$,

$$e^{j\theta} = \cos\theta + j\sin\theta,$$

where e *is the Euler's number 2.71828, base of natural logarithm.*

Proof. Let us consider a complex number $z = a + jb$ on the unitary circle of the complex plane. The number can be re-written, taking into account the polar coordinated, as

$$z = \rho\left(\cos\theta + j\sin\theta\right).$$

where $\rho = 1$. Hence,

$$z = \cos\theta + j\sin\theta.$$

Let us compute the derivative of z with respect to θ:

$$\frac{dz}{d\theta} = -\sin\theta + j\cos\theta = j\left(\cos\theta + j\sin\theta\right) = jz.$$

This means that the derivative operator is equivalent to the multiplication by the imaginary unit. Let us rearrange the equation as

$$\frac{dz}{z} = jd\theta.$$

If we integrate this differential equation we obtain:

$$\int \frac{dz}{z} = \int jd\theta \Rightarrow \ln\left(z\right) = j\theta \Rightarrow z = e^{j\theta}.$$

Considering that $z = \cos\theta + j\sin\theta$ we obtain

$$e^{j\theta} = \cos\theta + j\sin\theta. \qquad \square$$

Theorem B.3. Fundamental Theorem of Algebra. *If* $p(z) = \sum_{k=0}^{n} a_k z^k$ *is a complex polynomial having order* $n \geq 1$, *then this polynomial has at least one root.*

Proof. Let us assume, by contradiction, that a complex polynomial $p(z)$ with no roots exists. More specifically, let $p(z)$ be a complex polynomial such that $p(z) \neq 0$ $\forall z$, i.e. $p(z)$ is a polynomial that has no roots.
   The function

$$f(z) = \frac{1}{p(z)}.$$

is defined since it always occurs that $p(z) \neq 0$.
   Then, let us calculate

$$\lim_{|z| \to +\infty} |p(z)| = +\infty.$$

Thus,

$$\lim_{|z| \to +\infty} \left| \frac{1}{p(z)} \right| = 0.$$

This means that the function $f(z)$ is limited. For the Liouville's Theorem, see [32] and [33], the function $f(z)$ is constant. Hence, also $p(z)$ is constant. In other words, the only polynomials that have no roots are the constant polynomials. If the order of the polynomial is at least one there will be roots. $\square$

Theorem B.4. *If the vectors* $\mathbf{a_1}, \mathbf{a_2}, \dots, \mathbf{a_n}$ *are linearly independent the Gram determinant* $G(\mathbf{a_1}, \mathbf{a_2}, \dots, \mathbf{a_n}) > 0$.

Proof. If the vectors are linearly independent, let us assume that $\mathbf{x} = \lambda_1 \mathbf{a_1} + \lambda_2 \mathbf{a_2} + \dots + \lambda_n \mathbf{a_n}$ with $\lambda_1, \lambda_2, \dots, \lambda_n \in \mathbb{R}$. We can thus write that

$$\mathbf{x}^2 = (\lambda_1 \mathbf{a_1} + \lambda_2 \mathbf{a_2} + \dots + \lambda_n \mathbf{a_n})^2 = \sum_{i=1}^{n} \sum_{j=1}^{n} \lambda_i \lambda_j \mathbf{a_i} \mathbf{a_j}.$$

This polynomial is a *quadratic form*, see [34], whose discriminant is equal to the Gram determinant. It can be verified that this quadratic form is positive definite. Hence, $G(\mathbf{a_1}, \mathbf{a_2}, \dots, \mathbf{a_n}) > 0$.

Solutions to the Exercises

Exercises of Chap. 1

1.1

Proof. Let us consider the element $x \in A \cup (A \cap B)$ that is

$$x \in A \text{ OR } x \in (A \cap B).$$

All the elements $x \in (A \cap B)$ are elements of both A AND B. Hence the elements

$$x \in (A \cap B)$$

are elements of a subset of A, that is

$$(A \cap B) \subset A.$$

This means that the generic element x belong to either A or to its subset. In other words $x \in A$. This can be repeated for all the elements of $A \cup (A \cap B)$. Hence,

$$A \cup (A \cap B) = A. \square$$

1.2

Proof. Let us consider the element $x \in (A \cup B) \cup C$ that is

$$x \in (A \cup B) \text{ OR } x \in C,$$

that is

$$x \in A \text{ OR } x \in B \text{ OR } x \in C.$$

In other words x belongs to at least one of the three sets. Since x belongs to either B or C (or to both of them) then

$$x \in (B \cup C).$$

© Springer Nature Switzerland AG 2019

F. Neri, *Linear Algebra for Computational Sciences and Engineering*,
https://doi.org/10.1007/978-3-030-21321-3

Since x could also belong to A then

$$x \in A \cup (B \cup C).$$

Since we can repeat this considerations for all the elements of the set $(A \cup B) \cup C$ then

$$(A \cup B) \cup C = A \cup (B \cup C). \square$$

1.3 The roots of

$$x^2 - 2x - 8 = 0$$

are -2 and 4. Hence,

$$B = \{-2, 4\}.$$

The Cartesian product $A \times B$ is

$$A \times B = \{(a, -2), (a, 4), (b, -2), (b, 4), (c, -2), (c, 4)\}$$

1.4 Let us check the properties of the relations.

- Reflexivity: since $(1, 1), (2, 2), (3, 3) \in \mathcal{R}$, the relation is reflexive
- Transitivity: we have $(1, 2) \in \mathcal{R}$ and $(2, 3) \in \mathcal{R}$. It also happens that $(1, 3) \in \mathcal{R}$. There are no other elements to consider. The relation is transitive.
- Symmetry: since $(1, 2) \in \mathcal{R}$ and $(2, 1) \notin \mathcal{R}$ the relation is not symmetric. Since there are no symmetric elements, the relations is antisymmetric.

  Since the relation is reflexive, transitive and antisymmetric, it is an order relation.

1.5 The set f is a function since $\forall x \in A$ there exists only one $f(x) \in B$.

This function is not injective because its contains $(0, 1)$ and $(1, 1)$, i.e. $\exists$ two elements $x_1 \neq x_2$ such that $f(x_1) = f(x_2)$.

Exercises of Chap. 2

2.1

$$\mathbf{AB} = \begin{pmatrix} 5 & 0 & 0 & 5 \\ 0 & -15 & 19 & 13 \end{pmatrix}$$

2.2

$$\mathbf{xy} = 6 - 9 + 4 - 1 = 0$$

2.3

$$\mathbf{AB} = \begin{pmatrix} 5 & 0 & 0 & 5 \\ 0 & -15 & 19 & 13 \end{pmatrix}$$

2.4

$$\mathbf{AB} = \begin{pmatrix} 26 & -22 & 17 \\ 12 & -3 & 1 \\ 31 & 4 & -3 \end{pmatrix}$$

2.5

$$\det(\mathbf{A}) = 1$$
$$\det(\mathbf{B}) = 0$$

2.6

$$\det(\mathbf{A}) = 2k + 28.$$

This means that $\mathbf{A}$ is singular for $k = -7$.

$$\det(\mathbf{B}) = -2k^2 - k + 10$$

whose roots are 2 and $-\frac{5}{2}$. For these two values $\mathbf{B}$ is singular.

$$\det(\mathbf{C}) = 0$$

regardless of k since the third row is linear combination of the first and second rows. Hence, $\mathbf{C}$ is singular $\forall k \in \mathbb{R}$.

2.7

$$\mathrm{adj}(\mathbf{A}) = \begin{pmatrix} 0 & 4 & -2 \\ 0 & -6 & 4 \\ 1 & -8 & 5 \end{pmatrix}$$

2.8

$$\mathbf{A}^{-1} = \frac{1}{30} \begin{pmatrix} 2 & 3 \\ -8 & 3 \end{pmatrix}.$$

$$\mathbf{B}^{-1} = \frac{1}{8} \begin{pmatrix} 3 & 2 & 4 \\ -1 & 2 & 12 \\ 1 & -2 & -4 \end{pmatrix}.$$

2.9

1. Since $\det(\mathbf{A}) = 21$ the matrix $\mathbf{A}$ is invertible.
2.

$$\mathbf{A}^{-1} = \frac{1}{21} \begin{pmatrix} 7 & -1 & -4 \\ 0 & 9 & -6 \\ 0 & 6 & 3 \end{pmatrix}$$

3.

$$\mathbf{A}\mathbf{A}^{-1} = \frac{1}{21} \begin{pmatrix} 3 & -1 & 2 \\ 0 & 1 & 2 \\ 0 & -2 & 3 \end{pmatrix} \begin{pmatrix} 7 & -1 & -4 \\ 0 & 9 & -6 \\ 0 & 6 & 3 \end{pmatrix} = \frac{1}{21} \begin{pmatrix} 21 & 0 & 0 \\ 0 & 21 & 0 \\ 0 & 0 & 21 \end{pmatrix} = \begin{pmatrix} 1 & 0 & 0 \\ 0 & 1 & 0 \\ 0 & 0 & 1 \end{pmatrix}.$$

2.10 The $\det(\mathbf{A}) = 0$. Hence, the rank of the matrix cannot be 3. By cancelling the third row and the third column we can identify the nonsingular matrix

$$\begin{pmatrix} 2 & 1 \\ 0 & 1 \end{pmatrix}.$$

Hence, the rank of $\mathbf{A}$ is 2.

2.11 The determinant of the matrix $\mathbf{A}$ is equal to 0, i.e. the matrix $\mathbf{A}$ is singular. Thus its rank $\rho < 3$. It can be noticed that the second row is twice the first one and the third row is the sum of the first two. There is no 2×2 non-singular sub-matrices. Thus, $\rho = 1$.

Since the matrix is singular, it cannot be inverted.

Exercises of Chap. 3

3.1 The incomplete matrix is non-singular:

$$\det \begin{pmatrix} 1 & -2 & 1 \\ 1 & 5 & 0 \\ 0 & -3 & 1 \end{pmatrix} = 4.$$

Hence, the rank $\rho_A = \rho_{A_c} = 3$ and the system can be solved by the Cramer's Method. The solution is

$$x = \frac{\det \begin{pmatrix} 2 & -2 & 1 \\ 1 & 5 & 0 \\ 1 & -3 & 1 \end{pmatrix}}{4} = \frac{4}{4} = 1,$$

$$y = \frac{\det \begin{pmatrix} 1 & 2 & 1 \\ 1 & 1 & 0 \\ 0 & 1 & 1 \end{pmatrix}}{4} = \frac{0}{4} = 0$$

and

$$z = \frac{\det \begin{pmatrix} 1 & -2 & 2 \\ 1 & 5 & 1 \\ 0 & -3 & 1 \end{pmatrix}}{4} = \frac{4}{4} = 1.$$

3.2 The incomplete and complete matrices associated with the system are, respectively:

$$\mathbf{A} = \begin{pmatrix} (k+2) & (k-1) & -1 \\ k & -k & 0 \\ 4 & -1 & 0 \end{pmatrix}$$

and

$$\mathbf{A} = \begin{pmatrix} (k+2) & (k-1) & -1 & k-2 \\ k & -k & 0 & 2 \\ 4 & -1 & 0 & 1 \end{pmatrix}.$$

The $det(\mathbf{A}) = k - 4k = -3k$. Hence, the matrix $\mathbf{A}$ is non-singular when $k \neq 0$. Under this condition, $\rho_A = \rho_{A^c} = 3$, i.e. the system is compatible. Since $n = 3$, the system is also determined.

If $k = 0$ the incomplete matrix is singular and its rank is $\rho_A = 2$ because at least one non-singular submatrix of order 2 can be extracted. For example

$$\begin{pmatrix} -1 & -1 \\ -1 & 0 \end{pmatrix}.$$

is non-singular.

The complete matrix would be

$$A^c = \begin{pmatrix} 2 & -1 & -1 & -2 \\ 0 & 0 & 0 & 2 \\ 4 & -1 & 0 & 1 \end{pmatrix}.$$

whose rank is 3 because, e.g., the submatrix obtained by cancelling the third column

$$\begin{pmatrix} 2 & -1 & -2 \\ 0 & 0 & 2 \\ 4 & -1 & 1 \end{pmatrix}.$$

is non-singular. Hence, $\rho_A = 2 \neq \rho_{A^c} = 3$. If $k = 0$ the system is incompatible. There is no value of k that makes the system undetermined.

3.3 The incomplete matrix

$$\begin{pmatrix} 1 & 1 & -1 \\ 0 & 1 & -1 \\ 1 & 2 & -2 \end{pmatrix}$$

has null determinant as the third row is the sum of the first and second rows. Cramer's method cannot be applied. It follows that the rank of this matrix is $\rho_A = 2$.

The rank of the complete matrix is ρ_{A^c} is also 2 since the system is homogeneous. Thus, the system has ∞^1 solutions.

The general solution can be found posing $y = z = \alpha$. It then results that $x = 0$. Hence the general solution is $(0, \alpha, \alpha)$.

3.4 The incomplete matrix

$$\begin{pmatrix} 1 & 2 & 3 \\ 4 & 4 & 8 \\ 3 & -1 & 2 \end{pmatrix}$$

has null determinant as the third column is linear combination of the other two columns. It can be seen that there are non-singular order-2 submatrices. Hence, the rank of the matrix $\rho_A = 2$. Cramer method cannot be applied. On the contrary, the rank of the complete matrix is ρ_{A^c} is 3 as at least non-singular sub-matrix having order 3 could be extracted. For example the matrix

$$\begin{pmatrix} 1 & 2 & 1 \\ 4 & 4 & 2 \\ 3 & -1 & 1 \end{pmatrix}$$

is non-singular since its determinant is -6. This means that $\rho_A = 2 < \rho_{A^c} = 3$. The system is impossible. Thus it does not allow any solution.

3.5 The incomplete matrix

$$\begin{pmatrix} 1 & 2 & 3 \\ 2 & 4 & 6 \\ 3 & 6 & 9 \end{pmatrix}$$

has null determinant as the second row is twice the first row and third row is three times the first row. Cramer's method cannot be applied. It follows that the rank of this matrix is $\rho_A = 1$.

The rank of the complete matrix is ρ_{A^c} is also 1 since the known term of the second equation is twice the known term of the first equation and the known term of the third equation is three times the known term of the first equation.

This means that $\rho_A = 1\rho_{A^c}$. The system is undetermined and has $\infty^{n-\rho} = \infty^{3-1} = \infty^2$ solutions.

In order to find the general solution let us pose $y = \alpha$ and $z = \beta$:

$$x + 2\alpha + 3\beta = 1$$

which leads to

$$x = 1 - 2\alpha - 3\beta.$$

Hence, the general solution is

$$(1 - 2\alpha - 3\beta, \alpha, \beta) = \alpha \left(\frac{1}{\alpha} - 2, 1, 0 \right) + \beta (-3, 0, 1).$$

3.6 The associated complete matrix is

$$\mathbf{A}^c = (\mathbf{A}|\mathbf{b}) = \begin{pmatrix} 1 & -1 & 1 & | & 1 \\ 1 & 1 & 0 & | & 4 \\ 2 & 2 & 2 & | & 9 \end{pmatrix}.$$

Let us at first apply the following row operations:

$$\mathbf{r}^2 = \mathbf{r}^2 - \mathbf{r}^1$$

$$\mathbf{r}^3 = \mathbf{r}^3 - 2\mathbf{r}^1.$$

We obtain

$$\tilde{A}^c = (A|b) = \begin{pmatrix} 1 & -1 & 1 & | & 1 \\ 0 & 2 & -1 & | & 3 \\ 0 & 4 & 0 & | & 7 \end{pmatrix}.$$

Now, let us apply the following row transformation

$$r^3 = r^3 - 2r^2.$$

Thus, we obtain the triangular matrix

$$\tilde{A}^c = (A|b) = \begin{pmatrix} 1 & -1 & 1 & | & 1 \\ 0 & 2 & -1 & | & 3 \\ 0 & 0 & 2 & | & 1 \end{pmatrix}.$$

The matrix corresponds to the system

$$\begin{cases} x - y + z = 1 \\ 2y - z = 3 \\ 2z = 1 \end{cases}.$$

3.7 Considering that the generic formulas to fill U and L matrices are for $i \le j$

$$u_{i,j} = a_{i,j} - \sum_{k=1}^{i-1} l_{i,k} u_{k,j}$$

and for $j < i$

$$l_{i,j} = \frac{1}{u_{j,j}} \left(a_{i,j} - \sum_{k=1}^{j-1} l_{i,k} u_{k,j} \right)$$

we obtain:

$$u_{1,1} = a_{1,1} = 5$$
$$u_{1,2} = a_{1,2} = 0$$
$$u_{1,3} = a_{1,3} = 5$$
$$l_{2,1} = \frac{1}{u_{1,1}} (a_{2,1}) = 2$$
$$u_{2,2} = a_{2,2} - l_{2,1} u_{1,2} = 1$$
$$u_{2,3} = a_{2,3} - l_{2,1} u_{1,3} = 3$$
$$l_{3,1} = \frac{1}{u_{1,1}} (a_{3,1}) = 3$$
$$l_{3,2} = \frac{1}{u_{2,2}} (a_{3,2} - l_{3,1} u_{1,2}) = 2$$
$$u_{3,3} = a_{3,3} - l_{3,1} u_{1,3} - l_{3,2} u_{2,3} = 2.$$

Hence,

$$L = \begin{pmatrix} 1 & 0 & 0 \\ 2 & 1 & 0 \\ 3 & 2 & 1 \end{pmatrix}$$

and

$$\mathbf{U} = \begin{pmatrix} 5 & 0 & 5 \\ 0 & 1 & 3 \\ 0 & 0 & 2 \end{pmatrix}$$

3.8

1. From $x^{(0)} = 0, y^{(0)} = 0, z^{(0)} = 0$,

$$x^{(1)} = -2(0) = 0$$
$$y^{(1)} = -2 + 2(0) + 6(0) = -2$$
$$z^{(1)} = 8 - 4(0) = 8.$$

2. From $x^{(0)} = 0, y^{(0)} = 0, z^{(0)} = 0$,

$$x^{(1)} = -2(0) = 0$$
$$y^{(1)} = -2 + 2(0) + 6(0) = -2$$
$$z^{(1)} = 8 - 4(-2) = 16.$$

Exercises of Chap. 4

4.1 The two vectors are parallel if the rank of the associated matrix is less than 2. In this case the associated matrix is

$$\mathbf{A} = \begin{pmatrix} 2 & 1 & -2 \\ -8 & -4 & 8 \end{pmatrix}.$$

The rank of this matrix is 1 because each order 2 sub-matrix is singular. Thus, the vectors are parallel.

4.2

1. The perpendicularity occurs when the scalar product is null. Hence,

$$2 \cdot 1 + 0 \cdot 0 + 1 \cdot (1 - k) = 0.$$

This means

$$2 + 1 - k = 0$$

that is $k = 3$.

2. These two vectors are parallel if the rank of the matrix

$$\begin{pmatrix} 1 & 0 & 1 - k \\ 2 & 0 & 1 \end{pmatrix}$$

is < 2

This means that the vectors are parallel when $1 - 2 + 2k = 0$. This means $-1 + 2k = 0$ that is $k = \frac{1}{2}$.

4.3 In order to check whether or not the vectors are linearly independent let us pose

$$\vec{o} = \lambda \vec{u} + \mu \vec{v} + v \vec{w}.$$

If the only way to verify the equation is by means of

$$(\lambda, \mu, v) = (0, 0, 0)$$

then the vectors are linearly independent.

Thus,

$$\begin{pmatrix} 0 \\ 0 \\ 0 \end{pmatrix} = \lambda \begin{pmatrix} 2 \\ -3 \\ 2 \end{pmatrix} + \mu \begin{pmatrix} 3 \\ 0 \\ -1 \end{pmatrix} + v \begin{pmatrix} 1 \\ 0 \\ 2 \end{pmatrix}$$

which yields to a homogeneous system of linear equations that is determined only if the matrix associated with the system is non-singular. In this case,

$$\det \begin{pmatrix} 2 & 3 & 1 \\ -3 & 0 & 0 \\ 2 & 1 & 2 \end{pmatrix} = 15.$$

The vectors are linearly independent.

4.4

1. The associated matrix is

$$\mathbf{A} = \begin{pmatrix} 6 & 2 & 3 \\ 1 & 0 & 1 \\ 0 & 0 & 1 \end{pmatrix}$$

is non-singular since $\det(\mathbf{A}) = -2$. Hence, the vectors are linearly independent. The only way the null vector can be expressed as a linear combination of $\vec{u}$, $\vec{v}$, $\vec{w}$ is by means of all null scalars.

2. Since the vectors are linearly independent they are basis in $\mathbb{V}_3$. Hence, the vector $\vec{t}$ can be expressed in this new basis by imposing:

$$\vec{t} = \lambda \vec{u} + \mu \vec{v} + v \vec{w}.$$

This means

$$(1,1,1) = \lambda (6,2,3) + \mu (1,0,1) + v (0,0,1).$$

From this equation a system of linear equation is identified:

$$\begin{cases} 6\lambda + \mu = 1 \\ 2\lambda = 1 \\ 3\lambda + \mu + v = 1 \end{cases}.$$

The solution is $\lambda = \frac{1}{2}$, $\mu = -2$, and $v = \frac{3}{2}$. This means

$$\vec{t} = \frac{1}{2}\vec{u} - 2\vec{v} + \frac{3}{2}\vec{w}.$$

4.5

1. The vectors $\vec{u}$, $\vec{v}$, $\vec{w}$ are linearly dependent, the determinant of the matrix associated to them is singular. These vectors are not a basis.
2. Since the vectors $\vec{u}$, $\vec{v}$, $\vec{w}$ are not a basis the vector $\vec{t}$ cannot be expressed in that basis. The system of equations resulting from these vectors would be impossible.

4.6 For the three vectors, the third component is the second. In other words,

$$\vec{o} = \lambda \vec{u} + \mu \vec{v} + v \vec{w}.$$

is verified for

$$(\lambda, \mu, v) = (0, -2, 1)$$

The vectors are coplanar. Equivalently the vectors are linearly dependent.

This statement is equivalent to say that the matrix associated with the vectors is singular or that the mixed product is zero.

4.7 Let us calculate the vector product

$$\vec{u} \otimes \vec{v} = \det \begin{pmatrix} \vec{i} & \vec{j} & \vec{k} \\ (3h-5) & (2h-1) & 3 \\ 1 & -1 & 3 \end{pmatrix} =$$

$$= 3(2h-1)\,\vec{i} + 3\vec{j} - (3h-5) + 3\,\vec{i} - 3(3h-5)\,\vec{j} - (2h-1)\,\vec{k} =$$

$$= 6h\,\vec{i} + (-9h+18)\,\vec{j} + (-5h+6)\,\vec{k}$$

The two vectors are parallel if the vector product is equal to $\vec{0}$, i.e.

$$\begin{cases} 6h = 0 \\ -9h + 18 = 0 \\ -5h + 6 = 0 \end{cases}$$

that is impossible. Hence, there is no value of h that makes $\vec{u}$ and $\vec{v}$ parallel.

4.8 Let us calculate the mixed product

$$\det \begin{pmatrix} 2 & -1 & 3 \\ 1 & 1 & -2 \\ (h) & -1 & (h-1) \end{pmatrix} = 2h - 2 + 2h - 3 - 3h + 4 + h - 1 = 2h - 10.$$

By imposing the determinant to be null, the vectors are coplanar if $h = 5$.

Exercises of Chap. 5

5.1

$$\frac{1}{z} = \frac{1}{a+jb} = \frac{a-jb}{(a-jb)(a+jb)} = \frac{a-jb}{a^2+b^2}.$$

5.2 The module $\rho = \sqrt{1^2+1^2} = \sqrt{2}$. The phase $\arctan(-1) = -45° = 315°$

5.3

$$z = 4\cos(90°) + 4j\sin(90°) = 4j.$$

5.4 Let us represent the complex number in polar coordinates:

$$5 + j5 = \sqrt{50}\angle 45° = \sqrt{50}\left(\cos(45°) + j\sin 45°\right).$$

Let us now calculate by De Moivre's Formula

$$\sqrt[3]{5+j5} = \sqrt[3]{\sqrt{50}\left(\cos(45°) + j\sin 45°\right)} = \sqrt[6]{50}\left(\cos(15°) + j\sin(15°)\right)$$

5.5 By Ruffini's theorem, since $1^3 - 3 \times 1^2 - 13 \times 1 + 15 = 0$, it follows that $z^3 - 3z^2 - 13z + 15$ is divisible by $(z-1)$.

5.6 The third row is the sum of the other two rows. Thus, without the need of performing any calculation we can conclude that the determinant is null, i.e. the matrix is singular and thus cannot be inverted. The inverse of the matrix **A** does not exist.

5.7 For the Little Berzout's Theorem the remainder r is

$$r = p(2j) = (2j)^3 + 2(2j)^2 + 4(2j) - 8 = 8(j)^3 + 8(j)^2 + 8j - 8 = -8j - 8 + 8j - 8 = -16.$$

5.8

$$\frac{-9z+9}{2z^2+7z-4} = \frac{1}{(2z-1)} - \frac{5}{(z+4)}.$$

5.9 Expand in partial fractions the following rational fraction

$$\frac{3z+1}{(z-1)^2(z+2)}.$$

Solution

$$\frac{3z+1}{(z-1)^2(z+2)} = \frac{5}{9(z-1)} + \frac{4}{3(z-1)^2} - \frac{5}{9(z+2)}.$$

5.10

$$\frac{5z}{(z^3 - 3z^2 - 3z - 2)} = \frac{5z}{(z-2)(z^2 + z + 1)} = \frac{10}{7(z-2)} - \frac{10z + 5}{7(z^2 + z + 1)}.$$

Exercises of Chap. 6

6.1 If we consider the parallel line passing through the origin

$$4x - 3y = 0$$

and we pose $y = \alpha$ with the parameter $\alpha \in \mathbb{R}$, we have

$$x = \frac{3}{4}\alpha.$$

This line is parallel to the vectors $\alpha\left(\frac{3}{4}, 1\right)$. The direction $\left(\frac{3}{4}, 1\right)$ is the direction of the line. Equivalently, the direction of the line is $(3, 4)$.

6.2 The matrix

$$\begin{pmatrix} 3 & -2 \\ 4 & 1 \end{pmatrix}$$

is non-singular. Hence, an intersection point exists.

The coordinates of the intersection point are found by means of the Cramer's method:

$$x = \frac{\det \begin{pmatrix} -4 & -2 \\ -1 & 1 \end{pmatrix}}{\det \begin{pmatrix} 3 & -2 \\ 4 & 1 \end{pmatrix}} = -\frac{6}{11}$$

and

$$y = \frac{\det \begin{pmatrix} 3 & -4 \\ 4 & -1 \end{pmatrix}}{\det \begin{pmatrix} 3 & -2 \\ 4 & 1 \end{pmatrix}} = \frac{13}{11}.$$

6.3 The matrix

$$\begin{pmatrix} 3 & -2 \\ 9 & -6 \end{pmatrix}$$

is singular and its rank is 1. Hence, the two lines have the same direction.

On the other hand, the rank of the complete matrix

$$\begin{pmatrix} 3 & -2 & -4 \\ 9 & -6 & -1 \end{pmatrix}$$

is 2. The system of linear equations is impossible and thus has no solutions, i.e. the two lines do not intersect (the lines are parallel).

6.4 The matrix associated to the conic

$$\mathbf{A}^c = \begin{pmatrix} 4 & 1 & 0 \\ 1 & -2 & -4 \\ 0 & -4 & 8 \end{pmatrix}$$

is non-singular. Hence, the conic is non-degenerate. Since

$$\det(\mathbf{I}_{3,3}) = \det \begin{pmatrix} 4 & 1 \\ 1 & -2 \end{pmatrix} = -9.$$

Thus, the conic is a hyperbola.

6.5 The matrix associated to the conic

$$\mathbf{A}^c = \begin{pmatrix} 4 & 1 & 0 \\ 1 & 2 & -4 \\ 0 & -4 & -6 \end{pmatrix}$$

is non-singular. Hence, the conic is non-degenerate. Since

$$\det(\mathbf{I}_{3,3}) = \det \begin{pmatrix} 4 & 1 \\ 1 & 2 \end{pmatrix} = 7.$$

Thus, the conic is an ellipse.

6.6 The matrix associated to the conic

$$\mathbf{A}^c = \begin{pmatrix} 1 & 1 & -8 \\ 1 & 1 & 0 \\ -8 & 0 & -6 \end{pmatrix}$$

is non-singular. Hence, the conic is non-degenerate. Since

$$\det(\mathbf{I}_{3,3}) = \det \begin{pmatrix} 1 & 1 \\ 1 & 1 \end{pmatrix} = 0.$$

Thus, the conic is a parabola.

6.7 The matrix associated to the conic

$$\mathbf{A}^c = \begin{pmatrix} 1 & 1 & -7 \\ 1 & 0 & -8 \\ -7 & -8 & 12 \end{pmatrix}$$

is singular. Hence, the conic is degenerate. Since

$$\det(\mathbf{I}_{3,3}) = \det \begin{pmatrix} 1 & 1 \\ 1 & 0 \end{pmatrix} = -1.$$

Thus, the conic is a degenerate hyperbola, that is a pair of intersecting lines.

6.8 The matrix associated to the conic

$$\mathbf{A}^c = \begin{pmatrix} 1 & 3 & 5 \\ 3 & -16 & -40 \\ -16 & -40 & 24 \end{pmatrix}$$

is non-singular. Hence, the conic is non-degenerate. Since

$$\det(\mathbf{I}_{3,3}) = \det\begin{pmatrix} 1 & 3 \\ 3 & -16 \end{pmatrix} = -25.$$

Thus, the conic is a hyperbola.

Exercises of Chap. 7

7.1 $(A, +)$ is not an algebraic structure since the set is not closed with respect to the operation. For example $6 + 4 = 10 \notin A$.

7.2 Since the matrix product is an internal composition law, $(\mathbb{R}_{n,n}, \cdot)$ is an algebraic structure. Since the matrix product is associative then $(\mathbb{R}_{n,n}, \cdot)$ is a semigroup. Since a neutral element exists, that is the identity matrix $\mathbf{I}$ then $(\mathbb{R}_{3,3}, \cdot)$ is a monoid. d) Since there is no general inverse element (only non-singular matrices are invertible) then $(\mathbb{R}_{n,n}, \cdot)$ is not a group.

7.3 $(H, +_8)$ is a subgroup because it is a group. The operator $+_8$ is an associative internal composition law, every element of H has an inverse, i.e. $\{0, 8, 4, 2\}$.

The cosets are $H +_8 0 = H$ and $H_8 + 1 = \{1, 3, 5, 7\}$. All the other cosets are like these two, e.g. $H +_8 2 = \{2, 4, 6, 0\} = H +_8 0 = H$.

All the cosets have the same cardinality that is 4. The cardinality of Z_8 is 8. According to the Lagrange Theorem

$$\frac{|Z_8|}{|H|} = 2$$

that is an integer number.

7.4 Let us check the associative property. Let us calculate

$$(a * b) * c = (a + 5b) * c = a + 5b + 5c.$$

Let us calculate now

$$a * (b * c) = a * (b + 5c) = a + 5b + 25c.$$

Since the operator is not associative $(\mathbb{Q}, *)$ is not a semigroup. Thus, it cannot be a monoid.

7.5 This mapping is an homomorphism since

$$f(x + y) = e^{(x+y)} = e^{(x)} e^{(y)} = f(x) f(y).$$

The mapping is an isomorphism because it is injective as if $x_1 \neq x_2$ then $e^{(x_1)} \neq e^{(x_2)}$. The mapping is surjective since every positive number can be expressed as an exponential of a real number:

$$\forall t \in \mathbb{R}^+ \exists x \in \mathbb{R} \ni `t = e^x.$$

Exercises of Chap. 8

8.1 In order to check whether or not $(U,+,\cdot)$ and $(V,+,\cdot)$ are vector spaces, we have to prove the closure with respect to the two composition laws.

1. Let us consider two arbitrary vectors belonging to U, $\mathbf{u_1} = (x_1, y_1, z_1)$ and $\mathbf{u_2} = (x_2, y_2, z_2)$. These two vectors are such that

$$5x_1 + 5y_1 + 5z_1 = 0$$
$$5x_2 + 5y_2 + 5z_2 = 0.$$

Let us calculate

$$\mathbf{u_1} + \mathbf{u_2} = (x_1 + x_2, y_1 + y_2, z_1 + z_2).$$

In correspondence to the vector $\mathbf{u_1} + \mathbf{u_2}$,

$$5(x_1 + x_2) + 5(y_1 + y_2) + 5(z_1 + z_2) =$$
$$= 5x_1 + 5y_1 + 5z_1 + 5x_2 + 5y_2 + 5z_2 = 0 + 0 = 0.$$

This means that $\forall \mathbf{u_1}, \mathbf{u_2} \in U : \mathbf{u_1} + \mathbf{u_2} \in U$.

2. Let us consider an arbitrary vector $\mathbf{u} = (x, y, z) \in U$ and an arbitrary scalar $\lambda \in \mathbb{R}$. We know that $5x + 5y + 5z = 0$. Let us calculate

$$\lambda \mathbf{u} = (\lambda x, \lambda y, \lambda z).$$

In correspondence to the vector $\lambda \mathbf{u}$,

$$5\lambda x + 5\lambda y + 5\lambda z =$$
$$= \lambda(5x + 5y + 5z) = \lambda 0 = 0.$$

This means that $\forall \lambda \in \mathbb{R}$ and $\forall \mathbf{u} \in U : \lambda \mathbf{u} \in U$.
Thus, $(U,+,\cdot)$ is a vector space.
As for V we have

3. Let us check the closure with respect to the internal composition law:

$$5(x_1 + x_2) + 5(y_1 + y_2) + 5(z_1 + z_2) + 5 =$$
$$= 5x_1 + 5y_1 + 5z_1 + 5x_2 + 5y_2 + 5z_2 + 5 \neq 0.$$

Since there is no closure $(V,+,\cdot)$ is not a vector space.
We can easily check that also $(U \cap V, +, \cdot)$ is not a vector space since it would not contain the null vector $\mathbf{o}$. In this specific case, the set $U \cap V$ is given by

$$\begin{cases} 5x + 5y + 5z = 0 \\ 5x + 5y + 5z = -5 \end{cases}$$

which is impossible. In other words, the intersection is the empty set . The geometrical meaning of this problem is two parallel lines of the space.

8.2

Proof. Let us assume that $U \subset V$. It follows that

$$U \cup V = V.$$

Since $(V, +, \cdot)$ is a vector space then $(U \cup V, +, \cdot)$ is a vector subspace of $(E, +, \cdot)$. $\square$

The other case $(V \subset U)$ is analogous.

8.3 The intersection set $U \cap V$ is given by

$$\begin{cases} x - y + 4z = 0 \\ y - z = 0 \end{cases}$$

which has ∞^1 solutions of the type $(-3\alpha, \alpha, \alpha)$ with $\alpha \in \mathbb{R}$.

The sum set S is obtained by finding, at first, the general vector representation of each set

$$U = \{(\beta - 4\gamma, \beta, \gamma) \,|\, \beta, \gamma \in \mathbb{R}\}$$

and

$$V = \{(\delta, \alpha, \alpha) \,|\, \alpha \in \mathbb{R}\}.$$

Then, the sum set S is

$$S = U + V = \{(\beta - 4\gamma + \delta, \alpha + \beta, \alpha + \gamma) \,|\, \alpha, \beta, \gamma, \delta \in \mathbb{R}\}$$

which is $\mathbb{R}^3$ since every vector of $\mathbb{R}^3$ can be represented by arbitrarily choosing $\alpha, \beta, \gamma, \delta$.

The sum set is not direct sum since $U \cap V \neq \mathbf{o}$.

As shown above $\dim(U) = \dim(V) = 2$, the $\dim(U \cap V) = 1$ and $\dim(U + V) = 3$, i.e. $2 + 2 = 3 + 1$.

8.4 The matrix associated to this system of linear equations

$$\begin{pmatrix} 1 & 2 & 3 \\ 2 & 4 & 6 \\ 3 & 6 & 9 \end{pmatrix}$$

is singular as well as all its order 2 submatrices. Hence the rank of the matrix is 1 and the system has ∞^2 solutions. If we pose $y = \alpha$ and $z = \beta$ we have $x = -2\alpha - 3\beta$. Hence, the solutions are proportional to

$$(-2\alpha - 3\beta, \alpha, \beta)$$

which can written as

$$(-2\alpha - 3\beta, \alpha, \beta) = (-2\alpha, \alpha, 0) + (-3\beta, 0, \beta) = \alpha(-2, 1, 0) + \beta(-3, 0, 1).$$

$B = \{(-2, 1, 0), (-3, 0, 1)\}$. The dimension is then $\dim(E, +, \cdot) = 2$.

8.5 The incomplete matrix associated to the system is singular:

$$\det \begin{pmatrix} 1 & 1 & 2 \\ 2 & 1 & 3 \\ 0 & 1 & 1 \end{pmatrix} = 0.$$

The system is undetermined and has ∞^1 solutions. The general solution is $(\alpha, \alpha, -\alpha)$ with $\alpha \in \mathbb{R}$. A basis is then $B = \{(1, 1, -1)\}$ and the dimension of the space is 1.

8.6 The matrix associated to this system of linear equations

$$\begin{pmatrix} 1 & 2 & 3 \\ 2 & 4 & 6 \\ 3 & 6 & 9 \end{pmatrix}$$

is singular as well as all its order 2 submatrices. Hence the rank of the matrix is 1 and the system has ∞^2 solutions. If we pose $y = \alpha$ and $z = \beta$ we have $x = -2\alpha - 3\beta$. Hence, the solutions are proportional to

$$(-2\alpha - 3\beta, \alpha, \beta)$$

which can written as

$$(-2\alpha - 3\beta, \alpha, \beta) = (-2\alpha, \alpha, 0) + (-3\beta, 0, \beta) = \alpha(-2, 1, 0) + \beta(-3, 0, 1).$$

These couples of vectors solve the system of linear equations. In order to show that these two vectors compose a basis we need to verify that they are linearly independent. By definition, this means that

$$\lambda_1 \mathbf{v_1} + \lambda_2 \mathbf{v_2} = \mathbf{0}$$

only if $\lambda_1, \lambda_2 = 0, 0$.

In our case this means that

$$\lambda_1(-2, 1, 0) + \lambda_2(-3, 0, 1) = \mathbf{0}$$

only if $\lambda_2, \lambda_2 = 0, 0$. This is equivalent to say that the system of linear equations

$$\begin{cases} -2\lambda_1 - 3\lambda_2 = 0 \\ \lambda_1 = 0 \\ \lambda_2 = 0 \end{cases}$$

is determined. The only values that satisfy the equations are $\lambda_2, \lambda_2 = 0, 0$. This means that the vectors are linearly independent and hence compose a basis $B = \{(-2,1,0),(-3,0,1)\}$. The dimension is then dim $(E,+,\cdot) = 2$.

8.7 Let us name B the basis we want to find. At first we inspect the vectors.

Since $\mathbf{u}$ is not the null vector, it can be included in the basis: $B = \{\mathbf{u}\}$.

Let us check whether or not $\mathbf{u}$ and $\mathbf{v}$ are linearly independent.

$$\mathbf{o} = \lambda_1 \mathbf{u} + \lambda_2 \mathbf{v}$$

$$\begin{pmatrix} 0 \\ 0 \\ 0 \\ 0 \end{pmatrix} = \lambda_1 \begin{pmatrix} 2 \\ 2 \\ 1 \\ 3 \end{pmatrix} + \lambda_2 \begin{pmatrix} 0 \\ 2 \\ 1 \\ -1 \end{pmatrix}.$$

The associated system of linear equation is determined. Hence, $\lambda_1 = \lambda_2 = 0$ and the vectors are linearly independent. We can insert $\mathbf{v}$ in the basis B: $B = \{\mathbf{u}, \mathbf{v}\}$.

Let us now attempt to insert w

$$\mathbf{o} = \lambda_1 \mathbf{u} + \lambda_2 \mathbf{v} + \lambda_3 \mathbf{w}$$

that is

$$\begin{pmatrix} 0 \\ 0 \\ 0 \\ 0 \end{pmatrix} = \lambda_1 \begin{pmatrix} 2 \\ 2 \\ 1 \\ 3 \end{pmatrix} + \lambda_2 \begin{pmatrix} 0 \\ 2 \\ 1 \\ -1 \end{pmatrix} + \lambda_3 \begin{pmatrix} 2 \\ 4 \\ 2 \\ 2 \end{pmatrix}.$$

The matrix associated to the system is singular. Thus, the vectors $\mathbf{u}$, $\mathbf{v}$, and $\mathbf{w}$ are linearly dependent. This can be seen also by noticing that $\mathbf{w} = \mathbf{v} + \mathbf{u}$. Hence, we cannot include $\mathbf{w}$ in the basis B.

We can now attempt to insert $\mathbf{a}$ and verify

$$\begin{pmatrix} 0 \\ 0 \\ 0 \\ 0 \end{pmatrix} = \lambda_1 \begin{pmatrix} 2 \\ 2 \\ 1 \\ 3 \end{pmatrix} + \lambda_2 \begin{pmatrix} 0 \\ 2 \\ 1 \\ -1 \end{pmatrix} + \lambda_3 \begin{pmatrix} 3 \\ 1 \\ 2 \\ 1 \end{pmatrix}.$$

The associated matrix has rank 3 and thus $\mathbf{a}$ can be inserted into the basis B.

We can now attempt to add $\mathbf{b}$ to the basis:

$$\begin{pmatrix} 0 \\ 0 \\ 0 \\ 0 \end{pmatrix} = \lambda_1 \begin{pmatrix} 2 \\ 2 \\ 1 \\ 3 \end{pmatrix} + \lambda_2 \begin{pmatrix} 0 \\ 2 \\ 1 \\ -1 \end{pmatrix} + \lambda_3 \begin{pmatrix} 3 \\ 1 \\ 2 \\ 1 \end{pmatrix} + \lambda_4 \begin{pmatrix} 0 \\ 5 \\ 0 \\ 2 \end{pmatrix}.$$

The matrix associated to the system is non-singular. Hence we have found by elimination a basis of $\mathbb{R}^4$:

$$B = \{\mathbf{u}, \mathbf{v}, \mathbf{a}, \mathbf{b}\}.$$

Exercises of Chap. 9

9.1 Let us calculate
$$\| \mathbf{xy} \| = \mathbf{xy} = 76$$
and the respective modules
$$\| \mathbf{x} \| = 8.832$$
$$\| \mathbf{y} \| = 12.649.$$

Let us now calculate
$$\| \mathbf{x} \| \| \mathbf{y} \| = 111.71.$$

Since $71 < 111.71$, the Cauchy-Schwarz inequality is verified.
Let us calculate
$$\| \mathbf{x} + \mathbf{y} \| = \| (-2, 5, 19) \| = 19.748.$$

Considering that $19.748 < 8.832 + 12.649 = 21.481$, Minkowski's inequality is verified.

9.2 Let us now apply the Gram-Schmidt method to find an orthonormal basis $B_U = \{\mathbf{e}_1, \mathbf{e}_2\}$. The first vector is

$$\mathbf{e}_1 = \frac{\mathbf{x}_1}{\| \mathbf{x}_1 \|} = \frac{(2,1)}{\sqrt{5}} = \left(\frac{2}{\sqrt{5}}, \frac{1}{\sqrt{5}} \right) = (0.894 \, 0.447).$$

For the calculation of the second vector the direction must be detected at first. The direction is that given by

$$\mathbf{y}_2 = \mathbf{x}_2 + \lambda_1 \mathbf{e}_1 = (0,2) + \lambda_1 \left(\frac{2}{\sqrt{5}}, \frac{1}{\sqrt{5}} \right).$$

Let us find the orthogonal direction by imposing that $\mathbf{y}_2 \mathbf{e}_1 = 0$. Hence,

$$\mathbf{y}_2 \mathbf{e}_1 = 0 = \mathbf{x}_2 \mathbf{e}_1 + \lambda_1 \mathbf{e}_1 \mathbf{e}_1 = (0,2) \left(\frac{2}{\sqrt{5}}, \frac{1}{\sqrt{5}} \right) + \lambda_1,$$

which leads to
$$\lambda_1 = -\frac{2}{\sqrt{5}}.$$

The vector $\mathbf{y}_2$ is

$$\mathbf{y}_2 = \mathbf{x}_2 + \lambda_1 \mathbf{e}_1 = (0,2) + \left(-\frac{2}{\sqrt{5}} \right) \left(\frac{2}{\sqrt{5}}, \frac{1}{\sqrt{5}} \right) = (0,2) - \left(\frac{4}{5}, \frac{2}{5} \right) = \left(-\frac{4}{5}, \frac{8}{5} \right).$$

Finally, we can calculate the versor

$$\mathbf{e}_2 = \frac{\mathbf{y}_2}{\| \mathbf{y}_2 \|} = \frac{\left(-\frac{4}{5}, \frac{8}{5} \right)}{\sqrt{\frac{16}{25} + \frac{64}{25}}} = (-0.447, 0.894).$$

The vectors $\mathbf{e}_1, \mathbf{e}_2$ are an orthonormal basis of $\left(\mathbb{R}^2, +, \cdot \right)$.

Exercises of Chap. 10

10.1 The rank is the dimension of the image vector space and the rank of the matrix associated to the mapping. Since the image is only the null vector, $\mathbf{o_F}$, it has dimension 0.

The nullity is the dimension of the kernel. Since the mapped vector of all the vectors of the domain are equal to the null vector, then the kernel is the entire $\mathbb{R}^n$. Since the mapping is not injective then it is not invertible.

10.2 The kernel of a linear mapping is the set of vectors such that $f(\mathbf{x}) = \mathbf{o}$. This means that the kernel of this mapping is the solution of the system of linear equations:

$$\begin{cases} x + y + z = 0 \\ x - y - z = 0 \\ 2x + 2y + 2z = 0 \end{cases}$$

This system has rank $\rho = 2$ and thus has ∞^1 solutions. These solutions are proportional to $(0, 1, -1)$ and the kernel is

$$\ker(f) = \{\alpha(0, 1, -1), \alpha \in \mathbb{R}\}.$$

The nullity is the dimensions of $(\ker(f), +, \cdot)$ which is 1. To find the rank of this endomorphism, let us consider that $\dim(\mathbb{R}^3, +, \cdot) = 3$ and let us apply the rank-nullity theorem: the rank of the endomorphism is 2.

Since the kernel $\ker(f) \neq \mathbf{o}$, the mapping is not injective. Since the mapping is not injective, it is not surjective.

The image of the mapping is given by

$$L\left(\begin{pmatrix} 1 \\ 1 \\ 2 \end{pmatrix}, \begin{pmatrix} 1 \\ -1 \\ 2 \end{pmatrix}, \begin{pmatrix} 1 \\ 1 \\ 2 \end{pmatrix}\right).$$

Only two of them are linearly independent hence the image is

$$L\left(\begin{pmatrix} 1 \\ 1 \\ 2 \end{pmatrix}, \begin{pmatrix} 1 \\ -1 \\ 2 \end{pmatrix}\right).$$

This mapping transforms vectors of space into vectors of a plane (the image) in the space.

10.3 The kernel of this linear mapping is the set of points (x, y, z) such that

$$\begin{cases} x + 2y + z = 0 \\ 3x + 6y + 3z = 0 \\ 5x + 10y + 5z = 0. \end{cases}$$

It can be checked that the rank of this homogeneous system of linear equations is $\rho = 1$. Thus ∞^2 solutions exists. If we pose $x = \alpha$ and $z = \gamma$ with $\alpha, \gamma \in \mathbb{R}$ we have that the solution of the system of linear equations is

$$(x, y, z) = \left(\alpha, -\frac{\alpha + \gamma}{2}, \gamma \right),$$

that is also the kernel of the mapping:

$$\ker (f) = \left(\alpha, -\frac{\alpha + \gamma}{2}, \gamma \right).$$

It follows that $\dim (\ker (f), +, \cdot) = 2$. Since $\dim (\mathbb{R}^3, +, \cdot) = 3$, it follows from the rank-nullity theorem that $\dim (\operatorname{Im} (f)) = 1$.

Since $\ker (f) \neq \mathbf{o_E}$, the mapping is not injecting and since it is an endomorphism it is not surjective.

The image is given by

$$L \begin{pmatrix} 1 \\ 3 \\ 5 \end{pmatrix}.$$

We can conclude that the mapping f transforms the points of the space $(\mathbb{R}^3)$ into the points of a line of the space.

10.4

$$A = \begin{pmatrix} 1 & 2 & 0 \\ 0 & 3 & 0 \\ 2 & -4 & 2 \end{pmatrix}$$

Let us find the eigenvalues

$$\det (A - \lambda I) = \det \begin{pmatrix} 1 - \lambda & 2 & 0 \\ 0 & 3 - \lambda & 0 \\ 2 & -4 & 2 - \lambda \end{pmatrix} = (1 - \lambda)(2 - \lambda)(3 - \lambda).$$

The roots of the characteristic polynomial, i.e. the eigenvalues, are $\lambda_1 = 3, \lambda_2 = 2$, and $\lambda_3 = 1$.

In order to find an eigenvector let us take $\lambda_1 = 3$ and let us solve the system

$$(A - \lambda_1 I) x = \mathbf{o},$$

$$\begin{pmatrix} -2 & 2 & 0 \\ 0 & 0 & 0 \\ 2 & -4 & -1 \end{pmatrix} \begin{pmatrix} x \\ y \\ z \end{pmatrix} = \begin{pmatrix} 0 \\ 0 \\ 0 \end{pmatrix}$$

whose ∞^1 solutions are proportional to the eigenvector

$$\begin{pmatrix} -1 \\ -1 \\ 2 \end{pmatrix}$$

Since the eigenvalues are all distinct, the matrix is diagonalizable.

10.5 To find the eigenvalues means to find those values of λ such that

$$\begin{cases} x+y=\lambda x \\ 2x=\lambda y \end{cases} \Rightarrow \begin{cases} (1-\lambda)x+y=0 \\ 2x-\lambda y=0. \end{cases}$$

This means to find those values of λ such that

$$\det \begin{pmatrix} (1-\lambda) & 1 \\ 2 & -\lambda \end{pmatrix} = 0.$$

This means that

$$(1-\lambda)(-\lambda)-2=0 \Rightarrow \lambda^2 - \lambda - 2 = 0.$$

The solutions of this polynomial would be the eigenvalues of this endomorphism. The solutions are $\lambda_1 = -1$ and $\lambda_2 = 2$, that are the eigenvalues of the endomorphism.

Since the endomorphism has two distinct eigenvalues, it is diagonalizable. Let us find the eigenvectors. We have to solve

$$\begin{cases} 2x+y=0 \\ 2x+y=0 \end{cases}$$

whose solution is $\alpha(1,-2)$ and

$$\begin{cases} -x+y=0 \\ 2x-2y=0 \end{cases}$$

whose solution is $\alpha(1,1)$.

The transformation matrix

$$\mathbf{P} = \begin{pmatrix} 1 & 1 \\ -2 & 1 \end{pmatrix}$$

enables the diagonalization of mapping into

$$\mathbf{D} = \begin{pmatrix} -1 & 0 \\ 0 & 2 \end{pmatrix}.$$

10.6 The eigenvalues associated with the matrix $\mathbf{A}$ are

$$\lambda_1 = -1$$
$$\lambda_2 = 5$$
$$\lambda_3 = 1.$$

The dominant eigenvalue is $\lambda_2 = 5$.
Let us find it by Power method with starting vector $\mathbf{x}^\mathbf{T} = (1, 1, 1)$:

$$\mathbf{x} = \mathbf{Ax} = \begin{pmatrix} -7 \\ 9 \\ 2 \end{pmatrix}.$$

At the following iterations we obtain

$$\begin{pmatrix} -47 \\ 49 \\ 9 \end{pmatrix}; \begin{pmatrix} -247 \\ 249 \\ 42 \end{pmatrix}; \begin{pmatrix} -1247 \\ 1249 \\ 209 \end{pmatrix}; \begin{pmatrix} -6247 \\ 6249 \\ 1042 \end{pmatrix};$$

$$\begin{pmatrix} -31247 \\ 31249 \\ 5209 \end{pmatrix}; \begin{pmatrix} -156247 \\ 156249 \\ 26042 \end{pmatrix}; \begin{pmatrix} -781247 \\ 781249 \\ 130209 \end{pmatrix}; \begin{pmatrix} -3906247 \\ 3906249 \\ 651042 \end{pmatrix}.$$

We can now apply Rayleigh Theorem

$$\lambda_2 = \frac{\mathbf{x}^\mathbf{T} \mathbf{Ax}}{\mathbf{x}^\mathbf{T} \mathbf{x}} \approx 5.$$

10.7 Four iterations of Power Method starting from $\mathbf{x}^\mathbf{T} = (1, 1, 1)$ are

$$\begin{pmatrix} 2 \\ 2 \\ -2 \end{pmatrix}; \begin{pmatrix} 8 \\ 8 \\ -8 \end{pmatrix}; \begin{pmatrix} 16 \\ 16 \\ 16 \end{pmatrix}; \begin{pmatrix} 32 \\ 32 \\ -32 \end{pmatrix}; \begin{pmatrix} 64 \\ 64 \\ 64 \end{pmatrix}; \begin{pmatrix} 128 \\ 128 \\ -128 \end{pmatrix}; \begin{pmatrix} 256 \\ 256 \\ 256 \end{pmatrix}; \begin{pmatrix} 512 \\ 512 \\ 512 \end{pmatrix}.$$

If we apply the Rayleigh Theorem we obtain

$$\lambda = \frac{\mathbf{x}^\mathbf{T} \mathbf{Ax}}{\mathbf{x}^\mathbf{T} \mathbf{x}} \approx \frac{2}{3}.$$

However if we calculate the eigenvalues we find

$$\lambda_1 = 2$$
$$\lambda_2 = -2$$
$$\lambda_3 = -2.$$

Hence, since there is no dominant eigenvalue, the Power method does not converge to the dominant eigenvector.

Exercises of Chap. 12

12.1 Let us draw the graph

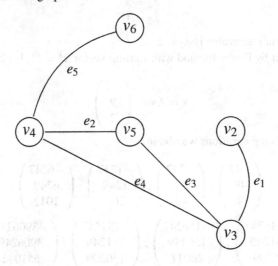

The graph is connected.

12.2 The degree of each note is

$$
\begin{aligned}
v_1 &: 2 \\
v_2 &: 3 \\
v_3 &: 2 \\
v_4 &: 3 \\
v_5 &: 2 \\
v_6 &: 3 \\
v_7 &: 2 \\
v_8 &: 3 \\
v_9 &: 2 \\
v_{10} &: 2.
\end{aligned}
$$

The degree of the graph is

$$
\sum_{i=1}^{10} \deg v_i = 24.
$$

12.3 The graph is connected. Let us check the degree of the vertices:

$$v_1 : 4$$
$$v_2 : 4$$
$$v_3 : 4$$
$$v_4 : 4$$
$$v_5 : 2$$
$$v_6 : 2$$
$$v_7 : 2$$
$$v_8 : 2.$$

Since the degree of each vertex is even the graph is Eulerian. An Eulerian cycle is:

$$(v_1, v_5, v_2, v_6, v_3, v_7, v_4, v_8, v_1, v_2, v_3, v_4, v_1).$$

12.4 If the vertex v_3 is the reference vertex the reduced incidence matrix is

$$\mathbf{A_f} = \begin{pmatrix} 1\,0\,0\,1\,1 \\ 1\,1\,0\,0\,0 \\ 0\,0\,1\,1\,0 \end{pmatrix}.$$

A spanning tree must be a tree (no cycles and connected) containing all the vertices. For example

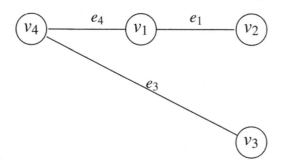

For the spanning tree above the corresponding fundamental cycle matrix is

$$\mathbf{B_f} = \begin{pmatrix} 0\,0\,1\,1\,1 \\ 1\,1\,0\,0\,1 \end{pmatrix}.$$

12.5 The graph is planar since it can be drawn without crossing edges:

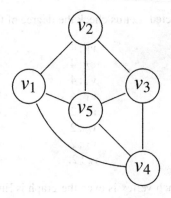

We have $n = 5$, $m = 8$, $f = 5$
The Euler's formula

$$n - m + f = 5 - 8 + 5 = 2$$

is then verified.

Exercises of Chap. 13

13.1 Let us represent the network in its minimal topology:

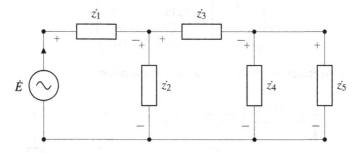

where

$$\dot{E} = (220\angle 0°)$$
$$\dot{z}_1 = 4\Omega$$
$$\dot{z}_2 = 4\Omega$$
$$\dot{z}_3 = 2 + j\left(314 \times 0.1 - \frac{1}{314 \times 50 \times 10^{-6}}\right)\Omega = 2 + j(31.4 - 63.38)\Omega = 2 - j31.98$$
$$\dot{z}_4 = 2\Omega$$
$$\dot{z}_5 = 2\Omega$$

We can now write the equations

$$\begin{cases} \dot{I}_1 = \dot{I}_2 + \dot{I}_3 \\ \dot{I}_3 = \dot{I}_4 + \dot{I}_5 \\ \dot{E} = \dot{z}_1 \dot{I}_1 + \dot{z}_2 \dot{I}_2 \\ \dot{z}_2 \dot{I}_2 = \dot{z}_3 \dot{I}_3 + \dot{z}_4 \dot{I}_4 \\ \dot{z}_4 \dot{I}_4 = \dot{z}_5 \dot{I}_5 \end{cases}$$

which can be rearranged as

$$\begin{pmatrix} -1 & 1 & 1 & 0 & 0 \\ 0 & 0 & -1 & 1 & 1 \\ \dot{z}_1 & \dot{z}_2 & 0 & 0 & 0 \\ 0 & -\dot{z}_2 & \dot{z}_3 & \dot{z}_4 & 0 \\ 0 & 0 & 0 & -\dot{z}_4 & \dot{z}_5 \end{pmatrix} \begin{pmatrix} \dot{I}_1 \\ \dot{I}_2 \\ \dot{I}_3 \\ \dot{I}_4 \\ \dot{I}_5 \end{pmatrix} = \begin{pmatrix} 0 \\ 0 \\ \dot{E} \\ 0 \\ 0 \end{pmatrix}.$$

By substituting the numbers

$$\begin{pmatrix} -1 & 1 & 1 & 0 & 0 \\ 0 & 0 & -1 & 1 & 1 \\ 4 & 4 & 0 & 0 & 0 \\ 0 & -4 & (2 - j31.98) & 2 & 0 \\ 0 & 0 & 0 & -2 & 2 \end{pmatrix} \begin{pmatrix} \dot{I}_1 \\ \dot{I}_2 \\ \dot{I}_3 \\ \dot{I}_4 \\ \dot{I}_5 \end{pmatrix} = \begin{pmatrix} 0 \\ 0 \\ 220 \\ 0 \\ 0 \end{pmatrix}.$$

The solution of the system of linear equations is

$$\begin{pmatrix} \dot{I}_1 \\ \dot{I}_2 \\ \dot{I}_3 \\ \dot{I}_4 \\ \dot{I}_5 \end{pmatrix} = \begin{pmatrix} 27.7625 + j1.6788 \\ 27.2375 - j1.6788 \\ 0.5249 + j3.3576 \\ 0.2625 + j1.6788 \\ 0.2625 + j1.6788 \end{pmatrix}.$$

13.2 Let us represent the network in its minimal topology:

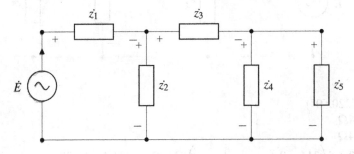

where

$$\dot{E} = (220\angle 0°)$$
$$\dot{z}_1 = 2 + j(314 * 300 * 10^-3)\Omega = 2 + j94.2\Omega$$
$$\dot{z}_2 = 4\Omega$$
$$\dot{z}_3 = 2 + j\left(314 \times 0.1 - \frac{1}{314\times 50\times 10^{-6}}\right)\Omega = 2 + j\left(31.4 - 63.38\right)\Omega = 2 - j31.980$$
$$\dot{z}_4 = 2\Omega$$
$$\dot{z}_5 = 2\Omega$$

We can now write the equations

$$\begin{cases} \dot{I}_1 = \dot{I}_2 + \dot{I}_3 \\ \dot{I}_3 = \dot{I}_4 + \dot{I}_5 \\ \dot{E} = \dot{z}_1\dot{I}_1 + \dot{z}_2\dot{I}_2 \\ \dot{z}_2\dot{I}_2 = \dot{z}_3\dot{I}_3 + \dot{z}_4\dot{I}_4 \\ \dot{z}_4\dot{I}_4 = \dot{z}_5\dot{I}_5 \end{cases}$$

which can be rearranged as

$$\begin{pmatrix} -1 & 1 & 1 & 0 & 0 \\ 0 & 0 & -1 & 1 & 1 \\ \dot{z}_1 & \dot{z}_2 & 0 & 0 & 0 \\ 0 & -\dot{z}_2 & \dot{z}_3 & \dot{z}_4 & 0 \\ 0 & 0 & 0 & -\dot{z}_4 & \dot{z}_5 \end{pmatrix} \begin{pmatrix} \dot{I}_1 \\ \dot{I}_2 \\ \dot{I}_3 \\ \dot{I}_4 \\ \dot{I}_5 \end{pmatrix} = \begin{pmatrix} 0 \\ 0 \\ \dot{E} \\ 0 \\ 0 \end{pmatrix}.$$

By substituting the numbers

$$
\begin{pmatrix}
-1 & 1 & 1 & 0 & 0 \\
0 & 0 & -1 & 1 & 1 \\
(2+j94.2) & 4 & 0 & 0 & 0 \\
0 & -4 & (2-j31.98) & 2 & 0 \\
0 & 0 & 0 & -2 & 2
\end{pmatrix}
\begin{pmatrix}
\dot{I}_1 \\ \dot{I}_2 \\ \dot{I}_3 \\ \dot{I}_4 \\ \dot{I}_5
\end{pmatrix}
=
\begin{pmatrix}
0 \\ 0 \\ 220 \\ 0 \\ 0
\end{pmatrix} .
$$

The solution of the system of linear equations is

$$
\begin{pmatrix}
\dot{I}_1 \\ \dot{I}_2 \\ \dot{I}_3 \\ \dot{I}_4 \\ \dot{I}_5
\end{pmatrix}
=
\begin{pmatrix}
0.14708 - j2.33810 \\
-0.13584 - j2.29457 \\
0.28292 - j0.04353 \\
0.14146 - j0.02177 \\
0.14146 - j0.02177
\end{pmatrix} .
$$

Exercises of Appendix A

A.1 (a) $x \wedge y$

A.2 $\begin{aligned} x &= 1 \\ y &= 0 \\ z &= 1 \\ v &= 0 \end{aligned}$

A.3 (b) $z(x \vee y) \vee v$

A.4 (c) $\bar{x} \wedge \bar{y} \wedge z \vee \bar{x} \wedge y \wedge z \vee x \wedge y \wedge \bar{z}$

References

[1] J. Hefferon, *Linear Algebra*. Saint Michael's College, 2012.

[2] G. Cramer, "Introduction a l'analyse des lignes courbes algebriques," *Geneva: Europeana*, pp. 656–659, 1750.

[3] K. Hoffman and R. Kunze, *Linear Algebra*. Englewood Cliffs, New Jersey, Prentice - Hall, 1971.

[4] A. Schönhage, A. Grotefeld, and E. Vetter, *Fast Algorithms. A Multitape Turing Machine Implementation*. 1994.

[5] J. B. Fraleigh and R. A. Beauregard, *Linear Algebra*. Addison-Wesley Publishing Company, 1987.

[6] J. Grcar, "Mathematicians of Gaussian elimination," *Notices of the American Mathematical Society*, vol. 58, no. 6, pp. 782–792, 2011.

[7] N. Higham, *Accuracy and Stability of Numerical Algorithms*. SIAM, 2002.

[8] D. M. Young, *Iterative methods for solving partial difference equations of elliptical type*. PhD thesis, Harvard University, 1950.

[9] H. Coxeter, *Introduction to Geometry*. Wiley, 1961.

[10] T. Blyth and E. Robertson, *Basic Linear Algebra*. Springer Undergraduate Mathematics Series, Springer, 2002.

[11] D. M. Burton, *The History of Mathematics*. McGraw-Hill, 1995.

[12] M. A. Moskowitz, *A Course in Complex Analysis in One Variable*. World Scientific Publishing Co, 2002.

[13] M. R. Spiegel, *Mathematical Handbook of Formulas and Tables*. Schaum, 1968.

[14] A. Kaw and E. Kalu, *Numerical Methods with Applications*. 2008.

[15] J. P. Ballantine and A. R. Jerbert, "Distance from a line, or plane, to a point," *The American Mathematical Monthly*, vol. 59, no. 4, pp. 242–243, 1952.

[16] D. Riddle, *Analytic geometry*. Wadsworth Pub. Co., 1982.

[17] H. S. M. Coxeter, *Projective Geometry*. Springer.

[18] K. G. Binmore, *Mathematical Analysis: A Straightforward Approach*. New York, NY, USA: Cambridge University Press, 2nd ed., 1982.

© Springer Nature Switzerland AG 2019   565
F. Neri, *Linear Algebra for Computational Sciences and Engineering*,
https://doi.org/10.1007/978-3-030-21321-3

[19] D. V. Widder, *The Laplace Transform*, vol. 6 of *Princeton Mathematical Series*. Princeton University Press, 1941.

[20] R. Larson and D. C. Falvo, *Elementary Linear Algebra*. Houghton Mifflin, 2008.

[21] A. Cobham, "The intrinsic computational difficulty of functions," *Proc. Logic, Methodology, and Philosophy of Science II, North Holland*, 1965.

[22] D. Kozen, *Theory of computation*. Birkhäuser, 2006.

[23] S. Arora and B. Barak, *Computational Complexity: A Modern Approach*. New York, NY, USA: Cambridge University Press, 1st ed., 2009.

[24] D. Huffman, "A method for the construction of minimum-redundancy codes," *Proceedings of the IRE*, vol. 40, pp. 1098–1101, 1952.

[25] J. Łukasiewicz, *Aristotle's Syllogistic from the Standpoint of Modern Formal Logic*. Oxford University Press, 1957.

[26] L. Euler, "Solutio problematis ad geometriam situs pertinensis," *Comm. Acad. Sc. Imperialis Petropolitanae*, vol. 8, 1736.

[27] K. Appel and W. Haken, "Solution of the four color map problem," *Scientific American*, vol. 237, no. 4, pp. 108–121, 1977.

[28] K. Mei-Ko, "Graphic programming using odd or even points," *Chinese Math.*, vol. 1, pp. 273–277, 1962.

[29] C. A. Desoer and E. S. Kuh, *Basic Circuit Theory*. McGraw-Hill Education, 2009.

[30] G. Boole, *An Investigation of the Laws of Thought*. Prometheus Books, 1853.

[31] L. A. Zadeh, "Fuzzy sets," *Information and Control*, vol. 8, no. 3, pp. 338–353, 1965.

[32] J. Liouville, "Lecons sur les fonctions doublement periodiques," *Journal f"ur die Reine und Angewandte Mathematik*, vol. 88, pp. 277–310, 1879.

[33] V. Vladimirov, *Methods of the theory of functions of several complex variables*. 1966.

[34] O. T. O'Meara, *Introduction to Quadratic Forms*. Springer-Verlag, 2000.

Index

© Springer Nature Switzerland AG 2019
F. Neri, *Linear Algebra for Computational Sciences and Engineering*,
https://doi.org/10.1007/978-3-030-21321-3